S0-BKW-937

Progress in Mathematics
Vol. 14

Edited by
J. Coates and
S. Helgason

Birkhäuser
Boston · Basel · Stuttgart

Yozô Matsushima

Manifolds and Lie Groups

Papers in Honor of Yozô Matsushima

J. Hano, A. Morimoto,
S. Murakami, K. Okamoto,
H. Ozeki, editors

1981

Birkhäuser
Boston · Basel · Stuttgart

Corresponding editor:

Jun-ichi Hano
Department of Mathematics
Washington University
St. Louis, Missouri 63130

Library of Congress Cataloging in Publication Data

Manifolds and Lie groups.

(Progress in mathematics ; v. 14)
Bibliography: p.
1. Manifolds (Mathematics)--Addresses, essays, lectures. 2. Lie groups--Addresses, essays, lectures. I. Matsushima, Yozô, 1921- .
II. Hano, J. (Jun-ichi), 1926- .
III. Series: Progress in mathematics (Cambridge, Mass.) ; v. 14.
QA613.M33 516'.07 81-38542
ISBN 3-7643-3053-8 AACR2

CIP-Kurztitelaufnahme der Deutschen Bibliothek

Manifolds and Lie groups: papers in honor of Yozo Matsushima / J. Hano... eds. -
Boston ; Basel ; Stuttgart : Birkhäuser, 1981.
(Progress in mathematics ; Vol. 14)
ISBN 3-7643-3053-8

NE: Hano, Jun-ichi (Hrsg.); GT

ISBN: 3-7643-3053-8
Printed in USA

FOREWORD

This volume is the collection of papers dedicated to Yozô Matsushima on his 60th birthday, which took place on February 11, 1980. A conference in Geometry in honor of Professor Matsushima was held at the University of Notre Dame on May 14 and 15, 1980. Some of the papers in this volume were delivered on this occasion.

Matsushima Conference, May 14-15, 1980

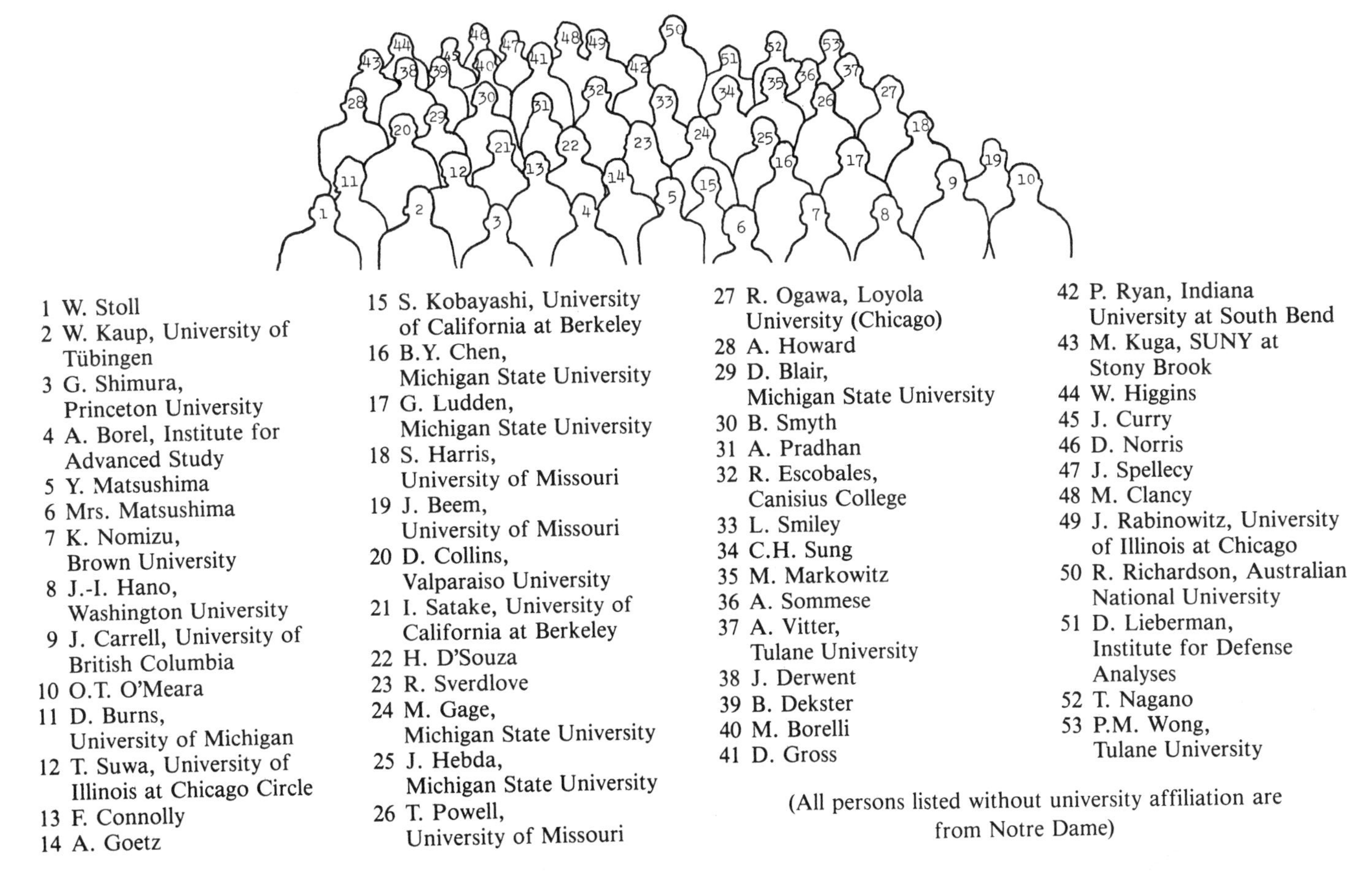

1 W. Stoll
2 W. Kaup, University of Tübingen
3 G. Shimura, Princeton University
4 A. Borel, Institute for Advanced Study
5 Y. Matsushima
6 Mrs. Matsushima
7 K. Nomizu, Brown University
8 J.-I. Hano, Washington University
9 J. Carrell, University of British Columbia
10 O.T. O'Meara
11 D. Burns, University of Michigan
12 T. Suwa, University of Illinois at Chicago Circle
13 F. Connolly
14 A. Goetz
15 S. Kobayashi, University of California at Berkeley
16 B.Y. Chen, Michigan State University
17 G. Ludden, Michigan State University
18 S. Harris, University of Missouri
19 J. Beem, University of Missouri
20 D. Collins, Valparaiso University
21 I. Satake, University of California at Berkeley
22 H. D'Souza
23 R. Sverdlove
24 M. Gage, Michigan State University
25 J. Hebda, Michigan State University
26 T. Powell, University of Missouri
27 R. Ogawa, Loyola University (Chicago)
28 A. Howard
29 D. Blair, Michigan State University
30 B. Smyth
31 A. Pradhan
32 R. Escobales, Canisius College
33 L. Smiley
34 C.H. Sung
35 M. Markowitz
36 A. Sommese
37 A. Vitter, Tulane University
38 J. Derwent
39 B. Dekster
40 M. Borelli
41 D. Gross
42 P. Ryan, Indiana University at South Bend
43 M. Kuga, SUNY at Stony Brook
44 W. Higgins
45 J. Curry
46 D. Norris
47 J. Spellecy
48 M. Clancy
49 J. Rabinowitz, University of Illinois at Chicago
50 R. Richardson, Australian National University
51 D. Lieberman, Institute for Defense Analyses
52 T. Nagano
53 P.M. Wong, Tulane University

(All persons listed without university affiliation are from Notre Dame)

A list of papers dedicated to Professor Matsushima on his 60th birthday, but not included in this volume.

Inoue, T., *Orthogonal projection onto spaces of holomorphic functions on bounded homogeneous domains.*

Koiso, N., *Rigidity and infinitesimal deformability of Einstein metrics.*

Konno, Y., *Multiplicity formulas for discrete series of spin* (1, 2m) *and* SU(1,n).

Miyanishi, M., *Regular subrings of a polynomial ring II.*

Nomura, Y., *Self-homotopy equivalences of Stiefel manifolds* $W_{n,2}$ *and* $V_{n,2}$.

Okamoto, K., Matsushita, H. and Sakurai, T., *On a certain class of irreducible unitary representations of the infinite dimensional rotation group I.*

Shimada, N., *A categorical-theoretical proof of the Garsten-Wagoner delooping theorem.*

Sumitomo, T., and Tandai, K., *Spectrum of Grassman manifold* $G_{2,n-1}(C)$.

Suwa, T., *Kupka-Reeb phenomena and universal unfoldings of certain foliation singularities.*

Tsujishita, T., *Characteristic classes for families of foliations.*

Uchida, F., *Action of special linear groups on a product manifold.*

CONTENTS

ON SOME GENERALIZATION OF B. KOSTANT'S PARTITION FUNCTION

Ichiro Amemiya, Nagayoshi Iwahori
and Kazuhiko Koike

0. Introduction

In this note we shall show some useful properties of the partition function given by B. Kostant [1] associated to complex semi-simple Lie algebras. Some of these properties are shown to be valid also for some generalized versions of Kostant's partition function (see Theorem 1 below). As an application of these properties we give (Theorem 4) an explicit formula for the multiplicity of the zero-weight in a given irreducible representation of the simple Lie algebra of type (G_2).

1. Preliminaries, Notations and Theorems

Let $\mathbb{Z}^\ell$ be the lattice consisting of all ℓ dimensional column vectors with integral components. Let $\mathbb{Z}^\ell_+$ be the subset of $\mathbb{Z}^\ell$ consisting of vectors with non-negative integral components. Let $A=(a_1,\dots,a_n)$ be an ordered sequence consisting of non-zero elements of $\mathbb{Z}^\ell_+$. A may be regarded to be an $\ell\times n$ matrix $A=(a_{ij})$ $(1\leqslant i\leqslant \ell,$ $1\leqslant j\leqslant n)$ where $a_j={}^t(a_{1j},a_{2j},\dots,a_{\ell j})$ for $j=1,\dots,n$. We define a function $P_A\colon \mathbb{Z}^\ell\to\mathbb{Z}$ as follows. Let $b\in\mathbb{Z}^\ell$. Then $P_A(b)$ is the number of the solutions $z={}^t(z_1,\dots,z_n)\in\mathbb{Z}^n_+$ of $a_1z_1+\dots+a_nz_n=b$. Thus $P_A(b)=0$ if $b\notin\mathbb{Z}^\ell_+$. We call P_A the partition function associated to the matrix A.

<u>Example 1</u>. Let Δ be a reduced root system (Bourbaki [2]). Let $\Pi=\{\alpha_1,\dots,\alpha_\ell\}$ be a base of Δ. Identify $\mathbb{Z}^\ell$ with $\mathbb{Z}\alpha_1+\dots+\mathbb{Z}\alpha_\ell$ via the map $(c_1,\dots,c_\ell)\mapsto c_1\alpha_1+\dots+c_\ell\alpha_\ell$. Let $\Delta^+=\{\beta_1,\dots,\beta_n\}$ be the set of all positive roots of Δ w.r.t. Π and let $A=(\beta_1,\dots,\beta_n)$. Then P_A is Kostant's partition function.

<u>Example 2</u>. Let $A=\begin{pmatrix}1&0&1&2&\dots&k\\0&1&1&1&\dots&1\end{pmatrix}$. Then it is easy to see that $P_A({}^t(i,j))=b_{j,j+k}(0)+b_{j,j+k}(1)+\cdots b_{j,j+k}(i)$ where $b_{p,q}(s)$ means the 2s-th Betti number of the complex Grassmann manifold $G_p(\mathbb{C}^q)$ consisting of all p-dimensional linear subspaces of $\mathbb{C}^q$.

Now, given such a partition function P_A, we define a power series $g(t;A,b,c)$ in the single variable t associated to $b \in \mathbb{Z}_+^\ell$, $c \in \mathbb{Z}_+^\ell$ as follows:

$$g(t;A,b,c) = \sum_{k=0}^{\infty} P_A(b+kc)\,t^k .$$

We call $g(t;A,b,c)$ the single-variable power series (to be abbreviated as s.v.p.s.) associated to the triple (A,b,c).

In order to state Theorem 1, we need one more definition:

Definition. *Let* $A=(a_{ij})$ *be an* $\ell \times n$ *matrix with integral entries* a_{ij}. *Let* r *be the rank of* A. *Let* $B_1,\dots,B_k$ *be the totality of the submatrices* B *of* A *of size* $\ell \times r$ *satisfying* $\operatorname{rank}(B) = r$. *Let* b_j *be the* r*-th elementary divisor of* B_j, *i.e.,* b_j *is the G.C.D. of the all minors of size* $r \times r$ *of* B_j. *We denote by* $c(A)$ *the L.C.M. of the* $b_1,\dots,b_k$ *and call* $c(A)$ *the characteristic content of the matrix* A.

Theorem 1. *Let* A *be an* $\ell \times n$ *matrix with non-negative integral entries containing no zero columns. Let* N *be the characteristic content of* A. *Then for any* b, c *in* $\mathbb{Z}_+^\ell$, *the s.v.p.s.* $g(t;A,b,c)$ *is a rational function in* t. *Furthermore the denominator of* $g(t;A,b,c)$ *is a divisor of* $(1-t^N)^n$, *i.e.,* $(1-t^N)^n \cdot g(t;A,b,c)$ *is a polynomial in* t.

Remark. The generating function of the multi-indexed sequence $P_A : \mathbb{Z}_+^\ell \to \mathbb{Z}$ is obviously given by

$$f_A(x_1,\dots,x_\ell) = \prod_{j=1}^{n} \left(1 - x_1^{a_{1j}} x_2^{a_{2j}} \cdots x_\ell^{a_{\ell j}}\right)^{-1} . \qquad (1.1)$$

Such a form of the generating function is the main reason for the validity of Theorem 1. If one starts from a Taylor expansion $\Sigma_b P(b)x^b$ (where $x^b = x_1^{b_1} \cdots x_\ell^{b_\ell}$) at the origin of a rational function $F(x_1,\dots,x_\ell)$, one does not always get a rational function as a s.v.p.s. For example, starting from

$$F(x,y) = (1-x-y)^{-1} = \sum \binom{i+j}{i} x^i y^j ,$$

for $b = {}^t(0,0)$, $c = {}^t(1,1)$ one gets

$$g(t;b,c) = \sum_{i=0}^{\infty} \binom{2i}{i} t^i = (1-4t)^{-1/2}$$

as s.v.p.s.

Now let us consider the case of Kostant's partition function P_A associated with a root system Δ. Let $c_{ij} = 2(\alpha_i,\alpha_j)/(\alpha_j,\alpha_j)$ be the Cartan integers w.r.t. a base $\Pi = \{\alpha_1,\ldots,\alpha_\ell\}$ of Δ. Let $f_1,\ldots,f_{j-1},f_{j+1},\ldots,f_\ell \in \mathbf{Z}_+$ be given. We consider the subsequence $\{\zeta_k^{(j)}\}$ of P_A restricted in the j-th direction starting from $P_A({}^t(f_1,\ldots,f_{j-1},0,f_{j+1},\ldots,f_\ell))$, i.e., we put

$$\zeta_k^{(j)} = P_A({}^t(f_1,\ldots,f_{j-1},k,f_{j+1},\ldots,f_\ell)) \quad .$$

Theorem 2. *Kostant's partition function satisfies for any direction* j *and for any starting values* $f_1,\ldots,f_{j-1},f_{j+1},\ldots,f_\ell$ *the following equalities.*

(i) *Stability:* $\zeta_N^{(j)} = \zeta_{N+1}^{(j)} = \ldots,$ *where*

$$N = \sum_{m \neq j} |c_{mj}| f_m \quad .$$

(ii) *Sum-formula: with respect to the value* N *in* (i),

$$\zeta_k^{(j)} + \zeta_{N-1-k}^{(j)} = \zeta_N^{(j)} \qquad \text{for} \quad k = 0,1,\ldots,N-1.$$

According to a remark due to S. Nakajima, these two properties (i), (ii) in Theorem 2 can be put into a single functional equation for the inverse $\varphi_A(x_1,\ldots,x_\ell)$ of the generating function $f_A(x_1,\ldots,x_\ell)$ of P_A. Namely (i) and (ii) mean

$$(S)_j: \ -x_j\varphi_A\left(x_1x_j^{-c_{1j}},\ldots,x_\ell\, x_j^{-c_{\ell j}}\right) = \varphi_A(x_1,\ldots,x_\ell) \quad .$$

(This will be proved in Lemma 3.1.)

We now characterize Kostant's partition function (or, its generating function) by the properties $(S)_j$ $(1 \leqslant j \leqslant \ell)$ as follows.

Theorem 3. *Let* $C = (c_{ij})$ *be an* $\ell \times \ell$ *matrix such that* $c_{ii} = 2$ $(1 \leqslant i \leqslant \ell)$ *and* $-c_{ij} \in \mathbb{Z}_+$ *for every pair* i,j *with* $i \neq j$. *Denote by* M_C *the complex linear subspace of the polynomial ring* $\mathbb{C}[x_1,\ldots,x_\ell]$ *defined by*

$$M_C = \{\varphi \in \mathbb{C}[x_1,\dots,x_\ell] \mid \varphi \text{ satisfies } (S)_j \text{ for } j=1,2,\dots,\ell\}.$$

Then $\dim_{\mathbb{C}} M_C \leqslant 1$. *Furthermore, if* $\dim_{\mathbb{C}} M_C = 1$, *then* C *is the Cartan matrix of some reduced root system and there exists an element* φ *in* M_C *such that* φ^{-1} *is the generating function of the corresponding Kostant's partition function.*

Finally, combining Theorems 1 and 2, we give an explicit formula for the multiplicity of zero weight in a given irreducible representation of complex simple Lie algebra of type (G_2) in Theorem 4. (The same method works in principle for every complex simple Lie algebra. However, the computation for the actual formula gets complicated with the rank of Lie algebras very rapidly.)

Theorem 4. *Let* $\mathfrak{g}$ *be a complex simple Lie algebra of type* (G_2). *Let* $\{\alpha_1,\alpha_2\}$ *be a base of the root system* Δ *of* $\mathfrak{g}$ *w.r.t. some Cartan subalgebra with* α_1 *as the short root. Let* $\{\Lambda_1,\Lambda_2\}$ *be the corresponding fundamental weights, i.e.,*

$$2(\Lambda_i,\alpha_j)/(\alpha_j,\alpha_j) = \delta_{ij} \qquad (1 \leqslant i,\ j \leqslant 2).$$

We denote by $m(i,j)$ *the multiplicity of the zero weight in the irreducible representation of* $\mathfrak{g}$ *with the highest weight* $\lambda = i\Lambda_1 + j\Lambda_2$. *Now define a* 4×4 *matrix* $B^{(i,j)} = (b_{pq}^{(i,j)})$ $(0 \leqslant p,\ q \leqslant 3)$ *as follows.*

(i) *If* $i = 6x$, $j = 2y$ *for some* $x,y \in \mathbb{Z}$, *then*

$$B^{(i,j)} = \begin{pmatrix} 1 & 4 & 7 & 4 \\ 17 & 82 & 84 & 24 \\ 51 & 156 & 72 & 0 \\ 36 & 72 & 0 & 0 \end{pmatrix}$$

(ii) *If* $i = 6x$, $j = 2y+1$ *for some* $x,y \in \mathbb{Z}$, *then*

$$B^{(i,j)} = \begin{pmatrix} 2 & 7 & 9 & 4 \\ 49 & 121 & 96 & 24 \\ 120 & 192 & 72 & 0 \\ 72 & 72 & 0 & 0 \end{pmatrix}$$

(iii) *If* $i = 6x+2$, $j = 2y$ *for some* $x,y \in \mathbb{Z}$, *then*

$$B^{(i,j)} = \begin{pmatrix} 3 & 18 & 27 & 12 \\ 30 & 126 & 108 & 24 \\ 63 & 180 & 72 & 0 \\ 36 & 72 & 0 & 0 \end{pmatrix} .$$

(iv) *If* $i = 6x + 2$, $j = 2y + 1$ *for some* $x, y \in \mathbb{Z}$, *then*

$$B^{(i,j)} = \begin{pmatrix} 9 & 30 & 33 & 12 \\ 81 & 177 & 120 & 24 \\ 144 & 216 & 72 & 0 \\ 72 & 72 & 0 & 0 \end{pmatrix} .$$

(v) *If* $i = 6x + 4$, $j = 2y$ *for some* $x, y \in \mathbb{Z}$, *then*

$$B^{(i,j)} = \begin{pmatrix} 8 & 44 & 55 & 20 \\ 47 & 178 & 132 & 24 \\ 75 & 204 & 72 & 0 \\ 36 & 72 & 0 & 0 \end{pmatrix} .$$

(vi) *If* $i = 6x + 4$, $j = 2y + 1$ *for some* $x, y \in \mathbb{Z}$, *then*

$$B^{(i,j)} = \begin{pmatrix} 24 & 69 & 65 & 20 \\ 121 & 241 & 144 & 24 \\ 168 & 240 & 72 & 0 \\ 72 & 72 & 0 & 0 \end{pmatrix} .$$

(vii) *If* $i = 6x + 1$, $i = y$ *for some* $x, y \in \mathbb{Z}$, *then*

$$B^{(i,j)} = \begin{pmatrix} 1 & 3 & 3 & 1 \\ 23 & 41 & 21 & 3 \\ 57 & 75 & 18 & 0 \\ 36 & 36 & 0 & 0 \end{pmatrix} .$$

(viii) *If* $i = 6x + 3$, $j = y$ *for some* $x, y \in \mathbb{Z}$, *then*

$$B^{(i,j)} = \begin{pmatrix} 5 & 11 & 8 & 1 \\ 38 & 62 & 27 & 3 \\ 69 & 87 & 18 & 0 \\ 36 & 36 & 0 & 0 \end{pmatrix} .$$

(ix) *If* $i = 6x + 5$, $j = y$ *for some* $x, y \in \mathbb{Z}$, *then*

$$B^{(i,j)} = \begin{pmatrix} 12 & 24 & 15 & 3 \\ 57 & 87 & 33 & 3 \\ 81 & 99 & 18 & 0 \\ 36 & 36 & 0 & 0 \end{pmatrix} .$$

Then $m(i,j)$ *is given by using the above* x,y *as follows:*

$$m(i,j) = \sum_{p,q=0}^{3} b_{p,q}^{(i,j)} \binom{x}{p} \binom{y}{q} ,$$

where $\binom{x}{p}$ *means as usual*

$$x(x-1)\dots(x-p+1)/p!$$

2. Proof of Theorem 1

Lemma 2.1. *Let* A *be an* $\ell \times n$ *matrix with integral entries. Let* A_1 *be a matrix which is obtained from* A *by deleting several columns of* A. *Then the characteristic content* $c(A_1)$ *is a divisor of the characteristic content* $c(A)$.

Proof. One has only to consider the case where A_1 is obtained from A by deleting a single column. Then the proof is immediate by the definition of characteristic content given in §1. ∎

Lemma 2.2. *Let* $b \in \mathbb{Z}_+^{\ell}$. *If there is a positive term in the sequence* $P_A(b)$, $P_A(2b)$, $P_A(3b), \dots,$ *then* $P_A(c(A)\cdot b) > 0$.

Proof. Let us prove our assertion by induction on n (= the column size of A). If $n=1$, our assertion is obvious. Now suppose that $n > 1$ and our assertion is valid up to $n-1$. Suppose $P_A(kb) > 0$ and take a solution z of $Az = kb$. Then $x = 1/k\ z \in \mathbb{Q}_+^n$ satisfies $Ax = b$. Put $x = {}^t(x_1,\dots,x_n)$. If some component x_j of x is 0, then our case is reduced to the case with smaller n. So we may assume that all the components x_j of x are positive.

Suppose now the column vectors $a_1,\dots,a_n$ of A are linearly dependent: $\gamma_1 a_1 + \dots + \gamma_n a_n = 0$ for some non-zero rational vector $(\gamma_1,\dots,\gamma_n)$. One can assume that some $\gamma_j > 0$. Then there exists a rational number $\varepsilon \in \mathbb{Q}$ such that $\mathrm{Min}_{1\leqslant j\leqslant n} \{x_j - \varepsilon\gamma_j\} = 0$. In fact, one can put $\varepsilon = \mathrm{Min}_{\gamma_j>0} \{x_j/\gamma_j\}$. Then $y = {}^t(x_1 - \varepsilon\gamma_1,\dots,x_n - \varepsilon\gamma_n) \in \mathbb{Q}_+^n$ satisfies $Ay = b$ having a zero component. Thus we are reduced to the case with smaller n.

Thus we may assume that the column vectors $a_1,\dots,a_n$ are linearly independent. Let $A_1,\dots,A_m$ be the totality of submatrices of A of size $n\times n$ and having nonvanishing determinant. In the equality $Az = kb$, let b_j be the n-dimensional integral vector obtained from b

by choosing the components of b corresponding to indices of the rows of the submatrix A_j. Since $Az = kb$, there exists a vector $y \in \mathbb{Q}_+^n$ such that $Ay = c(A)b$. Then we have $A_j y = c(A)b_j$ for $j = 1,\ldots,m$. Let A_j^* be the adjoint matrix of A_j: $A_j^* = \det(A_j)\cdot A_j^{-1}$. Put $A_j^* b_j = b_j^*$. Then $b_j^* \in \mathbb{Z}^n$ and we have $\det(A_j)y = c(A)b_j^*$. Now $c(A)$ is an integral linear combination of the $\det(A_j)$: $c(A) = \Sigma\, u_j \cdot \det(A_j)$. Hence $c(A)y =$ $\Sigma\, u_j \det(A_j)y = c(A) \times \Sigma\, u_j b_j^*$. Thus $y = \Sigma\, u_j b_j^*$ belongs to $\mathbb{Z}^n$, i.e., $P_A(c(A)b) > 0$. ∎

Let us now prove Theorem 1 by induction on n. The case $n = 1$ is easily proved. So assume that $n > 1$ and our assertion is true up to $n - 1$.

Let S be a subset of $\mathbb{Z}_+^n$. We denote by $P_A(b,S)$ the number of $z \in S$ satisfying $Az = b$. Thus if $S_1,\ldots,S_p$ are mutually disjoint subsets of $\mathbb{Z}_+^n$ one has $P_A(b, S_1 \cup \ldots \cup S_p) = \Sigma_{j=1}^p P_A(b,S_j)$.

Now we may assume that there exist $k_0 \in \mathbb{Z}_+$ and $z_0 = {}^t(f_1,\ldots,f_n) \in \mathbb{Z}_+^n$ such that $Az_0 = b + k_0 c$ (since otherwise one has $g(t;A,b,c) = 0$). We divide $\mathbb{Z}_+^n$ into 2^n subsets $\Omega(\varepsilon_1,\ldots,\varepsilon_n)$ (where each ε_j is $+1$ or -1) defined by

$$\Omega(\varepsilon_1,\ldots,\varepsilon_n) = \left\{ z = {}^t(z_1,\ldots,z_n) \in \mathbb{Z}_+^n \;\middle|\; \begin{array}{ll} z_i - f_i \geqslant 0 & (\text{for } \varepsilon_i = 1) \\ z_j - f_j < 0 & (\text{for } \varepsilon_j = -1) \end{array} \right\}$$

Then $\mathbb{Z}_+^n$ is a disjoint union of the $\Omega(\varepsilon_1,\ldots,\varepsilon_n)$ and we have

$$P_A(b) = \sum_{\varepsilon_1,\ldots,\varepsilon_n} P_A(b,\Omega(\varepsilon_1,\ldots,\varepsilon_n)) \quad .$$

Now, since there is a bijection $\{z \in \Omega(1,\ldots,1) \mid Az = b + kc\} \to \{w \in \mathbb{Z}_+^n \mid Aw = (k - k_0)c\}$ given by $z \mapsto w = z - z_0$, one has

$$P_A(b + kc, \Omega(1,\ldots,1)) = P_A((k - k_0)c) \quad .$$

Similarly we get (denoting by a_1 the 1st column of A),

$$P_A(b + kc, \Omega(-1,1,\ldots,1)) = \sum_{j=0}^{f_1 - 1} P_{A_1}((f_1 - j)a_1 + (k - k_0)c) \quad .$$

For general $(\varepsilon_1,\ldots,\varepsilon_n) \neq (1,\ldots,1)$, one gets by a similar consideration the following equality: suppose that $\varepsilon_{i_1} = \cdots = \varepsilon_{i_s} = -1$ $i_1 < \cdots < i_s$ and suppose that all other ε_j are equal to 1. Then, denoting by $A_{i_1,\ldots i_s}$ the matrix obtained from A by deleting the $i_1, i_2, \ldots, i_s$-th columns, one has

$$P_A(b + kc, \Omega(\varepsilon_1, \ldots, \varepsilon_n))$$

$$= \sum_{k_1=1}^{f_{i_1}} \cdots \sum_{k_s=1}^{f_{i_s}} P_{A_{i_1,\ldots,i_s}} \left(\sum_{p=1}^{s} k_p a_{i_p} + (k - k_0)c \right) .$$

Thus, if $(\varepsilon_1, \ldots, \varepsilon_n) \neq (1, \ldots, 1)$, $P_A(b + kc, \Omega(\varepsilon_1, \ldots, \varepsilon_n))$ is expressed as a sum of values of the partition function associated to the matrices obtained from A by deleting several columns. Hence

$$g(t;A,b,c) = \sum_k P_A(b + kc)\, t^k$$

$$= t^{k_0} \sum_k P_A((k - k_0)c)\, t^{k-k_0} + T(t)$$

i.e.,

$$g(t;A,b,c) = t^{k_0} g(t;A,0,c) + T(t) \tag{2.1}$$

where, by induction assumption and Lemma 2.1, $T(t)$ is a rational function in t whose denominator is a divisor of $(1 - t^{c(A)})^{n-1}$.

Thus in order to complete the proof it is enough to show that $g(t;A,0,c) \cdot (1 - t^{c(A)})^n \in \mathbb{C}[t]$. Now if $P_A(c) = P_A(2c) = \cdots = 0$, then $g(t;A,0,c)=1$ and our assertion is true. If some $P_A(kc) > 0$, then by Lemma 2.2, there exists a vector $z \in \mathbb{Z}_+^n$ with $Az = c(A)c$. Then, by putting $b = 0$, $k_0 = c(A)$ in (2.1), we see that

$$g(t;A,0,c) = t^{c(A)} g(t;A,0,c) + \tilde{T}(t)$$

with some $\tilde{T}(t) \in \mathbb{C}(t)$ satisfying $(1 - t^{c(A)})^{n-1} \tilde{T}(t) \in \mathbb{C}[t]$. Thus $(1 - t^{c(A)})^n g(t;A,0,c) = (1 - t^{c(A)})^{n-1} \tilde{T}(t) \in \mathbb{C}[t]$. ∎

3. Proof of Theorem 2

Let Δ be a reduced root system and $\Pi = \{\alpha_1, \ldots, \alpha_\ell\}$ a base of Δ. Denote by w_j the reflection w.r.t. α_j. Then $w_j(\alpha_i) = \alpha_i - c_{ij}\alpha_j$ where c_{ij}'s are Cartan integers. Let $\Delta^+ = \{\beta_1, \ldots, \beta_n\}$ be the positive roots and put

$$\beta_j = a_{1j}\alpha_1 + \cdots + a_{\ell j}\alpha_\ell \qquad (1 \leqslant j \leqslant n)$$

with $a_{ij} \in \mathbf{Z}_+$. Then the generating function $f_A(x_1, \ldots, x_\ell)$ of Kostant's partition function is given by (1.1). Now as is well-known, one has $w_j(\Delta^+ - \{\alpha_j\}) = \Delta^+ - \{\alpha_j\}$ and $w_j(\alpha_j) = -\alpha_j$. These two equalities immediately imply the equality $(S)_j$ in §1 for $\varphi_A(x_1, \ldots, x_\ell) = f_A(x_1, \ldots, x_\ell)^{-1}$. In fact, using the notations

$$x^{\beta_j} = x_1^{a_{1j}} x_2^{a_{2j}} \cdots x_\ell^{a_{\ell j}} \qquad ,$$

one has $f_A(x) = \prod_{\beta \in \Delta^+} (1 - x^\beta)^{-1}$. So

$$\begin{aligned}
\varphi_A\left(x_1 x_j^{-c_{1j}}, \ldots, x_\ell x_j^{-c_{\ell j}}\right) &= \prod_{\beta \in \Delta^+} \left(1 - x^{w_j(\beta)}\right) \\
&= \varphi_A(x_1, \ldots, x_\ell) \cdot (1 - x_j)^{-1}(1 - x_j^{-1}) \\
&= -\frac{1}{x_j} \varphi_A(x)
\end{aligned}$$

Let us now prove Theorem 2 for the direction $j = 1$ since other directions can be treated similarly. Put

$$f_A(x) = \sum_{f_2, \ldots, f_\ell \geqslant 0} H_{f_2, \ldots, f_\ell}(x_1) x_2^{f_2} \cdots x_\ell^{f_\ell}$$

where $H_{f_2, \ldots, f_\ell}(x_1)$ is a power series in x_1. Then in the notations of Theorem 2 we have

$$H_{f_2, \ldots, f_\ell}(x_1) = \zeta_0^{(1)} + \zeta_1^{(1)} x_1 + \zeta_2^{(1)} x_1^2 + \cdots \qquad .$$

We note that $H_{f_2,\dots,f_\ell}(x_1)$ is a rational function in x_1 by Theorem 1.

Lemma 3.1. *Let* $\psi(t) = c_0 + c_1 t + c_2 t^2 + \cdots$ *be a formal power series in* t *with coefficients* c_j *in a given commutative field* k. *Then a positive integer* M *satisfies*

$$(\alpha) \qquad c_M = c_{M+1} = \cdots$$

and

$$(\beta) \qquad c_i + c_{M-1-i} = c_M \qquad \text{for} \quad i = 0,1,\dots,M-1$$

if and only if $\psi(t)$ *is a rational function in* t *and satisfies*

$$(\gamma) \qquad -t^{M-1}\psi\left(\frac{1}{t}\right) = \psi(t) \quad .$$

Proof. Put $\theta(t) = (1-t)\psi(t)$. Then (γ) is equivalent to

$$(\gamma') \qquad t^M\theta\left(\frac{1}{t}\right) = \theta(t) \quad .$$

Now (γ') means that $\theta(t) = d_0 + d_1 t + \dots$ is a polynomial in t of degree at most M and satisfies the reciprocal condition $d_0 = d_M$, $d_1 = d_{M-1}, \dots$. However, since $d_j = c_j - c_{j-1}$ or $c_j = d_0 + d_1 + \dots + d_j$ for every j, a simple computation shows that (γ') is equivalent to the validity of (α) and (β). ∎

Let us now prove Theorem 2. Since $\varphi_A(x) = f_A(x)^{-1}$ satisfies $(S)_1$, we get

$$\sum_{f_2,\dots,f_\ell} H_{f_2,\dots,f_\ell}(x_1^{-1})\left(x_2 x_1^{-c_{21}}\right)^{f_2}\cdots\left(x_\ell x_1^{-c_{\ell 1}}\right)^{f_\ell}$$

$$= -x_1 \sum_{f_2,\dots,f_\ell} H_{f_2,\dots,f_\ell}(x_1) x_2^{f_2}\cdots x_\ell^{f_\ell} \quad .$$

Thus, comparing the coefficients of $x_2^{f_2}\cdots x_\ell^{f_\ell}$, we get

$$H_{f_2,\dots,f_\ell}(x_1) = -x_1^{N-1} H_{f_2,\dots,f_\ell}\left(\frac{1}{x_1}\right) \quad .$$

Thus, by Lemma 3.1, the proof is complete. ∎

4. Proof of Theorem 3

Let $C=(c_{ij})$ be an $\ell\times\ell$ integral matrix satisfying $c_{ii}=2$ $(1\leqslant i\leqslant\ell)$ and $c_{ij}\leqslant 0$ for $i\neq j$. We denote by R the ring of Laurent polynomials with complex coefficients in the variables $x_1,\ldots,x_\ell$: $R=\mathbb{C}[x_1,\ldots,x_\ell,x_1^{-1},\ldots,x_\ell^{-1}]$. Then R is isomorphic as an algebra over $\mathbb{C}$ with the group ring $\tilde{R}=\mathbb{C}[L]$ of the free abelian group $L=\mathbb{Z}e_1\oplus\cdots\oplus\mathbb{Z}e_\ell$ of rank ℓ via the isomorphism $x_1^{k_1}\ldots x_\ell^{k_\ell}\mapsto k_1e_1+\cdots+k_\ell e_\ell$ where $k_j\in\mathbb{Z}$ $(1\leqslant j\leqslant\ell)$. Now define an automorphism w_j of L by $w_j(e_s)=e_s-c_{sj}e_j$ $(1\leqslant s\leqslant\ell)$. Then $w_j^2=1$ and w_j induces an algebra automorphism of R and $\tilde{R}$ in a natural manner. Note that $w_j(x_s)=x_sx_j^{-c_{sj}}$ $(1\leqslant s\leqslant\ell)$. We denote by W the subgroup generated by $w_1,\ldots,w_\ell$ of the full automorphism group $\mathrm{Aut}(L)$ $(\cong GL(n,\mathbb{Z}))$ of the group L. Then W may be regarded as a subgroup of $GL(V)$ where $V=L\otimes_{\mathbb{Z}}\mathbb{R}$ is an ℓ-dimensional real vector space. Furthermore each w_j is a reflection of V.

Now let us prove the assertion that "$M_C\neq\{0\}$ implies that W is a finite reflection group."

For each element $w\in W$ we define a matrix $A(w)=(A_{ij}(w))\in GL(\ell,\mathbb{Z})$ by $w(e_j)=A_{1j}(w)e_1+A_{2j}(w)e_2+\cdots+A_{\ell j}(w)e_\ell$ $(1\leqslant j\leqslant\ell)$. Suppose now $M_C\neq\{0\}$. Fix a non-zero element $f(x_1,\ldots,x_\ell)$ of M_C. Then by $(S)_j$, $f(x_1,\ldots,x_\ell)=-x_jf(x_1x_j^{-c_{1j}},\ldots,x_\ell x_j^{-c_{\ell j}})$. Hence, by putting $x_j=1$, one gets $f(x_1,\ldots,x_{j-1},1,x_{j+1},\ldots,x_\ell)=0$, i.e., x_i-1 divides f in $\mathbb{C}[x_1,\ldots,x_\ell]$. Now $(S)_j$ means that $f(w_j(x_1),\ldots,w_j(x_\ell))=-x_j^{-1}f$, i.e., $w_j(f)=-x_j^{-1}f$. Thus for every element $w\in W$, $w(f)$ must have the following form: $w(f)=J_w\cdot f$ where J_w is an element of R of the form $J_w=\varepsilon x_1^{k_1}\ldots x_\ell^{k_\ell}$ with $\varepsilon=\pm1$, $k_i\in\mathbb{Z}$ $(1\leqslant i\leqslant\ell)$. Hence J_w is a unit of the ring R. Since $w(x_j)=x_1^{A_{1j}(w)}\ldots x_\ell^{A_{\ell j}(w)}$ and $x_j-1|f$ in $\mathbb{C}[x_1,\ldots,x_\ell]$, we get $x_1^{A_{1j}(w)}\ldots x_\ell^{A_{\ell j}(w)}-1|f$ in R.

Let us now introduce a notation for convenience. Given an element h of R of the form $h=x_1^{k_1}\ldots x_\ell^{k_\ell}-1$ with $k_j\in\mathbb{Z}$ $(1\leqslant j\leqslant\ell)$ and $k_1^2+\cdots+k_\ell^2>0$, define h^* by

$$h^*=x_1^{p_1}\ldots x_\ell^{p_\ell}-x_1^{q_1}\ldots x_\ell^{q_\ell},$$

where the p_j and q_j are given by

$$\begin{cases} p_j = k_j, \quad q_j = 0, & \text{if } k_j \geqslant 0, \\ p_j = 0, \quad q_j = -k_j, & \text{if } k_j < 0. \end{cases}$$

Thus $h^* \in \mathbb{C}[x_1,\ldots,x_\ell]$; $x_1^{p_1}\ldots x_\ell^{p_\ell}$ and $x_1^{q_1}\ldots x_\ell^{q_\ell}$ are relatively prime in $\mathbb{C}[x_1,\ldots,x_\ell]$. Furthermore h^* differs from h in R only by a unit factor.

Thus, $w(x_j) - 1 | f$ in R implies $(w(x_j) - 1)^* | f$ in R. So we may put $f = u \cdot (w(x_j) - 1)^* \cdot g$ for some g in $\mathbb{C}[x_1,\ldots,x_\ell]$ and a unit u in R. Then u is of the form $u = v/z$ with monomials v, z in $\mathbb{C}[x_1,\ldots,x_\ell]$ with positive exponents. Now because of the shape of $(w(x_j) - 1)^*$, z must divide g in $\mathbb{C}[x_1,\ldots,x_\ell]$. Thus we get $(w(x_j) - 1)^* | f$ in the polynomial ring $\mathbb{C}[x_1,\ldots,x_\ell]$. Hence the degree of $(w(x_j) - 1)^*$ is $\leqslant \deg(f)$. On the other hand $\deg(w(x_j) - 1)^*$ obviously satisfies $\deg(w(x_j) - 1)^* \geqslant \operatorname{Max}\{|A_{1j}(w)|,\ldots,|A_{\ell j}(w)|\}$. Therefore $\operatorname{Max}_{i,j} |A_{ij}(w)| \leqslant \deg(f)$. Thus W is a bounded subgroup of $GL(n,\mathbb{Z})$. Therefore W is a finite group.

Hence there exists a W-invariant symmetric, positive definite bilinear form $(\ ,\ )$ on V. Since w_j is a reflection of V, $w_j(e_i) = e_i - c_{ij}e_j$ implies that $c_{ij} = 2(e_i,e_j)/(e_j,e_j)$ for every i,j. Since $c_{ij} \in \mathbb{Z}$ satisfies $c_{ij} \leqslant 0$ for $i \neq j$, $\{e_1,\ldots,e_\ell\}$ forms an admissible system of roots and it is well-known (see [2]) that $\Delta = \cup_{i=1}^{\ell} W(e_i)$ forms a reduced root system having $\Pi = \{e_1,\ldots,e_\ell\}$ as a base. Furthermore W is the Weyl group of Δ.

Now let Δ^+ be the set of all positive roots of Δ w.r.t. Π. For $\beta = \Sigma k_i e_i$ in Δ, we put $x^\beta = x_1^{k_1}\ldots x_\ell^{k_\ell}$. Since $\beta \in \Delta$ is of the form $\beta = w(e_j)$ for some $w \in W$ and some j, $w(x_j) - 1 | f$ in R implies $x^\beta - 1 | f$ in R for every $\beta \in \Delta$. So by previous argument $(x^\beta - 1)^* | f$ in $\mathbb{C}[x_1,\ldots,x_\ell]$. In particular, if $\beta \in \Delta^+$, then $(x^\beta - 1)^* = x^\beta - 1$. Hence $x^\beta - 1 | f$ in $\mathbb{C}[x_1,\ldots,x_\ell]$. Now if $\beta \in \Delta^+$, $\gamma \in \Delta^+$ and $\beta \neq \gamma$, then $x^\beta - 1$ and $x^\gamma - 1$ are relatively prime in $\mathbb{C}[x_1,\ldots,x_\ell]$. Then, f is divisible in $\mathbb{C}[x_1,\ldots,x_\ell]$ by $f_0 = \prod_{\beta\in\Delta^+}(x^\beta - 1)$ in $\mathbb{C}[x_1,\ldots,x_\ell]$. On the other hand f_0 satisfies $(S)_j$ $(1 \leqslant j \leqslant \ell)$ as we have shown in §3. Therefore the polynomial $h = f/f_0$ is W-invariant. Now any W-invariant polynomial h is a constant. In fact, if h contains a monomial $\alpha x_1^{k_1}\ldots x_\ell^{k_\ell}$ with $\alpha \in \mathbb{C} - \{0\}$, $k_j \in \mathbb{Z}_+$ $(1 \leqslant j \leqslant \ell)$, $k_1^2 + \cdots + k_\ell^2 > 0$, h must contain a monomial with negative exponents, because the set $\Sigma_{j=1}^{\ell} \mathbb{Z}_+ e_j$ is not W-stable. Thus $M_C \neq \{0\}$ implies $M_C = \mathbb{C} f_0$ and C is a Cartan matrix of Δ. This completes the proof of Theorem 3. ∎

5. Proof of Theorem 4

Let Δ be a root system of type (G_2) and $\Pi = \{\alpha_1, \alpha_2\}$ a base of Δ with α_1 as the short root. Let $\{\Lambda_1, \Lambda_2\}$ be the corresponding fundamental weights, i.e.,

$$2(\Lambda_i, \alpha_j)/(\alpha_j, \alpha_j) = \delta_{ij} \qquad (1 \leqslant i,\ j \leqslant 2) \quad .$$

Thus $\Lambda_1 = 2\alpha_1 + \alpha_2$ and $\Lambda_2 = 3\alpha_1 + 2\alpha_2$. Let $m(i,j)$ be the multiplicity of the zero weight in the irreducible representation of the simple Lie algebra of type (G_2) with the highest weight $\lambda = i\Lambda_1 + j\Lambda_2$. Then by Kostant's formula [1] one has

$$m(i,j) = \sum_{w \in W} \det(w) P(w(\lambda + \delta) - \delta) \ , \tag{5.1}$$

where W is the Weyl group of Δ, $\delta = \Lambda_1 + \Lambda_2$ and P is the partition function associated to Δ, i.e., $P = P_A$ with $A = \begin{pmatrix} 1 & 0 & 1 & 2 & 3 & 3 \\ 0 & 1 & 1 & 1 & 1 & 2 \end{pmatrix}$. Put $P(i\alpha_1 + j\alpha_2) = P(i,j)$. Then $P(i,j) = 0$ if either i or j is negative. Thus by computing the sum (5.1) one has

$$\begin{aligned} m(i,j) = {} & P(2i+3j, i+2j) - P(2i+3j, i+j-1) \\ & + P(i+3j-1, j-2) - P(i+3j-1, i+2j) \\ & + P(i-4, i+j-1) \end{aligned} \tag{5.2}$$

Now the characteristic content $c(A)$ of the matrix A is 6. Hence by Theorem 1, every s.v.p.s. associated to A is of the form $h(t)/(1-t^6)^6$ where $h(t)$ is a polynomial in t. Since we need later several s.v.p.s.'s associated to A for the explicit formula of the $m(i,j)$, we give here some s.v.p.s.'s.

<u>Lemma 5.1</u>. *Let* $P(3j,j) = d_j$, $P(2j,j) = \Gamma_j$, $P(j,j) = \Delta_j$ *for* $j = 0,1,2,\ldots$. *Then*

$$\sum_{j=0}^{\infty} d_j t^j = (1-t)^{-4}(1-t^2)^{-1} , \tag{5.3}$$

$$\sum_{j=0}^{\infty} \Gamma_j t^j = (1+t+3t^2+2t^3+t^4)(1-t)^{-2}(1-t^2)^{-1}(1-t^3)^{-2} , \tag{5.4}$$

$$\sum_{j=0}^{\infty} \Delta_j t^j = (1-t)^{-2}(1-t^2)^{-1}(1-t^3)^{-2} . \tag{5.5}$$

Proof. We know by Theorem 1 that left hand sides of (5.3) ~ (5.5) are all rational functions of the form $h(t)/(1-t^6)^6$ with $h(t) \in \mathbb{C}[t]$. So we get the recursive relations for the sequences $\{d_j\}$, $\{\Gamma_j\}$, $\{\Delta_j\}$ which are obtained by expanding the polynomial $(1-t^6)^{6^j}$. On the other hand, for any given positive integer m, we can get the actual values for $P(i,j)$ $(0 \leqslant i, j \leqslant m)$ by using Kostant's recursive formula for the partition function P. It is enough to take $m = 36$ for our purpose. Thus we can actually determine the value of the index j from which the recursion formula is valid. Hence we can get a upper bound for the degree of the $h(t)$'s for (5.3) ~ (5.5). Then using the values of $P(i,j)$ $(0 \leqslant i, j \leqslant m)$, we can determine the $h(t)$'s, i.e., the right hand sides of (5.3) ~ (5.5). ∎

Remark. It is not difficult to get the generating functions (5.3) (resp. (5.5)) of "the boundary sequences" (d_j), (Δ_j) by rather simple considerations directly. (See the Table in Appendix 1 for the reason of the name "boundary sequence".) Namely, for the generating function $f_A(x,y) = \{(1-x)(1-y)(1-xy)(1-x^2y)(1-x^3y)(1-x^3y^2)\}^{-1}$ of (G_2), put $\psi(x,y) = f_A(x,y)^{-1}$. Put also

$$\psi_1(x,y) = \frac{\psi(x,y)}{1-x} , \qquad \psi_2(x,y) = \frac{\psi(x,y)}{1-y} .$$

Then $\psi_1(1,t)^{-1}$ (resp. $\psi_2(t,1)^{-1}$ gives the right hand side of (5.3) (resp. (5.5)).

However, we do not find yet to get a simple method to obtain generating functions of "the inner sequences" such as (Γ_j). ("Inner" is in the sense that the direction of (Γ_j) lies between two boundary sequences.)

Now let us introduce a notation. Let $\gamma_0, \gamma_1, \ldots, \gamma_m$ be complex numbers. Then we denote by $[\gamma_0, \gamma_1, \ldots, \gamma_m]_s$ the value at $t = s$ of the polynomial $\gamma_0 + \gamma_1 \cdot \binom{t}{1} + \cdots + \gamma_m \binom{t}{m}$ in t, where $\binom{t}{j} = t(t-1)\cdots(t-m+1)/j!$. Then one easily rewrites the results in Lemma 5.1 using the above notation. For Δ_j, Γ_j we get the following formulas according to the value of j modulo 6:

$$\Delta_{6n} = [1, 30, 125, 168, 72]_n ,$$

$$\Delta_{6n+1} = [2, 42, 148, 180, 72]_n ,$$

$$\Delta_{6n+2} = [4, 57, 173, 192, 72]_n ,$$

$$\Delta_{6n+3} = [8, 76, 200, 204, 72]_n ,$$

$$\Delta_{6n+4} = [13, 98, 229, 216, 72]_n ,$$

$$\Delta_{6n+5} = [20, 124, 260, 228, 72]_n ,$$

$$\Gamma_{6n} = [1, 109, 673, 1140, 576]_n ,$$

$$\Gamma_{6n+1} = [3, 166, 823, 1236, 576]_n ,$$

$$\Gamma_{6n+2} = [9, 242, 989, 1332, 576]_n ,$$

$$\Gamma_{6n+3} = [20, 339, 1171, 1428, 576]_n ,$$

$$\Gamma_{6n+4} = [38, 459, 1369, 1524, 576]_n ,$$

$$\Gamma_{6n+5} = [67, 606, 1583, 1620, 576]_n .$$

For d_j we get the following formulas according to the value of j modulo 2:

$$d_{2n} = [1, 10, 25, 24, 8]_n ,$$

$$d_{2n+1} = [4, 20, 36, 28\ 8]_n .$$

Now the stability and sum-formula in Theorem 2 become as follows for the partition function of (G_2).

(stability) $\quad j \geqslant i$ implies $P(i,j) = \Delta_i$,

$\qquad i \geqslant 3j$ implies $P(i,j) = d_j$,

$$\text{(sum-formula)}\qquad \alpha+\beta = j-1 \quad\text{implies}\quad P(j,\alpha)+P(j,\beta) = \Delta_j\ ,$$
$$\alpha+\beta = 3j-1 \quad\text{implies}\quad P(\alpha,j)+P(\beta,j) = d_j\ .$$

Using these properties, we get easily from (5.2) the following equality.

$$m(i,j) = \Delta_{2i+3j} - 2P(2i,i-1) - 2d_{i+j-1} + 2d_{i-1}$$
$$- \Delta_{i+3j-1} + 2d_{j-2} + \Delta_{i-4}\ .$$

Furthermore, since $P(2i,i-1) = \Delta_{2i} - P(2i,i) = \Delta_{2i} - \Gamma_i$ by sum-formula, we get finally an expression of $m(i,j)$ in terms of the $\{\Gamma_\nu\}$, $\{\Delta_\nu\}$, $\{d_\nu\}$. Thus using the previous expressions of the Γ_ν, Δ_ν, d_ν, we get, by a little computation, finally the explicit formula for $m(i,j)$. This completes the proof of Theorem 4. ■

As Appendices, we give tables of the values $P(i,j)$, $m(i,j)$ for (G_2). Especially from the table for $P(i,j)$ one can observe the stability and sum-formula very clearly.

Remark 1. Explicit formulas for the multiplicity of zero weight in a given irreducible representation for root systems of type (A_2), (B_2) are also available as in Theorem 4. But the computations for these cases are so easy that we give here only the results.

Let $\{\alpha_1,\alpha_2\}$ be simple roots such that $\|\alpha_1\| \leqslant \|\alpha_2\|$ and $\{\Lambda_1,\Lambda_2\}$ be the associated fundamental weights. Let $m(i,j)$ be the multiplicity of the zero weight in the irreducible representation of the corresponding simple Lie algebras with the highest weight $\lambda = i\Lambda_1 + j\Lambda_2$. Then the results are as follows.

Proposition 5.1. (i) *For the root system* (A_2),

$$m(i,j) = \begin{cases} 1+\mathrm{Min}(i,j), & \text{if } i \equiv j \pmod{3}\ , \\ 0, & \text{otherwise}\ . \end{cases}$$

(ii) *For the root system* (B_2),

$$m(i,j) = \begin{cases} \frac{ij}{2} + \left[\frac{i+j}{2}\right] + 1, & \text{if } i \equiv 0 \pmod{2}, \\ 0, & \text{otherwise}\ . \end{cases}$$

Remark 2. The property stated in Theorem 1 concerning a given power series $f(x,y) = \Sigma\, a_{ij}x^iy^j$ that all s.v.p.s. of f are rational functions does not imply in general that $f(x,y)$ is of the form given in Theorem 1; i.e., f is of the form $\prod_{i=1}^{m} (1 - x^{\alpha_i} y^{\beta_i})^{-1}$. For example, let

$$a_{ij} = \begin{cases} 1, & \text{if } j \text{ divides } i, \\ 0, & \text{otherwise} \end{cases}.$$

Then one can show that all s.v.p.s. of $f(x,y)$ are rational. But one can also show that $f(x,y)$ is not a rational function in x, y.

6. Acknowledgment

The authors are very grateful to Y. Ihara for suggesting the converse of Theorem 2, to S. Nakajima for the beautiful formulations of the stability and sum-formula, to K. Aomoto for suggesting the use of Cauchy integral formula for the coefficients of the power series $f_A(x_1,\ldots,x_\ell)$ in §1. Actually, our first proof of Theorem 1 was based on this method together with very lengthy calculations. The authors also thank T. Tokuyama and H. Yamaguchi for pointing out that in the formulation of stability and sum-formula, the "starting values" should satisfy the "smoothness condition" controlled by the (c_{ij}) in §4. They showed us the case $A = \begin{pmatrix} 1 & 0 & 2 \\ 0 & 1 & 2 \end{pmatrix}$. Then $f_A(x,y)$ satisfy stability and sum-formula; but the "starting values" lie on a graph of a discontinuous step function.

Appendix 1. Table for P(i, j).

i \ j	0	1	2	3	4	5	6	7	8	9	10	11	12	13	14	15	16	17	18	19
0	1	1	1	1	1	1	1	1	1	1	1	1	1	1	1	1	1	1	1	1
1	1	2	2	2	2	2	2	2	2	2	2	2	2	2	2	2	2	2	2	2
2	1	3	4	4	4	4	4	4	4	4	4	4	4	4	4	4	4	4	4	4
3	1	4	7	8	8	8	8	8	8	8	8	8	8	8	8	8	8	8	8	8
4	1	4	9	12	13	13	13	13	13	13	13	13	13	13	13	13	13	13	13	13
5	1	4	10	16	19	20	20	20	20	20	20	20	20	20	20	20	20	20	20	20
6	1	4	11	20	27	30	31	31	31	31	31	31	31	31	31	31	31	31	31	31
7	1	4	11	22	33	40	43	44	44	44	44	44	44	44	44	44	44	44	44	44
8	1	4	11	23	38	50	57	60	61	61	61	61	61	61	61	61	61	61	61	61
9	1	4	11	24	42	60	73	80	83	84	84	84	84	84	84	84	84	84	84	84
10	1	4	11	24	44	67	87	100	107	110	111	111	111	111	111	111	111	111	111	111
11	1	4	11	24	45	72	99	120	133	140	143	144	144	144	144	144	144	144	144	144
12	1	4	11	24	46	76	110	140	162	175	182	185	186	186	186	186	186	186	186	186
13	1	4	11	24	46	78	117	156	188	210	223	230	233	234	234	234	234	234	234	234
14	1	4	11	24	46	79	122	169	212	245	267	280	287	290	291	291	291	291	291	291
15	1	4	11	24	46	80	126	180	234	280	314	336	349	356	359	360	360	360	360	360
16	1	4	11	24	46	80	128	187	251	310	358	392	414	427	434	437	438	438	438	438
17	1	4	11	24	46	80	129	192	264	336	399	448	482	504	517	524	527	528	528	528
18	1	4	11	24	46	80	130	196	275	359	438	504	554	588	610	623	630	633	634	634
19	1	4	11	24	46	80	130	198	282	376	470	554	622	672	706	728	741	748	751	752
20	1	4	11	24	46	80	130	199	287	389	497	599	687	756	806	840	862	875	882	885
21	1	4	11	24	46	80	130	200	291	400	520	640	749	840	910	960	994	1016	1029	1036
22	1	4	11	24	46	80	130	200	293	407	537	673	803	917	1010	1080	1130	1164	1186	1199
23	1	4	11	24	46	80	130	200	294	412	550	700	850	988	1106	1200	1270	1320	1354	1376

Appendix 2. Table for $m(i,j)$

j \ i	0	1	2	3	4	5	6	7	8	9	10
0	1	1	3	5	8	12	18	24	33	43	55
1	2	4	9	16	24	36	51	68	90	116	145
2	5	10	21	35	52	75	104	136	177	224	277
3	9	20	39	64	93	132	179	232	297	372	455
4	16	35	66	105	151	210	281	360	456	565	686
5	25	56	102	160	227	312	412	524	657	808	974
6	38	84	150	231	325	441	577	728	906	1106	1326
7	54	120	210	320	446	600	778	976	1206	1464	1746
8	75	165	285	429	594	792	1020	1272	1563	1887	2241
9	100	220	375	560	770	1020	1305	1620	1980	2380	2815
10	131	286	483	715	978	1287	1638	2024	2463	2948	3475
11	167	364	609	896	1219	1596	2021	2488	3015	3596	4225
12	210	455	756	1105	1497	1950	2459	3016	3642	4329	5072
13	259	560	924	1344	1813	2352	2954	3612	4347	5152	6020
14	316	680	1116	1615	2171	2805	3511	4280	5136	6070	7076
15	380	816	1332	1920	2572	3312	4132	5024	6012	7088	8244

Appendix 3

The characteristic content $c(A)$ of the matrix A associated to the root system of a complex simple Lie algebra is given as follows.

$c(A_\ell) = 1$ $(\ell \geq 1)$, $c(G_2) = 3!$

$c(B_\ell) = 2^{[\ell/2]}$ $(\ell \geq 2)$, $c(F_4) = 4!$

$c(C_\ell) = 2^{\ell-1}$ $(\ell \geq 3)$, $c(E_6) = 3!$

$c(D_\ell) = 2^{[\ell/2]-1}$ $(\ell \geq 4)$, $c(E_7) = 4!$

$c(E_8) = 6!$

For exceptional types $(G_2) \sim (E_8)$, $c(A)$ is the product of the coefficients of the highest root w.r.t. simple roots taken without repetitions.

7. References

[1] B. Kostant, "A formula for the multiplicity of a weight," *Trans. A.M.S.*, vol. 93, 53-73, 1959.

[2] N. Bourbaki, *Groupes et algèbre de Lie*, Chap. 6, Hermann, Paris, 1968.

University of Tokyo
Bunkyoku, Tokyo 531
Japan

(Received January 22, 1981)

STABLE REAL COHOMOLOGY OF ARITHMETIC GROUPS II

Armand Borel

Given a discrete subgroup Γ of a connected real semisimple Lie group G with finite center there is a natural homomorphism

$$j_\Gamma^q\colon I_G^q \to H^q(\Gamma;\mathbf{C}) \;, \qquad (q=0,1,\ldots) \;, \tag{1}$$

where I_G^q denotes the space of G-invariant harmonic q-forms on the symmetric space quotient $X = G/K$ of G by a maximal compact subgroup K. If Γ is cocompact, this homomorphism is injective in all dimensions and the main objective of Matsushima in [19] is to give a range $m(G)$, independent of Γ, in which j_Γ^q is also surjective. The main argument there is to show that if a certain quadratic form depending on q is positive non-degenerate, then any Γ-invariant harmonic q-form is automatically G-invariant. In [3], we proved similarly the existence of a range in which j_Γ^q is bijective when Γ is arithmetic, but not necessarily cocompact. There are three main steps to the proof: (i) The cohomology of Γ can be computed by using differential forms which satisfy a certain growth condition, "logarithmic growth," at infinity; (ii) up to some range $c(G)$, these forms are all square integrable; and (iii) use the fact, pointed out in [16], that for $q \leqq m(G)$, Matsushima's arguments remain valid in the non-compact case for square integrable forms.

The first purpose of the present paper is to generalize and sharpen the results of [3] in several ways. First of all we shall also consider the case of non-trivial coefficients, at any rate when they are defined by a finite dimensional complex representation E of G. This extension could already have been easily carried out in [3], but was not chiefly for lack of applications. However, recent work on the rational homotopy type of diffeomorphism groups [15] shows that it may be useful. In fact, it is done there for $\mathbf{SL}_n\mathbf{Z}$ and the adjoint representation. Second we shall consider other growth conditions and show (3.4) that

$H^*(\Gamma;E)$ can also be computed by using forms which, together with their exterior differential, are either of moderate growth or weakly λ-bounded (where λ is a dominant linear form on the Lie algebra of a maximal $\mathbf{Q}$-split torus, see 3.2 for these notions). The proof is the same as that of the special case studied in [3: 7.4], and makes use of sheaf theory in the manifold with corners $\Gamma\backslash\bar{X}$ constructed in [8]. Those forms are square integrable up to a constant $C(G,\lambda,\tau^*)$ defined in §2, see (3.6). These growth conditions are expressed in terms of special frames in Siegel sets. In 3.10, we compare them with more usual notions of growth for functions on $\Gamma\backslash G$. Finally, as in [11], and following a development which has its origins in [20] and [22], we shall use relative Lie algebra cohomology and infinite dimensional unitary representations occurring in the spectrum of Γ, rather than Matsushima's original argument, and so can avail ourselves of some vanishing theorems recalled in 4.1. This leads again to the isomorphism of $I_G^{q\Gamma}$ and $H^q(\Gamma;\mathbf{C})$, but also to the vanishing of $H^q(\Gamma;E)$ in a certain range when E has no non-zero trivial subrepresentation (4.4). This range is the minimum of $C(G,\lambda,\tau^*)$ and of a constant defined by the vanishing of certain relative Lie algebra cohomology spaces (4.1). Since much information is known on the latter, this makes it worthwhile to study the former in more detail than in [3], and this is done in §2. Propositions 4.5 and 4.7 give two applications of these estimates.

Section 5 provides a counterpart to 3.4, which allows one to compute the cohomology with compact supports of $\Gamma\backslash X$ by means of forms which, together with their exterior differentials, are either fast decreasing (3.2) or weakly λ-bounded for $\lambda < 0$ (5.2). From this and 3.4 it follows that a non-zero fast decreasing Γ-invariant harmonic form is not cohomologous to zero in $H^*(\Gamma;E)$ and is cohomologous to a closed form with compact support (5.3). This applies in particular to harmonic cusp forms (5.5).

Finally, in §6, we show that 4.4 remains valid for S-arithmetic groups and groups of rational points, with essentially the same bounds (6.4). The main argument to effect this transition is contained in 6.2 and makes essential use of Bruhat-Tits buildings. As a result the stability theorems of [3] extend to S-arithmetic groups and groups of rational points and their consequences for the groups K_i of Quillen and ${}_{\varepsilon}L_i$ of Karoubi are also valid for rings of S-integers and number fields (6.5).

Some of the results proved here have been announced earlier, in particular in [2], [4: 6.1] and [6: Theorem 2].

Notation and Conventions. This paper is a sequel to [3], and we assume familiarity with it, in particular as regards Siegel sets and the representation of the invariant metric and differential forms with respect to special frames. However, we shall let K act on the right and G, Γ on the left, which causes some changes of signs and permutations of factors. For instance, A_t will now denote the subset of A where the simple roots are $\geqq t$ (1.4). This is understood in the sequel. The Lie algebra of a Lie group $H, U, \ldots$ is denoted either by $L(H)$, $L(U), \ldots$ or by the corresponding lower case German letter $\mathfrak{h}, \mathfrak{u}, \ldots$.

The group G of this introduction will be replaced by the group $G(\mathbf{R})$ of real points of a (Zariski)-connected reductive $\mathbf{Q}$-group. It is not necessarily connected (in the ordinary topology); this is why we have to replace I_G by I_G^Γ in certain statements.

G *is a connected isotropic reductive* $\mathbf{Q}$*-group without non-trivial rational characters defined over* $\mathbf{Q}$, K *a maximal compact subgroup of* $G(\mathbf{R})$, θ *the Cartan involution of* $G(\mathbf{R})$ *with respect to* K [8], $X = G(\mathbf{R})/K$. *Moreover,* (τ, E) *is a finite dimensional complex rational representation of* G. *We assume* E *to be endowed with an admissible scalar product* [11: II, 2.2]. *In* §§1 *to* 5, Γ *is an arithmetic subgroup of* G.

1. Preliminaries on Reductive Groups

1.1 Let P be a parabolic $\mathbf{Q}$-subgroup of G, U its unipotent radical and $\sigma_P \colon P \to P/U$ the canonical projection. The Levi subgroups defined over $\mathbf{R}$ of P are then mapped isomorphically on P/U by σ_P. Let Z_d be the greatest central $\mathbf{Q}$-split torus of P/U. We let A_P denote the (topological) identity component of $Z_d(\mathbf{R})$. Any subgroup belonging to a Levi subgroup of $P(\mathbf{R})$ and mapped isomorphically onto A_P by σ_P will be called a *split component relative to* $\mathbf{Q}$ *of* P. In particular, if S is a maximal $\mathbf{Q}$-split torus in the radical of P, then $S(\mathbf{R})^o$ is a split component rel.$\mathbf{Q}$. However, such a choice may be too restrictive. There is one and only one split component rel.$\mathbf{Q}$ which is stable under the Cartan involution θ associated to K (see, e.g., [8]). It is not necessarily contained in a $\mathbf{Q}$-split torus, though. We shall refer to it as the split component rel.$\mathbf{Q}$ of P. Unless otherwise stated, Levi subgroups and split components rel.$\mathbb{Q}$ are assumed to be θ-stable. The split components rel.$\mathbb{Q}$ are conjugate under $N(\mathbb{R})$.

1.2 Let now P_o be a minimal parabolic $\mathbb{Q}$-subgroup. The group P is conjugate to a unique parabolic $\mathbb{Q}$-subgroup containing P_o and, as is well-known, this yields a unique monomorphism $A_P \to A_{P_o}$. Using the projections σ_P and σ_{P_o}, we also get a unique monomorphism $A \to A'$, where A (resp. A') is any split component rel.$\mathbb{Q}$ of P (resp. P_o), whence also a canonical epimorphism $r_{PP_o}: X(A') \to X(A)$, where, as in [3], $X(H)$ denotes the commutative group of continuous homomorphisms of the real Lie group H into the multiplicative group $\mathbb{R}_+^*$ of strictly positive real numbers. In particular, an element $\lambda \in X(A_{P_o})$ defines an element $r_{P,P_o}(\lambda)$ of $X(A)$ for every parabolic $\mathbb{Q}$-subgroup P, which we shall also denote simply by λ.

1.3 We let $\Phi(P,A)$ denote the set of roots of P with respect to A and $\Delta(P,A) = \{\alpha_1,\ldots,\alpha_\ell\}$ $(\ell = \dim A)$ the set of simple roots of P with respect to A. An element

$$\lambda = \sum_{\alpha\in\Delta} c_\alpha(\lambda)\alpha \qquad ,$$

is dominant (resp. dominant regular) if $c_\alpha \geqq 0$ (resp. $c_\alpha > 0$) for all α. If $\lambda \in X(A_o)$ is dominant (resp. dominant regular), then $r_{PP_o}(\lambda)$ is dominant (resp. dominant regular) for any proper parabolic $\mathbb{Q}$-subgroup P. For $\lambda,\mu \in X(A)$ we write $\lambda \geqq \mu$ (resp. $\lambda > \mu$) if $\lambda - \mu \geqq 0$ (resp. $\lambda - \mu > 0$), and $\lambda \leqq 0$ (resp. $\lambda < 0$) if $-\lambda \geqq 0$ (resp. $-\lambda > 0$). As usual, ρ_P is defined by

$$a^{2\rho_P} = \det \operatorname{Ad} a|_{\mathfrak{n}} \qquad .$$

We have

$$\rho_P = r_{PP_o}\left(\rho_{P_o}\right) \qquad .$$

Let $\mathfrak{h}$ be a Cartan subalgebra of $\mathfrak{g}$ containing $L(A_o)$. We assume the root system $\Phi(\mathfrak{g}_c,\mathfrak{h}_c)$ be given an order compatible with $\Phi(P_o,A_o)$. Let

$$\rho = \frac{1}{2} \sum_{\beta>0} \beta \qquad .$$

Then

$$\rho|_{\mathfrak{a}_o} = \rho_{P_o} \quad .$$

1.4 In view of our shift from right to left for the action of Γ, the set A_t is defined here by

$$A_t = \{a \in A | a^{\alpha} \geqslant t , \qquad (\alpha \in \Delta(P,A)\} \quad . \tag{1}$$

1.5 Given $\lambda \in X(A)$ we denote by $C(G,P,\lambda)$ the greatest integer q such that

$$\rho_P > \lambda + \mu$$

for every weight μ of A in $\oplus \Lambda^q \mathfrak{n}$. It would of course be equivalent to letting μ run through the weights of A in $\oplus_{j \leq q} \Lambda^j \mathfrak{n}$.

We write simply $C(G,\lambda)$ for $C(G,P,\lambda)$ when $P = P_o$. It is immediate that

$$C(G,P,r_{PP_o}(\lambda)) \geq C(G,\lambda) \qquad (\lambda \in X(A_o)) \quad . \tag{1}$$

1.6 Let (σ,F) be a finite dimensional representation of G and $\lambda \in X(A_o)$. We let

$$C(G,\lambda,\sigma) = \inf_{\mu} C(G,\lambda+\mu) \quad ,$$

where μ runs through the restrictions to A_o of the weights of σ. It suffices of course to take the inf over the highest weights of the irreducible constituents of σ (for an ordering compatible with the one defined by P_o). For $\lambda = 0$, we denote this constant by $C(G,\sigma)$.

2. The Constant $C(G/k,\lambda)$

2.1 In this section we discuss the constant $C(G,\lambda)$, which we denote $C(G)$ or $C(G/\mathbf{Q})$ if $\lambda = 0$. It is equal to $C(\mathcal{D}G/\mathbf{Q},\lambda)$, where $\mathcal{D}G$ is the derived group of G, hence we may assume G to be semisimple. For the discussion it is also convenient to introduce this constant for more general groundfields than $\mathbf{Q}$. If H is a connected semisimple group defined over a field k, and $\lambda \in X(P_o)_k \otimes_{\mathbf{Z}} R$, we define then

$C(H/k,\lambda)$ as $C(G,P_o,\lambda)$ in 1.5. If k' is an extension of k, then any parabolic k-subgroup is a parabolic k' subgroup hence

$$C(H/k',\lambda') \leqq C(H/k,\lambda) \tag{1}$$

where λ is obtained from λ' by restriction to a maximal k-split torus. If L is a k-group which is k-isogeneous to H, then $C(H/k,\lambda) = C(L/k,\lambda)$. If H is k-isogeneous to a product H_i $(1 \leqq i \leqq s)$, then

$$C(H/k,\lambda) = \inf_i C(H_i/k,\lambda_i) \quad , \tag{2}$$

where λ_i is obtained from λ by restriction to a maximal k-split torus of H_i. In view of its definition, $C(H/k,\lambda)$ depends only on the relative root system ${}_k\Phi(H)$ and on the multiplicities of the roots. Given a root system Φ, a set of natural integers $\{m_\alpha\}$ $(\alpha \in \Phi)$, and $\lambda \in V[\Phi]$, where $V[\Phi]$ is the real vector space underlying the definition of Φ, fix an order on Φ, denote by ρ half the sum of the positive roots, each root α being counted m_α-times, and define $C(\Phi,\{m_\alpha\},\lambda)$ to be the greatest integer q such that $\rho > \mu + \lambda$ for any sum μ of q positive roots, where α occurs at most m_α times. Thus

$$C(H/k,\lambda) = C({}_k\Phi(H),\{m_\alpha\},\lambda) \quad , \tag{3}$$

where m_α is the dimension over k of the eigenspace in $L(H/k)$ corresponding to the root α and where $V[{}_k\Phi]$ is identified with $X(P_o)_k \otimes_{\mathbf{Z}} \mathbf{R}$ in the usual way.

If the m_α are all equal to one, we write $C(\Phi)$ for $C(\Phi,\{m_\alpha\})$, $C(\Phi,\lambda)$ for $C(\Phi,\{m_\alpha\},\lambda)$. In particular, if H splits over k, then

$$C(H/k,\lambda) = C(\Phi(H),\lambda).$$

2.2 Lemma. *Let* Φ *be a finite set,* C, d, m_α $(\alpha \in \Phi)$ *be strictly positive integers. Given a finite set* η *of elements belonging to* Φ, *let* $m_\eta(\alpha)$ *be the multiplicity of* α *in* η. *Let* ψ *be a finite set of elements of* Φ *of cardinality* $|\psi| \leqq d.C$ *and such that* $m_\psi(\alpha) \leqq d.m_\alpha$ *for all* $\alpha \in \Phi$. *Then* ψ *can be written as a disjoint union of subsets* ψ_i $(1 \leq i \leq d)$, *where* $|\psi_i| \leq C$ *and* $m_{\psi_i}(\alpha) \leq m_\alpha$ $(\alpha \in \Phi,\ 1 \leq i \leq d)$.

Proof by induction on d. There is nothing to prove if $d = 1$, so assume $d \geq 2$ and the lemma proved for $d - 1$. Let

$$\Phi_1 = \{\alpha \in \Phi \mid m_\psi(\alpha) > (d-1)m_\alpha\} \quad , \tag{1}$$

$$r_\alpha = m_\psi(\alpha) - (d-1)m_\alpha \quad , \quad (\alpha \in \Phi_1) \ . \tag{2}$$

We have then

$$r_\alpha \leq m_\alpha \quad , \quad (\alpha \in \Phi_1) \tag{3}$$

and also

$$\sum_{\alpha \in \Phi_1} r_\alpha \leq C \ . \tag{4}$$

Let θ be a subset of ψ which is maximal with respect to the following properties

$$|\theta| \leq C; \quad m_\theta(\alpha) \leq m_\alpha \quad (\alpha \in \Phi) \quad \text{and} \quad m_\theta(\alpha) \geq r_\alpha \ (\alpha \in \Phi_1). \tag{5}$$

Such subsets do exist in view of (3), (4). Set $\theta' = \psi - \theta$. By construction

$$m_{\theta'}(\alpha) \leq (d-1)m_\alpha \quad , \quad (\alpha \in \Phi) \ . \tag{6}$$

We claim moreover that $|\theta'| \leq (d-1)C$. This is clear if $\theta = |C|$. Assume now $|\theta| < C$. In view of the maximality assumption, we have then

$$m_\theta(\alpha) = \begin{cases} m_\alpha \quad , & (\alpha \in \Phi_1) \quad , \\ \min(m_\alpha, m_\psi(\alpha)), & (\alpha \in \Phi - \Phi_1) \end{cases} \tag{7}$$

$$m_{\theta'}(\alpha) = \begin{cases} m_\psi(\alpha) - m_\alpha & (\alpha \in \Phi - \Phi_1, m_\psi(\alpha) > m_\alpha) \\ 0 & (\alpha \in \Phi - \Phi_1, m_\psi(\alpha) \leq m_\alpha) \end{cases} \ . \tag{8}$$

It follows that

$$m_{\theta'}(\alpha) \leqq (d-1)m_\theta(\alpha) \quad , \qquad (\alpha \in \Phi) \quad , \tag{9}$$

whence

$$|\theta'| \leqslant (d-1)|\theta| < (d-1)C \quad . \tag{10}$$

We then take $\psi_1 = \theta$ and apply the induction assumption to θ'.

2.3 Lemma. *Let* Φ *be a root system* d, m_α $(\alpha \in \Phi)$ *strictly positive integers and let* $\lambda \in V[\Phi]$. *Then*

$$C(\Phi,\{dm_\alpha\},\lambda) \geqq d.C(\Phi,\{m_\alpha\},\lambda) \quad . \tag{1}$$

Given a set ψ of positive roots, let $\langle\psi\rangle$ be the sum of the elements in ψ and $m_\psi(\alpha)$ the multiplicity of α in ψ. Let $C' = C(\Phi,\{m_\alpha\},\lambda)$. Since $d\rho$ is half the sum of the positive roots with multiplicities dm_α, we have to prove

$$|\psi| \leqq d.C' \quad \text{and} \quad m_\psi(\alpha) \leqq d.m_\alpha \quad (\alpha \in \Phi) \Rightarrow d\rho > \lambda + \langle\psi\rangle \quad . \tag{2}$$

By 2.2 we can write ψ as a disjoint union of subsets ψ_i $(1 \leqslant i \leqslant d)$ such that $|\psi_i| \leqq C'$ and $m_{\psi_i}(\alpha) \leqq m_\alpha$ for all i and α. We have then

$$\rho > \lambda + \langle\psi_i\rangle \qquad (i = 1,\dots,d) \tag{3}$$

by assumption, whence

$$d\rho > d\lambda + \langle\psi\rangle \geqslant \lambda + \langle\psi\rangle \quad . \tag{4}$$

2.4 Remark. This estimate is not sharp. Assume for instance that $\Phi = \{\pm\alpha\}$ is of type $\mathbf{A}_1$ and that $m_\alpha = 1$. Then $C(\Phi,\{m_\alpha\}) = 0$. However

$$C(\mathbf{A}_1,d) = \left[\frac{d-1}{2}\right] \quad .$$

This is due to the fact that we may have $d\rho > \langle\psi\rangle$ for sets ψ with more than dC' elements, for which the proof of the inequality does not reduce to the case $d = 1$.

2.5 Proposition. *Let* k *be a field,* k' *a finite separable extension of* k. *Let* H^1 *be a connected semi-simple* k'*-group and*

$H = R_{k'/k}\, H'$, *where* $R_{k'/k}$ *refers to the restriction of scalars* [25: I]. *Then* $C(H/k) \geqq [k':k].C(H'/k')$.

Let $d = [k':k]$. There is a canonical isomorphism ${}_k\Phi(H) \to {}_{k'}\Phi(H')$ such that if α' corresponds to α, then $m_{\alpha'} = d.m_{\alpha}$ (cf. [10: 6.19, 6.21]). Our assertion then follows from 2.3, applied to H/k and H'/k', and from 2.1(3).

2.6 Remark. If P_o' is a minimal parabolic k'-subgroup of H', then $P_o = R_{k'/k}\, P_o$ is a minimal parabolic k-subgroup of H [10: loc. cit.], and there is a canonical isomorphism $X(P_o')_{k'} \otimes \mathbf{R} \to X(P_o)_k \otimes \mathbf{R}$ which is of course compatible with the isomorphism of relative root systems used above. If λ in the latter space corresponds to λ' in the former space by this isomorphism, then the same proof shows that we have

$$C(H/k,\lambda) \geqq [k':k]C(H'/k,\lambda') \quad . \tag{1}$$

2.7 Theorem. *Let* k *be a number field,* $\bar{k}$ *an algebraic closure of* k*, and* H *an almost simple* k*-group. Let* G_i $(1 \leqq i \leqq s)$ *be the simple factors of* $G/\bar{k}$*. Then* $C(H/k) \geqq \Sigma_i\, C(G_i/\bar{k})$.

There exists a finite extension k' of k contained in $\bar{k}$ and an absolutely almost simple k'-group H' such that H is isogeneous to $R_{k'/k}\, H'$ [10: 6.21]. Since $C(H/k)$ is the same for two k-isogeneous k-groups, we may replace H by $R_{k'/k}\, H'$. By 2.5

$$C(H/k) \geqq [k':k]C(H'/k') \quad . \tag{1}$$

We have $C(H'/k') \geqq C(H'/\bar{k})$. Now $H'/\bar{k}$ is one of the simple factors of $H/\bar{k}$. There are $[k':k]$ such factors, and they are isomorphic over $\bar{k}$. The theorem now follows from (1).

2.8 Remark. As a sequel to 2.4, we note that this is not necessarily sharp if H is not split over k. For instance, let $k = \mathbf{Q}$, $H' = \mathbf{SL}_2/k'$ and $H = R_{k/\mathbf{Q}}\, H'$. Then $C(H'/k') = 0$, but $C(H/k) = [(d-1)/2]$.

2.9 Using the tables of [12] one can compute $C(\Phi)$ for all irreducible reduced root systems. We get the following list:

Φ:	$\mathbf{A}_\ell (\ell \geqq 1)$,	$\mathbf{B}_\ell (\ell \geqq 3)$,	$\mathbf{C}_\ell (\ell \geqq 2)$,	$\mathbf{D}_\ell (\ell \geqq 4)$,	$\mathbf{E}_6$,	$\mathbf{E}_7$,	$\mathbf{E}_8$,	$\mathbf{F}_4$,	$\mathbf{G}_2$
$C(\Phi)$:	$[(\ell-1)/2]$,	$\ell - 1$,	$\ell - 2$,	$\ell - 2$,	7,	13,	25,	5,	1

2.10 Let k, k', H and H' be as in 2.7. Let (τ',E') be an absolutely irreducible rational representation of H' which is defined over k'. Then (τ,E), where $\tau = R_{k'/k}\tau'$, $E = R_{k'/k}E'$ is a rational representation of H defined over k. It is not irreducible, but a direct sum of irreducible representations of the simple $\bar{k}$-factors of H. If S' is a maximal k'-split torus of H', then the greatest k-split subtorus of $R_{k'/k}S$ is a maximal k-split torus of H, whence an isomorphism of $X(S')_{k'} \otimes \mathbf{R}$ onto $X(S)_k \otimes \mathbf{R}$ [10]. It maps the weights of τ' onto those of τ, but the multiplicity of the weight is multiplied by $[k':k]$. The restrictions to S of the weights of the constituents of τ are the same, with the original multiplicities. In particular there is only one highest weight. If $\lambda'_{\tau'}$ and λ_τ are the highest weights of τ' and τ, we have then with λ and λ' as before

$$C(H'/k',\lambda',\tau') = C({}_{k'}\Phi(H'),\lambda' + \lambda'_{\tau'}) \tag{1}$$

$$C(H/k,\lambda,\tau) = C({}_k\Phi(H),\lambda + \lambda_\tau) \quad . \tag{2}$$

Therefore 2.6(1) implies

$$C(H/k,\lambda,\tau) \geqq [k':k]C(H'/k',\lambda',\tau') \quad . \tag{3}$$

2.11 Consider in particular the case where $\lambda' = 0$ and $\tau' = Ad$ is the adjoint representation. Then τ is also the adjoint representation of H. Assume that H' is split over k'. The highest weight of τ' is the highest root δ_o. We have then

$$C(H/k,Ad) \geqq [k':k].C(\Phi(H'),\delta_o) \quad . \tag{4}$$

The values of $C(\Phi,\delta_o)$ can also be computed for all types by using the tables in [12]. One finds

Φ:	$\mathbf{A}_\ell$,	$\mathbf{B}_\ell (\ell \geqq 3)$,	$\mathbf{C}_\ell (\ell \geqq 2)$,	$\mathbf{D}_\ell (\ell \geqq 4)$,	$\mathbf{E}_6$,	$\mathbf{E}_7$,	$\mathbf{E}_8$,	$\mathbf{F}_4$,	$\mathbf{G}_2$
$C(\Phi,\delta_o)$:	$\left[\frac{\ell-3}{2}\right]$	$\ell - 2$,	$\ell - 4$,	$\ell - 4$,	6,	12,	23,	4,	0

3. Weak λ-boundedness of Differential Forms

3.1 If M is a manifold on which $G(\mathbb{R})$ operates (smoothly), then $\Omega(M;E)$ denotes the space of smooth differential E-valued forms on M. As usual, $G(\mathbb{R})$ acts on $\Omega^q(X;E)$ by the rule

$$(g\circ\omega)(x,Y_x) = \tau(g)(\omega(g^{-1}.x,g^{-1}.Y_x)),$$

$$(g\in G(\mathbb{R}),\ x\in M,\ Y_x \text{ a q-vector at } x).$$

For a subgroup H of $G(\mathbb{R})$, we let $\Omega(M;E)^H$ be the space of H-invariant elements in $\Omega(M;E)$. Our main case of interest is $M=X$, but we shall also on occasion take $M=G(\mathbb{R})$.

3.2 Growth Conditions for Differential Forms. Let P be a parabolic $\mathbb{Q}$-subgroup and A its split component. As usual $M(\mathbb{R}) = {}^oM(\mathbb{R})\times A$ is the (θ-stable) Levi subgroup of P, and $Z={}^oM(\mathbb{R})/(K\cap M)$. Let 0 be a fixed point of K in X and

$$\mu_o:\ N(\mathbb{R})\times Z\times A = Y\to X \tag{1}$$

be as in [3]. Let $\eta\in\Omega^j(X;E)$. With respect to a special frame, we write again

$$\mu_o^*(\eta) = \sum_I \eta_I\omega^I \quad , \tag{2}$$

where, however, the coefficients η_I are smooth E-valued functions. Fix $\lambda\in X(A_o)$. We say that η is weakly λ-bounded if, given P and a Siegel set $\mathfrak{S}_{t,\omega}=\mu_o(\omega\times A_t)$, there exists a polynomial P in $\ell=\dim A$ variables such that

$$|\mu_o^*(\eta)_I(q,a)| \prec a^\lambda|P(\ell n\ a^{\alpha_1},\dots,\ell n\ a^{\alpha_\ell})| \ , \quad (a\in A_t;\ q\in\omega)\ . \tag{3}$$

On the left-hand side $|\ |$ refers to the norm on E. on the right-hand side, $\alpha_1,\dots,\alpha_\ell$ are the simple roots of P with respect to A. Since the precise nature of P will not intervene, we shall also write this

$$|\mu_o^*(\eta)_I(q,a)| \prec_w a^\ell \quad , \qquad (a\in A_t;\ q\in\omega)\ . \tag{4}$$

We note that if $\lambda=0$, and $E=\mathbb{C}$, this is the condition of logarithmic growth in [3].

The form η is λ-*bounded* if

$$|\mu_o^*(\eta)_I(q,a)| \prec a^{\lambda} , \qquad (a \in A_t ;\ q \in \omega) . \tag{5}$$

It is of *moderate growth* if it is λ-bounded for some $\lambda \geqq 0$, and *fast decreasing* if it is λ-bounded for all λ's.

3.3 Notation. Let $\lambda \in X(A_o)$. We let $\Omega_{\lambda}^{w}(X;E)^{\Gamma}$ denote the spaces of smooth E-valued Γ-invariant forms η on X such that η and $d\eta$ are weakly λ-bounded. Moreover, $\Omega_{fd}(X;E)^{\Gamma}$ (resp. $\Omega_{mg}(X;E)^{\Gamma}$) will denote the complex of forms in $\Omega(X;E)^{\Gamma}$ which, together with their exterior differentials, are fast decreasing (resp. of moderate growth).

If $\lambda = 0$ and $E = \mathbb{C}$, this is just the complex C of Theorem 7.4 in [3], consisting of forms which, together with their exterior differentials, have logarithmic growth near the boundary.

3.4 Theorem. *Assume* $\lambda \in X(A_o)$ *to be dominant. Then the inclusions*

$$\Omega_{\lambda}^{w}(X;E)^{\Gamma} \to \Omega_{mg}(X;E)^{\Gamma} \to \Omega(X;E)^{\Gamma} \tag{1}$$

induces isomorphisms in cohomology.

The proof differs only in minor details from that of Theorem 7.4 in [3] and we shall refer to the latter to the extent possible.

Let Γ' be a normal subgroup of finite index of Γ. Then we have

$$\Omega_{\lambda}^{w}(X;E)^{\Gamma} = (\Omega_{\lambda}^{w}(X;E)^{\Gamma'})^{\Gamma/\Gamma'} , \tag{2}$$

$$H^*(\Omega_{\lambda}^{w}(X;E)^{\Gamma} = (H^*(\Omega_{\lambda}^{w}(X;E)^{\Gamma'})^{\Gamma/\Gamma'} , \tag{3}$$

and similarly for $\Omega_{mg}(X;E)^{\Gamma}$ and $\Omega(X;E)^{\Gamma}$. We may therefore replace Γ by Γ', hence assume Γ to be torsion-free. Then E gives rise to a locally constant sheaf $\tilde{E}$ on the manifold with corners $\Gamma\backslash\bar{X}$. We define a presheaf F on $\Gamma\backslash\bar{X}$ by assigning to an open set $U \subset \Gamma\backslash\bar{X}$ the space of $\tilde{E}$-valued forms on $U \cap (\Gamma\backslash X)$ which, together with their exterior differentials, are weakly λ-bounded (resp. of moderate growth) near the boundary. This is clearly a sheaf.

In the corner which is the closure of the Siegel set of 3.2, we may take as local coordinates $\beta^i = a^{-\alpha_i}$ $(1 \leqq i \leqq \ell)$ and local coordinates x^j $(\ell < j \leqq n = \dim X)$ in ω. Hence the $d\beta^i$ $(i \leqq \ell)$ and the ω^j $(j > \ell)$

form a local frame of the cotangent bundle which extends smoothly to a local frame on the corner. If φ is a smooth function on $\Gamma\backslash\bar{X}$, then $d\alpha$ is a linear combination of the $d\beta^i$, ω^j with bounded coefficients. Since $d\beta^i/\beta^i = -da^{\alpha_i}/a^{\alpha_i}$, it follows *a fortiori* that the coefficients $(d\varphi)_j$ of $d\varphi$, expressed as linear combination of the ω^j, are bounded. We even have

$$(d\varphi)_j \prec a^{-\alpha_j} \quad , \qquad (1 \leqq j \leqq \ell) \quad , \tag{4}$$

in $\mathfrak{S}_{t,\omega}$. It follows then that if η and $d\eta$ are weakly λ-bounded (resp. λ-bounded, resp. of moderate growth), then so are $\varphi\eta$ and $d\varphi.\eta$. As a consequence, F is fine. There remains to see that it provides a resolution of $\tilde{E}$. Since λ is *dominant*, $\Gamma(U)$ contains the smooth E-valued functions, hence $H^o(F) = \tilde{E}$. The main point is then again to check that $H^q(F) = 0$ for $q \geqq 1$. For this, it suffices to see that near a boundary point, the Poincaré lemma remains valid for weakly λ-bounded forms, or forms of moderate growth. This amounts to showing that the homotopy operator A of [3: p. 258] preserves these conditions. By (5), (7) *loc. cit.*, if η is of degree q, we have

$$A\eta = \sum_I c_I \omega^I \quad , \tag{5}$$

where

$$c_I = \sum_j \pm x^j \int_0^1 t^{q-1} \eta_{I\cup\{j\}}(tx)\, dt \quad , \tag{6}$$

the sum running through the indices j not belonging to I. By assumption,

$$\lambda = \sum_{1 \leqq i \leqq \ell_j} d_i \alpha_i \quad , \qquad \text{with } d_i \geqq 0 \text{ for } i = 1,\ldots,\ell. \tag{7}$$

Set

$$d(x) = \sum_{1 \leqq i \leqq \ell} d_i . x^i \quad . \tag{8}$$

We are interested only in $A\eta$ in some arbitrarily chosen smaller Siegel set, so we may assume $x^i \geqq D$ for $i = 1,\ldots,\ell$ and some $D \geqq 1$

and also $d(x) > 1$ unless $\lambda = 0$. Note also that the x^i for $i > \ell$ vary in a bounded interval. Assume now η to be weakly λ-bounded. There exist then $C, M > 0$ such that

$$|\eta_J(tx)| \leqq C.e^{td(x)}.(x^1 \dots x^\ell)^M \quad , \qquad (x \in \mathfrak{S},\ t \in [0,1]) \ . \tag{9}$$

It is then elementary that we can find $C', M' > 0$ such that

$$\left|\int_0^1 t^{q-1}\eta_J(tx)\ dt\right| \leqq C'.(x^1 \dots x^\ell)^{M'}.e^{d(x)} \quad , \qquad (x^i \geqq D,\ i = 1,\dots,\ell) \tag{10}$$

which implies that $A\eta$ is also weakly λ-bounded. It then also follows that $A\eta$ is of moderate growth if η is so, whence our assertion.

3.5 Scalar Product of Differential Forms. The elements of $\Omega(X;E)^\Gamma$ are smooth sections of a Γ-bundle whose fibre at $x \in X$ is $\Lambda T_x^*(X) \otimes E$. On the latter, we put hermitian product which is the tensor product of the product stemming from the metric on X by the admissible product on E, to be denoted $(\ ,\ )_x$.

As is well-known (cf. e.g. [11; 21]), we may identify $\Omega(X;E)^\Gamma$ with the space of smooth cross-sections of K-bundle over $\Gamma\backslash X$ by means of the map

$$\eta \mapsto \eta^o \ , \qquad \text{where} \qquad \eta^o(g) = \tau(g^{-1})(\eta\circ\pi)(g) \quad , \tag{1}$$

where $\pi: G(\mathbf{R}) \to X$ is the canonical projection. Since the hermitian product of the typical fibre is K-invariant, this allows one to define a scalar product on $\Omega(X;E)^\Gamma$. It is usually written

$$(\alpha,\beta) = \int_{\Gamma\backslash G(\mathbf{R})} (\alpha_g^o,\beta_g^o)\,dg \quad , \qquad (\alpha,\beta \in \Omega(X,E)^\Gamma) \ , \tag{2}$$

where dg is a Haar measure. The integrand is right invariant under K, so that the integral is in effect on $\Gamma\backslash X$. It is more convenient here for us to write it directly on $\Gamma\backslash X$, as

$$(\alpha,\beta) = \int_{\Gamma\backslash X} (\tau(g_x^{-1})\alpha_x, \tau(g_\alpha)^{-1}\beta_x)\ dv_x \quad , \tag{3}$$

where dv_x is the Riemannian volume element on X and g_x denotes an element in $G(\mathbf{R})$ which brings 0 onto x.

We let $H^*(\Gamma;E)_{(2)}$ denote the space of elements in $H^*(\Gamma;E)$ which are representable by a square integrable closed form. By a theorem of Kodaira, recalled in [3: 2.4], these elements are also representable by square integrable harmonic forms.

3.6 Lemma. *Assume that* (τ,E) *is irreducible. Let* $\lambda \in X(A_o)$. *Then, if* $j \leqq C(G,\lambda,\tau^*)$, *any weakly* λ-*bounded* E-*valued* Γ-*invariant* j-*form* η *is square integrable.*

It suffices to check that η is square integrable on any Siegel set $\mathfrak{S}_{t,\omega}$, with respect to any proper parabolic $\mathbf{Q}$-subgroup P. We use the notation of [3: 5.5]. We have then

$$(\eta,\eta)_{\mathfrak{S}} = \int_{A_t \times \omega} \sum_{I,J} g^{IJ}(\tau(a^{-1}q^{-1})\eta_I, \tau(a^{-1}q^{-1})\eta_J) a^{-2\rho_P} \, dv_A dv_Z dv_N \, . \tag{1}$$

Let (e_ν) be an orthonormal basis of E consisting of eigenvectors of A, where $\nu \in X(S)$ is the weight of e_ν. We can write

$$\tau(q^{-1})\eta_I(q,a) = \sum_\nu \eta_{I,\nu}(q,a) e_\nu \tag{2}$$

$$\tau(a^{-1})\tau(q^{-1})\eta_I(q,a) = \sum_\nu \eta_{I,\nu}(q,a) a^{-\nu} e_\nu \qquad (q \in \omega, \ a \in A) \, . \tag{3}$$

Since ω is compact, we get from (2)

$$|\tau(q^{-1})\eta_I(q,a)|^2 = \Sigma |\eta_{I,\nu}(q,a)|^2 \asymp |\eta_I(q,a)|^2 \, . \tag{4}$$

The linear form $-\nu$ is a weight of (τ^*,E^*), hence, if δ denotes the highest weight of τ^*:

$$\delta \geqslant -\nu \tag{5}$$

$$a^{-2\nu} \prec a^{2\delta} \qquad (a \in A_t) \, . \tag{6}$$

Together with (3), (4) this yields

$$|\tau(a^{-1}q^{-1})\eta_I(q,a)|^2 \qquad a^{2\delta}|\eta_I(q,a)|^2 \, , \qquad (a \in A_t; \ q \in \omega) \, . \tag{7}$$

This reduces the estimate of the sum on the right-hand side of (1) to the scalar case considered in [3]. We get then, using (11), (12) in [3: 5.5]:

$$\left|\sum_{I,J} g^{IJ}(\tau(a^{-1}q^{-1})\eta_I,\tau(a^{-1}q^{-1})\eta_J\right| \leqq \sum |\eta_J|^2 a^{2\alpha(J)+2\delta} \tag{8}$$

$$\left|\sum_{I,J} g^{IJ}(\tau(a^{-1}q^{-1})\eta_I,\tau(a^{-1}q^{-1})\eta_J)\right| \prec_w \sum_J a^{2(\alpha(J)+\lambda+\delta)} \quad (a \in A_t, q \in \omega) \;, \tag{9}$$

which, by [3: 5.4], is integrable on $A_t \times \omega$ if

$$\rho_P > \alpha(J) + \lambda + \delta \tag{10}$$

for all J with j-elements. Since the $\alpha(J)$ are the weights of A in $\Lambda^j\mathfrak{u}$, this condition is indeed fulfilled whenever $j \leqslant C(G,\lambda,\tau^*)$.

3.7 Proposition. *Let* $\lambda \in X(A)$ *be dominant.*

(i) *Let* $j \leqq C(G,\lambda,\tau^*)$. *Then any* j-*form in* $\Omega^w_\lambda(X;E)$ *is square integrable and any cohomology class in* $H^j(\Gamma;E)$ *is representable by a square integrable harmonic form.*

(ii) *For* $j \leqq C(G,\lambda,\tau^*) + 1$, *the space of harmonic square integrable* j-*forms contained in* $\Omega^w_\lambda(X;E)^\Gamma$ *maps injectively into* $H^j(\Gamma;E)$.

(i) The first assertion is a special case of 3.6, applied to each irreducible constituent of E. In view of 3.4, any element of $H^j(\Gamma;E)$ is then representable by a closed square integrable form, hence by a harmonic one (3.5).

(ii) Let now $j \leqq C(G,\lambda,\tau^*) + 1$ and η be a harmonic square integrable j-form contained in $\Omega^w_\lambda(X;E)^\Gamma$. [Of course, by (i), the assumption of square integrability is redundant if $j \leqq C(G,\lambda,\tau^*)$.] Assume η is cohomologous to zero in $\Omega(X;E)^\Gamma$. Then, by 3.4, it is already cohomologous to zero in $\Omega^w_\lambda(X;E)$, hence there exists $\mu \in \Omega^w_\lambda(X;E)$ of degree $j-1$ such that $\eta = d\mu$. But μ is square integrable by (i), hence $\eta = 0$ [3: 2.5].

3.8 Remark. Theorem 3.7, for $\lambda = 0$, shows that $H^j(\Gamma;E) = H^j(\Gamma;E)_{(2)}$ for $j \leqq C(G,\tau^*)$. Similar criteria were already given in [24] for $\mathbf{Q}$-rank 1 and announced in [17] in the general case. In fact the conditions given there are stronger. To compare, assume (τ,E) to be irreducible. Note first that our condition could be expressed by saying that $\rho > \mu + \nu$ for any weight μ of A in E^* and any weight ν of A in $\Lambda^j\mathfrak{n}$. These weights are the opposite of those of A in $\Lambda\mathfrak{n}^* \otimes E$, which is the complex used to compute $H^*(\mathfrak{n};E)$, so we can also phrase it by saying that $\rho + \mu > 0$ for any weight μ of A in $\Lambda^j\mathfrak{n}^* \otimes E$. Now the condition of [17;24] is that $s(\rho + \mu) > 0$ for certain elements s of $W(\mathfrak{g}_c, \mathfrak{h}_c)$. By a theorem of Kostant, these elements are such that $s(\rho + \mu) = \rho + \nu$ where ν runs through the weights of A in $H^*(\mathfrak{n};E)$. But $H^*(\mathfrak{n},E) = H(\Lambda\mathfrak{n}^* \otimes E)$. Thus the requirement in [17] is that $\rho + \mu > 0$ only for those μ which occur in $H^*(\mathfrak{n};E)$.

This is also contained in [27], where it is proved moreover that $H^*(\Gamma;E)_{(2)}$ is isomorphic to the space of L^2-harmonic forms up to that range.

3.9 We now relate the growth conditions introduced in 3.2 with conditions involving the coefficients of forms on $\Gamma\backslash G(\mathbf{R})$, written as linear combinations of Maurer-Cartan forms. Let then $\pi: G(\mathbf{R}) \to X$ be the canonical projection and (ω^i) be a basis of $L(\mathfrak{g}(\mathbf{R}))^*$. Any element $\eta \in \Omega(G;E)^\Gamma$ can be written

$$\mu = \sum_I \mu_I \omega^I \quad . \tag{1}$$

We say that μ is *weakly λ-bounded* (resp. *λ-bounded*) if 3.2(3) (resp. 3.2(5)) is true on any Siegel set on $G(\mathbf{R})$. This is clearly independent of the choice of the basis of $L(\mathfrak{g}(\mathbf{R}))^*$. Furthermore, since $G(\mathbf{R}) = P(\mathbf{R}).K$, it suffices to check this condition for the restriction of μ to $P(\mathbf{R})$, if μ is K-finite. The form μ is *fast decreasing* (resp. of *moderate growth*) if it is λ-bounded for all (resp. some) λ.

3.10 <u>Proposition</u>. *Let* $\eta \in \Omega(X;E)^\Gamma$ *and* $\lambda \in X(A_o)$.

(i) *Assume that* $\eta \circ \pi$ *is (weakly) λ-bounded on* $\Gamma\backslash G(\mathbf{R})$. *Then* η *is (weakly) λ-bounded on* $\Gamma\backslash X$.

(ii) $\eta \circ \pi$ *is of moderate growth (resp. fast decreasing), if and only* η *is so on* $\Gamma\backslash X$.

In 3.2 we have used an isomorphism of manifolds $\mu_o: Y = U(\mathbf{R}) \times Z \times A \to X$. Since $X = P(\mathbf{R})/(K \cap P)$, we could equally well have lifted our forms to $P(\mathbf{R})$ and considered instead an isomorphism

$$\mu_o^*: Y' = U(\mathbf{R}) \times {}^oM(\mathbf{R}) \times A \to P(\mathbf{R}) \quad , \tag{1}$$

given by $(u.m.a) \mapsto u.m.a$ $(u \in U(\mathbf{R}),\ m \in {}^oM(\mathbf{R}),\ a \in A)$.

On Y' we consider again special frames defined by the Maurer-Cartan forms ω^i on the three factors, those on $U(\mathbf{R})$ and A being chosen as before. The set of indices I_{-1}, I_o, I_β are as in [3: p. 250] except that I_o refers to the Lie algebra of ${}^oM(\mathbf{R})$. For a set of indices I, we define the sum $\alpha(I)$ as in [3], the summands being contributed only by indices in the I_β's $(\beta \in \Phi(P,A))$. The elements ω^i also define a basis of the space of left-invariant 1-forms on $P(\mathbf{R})$.

Let us denote by $\bar{\omega}^i$ (resp. $\bar{\omega}^I$) the left-invariant form on $P(\mathbf{R})$ which is equal to ω^i (resp. ω^I) at the identity. We want to express the $\mu_o^*(\bar{\omega}^I)$ in terms of the ω^I, and conversely.

For a set I of indices, we denote by (I) the collection of sets of indices J such that the cardinalities of $J \cap J_{-1}$, $J \cap J_o$ and $J \cap J_\beta$ $(\beta \in \Phi(P,A))$ are the same as those of $I \cap I_{-1}$, $I \cap I_o$ and $I \cap I_\beta$. We note that $\alpha(J) = \alpha(I)$ for $J \in (I)$. Let $j \in I_\beta$, $k \in I_o$ and $\ell \in I_{-1}$. We have

$$\mu_o^*(\bar{\omega}^j + \bar{\omega}^k + \bar{\omega}^\ell)_{(u,m,a)} = \mu_o^*(uma(\bar{\omega}_1^j + \bar{\omega}_1^k + \bar{\omega}_1^\ell)) = \mu_o^*(uma(\omega_1^j + \omega_1^k + \omega_1^\ell))$$

$$= u.\mathrm{Ad}ma(\omega_1^j) + m.\omega_1^k + a.\omega_1^\ell \tag{2}$$

$(u \in U(\mathbf{R}),\ m \in {}^oM(\mathbf{R}),\ a \in A)$. Since

$$\mathrm{Ad}a(\omega^j) = a^{-\beta}\omega^j \quad , \qquad (a \in A;\ j \in I_\beta) \ , \tag{3}$$

and ${}^oM(\mathbf{R})$ leaves $u(\mathbf{R})_\beta$ stable, there exist smooth functions c_i^j $(i,\ j \in I_\beta)$ on ${}^oM(\mathbf{R})$ such that

$$\mathrm{Ad}ma(\omega^j) = a^{-\beta} \sum_{i \in I_\beta} c_i^j(m)\omega^i \ , \qquad (m \in {}^oM(\mathbb{R})\ ;\ a \in A) \ . \tag{4}$$

From this it follows that we can write

$$\mu_o^*(\bar{\omega}^I)_{(u,m,a)} = a^{-\alpha(I)} \sum_{J \in (I)} c_J^I(m)\omega^J_{(u,m,a)} \quad , \tag{5}$$

where c_J^I is a smooth function on ${}^oM(\mathbf{R})$.

Conversely, we have

$$\mu_o^{*-1}(\omega_u^j + \omega_m^k + \omega_a^\ell) = \mu_o^{*-1}(u.\omega_1^j + m.\omega_1^k + a.\omega_1^\ell)$$

$$= uma(\mathrm{Ad}(ma)^{-1}\omega_1^j + \omega_1^k + \omega_1^\ell)$$

and a similar computation shows the existence of smooth functions d_I^J on ${}^oM(\mathbf{R})$ such that

$$\mu_o^{*-1}(\omega^J)_{uma} = a^{\alpha(J)} \sum_{I\in(J)} d_I^J(m)\omega^I_{uma} \quad . \tag{6}$$

For $\eta \in \Omega(X;E)^\Gamma$, let us write

$$(\eta \circ \pi)\big|_{P(R)} = \sum_I \bar{\eta}_I \bar{\omega}^I \quad , \qquad \mu_o^*(\eta) = \sum_J \eta_J \omega^J \quad . \tag{7}$$

It follows then from (5) that we have

$$\eta_J(u,m,a) = a^{-\alpha(J)} \sum_{I\in(J)} c_J^I(m)\overline{\eta_I}(uma) \quad , \tag{8}$$

$$\overline{\eta_I}(uma) = a^{\alpha(I)} \sum_{J\in(I)} d_I^J(m)\eta_J(u,m,a) \tag{9}$$

$(u \in U(\mathbf{R}),\ m \in {}^oM(\mathbf{R}),\ a \in A)$. If now u and m vary in relatively compact sets, we get

$$|\eta_J(u,m,a)| \prec a^{-\alpha(J)} \sum_{I\in(J)} |\overline{\eta_I}(um\dot{a})| \tag{10}$$

$$|\overline{\eta_I}(uma)| \prec a^{\alpha(J)} \sum |\eta_J(u,m,a)| \quad . \tag{11}$$

Then (i) follows from (10), and (ii) from (10) and (11). Note that the converse assertion to (i) need not be true since the right-hand side of (11) is not necessarily λ-bounded when the η_J's are so.

4. Some Vanishing and Isomorphism Theorems

4.1 Let H be a connected real semi-simple Lie group (with finite center, as usual), L a maximal compact subgroup of H and (μ,F) a finite dimensional complex representation of H. We let $M(H,\mu)$ or $M(H,F)$ denote the greatest integer such that $H^q(\mathfrak{h},L;V\otimes F)=0$ for all $q\leqq M(H,F)$ and all irreducible unitary representations of H with compact kernel, and $M(H,*)$ be the minimum of $M(H,F)$ over all F's. The study of $M(H,*)$ is one of the main objectives of [11].

Assume first H to be almost simple over $\mathbf{R}$. Then $M(H,*)\geqq \mathrm{rk}_{\mathbf{R}}\, H-1$ [11: V, 3.3] or [28], and $M(H,*)$ is also at least equal to Matsushima's constant $m(H)$, which, in a few cases, may be $\geqq \mathrm{rk}_{\mathbf{R}}\, H$ [11: II, 2.9]. If H is a product of non-compact $\mathbf{R}$-simple groups H_i $(1\leqq i\leqq s)$ by a compact group H_o, then V decomposes into a Hilbert tensor product $\otimes_i V_i$, where V_i is an irreducible unitary representation of H_i, which is non-trivial for $i\geqq 1$. If F decomposes similarly into a tensor product $\otimes F_i$, where F_i is a representation of H_i, then

$$M(H;F)+1 = \sum_{1\leqq i\leqq s} (M(H,F_i)+1) \quad .$$

If H is a connected complex simple group, viewed as a real Lie group, then $M(H,*)=r_H-1$, where r_H is the constant given by T. Enright in [14]. Except for $\mathbf{SL}_n$, this constant is at least equal to the $\mathbf{R}$-rank and to $m(H)$. It is given by the following table [14]:

H:	$\mathbf{A}_\ell(\ell\geqq 1)$,	$\mathbf{B}_\ell(\ell\geqq 2)$,	$\mathbf{C}_\ell(\ell\geqq 3)$,	$\mathbf{D}_\ell(\ell\geqq 4)$,	$\mathbf{E}_6$,	$\mathbf{E}_7$,	$\mathbf{E}_8$,	$\mathbf{F}_4$,	$\mathbf{G}_2$
r(H)-1:	$\ell-1$,	$2(\ell-1)$,	$2(\ell-1)$,	$2\ell-3$,	15,	26,	56,	14,	4

4.2 <u>Lemma</u>. *Let* H, (μ,F) *be as in* 4.1, H' *a connected semi-simple group and* $\sigma: H'\to H$ *a surjective homomorphism with finite kernel. Then* $M(H',\mu)=M(H,\mu)$.

Since any irreducible unitary representation with compact kernel of H is one of H' we have clearly $M(H',\mu)\leqq M(H,\mu)$.

Let (π,V) be an irreducible unitary representation of H' with compact kernel. Let C be the kernel of σ. By assumption, it is contained in the kernel of every direct summand of (μ,F). It follows therefore from [11: I, 5.3] that if C is not in the kernel of π, then

$$H^*(\mathfrak{h}',L;\ V\otimes F) = 0 .$$

As a consequence, the representations of H' which matter in the determination of $M(H',\mu)$ are only those which factor through H, whence the reverse inequality.

4.3 Let G be connected, almost simple and isotropic over $\mathbf{Q}$. Then $G(\mathbf{R})^o$ is isogeneous to a direct product of non-compact simple real Lie groups. In fact, up to isogeny, G is of the form $R_{k/\mathbf{Q}}G'$, where k is a finite extension of $\mathbf{Q}$ and G' an almost absolutely simple k-group [10: 6.21], the group $(R_{k/\mathbf{Q}}G')(\mathbf{R})$ is the direct product of the groups $G'(k_v)$, where k_v runs through the archimedean completions of k, and the group $G^1(k_v)$ is not compact and absolutely almost simple over k_v, hence almost simple over $\mathbf{R}$. If $H_1,\ldots,H_s$ are the normal simple subgroups of $G(\mathbf{R})^o$, then $G(\mathbf{R})^o$ is the quotient of $H = H_1\times\ldots\times H_s$ by a finite subgroup. Any irreducible unitary representation V of $G(\mathbf{R})^o$ is one of H, hence is a Hilbert tensor product $V_1\tilde{\otimes}\ldots\tilde{\otimes}V_s$, where V_i is an irreducible unitary representation of H_i $(1\leqq i\leqq s)$.

Let Γ be an arithmetic subgroup of G, contained in $G(R)^o$. It is then irreducible, so that 4.2 of [11: VII], applied to the inverse image of Γ in H, shows that if V occurs in $L^2(\Gamma\backslash G(\mathbb{R})^o)$, as a direct summand, then either V is trivial or no V_i is. In particular, *if* V *is not trivial, then its kernel is compact.*

4.4 Theorem. *Let* G *be connected and almost simple over* $\mathbf{Q}$.

(i) *Assume that* (τ,E) *does not contain any non-zero subspace on which* $G(R)^o$ *acts trivially. Then*

$$H^q(\Gamma;E) = 0 \qquad \textit{for} \quad q\leqq M(G(\mathbf{R})^o,E),\ C(G,\tau^*); \tag{1}$$

$$H^q(\Gamma;E)_{(2)} = 0 \qquad \textit{for} \quad q\leqq M(G(\mathbf{R})^o,E) . \tag{2}$$

(ii) *The natural homomorphism* $I_G^{q\Gamma}\to H^q(\Gamma;\mathbf{C})$ *is injective for* $q\leqq C(G)+1$, *surjective for* $q\leqq M(G(\mathbf{R})^o,\mathbf{C}),C(G)$, *and maps* $I_G^{q\Gamma}$ *onto* $H^q(\Gamma;\mathbf{C})_{(2)}$ *for* $q\leqq M(G(\mathbf{R})^o,\mathbf{C})$.

Let Γ' be a normal subgroup of finite index of Γ, contained in $G(\mathbf{R})^{o}$. We have

$$H^{*}(\Gamma;E) = (H^{*}(\Gamma';E))^{\Gamma/\Gamma'} , \qquad I_{G}^{\Gamma} = (I^{\Gamma'})^{\Gamma/\Gamma'} , \tag{3}$$

and the first isomorphism obviously induces an isomorphism

$$H^{*}(\Gamma;E)_{(2)} = (H^{*}(\Gamma';E)_{(2)})^{\Gamma/\Gamma'} . \tag{4}$$

This reduces us to proving the theorem for Γ'. We may therefore assume $\Gamma \subset G(\mathbf{R})^{o}$. In particular

$$I_{G}^{\Gamma} = I_{G} = H^{*}(\mathfrak{g},K^{o};\mathbf{C}) . \tag{5}$$

Let us write the discrete spectrum $L^{2}(\Gamma\backslash G(\mathbf{R})^{o})$ as a direct sum of irreducible invariant subspaces V_{i} $(i \in I)$. If V_{i} is not trivial, then it has compact kernel (4.3), hence

$$H^{q}(\mathfrak{g}(\mathbf{R}),K^{o};V_{i}\otimes E) = 0 \quad \text{for } q \leqq M(G(\mathbf{R})^{o},E),\ V_{i} \neq \mathbf{C} . \tag{6}$$

By [4; 7], $H^{*}(\Gamma;E)_{(2)}$ is a quotient of the sum of the spaces $H^{*}(\mathfrak{g}(\mathbf{R}), K^{o}; V_{i}\otimes E)$, hence $H^{q}(\Gamma;E)_{(2)}$ is a quotient of $H^{q}(\mathfrak{g}(\mathbf{R}),K^{o};E)$ for $q \leqq M(G(\mathbf{R})^{o},E)$. But we have $H^{*}(\mathfrak{g}(\mathbf{R}),K^{o};\mathbf{C}) = I_{G}$ and, in the case (i), $H^{*}(\mathfrak{g}(\mathbf{R}),K^{o};E) = 0$ (see e.g. [11: II, 3.2]). This proves (2) and the last part of (ii).

By 3.7, for $\lambda = 0$, we have $H^{q}(\Gamma;E) = H^{q}(\Gamma;E)_{(2)}$, for $q \leqq C(G,\tau^{*})$ so that now (1) and the second assertion of (ii) follows. Finally, I_{G} is a space of harmonic forms contained in the complex $\Omega_{0}^{w}(X;E)^{\Gamma}$, hence 3.7(ii) yields the first assertion of (ii).

Remark. For $E = \mathbf{C}$, 4.4 is basically 7.5 of [3], with however somewhat better bounds. As an illustration, let us mention the following proposition:

4.5 Proposition. *Let* k *be a quadratic imaginary field,* $\mathfrak{o}$ *the ring of integers of* k, G' *an almost simple* k*-split* k*-group and* Γ *an arithmetic subgroup of* G'. *Then* $H^{q}(\Gamma;\mathbf{C}) = H^{q}(\mathfrak{g}';\mathbf{C})$ *for* $q \leqq c(G')$, *where* $c(G')$ *is given by the following table:*

$\mathfrak{g}'$	$A_\ell(\ell \geq 1)$,	$B_\ell(\ell \geq 3)$,	$C_\ell(\ell \geq 2)$,	$D_\ell(\ell \geq 4)$,	E_6,	E_7,	E_8,	F_4,	G_2
$c(G')$	$2[(\ell-1)/2]$,	$2(\ell-1)$,	$2(\ell-2)$,	$2(\ell-2)$,	14,	26,	50,	10,	2

Let $G = R_{k/\mathbf{Q}}G'$. We view Γ as an arithmetic subgroup of G. By 2.5, $C(G) \geq 2.C(\Phi(G'))$. Moreover $G(\mathbf{R}) = G'(\mathbf{C})$, viewed as a real Lie group, so that $M(G(\mathbf{R}),\mathbf{C})$ is given by Enright's vanishing theorem (4.1). Finally, in that case, $H^*(\mathfrak{g}(\mathbf{R}),K;\mathbf{C})$ may be identified to the Lie algebra cohomology $H^*(\mathfrak{g}';\mathbf{C})$. Our assertion now follows from 4.4(ii) and from the tables given in 2.10 and 4.1.

4.6 Consider now a sequence $(G_n,\Gamma_n,(\tau_n,E_n))$, of almost simple $\mathbf{Q}$-groups, arithmetic subgroups and non-trivial irreducible representations. If $M(G_n(\mathbf{R})^o,\tau_n)$ and $C(G_n,\tau_n^*)$ tend to infinity with n, then, given j, there exists $n(j)$ such that

$$H^j(\Gamma_n;E_n) = 0 \qquad \text{for all} \quad n \geq n(j) \quad . \tag{1}$$

For $M(G_n(\mathbf{R})^o,\tau)$ to tend to infinity, it suffices that the $\mathbf{Q}$-rank of G_n does so (4.1). We deal then with a sequence of classical groups from some n on. Then the coefficients of ρ, expressed as sum of simple roots, tend to infinity. Therefore $C(G,\tau^*) \to \infty$ if there exists a constant k such that the coefficients of the highest weight of τ_n^* are $\leq k$ for all n's. As an example, let us mention the following proposition:

4.7 <u>Proposition</u>. *Let* k *be a number field,* d *the degree of* k *over* $\mathbf{Q}$, *and* $\mathfrak{o}$ *the ring of integers of* k. *Let* $G'_n = \mathbf{SL}_n$ *(resp.* $\mathbf{Sp}_{2n}$, *resp.* $\mathbf{SO}_{n,n}$*)* $(n \geq 4)$ *and* Γ'_n *be a subgroup of finite index of* $G'_n(\mathfrak{o})$. *Let* Ad_n *be the adjoint representation of* G'_n. *Then*

$$H^j(\Gamma'_n;\mathrm{Ad}_n) = 0 \qquad \textit{for} \quad j \leq d\left[\frac{n-4}{2}\right] \text{ (resp. } d(n-4)\text{, resp. } d(n-4)\text{)}. \tag{1}$$

Let $G_n = R_{k/\mathbf{Q}}G'_n$. Then the canonical isomorphism $G'_n(k) = G_n(\mathbf{Q})$ maps Γ'_n onto an arithmetic subgroup Γ_n of G_n. Let (τ'_n,E'_n) be the adjoint representation of G'. Then the representation of (τ_n,E_n) of G_n obtained by restriction of scalars from (τ'_n,E') is the adjoint representation of G_n. Over $\mathbf{C}$ it splits into d copies of (τ'_n,E'_n), hence $H^*(\Gamma_n;E_n)$ is the direct sum of d copies of

$H^*(\Gamma_n';E_n)$. It suffices to prove (1) for Γ_n and the adjoint representation of G_n.

The adjoint representation is self-contragradient and its highest weight is the highest root δ_o. Using 2.11 one sees that the upper bound given for j in (1) is $\leqq C(G_n,\tau_n)$. The group $G_n(\mathbf{R})$ is a product of r_1 copies of $G_n'(\mathbf{R})$ and of r_2 copies of $G_n'(\mathbf{C})$, where r_1 (resp. r_2) is the number of real (resp. complex) places of k. We have therefore as a consequence of 4.1(1):

$$M(G_n(\mathbb{R})^o,*) + 1 \geqq r_1 M(G_n'(\mathbb{R})^o,*) + r_2 M(G_n'(\mathbb{C})^*) + r_1 + r_2 \quad . \tag{2}$$

From this and the results recalled in 4.1, one sees easily that the upper bounds for j in (1) are also majorized by $M(G_n(\mathbb{R})^o,*)$, so that 4.7 follows from 4.4(i).

5. Decaying Forms and Cohomology with Compact Supports

5.1 We let $\Omega_c(X,E)^\Gamma$ denote the complex of forms in $\Omega(X,E)^\Gamma$ with compact support modulo Γ. Then $H^*(\Omega_c(X;E)^\Gamma) = H_c^*(\Gamma\backslash X;\tilde{E})$, where H_c^* refers to cohomology with compact supports. If Γ is torsion-free then $\tilde{E}$ is a locally constant sheaf and this follows from the de Rham theorem for cohomology with compact supports. If not, we can either reduce to that case by using a normal torsion-free subgroup of finite index of Γ, or by means of a mild extension of de Rham theorem. The discussion is the same as that given in [11: VII, 2.2].

5.2 Theorem. *Let* $\lambda \in X(A_o)$ *be* <0. *Then the inclusions*

$$\Omega_c(X;E)^\Gamma \to \Omega_{fd}(X;E)^\Gamma \to \Omega_\lambda^w(X;E)^\Gamma \quad ,$$

induce isomorphisms in cohomology.

The reduction to the case where Γ is torsion-free is carried out exactly as in 3.4, so we assume Γ to be torsion-free.

In this proof, we let ν range over c, fd, λ and set $\Omega_\nu = \Omega_\nu(X;E)^\Gamma$ if $\nu = c$, fd, and $\Omega_\nu = \Omega_\lambda^w(X;E)^\Gamma$ if $\nu = \lambda$. Let F_ν be the presheaf which associates to U open in $\Gamma\backslash\bar{X}$ the restrictions to $U \cap (\Gamma\backslash X)$ of elements in Ω_μ. Again this is a sheaf and, as in 3.4, it is seen to be fine. The space of sections of F_ν is Ω_ν. Therefore, by the comparison theorem in sheaf theory [18: II, 4.6.2], to prove the theorem, it suffices to show that inclusions $F_c \to F_\nu$ $(\nu = fd, \lambda)$ induce isomorphisms of the derived sheaves. If $z \in \Gamma\backslash X$, then the stalk $F_{\nu,z}$ of F at z is just the space of germs of differential forms around z, for all values of ν, hence the stalks themselves are equal. Let now $z \in \Gamma\backslash\partial\bar{X}$. Clearly, $F_{c,z} = 0$. Hence there remains to prove

$$H^q(F_{\nu,z}) = 0 \qquad (\nu = fd,\ \lambda,\ q = 0,1,2,\ldots) \ . \tag{1}$$

Let P be a parabolic **Q**-subgroup such that z is contained in the face $e'(P)$ associated to P [8]. The point z is then in the closure of a Siegel set $\mathfrak{S} = \mathfrak{S}_{t,\omega}$ associated to P. In the latter we use the notation and conventions of 3.4. A function f in Ω^o_ν tends to zero as the argument in $\mathfrak{S}$ tends to z. If its differential is zero near z, it is then zero in $\mathfrak{S}$, near z, hence $F^o_{\nu,z} = 0$ and (1) is clear for $q = 0$. To prove (1) for $q \geqq 1$ and $\nu = \lambda$, we have to show that if $\eta \in \Omega^w_\lambda(\mathfrak{S})$ is closed then its restriction to some smaller Siegel set $\mathfrak{S}'$ is the coboundary of an element in $\Omega^w_\lambda(\mathfrak{S}')$. We do this again by checking that the homotopy operator of the Poincaré lemma is compatible with our conditions. However, we have now to take the origin of our local coordinate system at z itself. We identify A_t to the strictly positive quadrant in the space with coordinates $\beta^i = a^{-\alpha_i}$ $(1 \leqq i \leqq \ell)$. Its closure $\bar{A}_t$ consists of the points with coordinates $\beta^i \in [0,1/t]$. The canonical isomorphism of $A_t \times \omega$ onto $\mathfrak{S}$ extends to one of $\bar{A}_t \times \omega$ onto the corner containing z, and in fact z is in the image of $(0,\omega)$. We have

$$d \log a^{\alpha_i} = -d\,\ell n\,\beta^i = -\omega^{i'} \qquad (1 \leqq i \leqq \ell) \ . \tag{2}$$

We take now as local coordinates in the corner the $y^i = -(\log \beta^i)^{-1}$ $(1 \leqq i \leqq \ell)$ and a set of local coordinates $(y^i)_{\ell < j \leqq n}$ in ω centered at z. We may assume that the coordinates y^j vary in $[-c,c]$ for $j > \ell$ and in $[0,c]$ for $j \leqq \ell$, where c is some small strictly positive constant. The intersection D of $\Gamma\backslash X$ with this neighborhood of z in the given corner is then the set of points

$$D = \{y = (y^i) \mid y^j \in (0,c] \quad \text{if} \quad j \leqq \ell,\ y^j \in [-c,c] \quad \text{if} \quad j > \ell\} \ . \tag{3}$$

By assumption

$$\lambda = -\sum_1^{\ell} d_i . \alpha_i \qquad (d_i > 0;\ i = 1,\dots,\ell) \quad . \tag{4}$$

The weak λ-boundedness of $\eta = \Sigma\, \eta_J \omega^J$ is originally expressed by the condition

$$|\eta_J(a,q)| \prec_w \prod_1^{\ell} (\beta^i)^{d_i} \qquad (a \in A_t,\ q \in \omega) \quad . \tag{5}$$

In D this is equivalent to requiring the existence of $d > 0$ and $M \in \mathbf{Z}$ such that

$$|\eta_J(y)| \leqq d . \prod_1^{\ell} (y^i)^M . e^{-d_i/y_i} \quad , \qquad (y \in D) \quad , \tag{6}$$

and, for all J's. Set

$$c(y) = \sum_{1 \leqq i \leqq \ell} d_i/y_i \quad . \tag{7}$$

Then, in D, and for c small enough

$$1 < c(y) < \infty \quad . \tag{8}$$

We shall now write (6) as

$$|\eta_J(y)| \prec_w e^{-c(y)} \quad , \qquad (y \in D) \quad . \tag{9}$$

We have

$$d \log \beta^i = (y^i)^{-2} . dy^i \quad , \qquad (1 \leqq i \leqq \ell) \quad . \tag{10}$$

On the other hand, the ω^j $(j > \ell)$ are linear combinations of the dy^j $(j > \ell)$ with bounded coefficients and *vice-versa*. Therefore if we write

$$\eta = \sum_J \eta_J \omega^J = \sum_J \eta'_J dy^J \quad , \tag{11}$$

then the condition (9) for the η_J's and for the η'_J's are equivalent, i.e., η is weakly λ-bounded if and only if we have for all J's:

$$|\eta'_J(y)| \prec_w e^{-c(y)} \quad , \qquad (y \in D) \quad . \tag{12}$$

Let $q \geqq 1$ and $J = \{j_1, \ldots, j_q\}$. Consider the form $\eta = f.dy^J$. The homotopy operator A of the Poincaré lemma transforms η onto $\Sigma\, c_I\, dy^I$, where, for I equal to J with j_i erased:

$$c_I(y) = (-1)^{i-1}\, y^{j_i} \int_0^1 f(ty).t^{q-1}.\, dt \quad . \tag{13}$$

Assume that f satisfies (6). Then

$$|c_I(y)| \leqq c.d \prod_1^{\ell} (y^i)^M \int_0^1 t^{M\ell+q-1}.e^{-c(y)/t}.\, dt \ , \ (y \in D) \ . \tag{14}$$

Going over to the variable $s = c(y)/t$, we see that the integral on the right-hand side of (14) is equal to

$$c(y)^{M\ell+q} \int_{c(y)}^{\infty} s^{-(M\ell+q+1)}.e^{-s}.\, ds \quad . \tag{15}$$

Let $N = M\ell+q+1$. It is easily seen that the integral in (15) is bounded by $C.c(y)^{-N}.e^{-c(y)}$, for some $C > 0$. Therefore c_I satisfies (9). As a consequence, $A(\sigma)$ is weakly λ-bounded if σ is so, for any form σ. If now σ is a closed form, then $\sigma = dA\sigma$, whence (1) for $\nu = \lambda$. But then it obviously follows also for $\nu = fd$ (in fact, the proof could be slightly simplified in that case since we need not keep the same exponent in the exponential).

5.3 Theorem. *Let* H_{fd} *be the space of fast decreasing harmonic forms contained in* $\Omega(X;E)^{\Gamma}$. *Then the natural map of* $H_{fd} \to H^*(\Gamma;E)$ *is injective. If* $\eta \in H_{fd}$, *then* η *can be written in the form* $\eta = \mu + d\nu$, *where* μ *has compact support* mod Γ *and* ν *is fast decreasing.*

The second assertion follows from 5.2. The first one depends only on the fact that cohomology can be computed by means of a complex of forms with moderate growth (3.4). We could also use 7.4 of [3].

Let $\eta \in H^q_{fd}$. Assume that it is cohomologous to zero in $\Omega(X;E)^\Gamma$. Then it is already so in Ω_{mg}, e.g., hence there exists $\sigma \in \Omega^{q-1}_{mg}(X;E)$ such that $\eta = d\sigma$. In a Siegel set, the coefficients g^{IJ} associated to a special frame have all moderate growth (see [3: 5.5]). Therefore, for a differential form τ which has moderate growth (resp. is fast decreasing) the function $|\tau|_y$ has moderate growth (resp. is fast decreasing). It follows that the functions $|\eta|_y \cdot |\sigma|_y$ and $(\eta_y, d\sigma_y)$ are fast decreasing, in particular are integrable. Since η is harmonic, $\partial\eta = 0$ hence $(\partial\eta, \sigma_y) = 0$. We have then, by [3: 2.2]:

$$(\eta,\eta) = (\eta, d\sigma) = (\partial\eta, \sigma) = 0 \quad ,$$

hence $\eta = 0$.

5.4 Let $^{o}L^2(\Gamma\backslash G)$ be the cuspidal spectrum of Γ, i.e., the subspace of $L^2(\Gamma\backslash G)$ consisting of cuspidal functions. The cohomology space $H^*(\mathfrak{g}(\mathbf{R}), K; {}^{o}L^2(\Gamma\backslash G)^\infty \otimes E)$ may be identified with the space of harmonic E-valued forms whose coefficients are E-valued cusp forms. Its image in $H^*(\Gamma;E)$ is, by definition, the cuspidal cohomology $H_{cusp}(\Gamma;E)$ of Γ. Since a cusp form is fast decreasing, these harmonic forms belong to H_{fd} by 3.10, hence 5.3 has the following corollary, whose first part was already announced in [4: 6.1] for $E = \mathbf{C}$.

5.5 Corollary. *The natural map of* $H^*(\mathfrak{g}(\mathbf{R}), K; {}^{o}L_2(\Gamma\backslash G)^\infty \otimes E)$ *onto* $H^*_{cusp}(\Gamma;E)$ *is injective. A cuspidal harmonic form* η *can be written in the form* $\eta = \mu + d\nu$, *where* μ *has compact support* mod Γ *and* ν *is fast decreasing.*

5.6 Assume Γ to be torsion-free and orientation preserving. Let n be the dimension of X. Then $H^n_c(\Gamma\backslash X;\mathbf{C}) \cong \mathbf{C}$ and this isomorphism is given by integration over $\Gamma\backslash X$. By 5.2, we have a canonical isomorphism $H^n(\Omega_{fd}(X;\mathbf{C})^\Gamma) = H^n_c(\Gamma\backslash X;\mathbf{C})$, whence also an isomorphism of $H^n(\Omega_{fd}(X;\mathbf{C})^\Gamma)$ onto $\mathbf{C}$. We claim that this isomorphism is also defined by integration. To see this it is enough to show that

$$\int_{\Gamma\backslash X} d\eta = 0 , \qquad \text{if} \quad \eta \in \Omega^{n-1}_{fd}(X;\mathbf{C})^\Gamma \quad . \tag{1}$$

But η tends to zero at infinity so it extends to a smooth form on $\Gamma\backslash\bar{X}$ which is identically zero on the boundary $\Gamma\backslash\partial\bar{X}$. We can then apply Stokes' theorem to η on $\Gamma\backslash\bar{X}$, and (1) follows. Since the product of a fast decreasing form by a form of moderate growth is fast decreasing, it follows immediately that the following diagram, where the horizontal arrows are defined by integration over $\Gamma\backslash X$, and the vertical arrows are the isomorphisms given by 3.4 and 5.2, is commutative:

$$\begin{array}{ccccc} H^p(\Omega_{fd}(X;\mathbf{C})^\Gamma) & \times & H^{n-p}(\Omega_{mg}(X;\mathbf{C})^\Gamma) & \to & \mathbf{C} \\ \uparrow & & \downarrow & & \\ H^p_c(\Gamma;\mathbf{C}) & \times & H^{n-p}(\Gamma;\mathbf{C}) & \to & \mathbf{C} \end{array} \qquad (0 \leqq p \leqq n)\ . \tag{2}$$

More generally, there are pairings

$$\Omega^p_c(X;E)^\Gamma \times \Omega^q(X;E^*)^\Gamma \to \Omega^{p+q}_c(X;\mathbf{C})^\Gamma \quad , \tag{3}$$

$$\Omega^p_{fd}(X;E)^\Gamma \times \Omega^q_{mg}(X;E^*)^\Gamma \to \Omega^{p+q}_{fd}(X;\mathbf{C})^\Gamma \quad , \qquad (p,q \geqq 0)\ , \tag{4}$$

defined by exterior product and the trace map $E \otimes E^* \to \mathbf{C}$. Composed with integration over $\Gamma\backslash X$, when $p+q=n$, they yield a commutative diagram, where the horizontal arrows are gain perfect pairings:

$$\begin{array}{ccccc} H^p(\Omega_{fd}(X;E)^\Gamma) & \times & H^{n-p}(\Omega_{mg}(X;E)^\Gamma) & \to & \mathbf{C} \\ \uparrow & & \downarrow & & \\ H^p_c(\Gamma\backslash X;E) & \times & H^{n-p}(\Gamma;E^*) & \to & \mathbf{C} \end{array} \qquad (0 \leqq p \leqq n)\ . \tag{5}$$

6. The Case of S-Arithmetic Groups

In this section, k *is a number field,* G' *a connected, almost* k-*simple isotropic* k-*group, and* $G = R_{k/\mathbf{Q}}G'$.

6.1 We let $\mathfrak{o}$ be the ring of integers of k, d the degree of k over $\mathbf{Q}$, V (resp. V_∞, resp. V_f) the set of (resp. archimedean, resp. finite) places of k and k_v the completion of k at v. For

$v \in V_f$, let $\mathfrak{o}_v$ denote the ring of integers of k_v.

If S is a finite set of places of k, set $S_\infty = S \cap V_\infty$ and $S_f = S \cap V_f$. Unless otherwise said, we assume that $S_\infty = V_\infty$. As usual, $\mathfrak{o}_S$ denotes the ring of S-integers of k (elements of k belonging to $\mathfrak{o}_v$ for all $v \in V - S$).

If H is a k-group, then we put

$$H_v = H(k_v), \qquad H_\infty = \prod_{v \in V_\infty} H_v, \qquad H_S = \prod_{v \in S} H_v \quad . \tag{1}$$

A subgroup $\Gamma \subset H(k)$ is S-arithmetic if, for any faithful k-morphism $\mu: H \to \mathbf{GL}_m$, the group $\mu(\Gamma)$ is commensurable with $\mu(H) \cap \mathbf{GL}_m \mathfrak{o}_S$. The group Γ, diagonally embedded in H_S is discrete. If L is a compact open subgroup of H_{S_f}, then $\Gamma \cap (H_\infty \times L)$ is an arithmetic subgroup of H.

If $H' = R_{k/\mathbf{Q}} H$, then we have canonical isomorphisms $H'(\mathbf{Q}) \cong H(k)$ and $H'(\mathbb{R}) = H_\infty$. The first isomorphism maps arithmetic subgroups of H onto arithmetic subgroups of H'. It also allows one to view any $H'(\mathbf{R})$ module as a $H(k)$-module. This is tacitly understood in the sequel.

6.2 Proposition. *Assume* G' *to be simply connected. Let* S *be a finite set of places of* V, Γ *a* S-*arithmetic subgroup of* G' *and* $N \in \mathbf{N}$.

(i) *Assume that for any pair* $\Gamma_1 \subset \Gamma_2$ *of arithmetic subgroups of* G' *contained in* Γ *the restriction map* $H^q(\Gamma_2;E) \to H^q(\Gamma_1;E)$ *is an isomorphism for* $q \leqq N$. *Then, for any arithmetic subgroup* Γ_o *of* G' *contained in* Γ, *the restriction map* $H^q(\Gamma;E) \to H^q(\Gamma_o;E)$ *is an isomorphism for* $q \leqq N$.

(ii) *If the assumption of* (i) *is satisfied for all finite* $S \subset V$ *and all* S-*arithmetic subgroups, then the restriction map* $H^q(G'(k),E) \to H^q(\Gamma_o;E)$ *is also an isomorphism for* $q \leqq N$.

The proof uses Bruhat-Tits buildings [13]; its framework is the same as that of 3.7 in [5] or 6.9 in [9]. See [5;9] for more references concerning Bruhat-Tits buildings.

For $v \in V_f$, let X_v be the Bruhat-Tits building of G' over k_v, set $X_f = \prod_{v \in S_f} X_v$ and $G_f = G_{S_f}$. Let $\bar{X}$ be the completion by corners of X constructed in [8] and $\bar{X}_S = \bar{X} \times X_f$. The group G'_S operates naturally on $\bar{X}_S$. The group Γ, embedded diagonally in G'_S, operates properly on $\bar{X}_S$ and the quotient $\bar{X}_S/\Gamma$ is compact [9: 6.9]. The group Γ also operates on X_f via its projection on G'_f and the natural projection $X_S \to X_f$ commutes with Γ, whence a projection

$$\pi\colon \bar{X}_S/\Gamma \to X_f/\Gamma \quad . \tag{1}$$

The space X_f is a product of simplicial complexes, a "polysimplicial complex" in the terminology of [13]. The group G_f operates on X_f as a group of automorphisms of the polysimplicial structure and X_f/G_f is a polysimplex C, of dimension the sum of the k_v-ranks of G'. Since G' is simply connected, and G'_∞ is not compact, the projection of Γ on each G'_v $(v \in V_f)$ is non-discrete hence Prop. 4.3 of [23] holds, and shows that Γ is *dense* in G'_f. Since the isotropy groups of G_f on X_f are compact open, it follows that $X_f/\Gamma = X_f/G_f = C$.

We consider now the Leray spectral sequence (E_r) of π. We have $E_2^{p,q} = H^p(C;F^q)$ where F^q is the Leray sheaf of π whose stalk at $c \in C$ is $H^q(\pi^{-1}(c);E)$. For a face σ of C let L_σ be its isotropy group in G_f. It is compact open. The isotropy group Γ_σ of σ in Γ may be identified with $\Gamma \cap (G'_\infty \times L_\sigma)$, hence is an arithmetic subgroup. This is also the isotropy group in Γ of any point c in the interior $\mathring{\sigma}$ of σ. We have then a homeomorphism:

$$\pi^{-1}(c) \cong \bar{X}_\infty/\Gamma_\sigma \ , \qquad (c \in \mathring{\sigma}) \tag{2}$$

and therefore

$$H^q(\pi^{-1}(c);E) \cong H^q(\Gamma_\sigma;E) \quad , \qquad (c \in \mathring{\sigma}) \quad . \tag{3}$$

If σ' is a face of σ, then $\Gamma_{\sigma'} \supset \Gamma_\sigma$ and we have a natural restriction map in cohomology. Thus the Leray sheaf is in this case just a system of coefficients, which assigns one graded space to each face, with natural mappings associated to inclusions of faces. By our assumption, there exists for $q \leqq N$ a finite dimensional vector space F^q and isomorphisms $H^q(\Gamma_\sigma;E) \to F^q$ compatible with restrictions. Therefore in those dimensions, our sheaf is just an ordinary constant system of coefficients and we have

$$E_2^{p,q} = H^p(C;F^q) \quad , \qquad (p \in \mathbf{N};\ q \leqq N) \quad . \tag{4}$$

Since C is a polysimplex, it is acyclic, hence

$$E_2^{o,q} \quad = \quad F^q \qquad\qquad (q \leqq N) \qquad (5)$$

$$E_2^{p,q} \quad = \quad 0 \qquad\qquad (p > 0;\ q \leqq N)\ . \qquad (6)$$

We have therefore, in total degree $m \leqq N$:

$$F^m \quad = \quad E_2^{o,m} \quad = \quad {}^mE_2 \quad = \quad {}^mE_\infty \quad = \quad H^m(\Gamma;E) \qquad , \qquad (7)$$

and (i) follows.

Assume now this is true for all finite $S \subset V$ and all S-arithmetic groups. Fix an increasing sequence S_n $(n = 1,2,\ldots)$ of subsets of V whose union is V. We can then view $G'(k)$ as the union of an increasing sequence of subgroups Γ_n, where Γ_n is S_n-arithmetic. There exists therefore an Eilenberg-MacLane space $K(G'(k),1)$ which is a union of subcomplexes K_n, where K_n is a $K(\Gamma_n,1)$ $(n = 1,2,\ldots)$. By Theorem 2.10*, p. 273 of [26: VI], we have an exact sequence

$$\to \varprojlim{}^1\ H^{q-1}(\Gamma_n;E) \to H^q(G'(k),E) \to \varprojlim\ H^q(\Gamma_n;E) \to 0, \qquad (q \in \mathbf{N})\ , \qquad (8)$$

where $\varprojlim{}^1$ is Milnor's first derived functor of $\varprojlim$ (*loc. cit.*). If $q \leqq N$, all the maps $H^q(\Gamma_{n+1};E) \to H^q(\Gamma_n;E)$ are isomorphisms, hence the $\varprojlim{}^1$ term is zero and the $\varprojlim$ term is just $H^q(\Gamma_1:E)$. The second assertion follows.

6.3 We let $\tilde{G}'$ denote the universal covering of G' and $\sigma: \tilde{G}' \to G'$ the canonical isogeny. Then $\tilde{G} = R_{k/G}\tilde{G}'$ is the universal covering of G. The canonical isogeny $\tilde{G} \to G$ will also be denoted by σ.

The group $\tilde{G}(\mathbf{R}) = \tilde{G}'_\infty$ is always connected, while $G(\mathbb{R}) = G'_\infty$ is not necessarily so. The homomorphism σ maps $\tilde{G}(\mathbf{R})$ onto $G(\mathbf{R})^o$ and has finite kernel.

As before, I_G denotes the space of $G(\mathbf{R})^o$ invariant forms on X and $j_\Gamma: I_G \to H^*(\Gamma;\mathbf{C})$ the natural homomorphism. If $G(\mathbf{R})$ is not connected, the space of $G(\mathbf{R})$-invariant harmonic forms may be of course $\neq I_G$, hence I_G^Γ may be $\neq I_G$, which introduces a minor complication. To state our next theorem we introduce still another constant, namely

$$i(G) \quad = \quad \max q\colon I_G^q \quad = \quad I_G^{G(\mathbf{R})} \qquad . \qquad (1)$$

If $G(\mathbb{R})$ is connected, hence in particular if G is simply connected, then $i(G) = \infty$.

6.4 Theorem. *Let* Γ *be equal either to* $G'(k)$ *or to a* S-*arithmetic subgroup of* G'.

(i) *Assume* E *does not contain any nonzero trivial* $G(\mathbb{R})^0$-*submodule. Then*

$$H^q(\Gamma;E) = 0 , \quad \textit{for} \quad q \leqq M(G(\mathbb{R})^o,\tau),C(G,\tau^*) \quad . \tag{1}$$

(ii) *If* Γ *is* S-*arithmetic and contained in* $G(\mathbb{R})^o$, *then*

$$H^q(\Gamma;\mathbb{C}) = I^q_{G'} \quad \textit{for} \quad q \leqq M(G(R)^o,\mathbb{C}),C(G) \quad . \tag{2}$$

(iii) *We have*

$$H^q(\Gamma;\mathbb{C}) = I^q_{G'} \quad \textit{for} \quad q \leqq M(G(\mathbb{R})^o,\mathbb{C}),C(G),i(G) . \tag{3}$$

[See 1.6, 4.1 and 6.3 for the definitions of $M(G(\mathbb{R})^o,\tau)$, $C(G,\tau^*)$, $C(G)$ and $i(G)$.]

In view of 4.4, our assumptions imply in each case that if $\Gamma_1 \subset \Gamma_2$ are arithmetic subgroups of G' contained in Γ, then $H^q(\Gamma_2;E) \to H^q(\Gamma_1;E)$ is an isomorphism in the range indicated, and furthermore that $H^q(\Gamma_1;E) = 0$ in case (i), and $H^q(\Gamma_1;\mathbb{C}) = I^q_G$ in cases (ii), (iii). Therefore, if G is simply connected, the theorem follows from 6.2.

In the general case, we first assume Γ to be S-arithmetic. As usual (see the beginning of 4.4), we may replace Γ by a subgroup of finite index. Since $\tilde{G}'(k) \to G'(k)$ maps S-arithmetic subgroups onto S-arithmetic subgroups [1: 8.12], we may then assume that $\Gamma = \sigma(\tilde{\Gamma})$, where $\tilde{\Gamma}$ is a torsion-free S-arithmetic subgroup of $\tilde{G}'(k)$. The map σ is then an isomorphism of $\tilde{\Gamma}$ onto Γ, hence $H^*(\tilde{\Gamma};E) = H^*(\Gamma;E)$. We are now back to the simply connected case. By our initial remark, this proves the theorem in this case, except that we have to replace G by $\tilde{G}$ in the constants giving the range for q. However $C(\tilde{G},\tau^*) = C(G,\tau^*)$ by definition, and $M(\tilde{G}(\mathbb{R}),E) = M(G(\mathbb{R})^o,E)$, $M(G(\mathbb{R}),\mathbb{C}) = M(G(\mathbb{R})^o,\mathbb{C})$ by 4.2. To prove (i) and (iii) when $\Gamma = G'(k)$, we then argue exactly as in the proof of 6.2(ii).

6.5 In [3], we applied 7.5 to sequences of arithmetic groups in classical groups. In view of 6.4, we can now do this for sequences of S-arithmetic groups, or also of groups of rational points. In particular, 11.1 remains valid if Γ_n stands for a S-arithmetic group or $G_n(\mathbf{Q})$. Similarly in 11.3, and in the examples which illustrate it, we may take for Γ'_n a S-arithmetic group or $G'_n(k)$ itself. Finally, the proofs of 12.2, and 12.3, giving the rank of the groups $K_i\mathfrak{o}\otimes\mathbf{Q}$ and ${}_\varepsilon L_i\mathfrak{o}\otimes\mathbf{Q}$ remain valid if $\mathfrak{o}$ is replaced by $\mathfrak{o}_S$ or by k. This then establishes the results announced in [2] and not proved in [3].

7. References

[1] A. Borel, "Some finiteness properties of adele groups over number fields," *Publ. Math. I.H.E.S.* 16, 5-30 (1963).

[2] A. Borel, "Cohomologie réelle stable de groupes S-arithmétiques," *C.R. Acad. Sci. Paris* 274, 1700-1702 (1972).

[3] A. Borel, "Stable real cohomology of arithmetic groups," *Annales Sci. E.N.S. Paris* (4)7, 235-272 (1974).

[4] A. Borel, "Cohomology of arithmetic groups,"Proc. Int. Congress of Math. Vancouver, Vol. 1, 435-442 (1974).

[5] A. Borel, "Cohomologie de sous-groupes discrets et représentations de groupes semi-simples," *Astéristique* 32-33, 73-111 (1976).

[6] A. Borel, "Stable and L^2-cohomology of arithmetic groups," *Bull. A.M.S. (N.S.)* 3, 1025-1027 (1980).

[7] A. Borel and H. Garland, "Laplacian and discrete spectrum of an arithmetic group" (in preparation).

[8] A. Borel and J-P. Serre, "Corners and arithmetic groups," *Comm. Math. Helv.* 48, 436-491 (1973).

[9] A. Borel and J-P. Serre, "Cohomologie d'immeubles et de groupes S-arithmétiques," *Topology* 15, 211-232 (1976).

[10] A. Borel and J. Tits, "Groupes réductifs," *Publ. Math. I.H.E.S.* 27, 55-150 (1965).

[11] A. Borel and N. Wallach, "Continuous cohomology, discrete subgroups and representations of reductive groups," *Annals of Mathematics Studies* 94; xvii + 387 p., Princeton University Press, 1980.

[12] N. Bourbaki, "Groupes et Algèbres de Lie," Chap. IV, V, VI, Act. Sci. Ind. 1337, Hermann, Paris, 1968.

[13] F. Bruhat and J. Tits, "Groupes réductifs sur un corps local I," *Publ. Math. I.H.E.S.* 41, 1-251 (1972).

[14] T. Enright, "Relative Lie algebra and unitary representations of complex Lie groups," *Duke M. J.* 46, 513-525 (1979).

[15] F.T. Farrell and W.C. Hsiang, "On the rational homotopy groups of the diffeomorphism groups of discs, spheres and aspherical manifolds," Proc. Symp. Pure Math. 32, Part 1 (1978), 403-415, A.M.S. Providence, RI.

[16] H. Garland, "A finiteness theorem for K_2 of a number field," *Annals of Math.* (2), 94, 534-548 (1971).

[17] H. Garland and W.C. Hsiang, "A square integrability criterion for the cohomology of an arithmetic group," Proc. Nat. Acad. Sci. USA, 59, 354-360 (1968).

[18] R. Godement, "Théorie des Faisceaux," Act. Sci. Ind. 1252, Hermann, Paris, 1958.

[19] Y. Matsushima, "On Betti numbers of compact, locally symmetric Riemannian manifolds," *Osaka Math. J.* 14, 1-20 (1962).

[20] Y. Matsushima, "A formula for the Betti numbers of compact locally symmetric Riemannian manifolds," *Jour. Diff. Geom.* 1, 99-109 (1967).

[21] Y. Matsushima and S. Murakami, "On vector bundle valued harmonic forms and automorphic forms on symmetric spaces," *Annals of Math.* (2) 78, 365-416 (1963).

[22] Y. Matsushima and S. Murakami, "On certain cohomology groups attached to hermitian symmetric spaces," *Osaka J. Math.* 2, 1-35, (1965).

[23] G. Prasad, "Strong approximation for semi-simple groups over function fields," *Annals of Math.* (2) 105, 553-572 (1977).

[24] M.S. Raghunathan, "Cohomology of arithmetic subgroups of algebraic groups II," *Annals of Math.* (2) 87, 279-304 (1968).

[25] A. Weil, Adeles and algebraic groups, Notes by M. Demazure and T. Ono, Institute for Advanced Study, Princeton, NJ, 1961.

[26] G. W. Whitehead, "Elements of homotopy theory," Grad. Texts in Math. 61, Springer Verlag, New York, 1978.

[27] S. Zucker, "L_2-cohomology of warped products and arithmetic groups," (to appear).

[28] S. Zuckermann, "Continuous cohomology and unitary representations of real reductive groups," *Annals of Math.* (2) 107, 495-516 (1978).

The Institute for Advanced Study
Princeton, NJ 08540 USA

(Received February 27, 1981)

VECTOR FIELDS AND COHOMOLOGY OF G/B

James B. Carrell

0. Introduction

The topic I will discuss today is one which arose from a question which I believe Professor Matsushima originally asked: namely, if one is given a holomorphic vector field V on a projective manifold X, is it true that X has no nontrivial holomorphic p-forms if $p > \dim_C$ zero (V)? Alan Howard answered this question affirmatively in [H] and later, D. Lieberman and I discovered other relationships between zeros of holomorphic vector fields and topology. Perhaps the most interesting of these is that if one has a holomorphic vector field V on a compact Kaehler manifold X with isolated zeros, then the whole cohomology ring of X can be calculated on the zeros of V. Although holomorphic vector fields with isolated zeros are not abundant, they do exist on a fundamental class of spaces, namely the algebraic homogeneous spaces. In the one example that has been carefully analyzed, the Grassmannians, the calculation of the cohomology ring on the zeros of V gives a new insight on the connection between Schubert calculus and the theory of symmetric functions [C].

What we shall discuss in this talk is the cohomology ring of a generalized flag manifold G/B from the viewpoint of zeros of a vector field. Rather surprisingly the classical description in terms of invariants of the Weyl group due to A. Borel [B] is arrived at as a limiting case. I would like to thank Bill Casselman for a suggestion which greatly simplified our presentation.

1. Review of G/B

Suppose G is a semi-simple complex Lie group, B a Borel subgroup, H a fixed maximal torus in B, and W the Weyl group of H in G. Let $X(H)$ denote the group of characters on H, and let $\mathfrak{h}$ denote the Lie algebra of H. For $w \in W$ and $v \in \mathfrak{h}$, $w \cdot v$ will denote the tangent action of W on $\mathfrak{h}$. W thus acts effectively on $\mathfrak{h}$ and on $\mathfrak{h}^*$ in the usual way: $w \cdot f(v) = f(w^{-1} \cdot v)$ for $f \in \mathfrak{h}^*$. To every character $\alpha \in X(H)$, one associates a line bundle L_α (obviously holomorphic) on G/B as

$$L_\alpha = (G \times C)/B$$

where $(g,z)b = (gb, \alpha(b^{-1})z)$ where α has been extended to B by the usual convention. Now $d\alpha \in \mathfrak{h}^*$ and since G is semi-simple, the $d\alpha$, for all $\alpha \in X(H)$, span $\mathfrak{h}^*$. Thus it can be checked that there is a well defined linear map

$$\beta \colon \mathfrak{h}^* \to H^2(G/B, C) \tag{1}$$

determined by the condition $\beta(d\alpha) = c_1(L_\alpha)$ for any $\alpha \in X(H)$. Let β also denote the algebra homomorphism

$$\beta \colon R = \mathrm{Sym}(\mathfrak{h}^*) \to H^\bullet(G/B, C)$$

extending (1), where $\mathrm{Sym}(\mathfrak{h}^*)$ is the symmetric algebra of $\mathfrak{h}^*$. Now W acts on R and on $H^\bullet(G/B,C)$ as well. Denote by I_W (resp. I_W^+) the ring of invariants of W acting on R (resp. $f \in I_W$ such that $f(0) = 0$). The following theorem is due to Borel [B].

<u>Theorem 1</u>. *β is a W-equivariant surjective homomorphism whose kernel is RI_W^+. Consequently, since RI_W^+ is a homogeneous ideal invariant under W, β induces a W-equivariant isomorphism of graded rings*

$$\bar{\beta} \colon R/RI_W^+ \to H^\bullet(G/B, C) \tag{2}$$

One obtains holomorphic vector fields with isolated zeros on G/B as follows: We call a vector $v \in \mathfrak{h}$ regular if the set of fixed points of the one parameter group $\exp(tv)$ of H acting on G/B by left translation is exactly $(G/B)^H$ (where H also acts on the left). It is well-known that $(G/B)^H = \{gB \colon g \in N_G(H)\}$ and that gB depends only on $\bar{g} \in N_G(H)/C_G(H) = W$. Thus the correspondence $\bar{g} \to gB$ sets up a one-to-one correspondence between W and $(G/B)^H$ and consequently between zero(V), $V = (d/dt)\exp(tv)\big|_{t=0}$, and W for any regular v in $\mathfrak{h}$. For $w \in W$, we may unambiguously refer to wB. It is well-known that the zeros of V are all simple.

2. A Resume on Vector Fields

A holomorphic vector field V on a complex manifold X defines, by way of the contraction operator $i(V)$, a complex of sheaves,

$$0 \to \Omega^n \xrightarrow{i(V)} \Omega^{n-1} \to \ldots \to \Omega^1 \to \mathcal{O} \to 0 \quad . \tag{3}$$

If V has only finitely many zeros, then this complex is exact except at $\mathcal{O}$, and in fact provides a locally free resolution of the sheaf $\mathcal{O}_Z = \mathcal{O}/i(V)\Omega^1$, which is, by definition, the structure sheaf of the variety Z of zeros of V. It follows from general facts that there exists a spectral sequence with $E_1^{-p,q} = H^q(X,\Omega^p)$ abutting to $H^0(X,\mathcal{O}_Z)$. The key fact proved in [C-L$_1$] is that if X is compact Kaehler, then this spectral sequence degenerates at E_1 as long as $Z \neq \emptyset$. As a consequence of the finiteness of Z and $i(V)$ being a derivation, we have the

Theorem [C-L$_2$]. *If* X *is a compact Kaehler manifold admitting a holomorphic vector field* V *with* $Z = \mathrm{zero}(V)$ *finite but nontrivial, then*

(i) $H^p(X,\Omega^q) = 0$ *if* $p \neq q$ *(consequently* $H^{2p}(X,\mathbb{C}) = H^p(X,\Omega^p)$ *and* $H^{2p+1}(X,\mathbb{C}) = 0$*), and*

(ii) *there exists a filtration*

$$H^0(X,\mathcal{O}_Z) \;=\; F_n \supset F_{n-1} \supset \ldots \supset F_1 \supset F_0 \quad , \tag{4}$$

where $n = \dim X$, *such that* $F_i F_j \subset F_{i+j}$ *and having the property that as graded rings*

$$\oplus_p F_p/F_{p-1} \;\cong\; \oplus_p H^{2p}(X,\mathbb{C}) \quad . \tag{5}$$

For example, if V has only simple zeros, in other words Z is nonsingular, then $H^0(X,\mathcal{O}_Z)$ is precisely the ring of complex valued functions on Z. Thus, algebraically, $H^0(X,\mathcal{O}_Z)$ can be quite simple. The difficulty in analyzing the cohomology ring is in describing the filtration F. Note that $H^0(G/B,\mathcal{O}_Z) = \mathbb{C}^W$ for any vector field on G/B with simple zeros, hence for any vector field generated by a regular vector in $\mathfrak{h}$. (At the other extreme, G/B always admits a vector field with exactly one zero by [A], but for these, the structure of $H^0(G/B,\mathcal{O}_Z)$ is not known).

3. The V-Equivariant Chern Class of L_α

The key to understanding $H^0(G/B,\mathcal{O}_Z)$ is in knowing how $c_1(L_\alpha)$ arises. To answer this we need to recall the theory of V-equivariant Chern classes. We say a holomorphic line bundle L on X is V-equivariant if the derivation $V\colon \mathcal{O}\to\mathcal{O}$ lifts to a derivation $\tilde{V}\colon \mathcal{O}(L)\to\mathcal{O}(L)$; i.e., a C-linear map satisfying $\tilde{V}(fs) = V(f)s + f\tilde{V}(s)$ if $f\in\mathcal{O}$, $s\in\mathcal{O}(L)$. Since $V(f) = i(V)df$, $\tilde{V}$ defines a global section of $\mathrm{End}(\mathcal{O}(L)\otimes_{\mathcal{O}}\mathcal{O}_Z)\cong\mathcal{O}_Z$; i.e., $\tilde{V}\in H^0(X,\mathcal{O}_Z)$. It is shown in [C-L$_2$] that

(1) $\tilde{V}\in F_1$ and has image $c_1(L)$ under the isomorphism (5), and

(2) every line bundle on X is V-equivariant if X is compact Kaehler and $Z\neq\emptyset$.

The calculation of V for L_α is originally due to E. Akyildiz.

Lemma 1. *Let* V *be the vector field on* G/B *generated by a regular vector* $v\in\mathfrak{h}$ *and let* $\alpha\in X(H)$. *Then there exists a lifting* $\tilde{V}_\alpha$ *of* V *such that in* $H^0(G/B,\mathcal{O}_Z)$, $\tilde{V}_\alpha(w) = -d\alpha(w^{-1}\cdot v)$.

Proof. Recall that by definition, $L_\alpha = G\times C/B$, where $(g,z)b = (gb,\alpha(b^{-1})z)$. Define a one-parameter group on $G\times C$ by $\phi_t(g,z) = (\exp(tv)g,\alpha(\exp(tv))z)$. Clearly ϕ_t commutes with the action of B. Now if $g\in N_G(H)$, $\exp(tv)g = g\exp(tw^{-1}\cdot v)$, where $w=\bar{g}$, so

$$(\exp(tv)g,\alpha(\exp(tv))z) = (g,\alpha(\exp t(v-w^{-1}\cdot v))z) .$$

It follows that for some lift $\tilde{V}'_\alpha$ of V,

$$\begin{aligned}\tilde{V}'_\alpha(w) &= \frac{d}{dt}\alpha(\exp(t(v-w^{-1}\cdot v)))\Big|_{t=0}\\ &= d\alpha(v-w^{-1}\cdot v) .\end{aligned}$$

Thus taking $\tilde{V}_\alpha = \tilde{V}'_\alpha - d\alpha(v)I$, where I denotes the identity in $\mathrm{End}(L)$, we get the desired result. ∎

The lemma implies that for each $\omega\in\mathfrak{h}^*$, one can define an element $s_\omega\in F_1$ by setting $s_\omega(w) = -\omega(w^{-1}\cdot v) = -w\cdot\omega(v)$. So define a linear map

Ψ_v: $\mathfrak{h}^*$ $H^0(G/B,\mathcal{O}_Z)$ by $\Psi_v(\omega)(w) = -w\cdot\omega(v)$. Then Ψ_v can be extended to an algebra homomorphism Ψ_v: $R \to H^0(G/B,\mathcal{O}_Z)$. Unfortunately, Ψ_v is not W-equivariant with respect to the natural action of W on $f \in H^0(G/B,\mathcal{O}_Z)$ given by $(u\cdot f)(w) = f(u^{-1}w)$. To obtain equivariance, one must force W to act on $Z(=W)$ on the right. Thus $u \in W$ acts on f according to $f^u(w) = f(wu^{-1})$. Then one gets that $\Psi_v(u\cdot\omega) = (\Psi_v(\omega))^{u^{-1}}$ for all $u \in W$.

Description of $H^0(G/B,\mathcal{O}_Z)$. For any $v \in \mathfrak{h}$, let I_v denote the ideal in R generated by all $\phi \in I_W$ such that $\phi(v) = 0$. The ring R/I_v is only graded when I_v is homogeneous, i.e., only when $v = 0$ and $R/I_v = R/RI_W^+$. However, R/I_v is always filtered by degree. Namely, if $p = 0,1,\ldots$, set $(R/I_v)_p = R_p/R_p \cap I_v$ where $R_p = \{f \in R\colon \deg f \leqslant p\}$. $I_v \subset \ker \Psi_v$, for if $\phi \in I_W$ and $\phi(v) = 0$, then for all $w \in W$,

$$\begin{aligned}\Psi_v(\phi)(w) &= (w\cdot\phi)(v)\\ &= \phi(v)\\ &= 0 \quad .\end{aligned}$$

This motivates

Theorem 2. *For* v *in a dense open set in* $\mathfrak{h}$, Ψ_v *induces an isomorphism*

$$\bar{\Psi}_v\colon\ R/I_v \to H^0(G/B,\mathcal{O}_Z)$$

preserving the filtrations, i.e., $\bar{\Psi}_v((R/I_v)_p) = F_p$. *Consequently, for each* p, $WF_p = F_p$ *and the natural morphism* $F_1^{\otimes p} \to F_p$ *is onto.*

This theorem was inspired by Bill Casselman's observation that $\ker \Psi_v = I_v$.

Proof. Let $A_v = \ker \Psi_v$. Clearly A_v is W-invariant. Since $\bar{\Psi}_v\colon R/A_v \to H^0(G/B,\mathcal{O}_Z)$ is injective, R/A_v is finite dimensional, so the Nullstellensatz implies that the variety $V(A_v)$ in $\mathfrak{h}$ defined by A_v is finite. It is moreover W-invariant (since A_v is) and contains v. Since W is finite and acts effectively on $\mathfrak{h}$, the set $\Sigma = \{v \in \Sigma\colon v$ is regular and all translates $w\cdot v$, $w \in W$, are distinct$\}$ is dense and open. Thus if $v \in \Sigma$, then $V(A_v)$ has cardinality at least $|W|$, while $\operatorname{card} V(A_v) \leqslant \dim R/A_v \leqslant |W|$ shows that $\operatorname{card} V(A_v) = \dim R/A_v = |W|$. Consequently, $\bar{\Psi}_v$ is an isomorphism.

We next show that $I_v = A_v$. But by a theorem of Chevalley [Ch], $\dim R/RI_W^+ = |W|$, and thus $\dim R/I_v = |W|$ as well. Since $I_v \subset A_v$, $R/A_v \cong R/I_v$, and since $v \in V(I_v)$, $V(I_v) = V(A_v)$. These facts imply that $I_v = A_v$. The assertion that $\bar{\Psi}_v$ preserves the filtrations follows from $\Psi_v((R/I_v)_p) \subset F_p$ because both spaces have the same dimension (Chevalley's theorem). The last assertion is now obvious.

4. Proof of Borel's Theorem

To summarize, we have, by Lemma 1, produced for each v in a dense open set in $\mathfrak{h}$, a commutative diagram

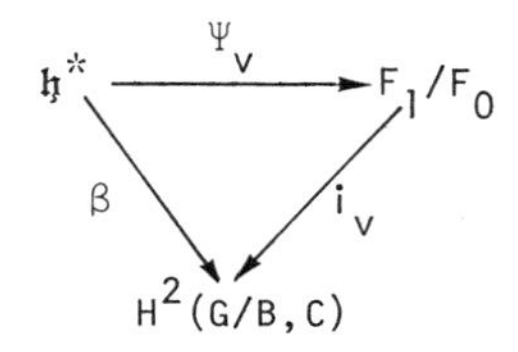

where i_v is an isomorphism, and Ψ_v is surjective. Consequently β is surjective. Moreover, by Theorem 2, this results in a commutative diagram for each $p \geqslant 1$

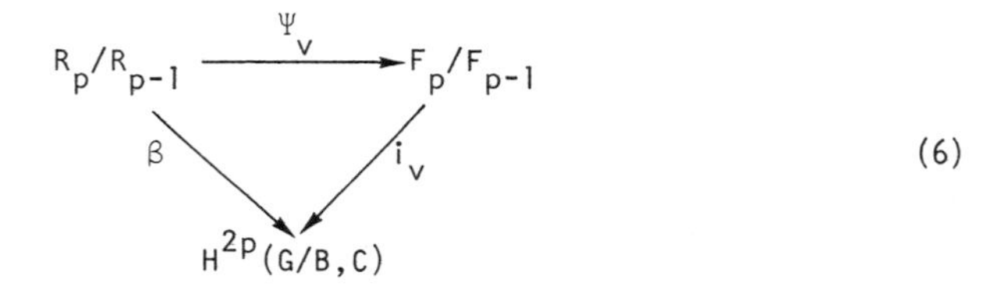

where Ψ_v is surjective and i_v is an isomorphism. Thus $\beta\colon R \to H^{\bullet}(G/B,C)$ is surjective. To complete the proof, one must show that $\ker \beta = RI_W^+$. But because $\dim R/RI_W^+ = \dim H^0(G/B,\mathcal{O}_Z) = |W|$, it suffices to show that $RI_W^+ \subset \ker \beta$, and this is surprisingly easy. In fact, if $f \in R_p \cap RI_W^+$, then $\Psi_v(f) \in F_{p-1}$ due to the fact that $\Psi_v(I_W^+) \subset F_0$. Hence, by commutativity of (6), $\beta(f) = 0$, and the result is proved.

Thus one gets the result that $H^{\bullet}(G/B,C)$ is a limit of filtered rings R/I_v, each R/I_v having the additional property that its associated graded ring is $H^{\bullet}(G/B,C)$.

5. Remarks on Computations in $H^{\bullet}(G/B,C)$

In effect, when one throws computations in $H^{\bullet}(G/B,C)$ back into $H^0(G/B,\mathcal{O}_Z)$, one reduces a problem involving invariants of the Weyl group and possibly roots of G in H into a problem in linear algebra in which knowledge of roots and invariants is not needed. For example, suppose $\alpha_1,\ldots,\alpha_k \in X(H)$ and one wishes to check a polynomial relation $p(c_1(L_{\alpha_1}),\ldots,c_1(L_{\alpha_k})) = 0$ where p is a homogeneous polynomial of degree q. Then one must check that for just one regular v,

$$\Psi_v(p(d\alpha_1,\ldots,d\alpha_k)) \in F_{q-1} \tag{7}$$

where F refers to the filtration induced by v. If $\omega_1,\ldots,\omega_n$ is a basis of $\mathfrak{h}^*$ and if $w_1,\ldots,w_N$ are the elements of W, then F_1 can be viewed as the row space of the matrix

$$\begin{pmatrix} \omega_1(w_1^{-1}\cdot v) & \omega_1(w_2^{-1}\cdot v) & \cdots & \omega_1(w_N^{-1}\cdot v) \\ \vdots & \vdots & & \vdots \\ \omega_n(w_1^{-1}\cdot v) & \omega_n(w_2^{-1}\cdot v) & \cdots & \omega_n(w_N^{-1}\cdot v) \end{pmatrix}$$

under the identification of $s \in H^0(G/B,\mathcal{O}_Z)$ with $(s(w_1),\ldots,s(w_N)) \in C^N$. Similarly, since $F_1F_1 = F_2$, F_2 is the row space of

$$\begin{pmatrix} \omega_1(w_1^{-1}\cdot v)^2 & \cdots & \omega_1(w_N^{-1}\cdot v)^2 \\ \vdots & & \vdots \\ \omega_1(w_1^{-1}\cdot v)^2 & \cdots & \omega_n(w_N^{-1}\cdot v)^2 \\ \omega_1(w_1^{-1}\cdot v)\omega_2(w_1^{-1}\cdot v) & \cdots & \omega_1(w_N^{-1}\cdot v)\omega_2(w_N^{-1}\cdot v) \\ \vdots & & \\ \omega_{n-1}(w_1^{-1}\cdot v)\omega_n(w_1^{-1}\cdot v) & \cdots & \omega_{n-1}(w_N^{-1}\cdot v)\omega_n(w_N^{-1}\cdot v) \end{pmatrix}$$

Thus to verify (7), one is reduced to checking whether

$$(p(d\alpha_1(w_1^{-1}\cdot v),\ldots,d\alpha_k(w_1^{-1}\cdot v)),\ldots,p(d\alpha_1(w_N^{-1}\cdot v),\ldots,d\alpha_k(w_N^{-1}\cdot v)))$$

is in the row space of the matrix whose typical row is

$$\left(\omega_{i_1}(w_1^{-1}\cdot v)\ldots\omega_{i_{q-1}}(w_1^{-1}\cdot v),\ldots,\omega_{i_1}(w_N^{-1}\cdot v)\ldots\omega_{i_{q-1}}(w_N^{-1}\cdot v)\right)$$

where $1 \leqslant i_1 \leqslant \ldots \leqslant i_{q-1} \leqslant n$.

Let us take a specific example. Suppose $G = GL_3$, so that B consists of the nonsingular upper triangular 3×3 matrices, and W is S_3. Take H to be the set of all 3×3 diagonal matrices.

$$\begin{pmatrix} x_{11} & & 0 \\ & x_{22} & \\ 0 & & x_{33} \end{pmatrix}$$

and $\omega_i = dx_{ii}$ $(i = 1, 2, 3)$ a basis of $\mathfrak{h}^*$. If we let $v = (1, 2, 3)$, then F_1 is the row space of

$$\begin{pmatrix} 1 & 1 & 2 & 2 & 3 & 3 \\ 2 & 3 & 1 & 3 & 1 & 2 \\ 3 & 2 & 3 & 1 & 2 & 1 \end{pmatrix} = \begin{pmatrix} a_1 \\ a_2 \\ a_3 \end{pmatrix}$$

and F_2 is the row space of

$$\begin{pmatrix} 1 & 1 & 4 & 4 & 9 & 9 \\ 4 & 9 & 1 & 9 & 1 & 4 \\ 9 & 4 & 9 & 1 & 4 & 1 \\ 2 & 3 & 2 & 6 & 3 & 6 \\ 3 & 2 & 6 & 2 & 6 & 3 \\ 6 & 6 & 3 & 3 & 2 & 2 \end{pmatrix} = \begin{pmatrix} a_1^2 \\ a_2^2 \\ a_3^2 \\ a_1a_2 \\ a_1a_3 \\ a_2a_3 \end{pmatrix} .$$

We conclude by mentioning a curious relation between weights and cohomology. Suppose that λ is a regular one parameter subgroup of H. Then λ defines a C^* action on G/B with $(G/B)^{C^*} = Z$ as usual. The induced representation of C^* on $T_w(G/B)$ for $w \in W$ (viewed as Z) is characterized by its weights which are in fact the integers $d\alpha(w^{-1}\cdot\lambda'(1))$ where α runs through the positive roots (with respect to B). Suppose we denote these positive roots by $\alpha_1,\ldots,\alpha_k$. Then the row space of

$$\begin{pmatrix} d\alpha_1(w_1^{-1}\cdot\lambda'(1)) & \cdots & d\alpha_1(w_N^{-1}\cdot\lambda'(1)) \\ \vdots & & \vdots \\ d\alpha_k(w_1^{-1}\cdot\lambda'(1)) & \cdots & d\alpha_k(w_N^{-1}\cdot\lambda'(1)) \end{pmatrix}$$

has dimension $n-1$. Consequently, exactly $(n-1)$ column vectors are linearly independent and this places a curious restriction on the weights of any C^* action on G/B with finitely many fixed points. By looking at F_2, F_3, etc. one can deduce further restrictions on the weights.

6. References

[A] Akyildiz, E., "A vector field on G/P with one zero," Proc. Amer. Math. Soc. 67, 32-34 (1977).

[B] Borel, A., "Sur la cohomologie des espaces fibrés principaux et des espaces homogenes des groupes de Lie compacts," *Ann. of Math.* (2), 57, 115-207 (1953).

[C] Carrell, J.B., "Chern classes of the Grassmannians and Schubert calculus," *Topology* 17, 177-182 (1978).

[C-L$_1$] Carrell, J.B. and Lieberman, D.I., "Holomorphic vector fields and Kaehler manifolds," *Invent. Math.* 21, 303-309 (1974).

[C-L$_2$] Carrell, J.B. and Lieberman, D.I., "Vector fields and Chern numbers," *Math. Annalen* 225, 263-273 (1977).

[Ch] Chevalley, C., "Invariants of finite groups generated by reflections," *Amer. J. Math.* 67, 778-782 (1955).

[H] Howard, A., "Holomorphic vector fields on projective manifolds," *J. Math.* 94, 1282-1290 (1972).

The University of British Columbia
Vancouver, Canada V6T 1Y4

Partially supported by a grant from NSERC of Canada

(Received April 26, 1980)

A SIMPLE PROOF OF FROBENIUS THEOREM

Shiing-shen Chern[†] and Jon G. Wolfson[‡]

Frobenius Theorem, as stated in Y. Matsushima, *Differential Manifolds*, Marcel Dekker, N.Y., 1972, p. 167, is the following:

Let D *be an* r*-dimensional differential system on an* n*-dimensional manifold* M. *Then* D *is completely integrable if and only if for every local basis* $\{X_1,\ldots,X_r\}$ *of* D *on any open set* V *of* M, *there are* C^∞*-functions* c^k_{ij} *on* V *such that we have*

$$[X_i,X_j] = \sum_k c^k_{ij}X_k \quad , \qquad 1 \leqq i,j,k \leqq r \quad . \tag{1}$$

We recall that D is called completely integrable if, at each point $p \in M$, there is a local coordinate system $(x^1,\ldots,x^n)$ such that $\partial/\partial x^1,\ldots,\partial/\partial x^r$ form a local basis of D.

The theorem is of course a fundamental one in differential geometry and every mathematician has his favorite proof. We wish to record the following proof, because it is surprisingly simple and we have not found it in the literature.

The "only if" part of the theorem being trivial, we will prove the "if" part.

For $r = 1$ condition (1) is automatically satisfied, and a stronger version of the theorem holds: *If a vector field* $X \neq 0$, *then there is at every point* $p \in M$ *a local coordinate system* $(x^1,\ldots,x^n)$ *such that* $X = \partial/\partial x^1$.

The proof of this statement is based on an existence theorem on ordinary differential equations. We will assume it. It turns out that this is the hardest part of the proof.

We suppose $r \geqq 2$. To apply induction we suppose the theorem be true for $r - 1$. By the above statement there is a coordinate system $(y^1,\ldots,y^n)$ at p such that

$$X_r = \frac{\partial}{\partial y^r} \quad . \tag{2}$$

Let

$$X'_\lambda = X_\lambda - (X_\lambda y^r)X_r \quad , \quad 1 \leqq \lambda,\mu,\nu \leqq r-1 \quad . \tag{3}$$

Then

$$X'_\lambda y^r = 0 \quad , \quad X_r y^r = 1 \quad . \tag{4}$$

Let

$$[X'_\lambda, X'_\mu] \equiv a_{\lambda\mu} X_r \quad , \quad \text{mod } X'_\nu \quad .$$

Applying both sides of the operator to the function y^r, we get $a_{\lambda\mu} = 0$. Hence the differential system $D' = \{X'_1, \ldots, X'_{r-1}\}$ satisfies the condition (1). By induction hypothesis there is a local coordinate system $(z^1, \ldots, z^n)$ at p such that

$$D' = \{X''_1, \ldots, X''_{r-1}\} \tag{5}$$

and

$$X''_\lambda = \frac{\partial}{\partial z^\lambda} \quad . \tag{6}$$

The X''_λ differ from X'_μ by a non-singular linear transformation. This has the consequence

$$X''_\lambda y^r = 0 \quad . \tag{7}$$

We have $D = \{X''_1, \ldots, X''_{r-1}, X_r\}$ and condition (1) remains satisfied. Put

$$[X''_\lambda, X_r] \equiv b_\lambda X_r \quad , \quad \text{mod } X''_\mu \quad .$$

Applying the operator on y^r, we get $b_\lambda = 0$. It follows that

$$[X''_\lambda, X_r] = \sum_\mu d^\mu_\lambda X''_\mu \quad . \tag{8}$$

In the z-coordinates let

$$X_r = \sum_A \xi^A \frac{\partial}{\partial z^A} \quad , \quad 1 \leqq A \leqq n \quad . \tag{9}$$

Then

$$[X''_\lambda, X_r] = \sum_A \frac{\partial \xi^A}{\partial z^\lambda} \frac{\partial}{\partial z^A}$$

and condition (8) becomes

$$\frac{\partial \xi^\rho}{\partial z^\lambda} = 0 \quad , \qquad 1 \leqq \lambda \leqq r-1, \quad r \leqq \rho \leqq n \quad , \tag{10}$$

which means that ξ^ρ are functions of $z^r,\dots,z^n$ only. But D is also spanned by $X''_1,\dots,X''_{r-1},X'_r$, where

$$X'_r = \sum_\rho \xi^\rho \frac{\partial}{\partial z^\rho} \qquad r \leqq \rho \leqq n \quad . \tag{11}$$

The last operator involves only the coordinates $z^r,\dots,z^n$. By a change of coordinates $(z^r,\dots,z^n) \to (w^r,\dots,w^n)$, which will not affect $z^1,\dots,z^{r-1}$, and hence not the equations (6), we can get

$$X'_r = \frac{\partial}{\partial w^r} \quad . \tag{12}$$

This completes the proof of the theorem. ■

University of California
Berkeley, California 94720, USA

† Work done under partial support of NSF Grant MC577-23579.

‡ Supported by a postgraduate scholarship of NSERC of Canada.

(Received December 19, 1980)

ON FLAT SURFACES IN S_1^3 AND H_1^3

Marcos Dajczer and Katsumi Nomizu

The main purpose of the present paper is to study isometric immersions of the Euclidean plane E^2 and the Lorentzian plane L^2 into the 3-dimensional Lorentzian manifolds S_1^3 and H_1^3 of constant sectional curvature 1 and -1, respectively.

The paper is organized as follows.

In Section 1 we give a brief survey of some known results concerning isometric immersions between space forms that will motivate the problems treated in this paper. Section 2 deals with a general principle on the correspondence between the shape operators for isometric immersions of different spaces and its application to the problem of determining isometric immersions of E^2 and L^2 into S_1^3.

In Section 3 we study the possibilities of the shape operator for isometric immersions of L^2 into H_1^3. In Section 4 we construct models of isometric immersions of L^2 into H_1^3 and compute their shape operators. One important tool is the notion of B-scroll of a Frenet curve (time-like, space-like or null) in H_1^3. In Section 5, we prove three theorems which characterize the models obtained in Section 4. In particular, we determine complete Lorentz surfaces with parallel shape operators in H_1^3 as well as complete Lorentz surfaces with constant mean curvature in H_1^3.

The remainder of the paper is devoted to the group-theoretic interpretation of our results on isometric immersions of L^2 into H_1^3 and an indication of a more general result. In Section 6 we show that the geometry of H_1^3 can be realized by a bi-invariant Lorentzian metric on the Lie group $SL(2,R)$. We prove Lemmas 5 and 6 on the group-theoretic meaning of the condition that the torsion of a Frenet curve in $SL(2,R)$ is equal to 1 or -1. In Section 7, we discuss flat Lorentz surfaces obtained as the product of two 1-parameter subgroups. This idea is then generalized. We show that the B-scroll of a Frenet curve in $H_1^3 = SL(2,R)$ can be obtained in the group-theoretic fashion. Finally, we show a way of obtaining a flat Lorentz surface as a product of two appropriate curves. Whether this Lorentz surface is complete or not remains an open problem.

The results in this paper are based on natural extensions of the ideas in [2], [3], [4] and [8].

The work of the second author was done while he was visiting professor at Instituto de Matemática Pura e Aplicada, although some of his ideas were developed earlier in the work supported by the National Science Foundation (MCS79-01310).

1. Introduction

We use the standard notation R_s^{n+1}, S_s^n, H_s^n as follows [10].

R_s^{n+1} is the $(n+1)$-dimensional vector space R^{n+1} with an inner product of signature $(s, n+1-s)$ given by

$$\langle x,y \rangle = -\sum_{i=1}^{s} x_i y_i + \sum_{j=s+1}^{n+1} x_j y_j$$

for $x = (x_1,\ldots,x_{n+1})$, $y = (y_1,\ldots,y_{n+1}) \in R^{n+1}$.

S_s^n is the hypersurface of R_s^{n+1}

$$\{x \in R_s^{n+1};\ \langle x,x \rangle = 1\} \quad ,$$

on which the restriction of $\langle , \rangle$ is an indefinite metric of signature $(s, n-s)$ with constant sectional curvature 1.

H_s^n is the hypersurface of R_{s+1}^{n+1}

$$\{x \in R_{s+1}^{n+1};\ \langle x,x \rangle = -1\} \quad ,$$

on which the restriction of $\langle , \rangle$ is an indefinite metric of signature $(s, n-s)$ with constant sectional curvature -1. Only when $s = 0$, we set

$$H_0^n = \{x \in R_s^{n+1};\ \langle x,x \rangle = -1,\ x_1 > \ \} \quad .$$

For $s = 0$, R_0^n is the Euclidean space E^n, S_0^n is the sphere S^n, and H_0^n is the hyperbolic space H^n. These are the standard models of Riemannian space forms of constant sectional curvature 0,1 and -1, respectively.

For $s = 1$, R_1^n is the Lorentz space L^n, S_1^n and H_1^n are the standard models of Lorentzian space forms of constant sectional curvature 1 and -1, respectively.

More generally, we denote by $M_s^n(c)$ a manifold with an indefinite metric of signature $(s, n-s)$ of constant sectional curvature c.

The following result can be found in [4] (also [5], p. 458).

Theorem 0. *Suppose*

(i) $M_0^n(c)$ *is isometrically immersed in* $M_1^{n+1}(\tilde{c})$ *or*

(ii) $M_1^n(c)$ *is isometrically immersed in* $M_1^{n+1}(\tilde{c})$ *or*

(iii) $M_1^n(c)$ *is isometrically immersed in* $M_2^{n+1}(\tilde{c})$.

If $c \neq \tilde{c}$ *and* $n > 2$, *then the immersion is umbilical and furthermore* $\tilde{c} > c$ *in case* (i), $\tilde{c} < c$ *in case* (ii), *and* $\tilde{c} > c$ *in case* (iii).

From this result we may derive the following facts and problems.

1.1 *If* $f: E^n \to S_1^{n+1}$ *is an isometric immersion with* $n \geqq 3$, *then it is umbilical and can be described by*

$$(x_1, \dots, x_n) \to \left(\frac{x_1^2 + \cdots + x_n^2}{2}, x_1, \dots, x_n, 1 - \frac{x_1^2 + \cdots + x_n^2}{2} \right) .$$

Problem 1. *Determine all isometric immersions* $E^2 \to S_1^3$.

1.2 *If there is an isometric immersion* $L^n \to S_1^{n+1}$, *then* $n = 2$.

Problem 2. *Determine all isometric immersions* $L^2 \to S_1^3$.

1.3 *If* $f: L^n \to H_1^{n+1}$ *is an isometric immersion with* $n \geqq 3$, *then it is umbilical and can be described by*

$$(x_1, \dots, x_n) \to \left(1 + \frac{-x_1^2 + x_2^2 + \cdots + x_n^2}{2}, x_1, \dots, x_n, \frac{-x_1^2 + x_2^2 + \cdots + x_n^2}{2} \right) .$$

Problem 3. *Determine all isometric immersions* $L^2 \to H_1^3$.

1.4 *If there is an isometric immersion* $E^n \to H_1^{n+1}$, *then* $n = 2$.

Problem 4. *Determine all isometric immersions* $E^2 \to H_1^3$.

1.5 *If there is an isometric immersion* $L^n \to H_2^{n+1}$, *then* $n = 2$.

Observe that there is a one-to-one correspondence between isometric immersions $L^2 \to H_2^3$ and $L^2 \to S_1^3$, because putting the minus sign in front of the inner product changes R^4_3 into R^4_1 and H^3_2 into S^3_1. Thus the determination of all isometric immersions $L^2 \to H^3_2$ is the same as Problem 3.

1.6 *If* $f: L^n \to S^{n+1}_2$ *is an isometric immersion with* $n \geq 3$, *then it is umbilical and can be described by*

$$(x_1, \ldots, x_n) \to \left(\frac{x_0^2 - x_1^2 - \cdots - x_n^2}{2},\ x_1, \ldots, x_n,\ 1 + \frac{x_0^2 - x_1^2 - \cdots - x_n^2}{2} \right).$$

Just like the correspondence between $L^2 \to H^3_2$ and $L^2 \to S^3_1$, we have a natural correspondence between $L^2 \to S^3_2$ and $L^2 \to H^3_1$. Thus the determination of all isometric immersions $L^2 \to S^3_2$ is the same as Problem 4.

Problems 1 and 2 can be reduced to the known result on isometric immersions $E^2 \to H^3$, [7], [8], [9], as is done in Section 2 by virtue of a general principle on shape operators. The same principle will also show that Problems 3 and 4 are equivalent. The study of Problem 3 is the primary purpose of the present paper.

We note that there are isometric immersions of other spaces to study, for example, $S^2_1 \to L^3$, $H^2_1 \to L^3$, etc. Isometric immersions between Lorentzian spaces of the same sectional curvature have been studied; see [2] for $L^n \to L^{n+1}$, [3] for $S^n_1 \to S^{n+1}_1$, [4] for $E^n \to L^{n+k}$, $L^2 \to L^4$, $L^2 \to R^4_2$, etc.

2. Correspondence of Shape Operators

Let M^3 be any Riemannian or indefinite Riemannian manifold. The shape operator for an isometric immersion $E^2 \to M^3$ is a field of symmetric transformations which, relative to a Euclidean coordinate system $\{x,y\}$ on E^2, is represented by a matrix

$$A = \begin{bmatrix} a & b \\ b & c \end{bmatrix}, \tag{1}$$

where a, b, c are differentiable functions of (x,y). The equation of Codazzi is

$$\nabla_x (A\, \partial/\partial y) = \nabla_y (A\, \partial/\partial x)$$

where ∇ denotes covariant differentiation on E^2. This equation is equivalent to the system of partial differential equations

$$b_x = a_y \quad \text{and} \quad b_y = c_x \quad . \tag{2}$$

Now let us consider

$$B = \begin{bmatrix} a & b \\ -b & -c \end{bmatrix} \tag{3}$$

as a tensor field of Lorentz-symmetric transformations on L^2 relative to a Lorentzian coordinate system $\{x,y\}$, that is, $\langle\partial/\partial x, \partial/\partial x\rangle = -1$, $\langle\partial/\partial y, \partial/\partial y\rangle = 1$, $\langle\partial/\partial x, \partial/\partial y\rangle = 0$. The equation of Codazzi for B is also equivalent to (2).

2.1 Now we take M^3 to be E^3 or L^3. Then the equation of Gauss for an isometric immersion is $\det A = 0$. Since $\det B = -\det A = 0$, we see that B satisfies the equation of Gauss for an isometric immersion $L^2 \to L^3$.

2.2 Let $M^3 = H^3$ or S_1^3. Then the equation of Gauss is $\det A = 1$. In this case, $\det B = -1$, which is the equation of Gauss for $L^2 \to S_1^3$. Conversely, if B is the shape operator for $L^2 \to S_1^3$, then the corresponding A is the shape operator for $E^2 \to H^3$ or $E^2 \to S_1^3$.

2.3 Let $M^3 = S^3$ or H_1^3. Then the equation of Gauss is $\det A = -1$. In this case, $\det B = 1$, which is the equation of Gauss for $L^2 \to H_1^3$. Conversely, if B is the shape operator for $L^2 \to H_1^3$, then the corresponding A is the shape operator for $E^2 \to S^3$ or $E^2 \to H_1^3$.

Summarizing these observations we have

Proposition 0 (General principle on correspondence of shape operators). *There is a one-to-one correspondence between the shape operators for isometric immersions as follows:*

(1) A *for* $E^2 \to E^3$, A *for* $E^2 \to L^3$ *and* B *for* $L^2 \to L^3$;

(2) A *for* $E^2 \to H^3$, A *for* $E^2 \to S_1^3$ *and* B *for* $L^2 \to S_1^3$;

(3) A *for* $E^2 \to S^3$, A *for* $E^2 \to H_1^3$ *and* B *for* $L^2 \to H_1^3$.

Here

$$A = \begin{bmatrix} a & b \\ b & c \end{bmatrix} \quad \textit{relative to a Euclidean coordinate system}$$

and

$$B = \begin{bmatrix} a & b \\ -b & -c \end{bmatrix}$$ *relative to a Lorentzian coordinate system.*

By this principle, the study of isometric immersions $E^2 \to S_1^3$ in Problem 1 and $L^2 \to S_1^3$ in Problem 2 can be reduced to that of $E^2 \to H^3$, which was done in [7] and [9]. Indeed, according to [8, pp. 163-171], the matrix for the shape operator A for any isometric immersion $E^2 \to H^3$ is a constant matrix, that is, A is covariant constant on L^2. Therefore, the shape operators for any isometric immersion $E^2 \to S_1^3$ and $L^2 \to S_1^3$ are covariant constant. Using this fact, we prove

Theorem 1. *Let* f *be an isometric immersion* $E^2 \to S_1^3$. *Then there is a Euclidean coordinate system* $\{x,y\}$ *on* E^2 *relative to which the shape operator* A *has the form*

$$A = \begin{bmatrix} \lambda & 0 \\ 0 & 1/\lambda \end{bmatrix} \quad (\lambda\text{: constant, } \geqq 1) \quad .$$

(1) *If* $\lambda = 1$, *then* f *is umbilical and congruent to*

$$f_1 : (x,y) \in E^2 \to \left(\frac{x^2+y^2}{2} \, , \; 1 - \frac{x^2+y^2}{2} \, , \; x,y \right) \in S_1^3 \subset R_1^4 \quad ,$$

where

$$f_1(E^2) = \left\{ x = (x_1,x_2,x_3,x_4) \in S_1^3 ; \; x_1 + x_2 = 1 \right\} \quad .$$

(2) *If* $\lambda > 1$, *then let*

$$r = \frac{1}{\sqrt{\lambda^2 - 1}} \quad .$$

f *is congruent to*

$$f_2 : (x,y) \in E^2 \to \left(r \cosh \frac{x}{r}, \; r \sinh \frac{x}{r} \, , \; \sqrt{1+r^2} \, \cos \frac{y}{\sqrt{1+r^2}} \, , \; \sqrt{1+r^2} \, \sin \frac{y}{\sqrt{1+r^2}} \right) \in S_1^3 \subset R_1^4 \quad ,$$

where

$$f_2(E^2) = \left\{ x = (x_1, x_2, x_3, x_4) \in S_1^3;\ x_2^2 + x_3^2 = 1 + r^2 \right\} .$$

Remark. Here and in all the classifications of isometric immersions, we determine the shape operator A up to a sign, because the shape operator $-A$ is obtained from the same immersion for the opposite choice of a field of unit normals.

Proof. We know that the shape operator A is covariant constant. If we diagonalize A at the origin $(0,0)$ of E_2:

$$A = \begin{bmatrix} \lambda & 0 \\ 0 & \mu \end{bmatrix} \quad (\lambda, \mu\text{: constants})$$

with respect to an orthonormal basis X and Y, then A is given by the same constant matrix relative to the Euclidean coordinate system $\{x,y\}$ such that $(\partial/\partial x)_{(0,0)} = X$, $(\partial/\partial y)_{(0,0)} = Y$. Since $\det A = 1$, we have $\mu = 1/\lambda$. We may, as in the Remark above, assume $\lambda, \mu > 0$ and hence $\lambda \geqq 1 \geqq \mu$.

We can easily verify that f_1 and f_2 are isometric immersions with shape operators

$$\begin{bmatrix} 1 & 0 \\ 0 & 1 \end{bmatrix} \quad \text{and} \quad \begin{bmatrix} \lambda & 0 \\ 0 & 1/\lambda \end{bmatrix} \quad (\lambda > 1), \quad \text{respectively,}$$

for the choice of time-like unit normal vector field

$$\xi = -\left(\frac{x^2 + y^2}{2} + 1,\ -\frac{x^2 + y^2}{2},\ x, y \right)$$

and

$$\xi = -\left(\sqrt{1 + r^2} \cosh \frac{x}{r},\ \sqrt{1 + r^2} \sinh \frac{x}{r},\ r \cos \frac{y}{\sqrt{1 + r^2}},\ r \sin \frac{y}{\sqrt{1 + r^2}} \right)$$

respectively.

Theorem 2. *Let* f *be an isometric immersion* $L^2 \to S_1^3$. *Then there is a Lorentzian coordinate system* $\{x,y\}$ *on* E^2 *relative to which the shape operator* B *has the form*

$$B = \begin{bmatrix} \lambda & 0 \\ 0 & -\frac{1}{\lambda} \end{bmatrix} \quad (\lambda\text{: } \textit{non-zero constant}) \quad .$$

f *is congruent to*

$$f_1 : (x,y) \in L^2 \to \left(r \sinh \frac{x}{r},\ r \cosh \frac{x}{r},\ \sqrt{1-r^2} \cos \frac{y}{\sqrt{1-\lambda^2}},\ \sqrt{1-r^2} \sin \frac{y}{\sqrt{1-r^2}} \right),$$

where

$$r = \frac{1}{\sqrt{1+\lambda^2}} \quad .$$

<u>Proof</u>. We know that the shape operator B is covariant constant. Let

$$B = \begin{bmatrix} a & b \\ -b & -c \end{bmatrix}$$

relative to a Lorentzian coordinate system. Then $\det B = -ac + b^2 = -1$. The characteristic polynomial of B is

$$\lambda^2 + (c-a)\lambda - 1 = 0 \quad .$$

Since the discriminant $(c-a)^2 + 4 > 0$, B has two distinct real eigenvalues. It follows that there is a new Lorentz coordinate system relative to which

$$B = \begin{bmatrix} \lambda & 0 \\ 0 & -\frac{1}{\lambda} \end{bmatrix} \quad .$$

The mapping f_1 is an isometric immersion whose shape operator for the field of space-like unit normals

$$\xi = -\left(\sqrt{1-r^2} \sinh \frac{x}{r},\ \sqrt{1-r^2} \cosh \frac{x}{r},\ -r \cos \frac{y}{\sqrt{1-r^2}},\ -r \sin \frac{y}{\sqrt{1-r^2}} \right)$$

coincides with B above. Thus f is congruent to f_1.

Theorems 1 and 2 are the solutions to Problems 1 and 2, respectively. By (3) of Proposition 0, we see that Problems 3 and 4 are equivalent. We study Problem 3, namely $L^2 \to H_1^3$, in the remainder of the paper.

3. Shape Operators for $L^2 \to H_1^3$

We recall our terminology. By a Lorentzian coordinate system $\{x,y\}$ or $\{t,s\}$ on L^2 we mean an affine coordinate system $\{x,y\}$ such that

$$\langle\partial/\partial x,\partial/\partial x\rangle = -1, \quad \langle\partial/\partial x,\partial/\partial y\rangle = 0, \quad \langle\partial/\partial y,\partial/\partial y\rangle = 1.$$

If $\{x,y\}$ is a Lorentzian coordinate system, we let

$$\begin{cases} u = \dfrac{1}{\sqrt{2}}(x+y) \\ v = \dfrac{1}{\sqrt{2}}(x-y) \end{cases} \quad \text{so that} \quad \begin{cases} x = \dfrac{1}{\sqrt{2}}(u+v) \\ y = \dfrac{1}{\sqrt{2}}(u-v) \end{cases} . \tag{4}$$

Then $\{u,v\}$ is an affine coordinate system such that

$$\langle\partial/\partial u,\partial/\partial u\rangle = \langle\partial/\partial v,\partial/\partial v\rangle = 0, \quad \langle\partial/\partial u,\partial/\partial v\rangle = -1 .$$

We call $\{u,v\}$ the null coordinate system associated to $\{x,y\}$. Remark that $(u,v) \to (v,u)$ is a Lorentz transformation.

Lemma 1. *If a shape operator* S *(not necessarily covariant constant) on* L^2 *is given by*

$$S = \begin{bmatrix} a & -b \\ b & c \end{bmatrix}$$

relative to a Lorentzian coordinate system, then

$$S = \begin{bmatrix} \dfrac{a+c}{2} & \dfrac{a+2b-c}{2} \\ \dfrac{a-2b-c}{2} & \dfrac{a+c}{2} \end{bmatrix}$$

relative to the null coordinate system $\{u,v\}$ *given in* (4).

Proof. Easy verification.

Lemma 2. *If*

$$S = \begin{bmatrix} a & -b \\ b & c \end{bmatrix}$$

relative to a Lorentzian coordinate system $\{x,y\}$, *then*

$$S = \begin{bmatrix} a & b \\ b & c \end{bmatrix}$$

relative to the coordinate system $\{-x,y\}$.

Proof. Easy verification.

It will be necessary to classify covariant constant shape operators S that arise from isometric immersions $L^2 \to H_1^3$. Such S is represented by a constant matrix

$$\begin{bmatrix} a & -b \\ b & c \end{bmatrix}, \qquad \text{where} \qquad ac + b^2 = 1$$

relative to a Lorentzian coordinate system $\{x,y\}$. We may assume trace $S \geqq 0$ by the Remark following Theorem 1. Although the following classification follows from [4], it can be directly proved by considering the characteristic roots.

Lemma 3. *Covariant constant shape operators* S *on* L^2 *with* $\det S = 1$ *and* trace $S \geqq 0$ *can be classified as follows.*

(1) *If* trace $S > 2$, *then there is a Lorentz coordinate system relative to which*

$$S = \begin{bmatrix} \lambda & 0 \\ 0 & \frac{1}{\lambda} \end{bmatrix}, \qquad \lambda > 0, \quad \lambda \neq 1 \ ;$$

(2) trace $S = 2$ *and*

$$S = \begin{bmatrix} 1 & 0 \\ 0 & 1 \end{bmatrix};$$

(3) trace $S = 2$ *and there is a null coordinate system* $\{u,v\}$ *relative to which*

$$S = \begin{bmatrix} 1 & 0 \\ 1 & 1 \end{bmatrix} \quad \text{or} \quad \begin{bmatrix} 1 & 0 \\ -1 & 1 \end{bmatrix} ;$$

(4) *If* trace $S < 2$, *then there is a Lorentz coordinate system relative to which*

$$S = \begin{bmatrix} \alpha & -\beta \\ \beta & \alpha \end{bmatrix} ,$$

where $\alpha^2 + \beta^2 = 1$, $\beta > 0$.

Later on, we shall use different matrix representations for S in cases 1 and 4 of the above lemma.

Lemma 4.

(1) *If*

$$S = \begin{bmatrix} \lambda & 0 \\ 0 & \frac{1}{\lambda} \end{bmatrix}, \ \lambda > 0, \ \lambda \neq 1 ,$$

then there is a new Lorentz coordinate system relative to which S *can be represented by*

$$\begin{bmatrix} k & -1 \\ 1 & 0 \end{bmatrix} \quad \text{if} \quad \lambda > \frac{1}{\lambda}$$

or

$$\begin{bmatrix} 0 & -1 \\ 1 & k \end{bmatrix} \quad \text{if} \quad \lambda < \frac{1}{\lambda} .$$

Proof. We have

$$\begin{bmatrix} \cosh\theta & \sinh\theta \\ \sinh\theta & \cosh\theta \end{bmatrix} \begin{bmatrix} \lambda & 0 \\ 0 & \frac{1}{\lambda} \end{bmatrix} \begin{bmatrix} \cosh\theta & \sinh\theta \\ \sinh\theta & \cosh\theta \end{bmatrix}^{-1}$$

$$= \begin{bmatrix} \lambda\cosh^2\theta - \frac{1}{\lambda}\sinh^2\theta & \left(\frac{1}{\lambda} - \lambda\right)\cosh\theta\sinh\theta \\ \left(\lambda - \frac{1}{\lambda}\right)\cosh\theta\sinh\theta & \frac{1}{\lambda}\cosh^2\theta - \lambda\sinh^2\theta \end{bmatrix}.$$

If $\lambda > 1/\lambda$, choose $\theta > 0$ such that $1/\lambda \cosh^2\theta - \lambda\sinh^2\theta = 0$, i.e., $\lambda = \cosh\theta/\sinh\theta$. Then

$$\left(\lambda - \frac{1}{\lambda}\right)\cosh\theta \sinh\theta = \left(\frac{\cosh\theta}{\sinh\theta} - \frac{\sinh\theta}{\cosh\theta}\right)\cosh\theta\sinh\theta = 1.$$

Thus the above matrix takes the form $\begin{bmatrix} k & -1 \\ 1 & 0 \end{bmatrix}$.

If $\lambda < 1/\lambda$, choose $\theta > 0$ such that $\lambda\cosh^2\theta - 1/\lambda\sinh^2\theta = 0$, i.e., $\lambda = \cosh\theta/\sinh\theta$. Then we obtain the matrix of the form $\begin{bmatrix} 0 & -1 \\ 1 & k \end{bmatrix}$.

Lemma 5. *If*

$$S = \begin{bmatrix} \alpha & -\beta \\ \beta & \alpha \end{bmatrix},$$

where $\alpha^2 + \beta^2 = 1$, $\beta > 0$, *then there is a new Lorentzian coordinate system relative to which* S *is represented by*

$$\begin{bmatrix} 0 & -1 \\ 1 & 2\alpha \end{bmatrix} \quad \textit{or} \quad \begin{bmatrix} 2\alpha & -1 \\ 1 & 0 \end{bmatrix} \quad \textit{(whichever)}.$$

Proof. Choose θ such that

$$\sinh 2\theta = \frac{-\alpha}{\beta} \quad \left(\text{or } \frac{\alpha}{\beta}\right).$$

Then

$$\begin{bmatrix} \cosh\theta & \sinh\theta \\ \sinh\theta & \cosh\theta \end{bmatrix} \begin{bmatrix} \alpha & -\beta \\ \beta & \alpha \end{bmatrix} \begin{bmatrix} \cosh\theta & \sinh\theta \\ \sinh\theta & \cosh\theta \end{bmatrix}^{-1}$$

$$= \begin{bmatrix} 0 & -1 \\ 1 & 2\alpha \end{bmatrix} \text{ or } \begin{bmatrix} 2\alpha & -1 \\ 1 & 0 \end{bmatrix}.$$

4. Models for $L^2 \to H_1^3$

In this section we construct models of isometric immersions $L^2 \to H_1^3$ and determine their shape operators.

Example 1. Let $\{t,s\}$ be a Lorentzian coordinate system and define $f: L^2 \to R_2^4$ by

$$f(t,s) = \left(\frac{s^2 - t^2}{2} + 1,\ t,\ s,\ \frac{t^2 - s^2}{2}\right) .$$

Then $f(L^2) \subset H_1^3$. Since

$$f_*(\partial/\partial t) = (-t,1,0,t), \quad f_*(\partial/\partial s) = (s,0,1,-s) ,$$

we have

$$<f_*(\partial/\partial t), f_*(\partial/\partial t)> = -1, \quad <f_*(\partial/\partial s), f_*(\partial/\partial s)> = 1$$

$$<f_*(\partial/\partial t), f_*(\partial/\partial s)> = 0 ,$$

namely, f is an isometric immersion of L^2 into H_1^3. We may choose a field of unit normal vectors

$$\xi = \left(\frac{t^2 - s^2}{2},\ t,\ s,\ \frac{s^2 - t^2}{2} + 1\right)$$

and the shape operator is $S = I$ (identity transformation).

Our immersion is umbilical. It follows that any umbilical isometric immersion is congruent to this example.

In order to construct other examples, we discuss curves in the space H_1^3.

For each point $x \in H_1^3 \subset R_2^4$, the tangent space $T_x(H_1^3)$ is a vector space with Lorentzian inner product. There is also a natural orientation defined as follows: An ordered basis $\{X,Y,Z\}$ in $T_x(H_1^3)$ is positively oriented if

$$\det[x\ X\ Y\ Z] > 0 ,$$

where $[x\ X\ Y\ Z]$ is the matrix with $x, X, Y, Z \in R_2^4$ as row vectors.

A natural volume element ω on H_1^2 is defined by

$$\omega(X,Y,Z) = \det[x\ X\ Y\ Z] \quad .$$

By means of the Lorentzian inner product and the volume element we define the notion of cross product in each $T_x(H_1^2)$. Given $X, Y \in T_x(H_1^2)$ the cross product $X \times Y$ is the unique vector in $T_x(H_1^2)$ such that

$$<X \times Y, Z> = \omega(X,Y,Z) \qquad \text{for every} \quad Z \in T_x(H_1^2) \quad .$$

Obviously, $Y \times X = -X \times Y$. We have also

$$<X \times Y, X \times Y> = <X,Y>^2 - <X,X><Y,Y>$$

as can be verified.

By a time-like curve in H_1^3 we mean a differentiable curve x_t, $t \in I$, where I is an interval containing 0 on the real line R, such that the tangent vector $T = dx/dt$ satisfies $<T,T> = -1$ for every t. We say that the curve is complete if $I = R$.

By a *time-like Frenet curve* x_t we mean a time-like curve which admits two space-like unit vector fields N and B along the curve such that $B = T \times N$ which satisfy the differential equations

$$\begin{aligned} \tilde{\nabla}_t T &= kN \\ \tilde{\nabla}_t N &= kT + wB \\ \tilde{\nabla}_t B &= -wN \end{aligned} \tag{5}$$

where $\tilde{\nabla}_t$ is covariant differentiation along x_t in H_1^3 and $k = k(t)$ and $w = w(t)$ are differentiable functions. As in the elementary theory of curves in E^3, k and w are called the curvature and the torsion, respectively. Note that the function k may change its sign.

The equations (5) can be written in R_2^4 as follows:

$$\frac{dT}{dt} = kN - x$$

$$\frac{dN}{dt} = kT + wB \tag{6}$$

$$\frac{dB}{dt} = -wN \quad .$$

We have the *fundamental theorem for time-like curves: Given differentiable functions* k *and* w *on* I, *an initial point* x_0 *in* H_1^3 *and an orthonormal frame* (T_0, N_0, B_0) *at* x_0, *where* T_0 *is time-like and* $B_0 = T_0 \times N_0$, *there exists a unique time-like Frenet curve with* (T_0, N_0, B_0) *as the initial frame at* x_0 *and* k *and* w *as curvature and torsion, respectively.*

We now define the B-scroll of a time-like curve.

Example 2. Let x_t be a complete time-like Frenet curve of constant torsion 1 (or -1). By the B-scroll of x_t we mean a mapping $f: L^2 \to H_1^3$ given by

$$f(t,s) = (\cosh s)x_t + (\sinh s)B_t \quad ,$$

where $\{t,s\}$ is a Lorentzian coordinate system on L^2. This mapping f is indeed an isometric immersion of L^2 into H_1^3 as we can easily verify. We compute the shape operator.

In the case $w \equiv 1$, we may take a field of unit normal vector field

$$\xi_{(t,s)} = \sinh s\, T - \cosh s\, N \quad .$$

Then

$$\frac{\partial \xi}{\partial t} = \sinh s(kN - x) - \cosh s(kT + B)$$
$$= f_*(-k\, \partial/\partial t - \partial/\partial s)$$

$$\frac{\partial \xi}{\partial s} = \cosh s\, T - \sinh s\, N$$
$$= f_*(\partial/\partial t)$$

so that the shape operator is

$$S = \begin{bmatrix} k & -1 \\ 1 & 0 \end{bmatrix}, \quad k = k(t) \tag{7}$$

relative to the coordinate system $\{t,s\}$ on L^2.

In the case $w \equiv -1$, we take

$$\xi(t,s) = -\sinh s\, T - \cosh s\, N$$

and find the shape operator

$$S = \begin{bmatrix} k & 1 \\ -1 & 0 \end{bmatrix}, \quad k = k(t) \quad . \tag{8}$$

Remark. The shape operator (7) changes to (8) by the Lorentz transformation $(t,s) \to (-t,s)$.

Example 3. (Special case of Example 2.) For a complete time-like geodesic x_t, we may associate the frame field T,N,B so that $k \equiv 0$ and $w \equiv 1$ (or -1). The B-scroll is an isometric immersion $L^2 \to H_1^3$ whose shape operator is

$$\begin{bmatrix} 0 & -1 \\ 1 & 0 \end{bmatrix} \quad .$$

By an isometry of H_1^3, we may assume that $x_t = (\cos t, \sin t, 0, 0)$. Then $T = (-\sin t, \cos t, 0, 0)$. We may take $N = (0, 0, \cos t, \sin t)$ and $B = (0, 0, -\sin t, \cos t)$. The B-scroll is then

$$(t,s) \to (\cos t \cosh s, \sin t \cosh s, -\sin t \sinh s, \cos t \sinh s) \quad .$$

By a *space-like Frenet curve* in H_1^3 we mean a differentiable curve y_s, $s \in I$, with the frame field, $T = dy/ds$, N, B such that

$$\langle T,T\rangle = 1, \langle N,N\rangle = 1, \quad B = T \times N \quad (\text{thus } \langle B,B\rangle = -1)$$

which satisfy the differential equations

$$\tilde{\nabla}_s T = kN$$

$$\tilde{\nabla}_s N = -kT + wB \tag{9}$$

$$\tilde{\nabla}_s B = wN$$

with functions $k = k(s)$ and $w = w(s)$, which are the curvature and the torsion of the curve. Equivalently we have

$$\frac{dT}{ds} = kN + y$$

$$\frac{dN}{ds} = -kT + wB \quad . \tag{10}$$

$$\frac{dB}{ds} = wN$$

As before, given k and w, and an appropriate initial frame (T_0, N_0, B_0) at a point y_0, there is a unique space-like Frenet curve with k and w as curvature and torsion.

Example 4. For a complete space-like Frenet curve y_s with torsion $w \equiv 1$ (or -1), the mapping f defined by

$$f(t,s) = (\cos t) y_s + (\sin t) B_s$$

is an isometric immersion of L^2 into H_1^3, called the B-scroll of y_s.

For $w \equiv 1$, we take a field of unit normal vectors

$$\xi = -\sin t\, T + \cos t\, N \quad .$$

Then the shape operator is

$$S = \begin{bmatrix} 0 & -1 \\ 1 & k \end{bmatrix}, \quad k = k(s) \tag{11}$$

relative to the coordinate system $\{t,s\}$.

For $w \equiv -1$, we have

$$\xi = \sin t\, T + \cos t\, N$$

and

$$S = \begin{bmatrix} 0 & 1 \\ -1 & k \end{bmatrix}, \quad k = k(s) \quad . \tag{12}$$

The same kind of remark as before applies to (11) and (12).

Example 5. (Special case of Example 4.) For a complete space-like geodesic y_s we may associate the frame field T, N, B such that $k \equiv 0$ and $w \equiv 1$ (or -1). The B-scroll has the shape operator

$$\begin{bmatrix} 0 & -1 \\ 1 & 0 \end{bmatrix} \quad ,$$

which is the same as that in Example 3.

Finally, we consider a null Frenet curve. A differentiable curve x_u, $u \in I$, is a *null Frenet curve* if it admits a frame field $(A = dx/du,\ B,\ C)$ such that

$$\langle A,A\rangle = \langle B,B\rangle = 0, \quad \langle A,B\rangle = -1$$

and

$$C = A \times B \quad (\text{thus } \langle C,C\rangle = 1, \quad \langle C,A\rangle = \langle C,B\rangle = 0)$$

satisfying the differential equations

$$\begin{aligned} \tilde{\nabla}_u A &= kC \\ \tilde{\nabla}_u B &= wC \\ \tilde{\nabla}_u C &= wA + kB \end{aligned} \quad , \tag{13}$$

where $k = k(u)$ and $w = w(u)$ are appropriate functions. We call k and w the curvature and the torison of the null curve $x(u)$. Equivalently, we have

$$\begin{aligned} \frac{dA}{du} &= kC \\ \frac{dB}{du} &= -x + wC \\ \frac{dC}{du} &= wA + kB \end{aligned} \quad . \tag{14}$$

As before, given two functions $k = k(u)$, $w = w(u)$ on I, and an appropriate initial frame (A_0, B_0, C_0) at x_0, there is a unique null

Frenet curve with k and w as curvature. The significance of the B-scroll of a null curve with $w \equiv 0$ in L^3 was first recognized in [2].

Example 6. Let x_u be a complete null Frenet curve with torsion $w \equiv 1$ or -1. Let $\{u,v\}$ be a null coordinate system on L^2. The B-scroll is defined by

$$f(u,v) = x_u + vB \quad .$$

For a fixed u, $x + vB$ is a null geodesic in H_1^3 starting at x in the direction of B.

Since

$$f_*(\partial/\partial u) = A + v(-x + wC)$$
$$f_*(\partial/\partial v) = B \quad ,$$

we get

$$\langle f_*(\partial/\partial u), f_*(\partial/\partial u)\rangle = v^2(w^2 - 1) = 0 \quad \text{for } w \equiv 1 \text{ or } -1$$
$$\langle f_*(\partial/\partial u), f_*(\partial/\partial v)\rangle = \langle A, B\rangle = -1$$
$$\langle f_*(\partial/\partial v), f_*(\partial/\partial v)\rangle = 0 \quad .$$

Since $\{u,v\}$ is a null coordinate system, this shows that f is an isometric immersion of L^2 into H_1^3.

In the case $w \equiv 1$, we take a field of unit normal vectors

$$\xi = -vB - C \quad .$$

The shape operator is given by

$$S = \begin{bmatrix} 1 & 0 \\ k & 1 \end{bmatrix}, \quad k = k(u) \ , \tag{15}$$

relative to the null coordinate system $\{u,v\}$. Relative to the Lorentzian coordinate system $\{x,y\}$ related to $\{u,v\}$ as in (4), the shape operator takes the form

$$S = \begin{bmatrix} 1 + m & m \\ -m & 1 - m \end{bmatrix}, \tag{16}$$

where

$$m = \frac{k\left(\frac{x+y}{\sqrt{2}}\right)}{2} .$$

We note that for any differentiable function $\phi(t)$ of one variable t, (16) with $m = \phi(x+y)$ satisfies the equations of Gauss and Codazzi, as can be easily verified. It is indeed the shape operator of the B-scroll of an appropriate null Frenet curve of torsion 1 and of curvature $k(u) = 2\phi(\sqrt{2}\,u)$.

In the case $w \equiv -1$, we take

$$\xi = vB - C$$

and get

$$S = \begin{bmatrix} -1 & 0 \\ k & -1 \end{bmatrix}, \quad k = k(u) \tag{17}$$

relative to the null coordinate system $\{u,v\}$.

Example 7. (Special case of Example 6.) If x_u is a null geodesic, we may associate a frame field (A,B,C), $A = dx/du$, so that the equations of a null Frenet curve are satisfied with $k \equiv 0$ and $w \equiv 1$ (or -1). We get the shape operator $S = I$ (or $-I$). If we choose appropriate initial conditions for $u = 0$, then the B-scroll is given by

$$f(u,v) = \left(1 - uv,\ u + \frac{1}{2}v,\ u - \frac{1}{2}v,\ uv\right) .$$

If we take a Lorentzian coordinate system $\{t,s\}$ with

$$t = u + \frac{1}{2}v, \qquad s = u - \frac{1}{2}v ,$$

then the B-scroll is expressed by

$$f_1(t,s) = \left(\frac{s^2 - t^2}{2} + 1,\ t,\ s,\ \frac{t^2 - s^2}{2}\right) .$$

This is the same isometric immersion as Example 1.

5. Characterization

In this section we shall characterize the models discussed in Section 4.

Theorem 1. *Let* $f: L^2 \to H_1^3$ *be an isometric immersion with shape operator* S. *Let* $\{t,s\}$ *be a Lorentzian coordinate system on* L^2 *and* $\{u,v\}$ *the associated null coordinate system.*

If the covariant derivative $\nabla_t S = 0$ *or* $\nabla_s S = 0$ *or* $\nabla_u S = 0$, *then* f *is the B-scroll of a suitable curve of torsion* $\equiv 1$ *(or* -1*).*

Proof. Let

$$S = \begin{bmatrix} a & -b \\ b & c \end{bmatrix}$$

relative to the coordinate system $\{t,s\}$, where a, b and c are functions of t and s. Then $\det S = ac + b^2 = 1$.

We shall prove Theorem 1 under the assumption $\nabla_t S = 0$. The other two cases are similar. We have

$$\begin{aligned}(\nabla_t S)(\partial/\partial t) &= \nabla_t(a\, \partial/\partial t + b\, \partial/\partial s) \\ &= \frac{\partial a}{\partial t}\, \partial/\partial t + \frac{\partial b}{\partial t}\, \partial/\partial s = 0\end{aligned}$$

and

$$\begin{aligned}(\nabla_t S)(\partial/\partial s) &= \nabla_t(-b\, \partial/\partial t + c\, \partial/\partial s) \\ &= -\frac{\partial b}{\partial t}\, \partial/\partial t + \frac{\partial c}{\partial t}\, \partial/\partial s = 0 .\end{aligned}$$

Thus

$$\frac{\partial a}{\partial t} = \frac{\partial b}{\partial t} = \frac{\partial c}{\partial t} = 0 \tag{18}$$

that is, a, b, and c are functions of s only.

From the equation of Codazzi

$$(\nabla_s S)(\partial/\partial t) = (\nabla_t S)(\partial/\partial s) = 0 ,$$

we get

$$\nabla_s(a\,\partial/\partial t + b\,\partial/\partial s) = \frac{\partial a}{\partial s}\,\partial/\partial t + \frac{\partial b}{\partial s}\,\partial/\partial s = 0,$$

thus

$$\frac{\partial a}{\partial s} = \frac{\partial b}{\partial s} = 0 \quad . \tag{19}$$

From (18) and (19), we see that a and b are constants. We consider Case I: $a \neq 0$ and Case II: $a = 0$.

Case I. $a \neq 0$. From $ac + b^2 = 1$, we see that a is also a constant. Thus the shape operator S is covariant constant. We may assume trace $S \geqq 0$. We have cases (1), (2), (3) and (4) in Lemma 3. In case (1), Lemma 4 says that there is a Lorentzian coordinate system in which either

$$S = \begin{bmatrix} k & -1 \\ 1 & 0 \end{bmatrix} \quad \text{or} \quad S = \begin{bmatrix} 0 & -1 \\ 1 & k \end{bmatrix} .$$

Then f is either the B-scroll of a complete time-like Frenet curve of torsion 1 (Example 2) or the B-scroll of a complete space-like Frenet curve of torsion 1 (Example 4).

In case (2) of Lemma 3, we have $S = I$ and f is umbilical. It can be considered as the B-scroll of a null geodesic in Example 7.

In case (3) of Lemma 3, there is a null coordinate system in which

$$S = \begin{bmatrix} 1 & 0 \\ 1 & 1 \end{bmatrix} ,$$

which is the shape operator (15) for $k \equiv 1$. Hence f is the B-scroll of a complete null Frenet curve of torsion 1.

Finally, in case (4) of Lemma 3, Lemma 5 says that there is a Lorentzian coordinate system in which

$$S = \begin{bmatrix} 0 & -1 \\ 1 & k \end{bmatrix} .$$

This is the shape operator (11) with constant k. Thus f is the B-scroll of a complete space-like Frenet curve of constant curvature k and torsion 1.

<u>Case II</u>. $a=0$. In this case, $b=\pm 1$ and $c=c(s)$ is an arbitrary function. Thus we may assume

$$S = \begin{bmatrix} 0 & -1 \\ 1 & c(s) \end{bmatrix}$$

(by Lemma 2). f is the B-scroll of a complete space-like Frenet curve of torsion 1 and curvature $c(s)$.

<u>Corollary</u>. *Let* $f: L^2 \to H_1^3$ *be an isometric immersion with parallel (i.e., covariant constant) shape operator* S. *Then* f *is the* B-*scroll of a complete curve of torsion* 1 *and constant curvature.*

Let us remark that isometric immersions $L^2 \to R_2^4$ with parallel second fundamental forms have been classified by Magid [4, Theorem 1.17]. Isometric immersions $L^2 \to H_1^3$ with parallel shape operators are those on his list whose images lie in H_1^3. According to the forms of shape operators (1), (2), (3) and (4) in Lemma 3, these immersions can be described explicitly. We omit the detail.

Next we prove

<u>Theorem 2</u>. *Let* $f: L^2 \to H_1^3$ *be an isometric immersion. Suppose that at each point of* L^2 *there is a null asymptotic direction* [*or, equivalently, a null principal direction*]. *Then* f *is the* B-*scroll of a null curve with torsion* 1 *(or* -1*).*

<u>Proof</u>. Suppose U is a null vector at x and V a null vector at x such that $\langle U,V\rangle = -1$. If U is asymptotic: $\langle SU,U\rangle = 0$, then writing $SU = aU + bV$ we find

$$\langle SU,U\rangle = -b = 0 \quad .$$

Thus

$$SU = aU \quad ,$$

that is, U is a principal vector. Conversely, if $SU = aU$, then obviously $\langle SU,U\rangle = 0$.

Write $SV = cU + dV$. Then

$$\langle SU,V\rangle = \langle U,SV\rangle \quad .$$

implies $a = d$. Hence

$$S = \begin{bmatrix} a & c \\ 0 & a \end{bmatrix}$$

relative to the basis $\{U,V\}$. From $\det S = 1$, we find $a = \pm 1$ so

$$\text{either} \quad S = \begin{bmatrix} 1 & c \\ 0 & 1 \end{bmatrix} \quad \text{or} \quad \begin{bmatrix} -1 & c \\ 0 & -1 \end{bmatrix} .$$

The characteristic polynomial of S is then

$$\text{either} \quad (\lambda - 1)^2 \quad \text{or} \quad (\lambda + 1)^2 .$$

This is valid at each point x.

Now let

$$S = \begin{bmatrix} \alpha & \gamma \\ \beta & \alpha \end{bmatrix}$$

relative to a fixed null coordinate system $\{u,v\}$ on L^2, where α,β,γ are functions of (u,v). We have $\det S = \alpha^2 - \beta\gamma = 1$. The characteristic polynomial of S is

$$(\lambda - \alpha)^2 - \beta\gamma = \lambda^2 - 2\alpha\lambda + 1 .$$

From our previous information we get $\alpha = 1$ or -1. Then $\beta\gamma = 0$ so $\beta = 0$ or $\gamma = 0$.

Thus S is one of the following four forms:

$$\text{(i)} \begin{bmatrix} 1 & \gamma \\ 0 & 1 \end{bmatrix} \quad \text{(ii)} \begin{bmatrix} 1 & 0 \\ \beta & 1 \end{bmatrix} \quad \text{(iii)} \begin{bmatrix} -1 & \gamma \\ 0 & -1 \end{bmatrix} \quad \text{(iv)} \begin{bmatrix} -1 & 0 \\ \beta & -1 \end{bmatrix} .$$

Using Codazzi's equation $\nabla_u(S\, \partial/\partial v) = \nabla_v(S\, \partial/\partial u)$, we find in each case

$$\text{(i)} \ \frac{\partial\gamma}{\partial u} = 0 \qquad \text{(ii)} \ \frac{\partial\beta}{\partial v} = 0 \qquad \text{(iii)} \ \frac{\partial\gamma}{\partial u} = 0 \qquad \text{(iv)} \ \frac{\partial\beta}{\partial v} = 0$$

that is, γ is a function of v only and β is a function of u only.

In case (i), we may interchange u and v so that

$$S = \begin{bmatrix} 1 & 0 \\ \gamma(u) & 1 \end{bmatrix}.$$

Thus f is the B-scroll of a complete null curve of curvature $\gamma(u)$ and torsion 1 (see (15)).

The same conclusion is valid for f in case (ii). In cases (iii) and (iv) we find that f is the B-scroll of a complete null curve of torsion -1 (see (17)).

We shall now establish

Theorem 3. *Let* $f: L^2 \to H_1^3$ *be an isometric immersion with constant mean curvature* H. *Then* f *is the* B-*scroll of a suitable curve of torsion* 1 *or* -1. *More precisely,* f *is an immersion with parallel shape operator* [*Corollary to Theorem 1 and the subsequent Remark*] *or the* B-*scroll of a null curve with torsion* 1 *or* -1.

Proof. Let

$$S = \begin{bmatrix} a & -b \\ b & c \end{bmatrix}$$

be the shape operator of f relative to a Lorentzian coordinate system $\{t,s\}$. The coefficients a, b, c are functions of t,s and, by assumption, $a+c$ is a constant $2H$.

The equation of Codazzi $\nabla_t(S(\partial/\partial s)) = \nabla_s(S(\partial/\partial t))$ implies

$$\frac{\partial a}{\partial s} = -\frac{\partial b}{\partial t}, \qquad \frac{\partial b}{\partial s} = -\frac{\partial a}{\partial t} \tag{20}$$

and thus

$$\left(\frac{\partial a}{\partial t}\right)^2 + \left(\frac{\partial a}{\partial s}\right)^2 = \left(\frac{\partial b}{\partial t}\right)^2 + \left(\frac{\partial b}{\partial s}\right)^2 . \tag{21}$$

From $\det S = a(2H-a) + b^2 = 2aH - a^2 + b^2 = 1$ we obtain

$$(H-a)\frac{\partial a}{\partial t} = -b\frac{\partial b}{\partial t}$$

$$(H-a)\frac{\partial a}{\partial s} = -b\frac{\partial b}{\partial s}$$

from which

$$(H - a)^2\left(\left(\frac{\partial a}{\partial t}\right)^2 + \left(\frac{\partial a}{\partial s}\right)^2\right) = b^2\left(\left(\frac{\partial b}{\partial t}\right)^2 + \left(\frac{\partial b}{\partial s}\right)^2\right)$$
$$= b^2\left(\left(\frac{\partial a}{\partial t}\right)^2 + \left(\frac{\partial a}{\partial s}\right)^2\right) \tag{22}$$

by virtue of (21).

If $b^2 = (H - a)^2$ at some point of L^2, then $H^2 = b^2 - a^2 + 2aH$. But $\det S = 2aH - a^2 + b^2 = 1$. Thus $H^2 = 1$, i.e., $H = \pm 1$. Conversely, if $H = 1$ [resp. -1], then from $\det S = 2aH - a^2 + b^2 = 1$ we get

$$b^2 = (a - 1)^2 \qquad [\text{resp. } b^2 = (a + 1)^2] \quad ,$$

that is, $b^2 = (H - a)^2$ at every point.

We have two cases to consider.

Case (i). $|H| \neq 1$. From the remark above, $b^2 \neq (H - a)^2$ at each point. From (22) we get

$$\frac{\partial a}{\partial t} = \frac{\partial a}{\partial s} = 0$$

and, by (21),

$$\frac{\partial b}{\partial t} = \frac{\partial b}{\partial s} = 0 \quad .$$

Thus a and b are constants and so is $c = 2H - a$. Therefore S is parallel and we have the description of such isometric immersions in Corollary to Theorem 1.

Case (ii). $H = \pm 1$. Then

$$S = \begin{bmatrix} a & -b \\ b & 2 - a \end{bmatrix} \quad \text{or} \quad \begin{bmatrix} a & -b \\ b & -2 - a \end{bmatrix} \quad .$$

From $\det S = 1$, we obtain $b = 1 - a$ or $a - 1$ for the first matrix and $b = a + 1$ or $-(a + 1)$ for the second matrix. Thus we have one of the following four cases:

$$S = \begin{bmatrix} a & a - 1 \\ 1 - a & 2 - a \end{bmatrix} \quad \text{or} \quad \begin{bmatrix} a & 1 - a \\ a - 1 & 2 - a \end{bmatrix} \quad \text{or} \quad \begin{bmatrix} a & -(a + 1) \\ a + 1 & -2 - a \end{bmatrix} \quad \text{or}$$

$$\begin{bmatrix} a & a+1 \\ -(a+1) & -2-a \end{bmatrix} .$$

We shall complete our proof for the first case; the others can be handled similarly.

The equation of Codazzi gives

$$\frac{\partial a}{\partial t} = \frac{\partial a}{\partial s}$$

so that a is a function of $t+s$, namely, of $u = (t+s)/\sqrt{2}$. Relative to the null coordinate system $\{u,v\}$ S takes the form

$$\begin{bmatrix} 1 & 0 \\ 2a-2 & 1 \end{bmatrix} .$$

Thus f is the B-scroll of a null curve of torsion 1 and curvature $2a-2$.

Corollary. *Let* $f: L^2 \to H_1^3$ *be an isometric immersion with mean curvature* 0. *Then* f *is congruent to the* B-*scroll in Example 3:*

$$(t,s) \to (\cos t \cosh s, \sin t \cosh s, -\sin t \sinh s, \cos t \sinh s) .$$

6. Geometry of SL(2,R)

In this section, we show that the geometry in H_1^3 can be realized on the group $SL(2,R)$ and thus admits a group-theoretic interpretation. We discuss the bi-invariant Lorentz metric on $SL(2,R)$, its geodesics, the structure of cross-product in each tangent space and its relation to covariant differentiation. Finally, we interpret the condition of torsion = 1 for a Frenet curve as left-invariance of the vector field B along the curve. This fact will be used in Section 7.

In the vector space $\mathfrak{gl}(2,R)$ of all 2×2 real matrices, we introduce a nondegenerate inner product by

$$\langle A,B \rangle = \frac{1}{2}\{tr(AB) - (tr\, A)(tr\, B)\} . \tag{23}$$

In particular,

$$\langle A,A\rangle = -\det A \quad .$$

This inner product corresponds to the standard inner product in R^4_2 by a linear isomorphism

$$A = \begin{bmatrix} a & b \\ c & d \end{bmatrix} \in \mathfrak{gl}(2,R) \leftrightarrow$$

$$x = \frac{1}{2}(a+d,\ c-b,\ a-d,\ b+c) \in R^4_2 \tag{24}$$

or, equivalently,

$$A = \begin{bmatrix} x_1+x_3 & x_4-x_2 \\ x_2+x_4 & x_1-x_3 \end{bmatrix} \in \mathfrak{gl}(2,R) \leftrightarrow$$

$$x = (x_1,x_2,x_3,x_4) \in R^4_2 \quad . \tag{25}$$

By this linear isomorphism, $SL(2,R) = \{A \in \mathfrak{gl}(2,R);\ \det A = 1\}$ corresponds to H^3_1. Thus the Lorentz metric induced on the hypersurface $SL(2,R)$ has constant sectional curvature -1. On the Lie algebra $\mathfrak{sl}(2,R) = \{X \in \mathfrak{gl}(2,R);\ \mathrm{tr}\ X = 0\}$, the metric is given by

$$\langle X,Y\rangle = \frac{1}{2}\,\mathrm{trace}\ XY \quad . \tag{26}$$

The matrices

$$E_0 = \begin{bmatrix} 0 & -1 \\ 1 & 0 \end{bmatrix}, \quad E_1 = \begin{bmatrix} 1 & 0 \\ 0 & -1 \end{bmatrix}, \quad E_2 = \begin{bmatrix} 0 & 1 \\ 1 & 0 \end{bmatrix} \tag{27}$$

form an orthonormal bais:

$$\langle E_0,E_0\rangle = -1, \quad \langle E_1,E_1\rangle = \langle E_2,E_2\rangle = 1 \quad .$$

For the bracket in $\mathfrak{sl}(2,R)$ we have

$$[E_0,E_1] = 2E_2, \quad [E_1,E_2] = -2E_0, \quad [E_2,E_0] = 2E_1 \ . \tag{28}$$

The metric on $SL(2,R)$ is bi-invariant. Covariant differentiation on $SL(2,R)$ can be expressed as a bilinear mapping

$$X,Y \in \mathfrak{sl}(2,R) \to \tilde{\nabla}_X Y \in \mathfrak{sl}(2,R) \quad ,$$

given by

$$\tilde{\nabla}_X Y = \frac{1}{2}[X,Y] \quad . \tag{29}$$

Thus

$$\tilde{\nabla}_{E_0} E_1 = E_2, \quad \tilde{\nabla}_{E_1} E_2 = -E_0, \quad \tilde{\nabla}_{E_2} E_0 = E_1 \ . \tag{30}$$

We have a natural volume element ω on $SL(2,R)$ such that $\omega(E_0,E_1,E_2) = 1$. It is bi-invariant together with the metric. We may define the structure of cross-product in $\mathfrak{sl}(2,R)$ as well as in the tangent space of each point by

$$\langle X \times Y, Z \rangle = \omega(X,Y,Z) \quad . \tag{31}$$

We have

$$E_0 \times E_1 = E_2, \quad E_1 \times E_2 = -E_0, \quad E_2 \times E_0 = E_1 \ . \tag{32}$$

From (30) and (32) we get, by using bilinearity,

$$\tilde{\nabla}_X Y = X \times Y$$

for all $X, Y \in \mathfrak{sl}(2,R)$. Indeed, more generally,

<u>Lemma 1</u>. *If* Y *is a left-invariant vector field, then*

$$\tilde{\nabla}_X Y = X \times Y$$

for any vector field X.

<u>Proof</u>. We write $X = \Sigma_{i=0}^{2} f_i E_i$, where f_i's are functions on $SL(2,R)$. Then

$$\begin{aligned}\tilde{\nabla}_X Y &= \sum f_i \tilde{\nabla}_{E_i} Y = \sum f_i (E_i \times Y) \\ &= \left(\sum f_i E_i\right) \times Y = X \times Y \quad .\end{aligned}$$

We shall prepare a few lemmas. The proofs are easy and hence omitted.

Lemma 2.

(i) *If* $\langle X,X\rangle = -1$, $\langle Y,Y\rangle = 1$ *and* $\langle X,Y\rangle = 0$, *then* $(X\times Y)\times Y = X$ *and* $(X\times Y)\times Y = Y$.

(ii) *If* $\langle U,U\rangle = \langle V,V\rangle = 0$ *and* $\langle U,V\rangle = -1$, *then* $(U\times V)\times V = V$ *and* $(U\times V)\times U = -U$.

Lemma 3. *For* $X,Y \in \mathfrak{sl}(2,R)$,

$$XY = X\times Y + \langle X,Y\rangle e \quad ,$$

where e *is the identity matrix.*

Lemma 4.

(i) *If* $\langle X,X\rangle = -1$, $\langle Y,Y\rangle = 1$ *and* $\langle X,Y\rangle = 0$, *then* $X(X\times Y) = -Y$ *and* $(X\times Y)Y = X$.

(ii) *If* $\langle U,U\rangle = \langle V,V\rangle = 0$ *and* $\langle U,V\rangle = -1$, *then* $U(U\times V) = U$ *and* $(U\times V)V = V$.

We shall now consider a Frenet curve (time-like or space-like or null) and prove.

Lemma 5. *The torsion* w *is identically equal to* 1 *if and only if the vector field* B *is left-invariant along the curve.*

Proof. We first consider a time-like Frenet curve x_t with frame field (T,N,B), where $B = T\times N$. Using left-invariant vector fields E_0, E_1 and E_2 obtained from (27), we write

$$B_t = \sum_{i=0}^{2} B^i(t)E_i \quad .$$

Then

$$\tilde{\nabla}_t B = \sum \frac{dB^i}{dt} E_i + \sum B^i(t)\tilde{\nabla}_T E_i \quad .$$

By Lemma 1, we have $\tilde{\nabla}_T E_i = T\times E_i$ so

$$\begin{aligned}\sum B^i(t)\tilde{\nabla}_T E_i &= \sum B^i(t)T\times E_i \\ &= T\times\Big(\sum_i B^i(t)E_i\Big) = T\times B = -N \quad .\end{aligned}$$

Thus

$$\tilde{\nabla}_t B = \sum \frac{dB^i}{dt} E_i - N \quad .$$

From one of the Frenet equations

$$\tilde{\nabla}_t B = -wN \quad ,$$

we see that $w \equiv 1$ if and only if $dB^i/dt \equiv 0$, i.e., B^i is constant for each i.

The proof in the case of a space-like Frenet curve y_s with frame field T,N,B, where $B = T \times N$ is time-like, is quite similar if we note $T \times B = N$ and $\tilde{\nabla}_s B = wN$.

Finally, let x_u be a null Frenet curve with frame field A,B,C, where $C = A \times B$. The proof is again similar if we use the equation $\tilde{\nabla}_u B = wC$.

Now let ϕ be the diffeomorphism $x \to x^{-1}$ of $SL(2,R)$ onto itself. If X is a left-invariant vector field, then $\phi_*(X)$ is the right-invariant vector field such that

$$(\phi_*(X))_e = -X_e \quad ,$$

where e is the identity matrix. We have

$$\phi_*[X,Y] = [\phi_* X, \phi_* Y]$$

for any left-invariant vector fields X and Y. Each side of the above equation is a right-invariant vector field.

Since the metric is bi-invariant, ϕ is an isometry:

$$\langle \phi_* X, \phi_* Y \rangle_{\phi(x)} = \langle X, Y \rangle_x \quad . \tag{33}$$

Thus ϕ preserves the Levi-Civita connection:

$$\tilde{\nabla}_{\phi_* X} \phi_* Y = \phi_*(\tilde{\nabla}_X Y)$$

for any left-invariant vector fields X and Y.

The transformation ϕ reverses the orientation given by the volume element ω:

$$\omega(\phi_* X, \phi_* Y, \phi_* Z) = -\omega(X,Y,Z) \quad . \tag{34}$$

From (33) and (34) we have

$$\phi_*(X \times Y) = -(\phi_* X \times \phi_* Y)$$

for any vector fields X and Y. We can now prove

Lemma 6. *The torsion* w *is identically equal to* -1 *if and only if the vector field* B *is right-invariant along the curve.*

Proof. We shall prove this for a time-like Frenet curve x_t with frame field $\{T,N,B\}$. The other cases are similar. We consider $x_t^* = \phi(x_t)$, which is a time-like curve. Let $T_t^* = \phi_*(T_t)$ and $N_t^* = \phi_*(N_t)$. Then

$$T_t^* \times N_t^* = \phi_*(T_t) \times \phi_*(N_t) = -\phi_*(T_t \times N_t)$$

$$= -\phi_*(B_t) \quad .$$

Thus if we let $B_t^* = -\phi_*(B_t)$, then $\{T^*,N^*,B^*\}$ is a frame field for x_t^* with $T^* \times N^* = B^*$. The Frenet equations for x_t now give rise to the Frenet equations for x_t^*:

$$\tilde{\nabla}_t T^* = kN^*$$

$$\tilde{\nabla}_t N^* = kT^* - wB^*$$

$$\tilde{\nabla}_t B^* = wN^* \quad .$$

Thus x_t^* has torsion $-w$. By Lemma 2, B^* is left-invariant if and only if $-w \equiv 1$. This means that B is right-invariant if and only if $w \equiv 1$.

We remark that Lemmas 5 and 6 are analogues of results given in [8, p. 148] for the group $SO(3)$ with bivariant Riemannian metric. Our Lemma 1 simplifies the proof. The approach we are taking in the next two sections follow that of Bianchi as explained in [8, pp. 140-157] for $SO(3)$.

7. Product of Curves

Let $\{X,Y\}$ be a pair of vectors in $\mathfrak{sl}(2,R)$ such that $\langle X,X\rangle = -1$, $\langle Y,Y\rangle = 1$, $\langle X,Y\rangle = 0$. We define a mapping $f: L^2 \to SL(2,R)$ by

$$f(t,s) = x_t y_s \quad ,$$

where the 1-parameter subgroup $x_t = \exp tX$ is a time-like geodesic, and $y_s = \exp sY$ is a space-like geodesic. We have

$$f_*(\partial/\partial t) = x_t \; X \; y_s$$

$$f_*(\partial/\partial s) = x_t \; Y \; y_s \quad .$$

Since the metric on $SL(2,R)$ is bi-invariant, we get

$$\langle f_*(\partial/\partial t), f_*(\partial/\partial t)\rangle = \langle X,X\rangle = -1$$

$$\langle f_*(\partial/\partial s), f_*(\partial/\partial s)\rangle = \langle Y,Y\rangle = 1$$

$$\langle f_*(\partial/\partial t), f_*(\partial/\partial s)\rangle = \langle X,Y\rangle = 0 \quad ,$$

showing that f is an isometric immersion of L^2 into $SL(2,R)$.

To find the shape operator S, we take a unit normal $\xi_0 \in \mathfrak{sl}(2,R)$ at $f(0,0) = e$, say, $\xi_0 = X \times Y$. Then

$$\xi_{(t,s)} = x_t \xi_0 y_s$$

is a field of unit normal vectors. We get

$$\left(\frac{\partial \xi}{\partial t}\right)_{t=s=0} = (x_t X \xi_0 y_s)_{t=s=0} = X\xi_0 \quad .$$

By Lemma 4(i), $X\xi_0 = -Y$. Thus we have

$$S(\partial/\partial t) = \partial/\partial s \qquad \text{at} \qquad (t,s) = (0,0) \quad .$$

Similarly

$$S(\partial/\partial s) = -\partial/\partial t \qquad \text{at} \qquad (t,s) = (0,0) \quad .$$

For an arbitrary point (t_0,s_0), note that the mapping $\tau : x \to x_{t_0} x y_{s_0}$ is an isometry of $SL(2,R)$ such that $f(t+t_0, s+s_0) = \tau f(t,s)$. We have

$$\tau(f_*(\partial/\partial t)_{(0,0)}) = f_*(\partial/\partial t)_{(t_0,s_0)}$$

$$\tau(f_*(\partial/\partial s)_{(0,0)}) = f_*(\partial/\partial s)_{(t_0,s_0)} .$$

Since $\xi_{(t+t_0,s+s_0)} = \tau\xi_{(t,s)}$, we see that the shape operators $S_{(t_0,s_0)}$ at (t_0,s_0) and $S_{(0,0)}$ are related by

$$f_*(S(\partial/\partial t)_{(t_0,s_0)}) = \tau f_*(S(\partial/\partial t)_{(0,0)}) .$$

Thus

$$S(\partial/\partial t)_{(t_0,s_0)} = (\partial/\partial s)_{(t_0,s_0)}$$

and, similarly,

$$S(\partial/\partial s)_{(t_0,s_0)} = -(\partial/\partial t)_{(t_0,s_0)} .$$

Hence the shape operator S is expressed by

$$S = \begin{bmatrix} 0 & -1 \\ 1 & 0 \end{bmatrix} \quad \text{relative to } \{t,s\}.$$

We observe that the shape operator S is the same for any choice of orthonormal vectors $\{X,Y\}$. Thus all immersions of this type are congruent to each other.

We also recognize that the shape operator S is the same as that in Example 3.

Actually, the expression $f(t,s) = x_t y_s$ represents the B-scroll of the time-like geodesic x_t in the following way. At $t=0$ $(x_0 = e)$, we choose $T_0 = X$, N_0 and $B_0 = Y$, where $N_0 = Y \times X$, as the initial frame. We can find the frame field T,N,B along the geodesic x_t so that the Frenet equations are valid with $k \equiv 0$ and $w \equiv 1$. By Lemma 5, B is left-invariant. This means that

$$B_{t_0} = x_{t_0} B_0 = x_{t_0} Y .$$

The geodesic starting from x_{t_0} in the direction of B_{t_0} is then $x_{t_0} y_s$, because this curve (with parameter s) is the geodesic with the initial tangent vector $x_{t_0} Y = x_{t_0} B_0 = B_{t_0}$.

To sum up, we state

Proposition 4. *Let* X,Y *be an orthonormal pair of vectors in* $\mathfrak{sl}(2,R)$, *where* X *is time-like and* Y *is space-like. Then the product of* 1-*parameter subgroups*

$$f(t,s) = \exp tX \exp sY$$

is an isometric immersion of L^2 *into* SL(2,R). *These immersions are congruent to each other for all choices of the pair* {X,Y}. *The shape operator is*

$$S = \begin{bmatrix} 0 & -1 \\ 1 & 0 \end{bmatrix}$$ *relative to a Lorentzian coordinate system.*

Such an isometric immersion can be viewed as the B-*scroll of a time-like geodesic, as in Example 3.*

We can also relate Proposition 4 to Example 5, but we omit the detail.

Now we consider a pair of null vectors {U,V} in $\mathfrak{sl}(2,R)$ such that $\langle U,V\rangle = -1$. We define

$$f: L^2 \to SL(2,R)$$

by

$$f(u,v) = \exp uU \exp vV ,$$

where {u,v} is a null coordinate system on L^2. We see that f is an isometric immersion of L^2 into SL(2,R). To find its shape operator S, we take a unit normal vector

$$\xi_0 = V \times u \qquad \text{at} \qquad (u,v) = (0,0)$$

and let

$$\xi_{(u,v)} = \exp uU\, \xi_0 \exp vV .$$

By using Lemma 4, (ii) we see as before that $S = I$ at (0,0) and in fact at every point (u,v) of L^2. Hence our immersion f is umbilical.

We obtain

<u>Proposition 5</u>. *Let* $\{U,V\}$ *be a pair of null vectors such that* $\langle U,V\rangle = -1$ *in* $\mathfrak{sl}(2,R)$. *Then the product of the 1-parameter subgroups*

$$f(u,v) = \exp uU \exp vV$$

is an isometric immersion of L^2 *into* $SL(2,R)$ *which is umbilical. It is thus congruent to the immersion in Example 1. It can be also described as the* B*-scroll of the null geodesic* x_u *as in Example 7* (where we take the initial vector B_0 of x_u to be V).

For the last assertion we use Lemma 5 for a null Frenet curve of torsion 1.

We first observe that the B-scroll of a time-like Frenet curve x_t of torsion $\equiv 1$ can be expressed in the form of the product $x_t y_s$, where y_s is a certain 1-parameter subgroup.

To show this, we may assume, without loss of generality, that $x_0 = e$. Let $\{T,N,B\}$ be the Frenet frame. By Lemma 5, B is left-invariant, i.e., $B_t = x_t B_0$. We let $y_s = \exp sB_0$, which is the geodesic from e with initial tangent vector B_0. For each t, $x_t y_s = x_t \exp sB_0$ is the geodesic from x_t in the direction of $B_t = x_t B_0$. Thus $f(t,s) = x_t y_s$ is the B-scroll of the curve x_t.

The same situations prevail for the B-scroll of a space-like Frenet curve of torsion $\equiv 1$ as well as of a null Frenet curve of torsion $\equiv 1$.

We also remark that if x_t is a time-like Frenet curve of torsion -1, then its B-scroll can be expressed in the form $f(t,s) = y_s x_t$, where y_s is a certain 1-parameter subgroup. The proof is similar to the case of torsion $\equiv 1$, except that for torsion $\equiv -1$ the vector field B is right-invariant along x_t, i.e., $B_t = B_0 x_t$, by Lemma 6. Hence $f(t,s) = \exp sB_0 x_t$.

Similar results are valid for space-like or null Frenet curves of torsion -1.

Finally, we consider a more general situation and prove

<u>Theorem 6</u>. *Let* x_t *be a time-like Frenet curve of torsion* 1 *with frame field* T,N,B *and let* y_s *be a space-like Frenet curve of torsion* -1 *with frame field* $\bar{T},\bar{N},\bar{B}$ *such that*

(1) $x_0 = y_0 = e$

(2) $B_0 = \bar{B}_0$

(3) T_0 *and* $\bar{T}_0$ *are linearly independent.*
Then the surface M *given by*

$$(t,s) \in D \to f(t,s) = x_t y_s$$

is a flat Lorentzian surface, where D *is a certain domain in* R^2.

<u>Proof</u>. By Lemma 5, we have $B_{t_0} = x_{t_0} B_0$. Since $u \to u y_s$, is an isometry of $SL(2,R)$ onto itself, the t-curve: $t \to x_t y_{s_0}$ is a time-like Frenet curve with frame field $\{Ty_{s_0}, Ny_{s_0}, By_{s_0}\}$. Thus the binormal vector at t_0 of the t-curve is $B_{t_0} y_{s_0} = x_{t_0} B_0 y_{s_0}$.

By Lemma 6, we have $\bar{B}_{s_0} = \bar{B}_0 y_{s_0}$. The s-curve: $s \to x_{t_0} y_s$ is a space-like Frenet curve with frame field $\{x_{t_0}\bar{T}, x_{t_0}\bar{N}, x_{t_0}\bar{B}\}$. Thus the binormal vector at s_0 of the s-curve is $x_{t_0}\bar{B}_{s_0} = x_{t_0}\bar{B}_0 y_{s_0}$.

Since $B_0 = \bar{B}_0$ by assumption, we see that for each (t_0, s_0) the t-curve $t \to f(t,s_0)$ and the s-curve $s \to f(t_0,s)$ have the same binormal hence the same osculating plane at $p = f(t_0,s_0)$, say, Σ_p.

We choose a domain D in R^2 so that the tangent vectors $f_*(\partial/\partial t)$ and $f_*(\partial/\partial s)$ are linearly independent for $(t,s) \in D$. They belong to Σ_p at (t_0,s_0). The tangent plane of M at p coincides with Σ_p. This implies that the principal normal vector $N_{t_0} y_{s_0}$ of the t-curve $f(t,s_0)$ at $t = t_0$ lies in the tangent space $T_p(M)$.

We can find the shape operator S of M at p as follows. The vector field $B_t y_{s_0}$ is normal to the surface along the curve $x_t y_{s_0}$. From one of the Frenet equations for x_t with torsion $\equiv 1$:

$$\tilde{\nabla}_t B_t = -N_t .$$

We get a similar equation

$$\tilde{\nabla}_t (B_t \cdot y_{s_0}) = -N_t y_{s_0} .$$

If we take $\{T_{t_0} y_{s_0}, N_{t_0} y_{s_0}\}$ as a basis of the tangent space $T_p(M)$, then the above equation implies that the shape operator S takes the form

$$S = \begin{bmatrix} 0 & -1 \\ 1 & * \end{bmatrix} .$$

Therefore $\det S = 1$. The equation of Gauss shows that the surface M is flat. This completes the proof of Theorem 6.

Let us remark that, unfortunately, we do not know whether the domain D can be the entire plane R^2 when the curves x_t and y_s are complete (that is, defined for all t and s). Even if $D = R^2$, we do not know whether the flat Lorentzian surface M is geodesically complete or not. This seems to be a much more difficult problem than the question of completeness of flat surfaces in S^3 treated in [1] and [6].

Finally, it is also an open problem to decide whether an arbitrary isometric immersion $L^2 \to SL(2,R) = H_1^3$ can be obtained as a product of two appropriate curves.

8. References

[1] T. Cecil, "On the completeness of flat surfaces in S^3," *Coll. Math.* 33 (1975), 139-143.

[2] L. Graves, "Codimension one isometric immersions between Lorentz spaces," Thesis, Brown University, 1977; also *Trans. AMS* 252 (1979), 367-392.

[3] L. Graves and K. Nomizu, "Isometric immersions of Lorentzian space forms," *Math. Ann.* 233 (1978), 125-136.

[4] M.M. Magid, "Isometric immersions between indefinite flat spaces with parallel second fundamental forms," Thesis, Brown Univ., 1978.

[5] J.D. Moore, " Submanifolds of constant positive curvature I," *Duke Math. J.* 44 (1977), 449-484.

[6] S. Sasaki, "On complete surfaces with Gaussian curvature zero in 3-sphere," *Coll. Math.* 26 (1972), 165-174.

[7] S. Sasaki, "On complete flat surfaces in hyperbolic 3-space," *Kodai Math. Sem. Rep.* 25 (1973), 449-457.

[8] M. Spivak, *A Comprehensive Introduction to Differential Geometry*, Vol. IV, Publish or Perish, Boston, 1975.

[9] Ju. A. Volkov and S.M. Vladimivova, "Isometric immersions of the Euclidean plane in Lobačevskiĭ space," *Mat. Zametki* 10 (1971), 327-332 = *Math. Notes* 10 (1971), 655-661.

[10] J.A. Wolf, *Spaces of Constant Curvature*, McGraw-Hill, New York, 1967.

Instituto de Matemática Pura e Aplicada
Rua Luiz de Camões 68, Rio de Janeiro, Brazil

Brown University, Providence, R.I. 02912, USA

(Received September 6, 1980)

THE COMPLEX LAPLACE-BELTRAMI OPERATOR CANONICALLY ASSOCIATED TO A POLARIZED ABELIAN VARIETY

Jun-ichi Hano

0. Introduction

A holomorphic line bundle over an abelian variety whose Chern class is positive definite admits a complex Laplace-Beltrami operator canonically determined by the Chern class. In this paper, we study some implications of the Selberg trace formula applied to this elliptic differential operator acting on the Hilbert space of sections of the bundle.

Making use of the trace formula, we determine first the spectrum of the operator (Theorem 1). Then, observing a relation of the operator with a unitary representation of the Heisenberg group, called "Lattice representation" by P. Cartier [3], we determine the eigenspaces of the operator (Theorem 2) and obtain a new proof of the theorem by Cartier (Theorem 2 in [3]) concerning the irreducible decomposition of the lattice representation (Theorem 3). For other approaches to this theorem, we refer to L. Auslander and R. Tolimieri [1] and Tolimieri [9]. We also show that conversely, using the theorem of Cartier, we can determine the eigenvalues and the eigenspaces of the operator in question (Remark in 8).

Choosing a suitable theta automorphy factor, let us identify the Hilbert space of sections with a Hilbert space of functions on a complex cartesian space in such a fashion that a holomorphic theta function of reduced type corresponds to a holomorphic section. Then, each eigenspace of the complex Laplace-Beltrami operator is spanned by those functions obtained from holomorphic theta functions by applying certain linear differential operators successively. This is the content of Theorem 2 and is pointed out by G. Shimura to the author. The investigation of those functions, namely, the eigenfunctions of the operator is one of the main themes in his paper [8]. Those linear differential operators are complex linear combinations of infinitesimal generators of the lattice representation.

The last section is devoted to exhibit an orthonormal basis of the Hilbert space of sections, consisting of eigenfunctions of the operator, and to give a concrete irreducible decomposition of the lattice representation. An intertwining operator from the Schresinger representation onto each irreducible factor is explicitly determined (Theorems 4 and 5).

As for the theta function theory, we refer to Chapter IV, Weil [9]. A description of a holomorphic line bundle over an abelian variety in terms of the Heisenberg groups, which is useful for our application of the Selberg trace formula, is in Cartier [3] and Igusa [6]. Formalism of the Selberg trace formula in Hejhal [5] is adopted here almost *verbatim*.

As usual, $\mathbb{Z}$, $\mathbb{C}$ and $\mathbb{R}$ denotes the set of all integers, the field of complex numbers and the field of real numbers, respectively.

The author would like to express his thanks to G. Shimura for his useful suggestion on the subject. The second section is motivated by his comments given to the author. The author also extends thanks to G. Weiss for his kind explanation of the Dirichlet series.

1. An Application of the Trace Formula

1. An abelian variety with a Riemann form A means the following setting:

a real vector space V of dimension $2n$,

a lattice group L in V,

a nondegenerate alternate bilinear form A on $V \times V$

and

an endomorphism J on V such that $J^2 = -1$,

satisfying the following conditions:

(i) $A(L,L) \in \mathbb{Z}$

(ii) $A(Ju,Jv) = A(u,v)$

(iii) $A(u,Ju) > 0$ unless $u = 0$.

We denote by $\mathbb{V}$ the complex vector space (V,J) and by $\mathbf{T}$ the complex torus $L\backslash\mathbb{V}$. From the above conditions (ii) and (iii), it follows that

$$H(u,v) = A(u,Jv) + iA(u,v), \qquad u,v \in \mathbb{V},$$

is a positive definite hermitian form.

On $\mathbb{V}\times\mathbb{C}^*$, we define a multiplication

$$(u,c)\cdot(v,d) = \left(u+v, e\left[\frac{1}{2i}H(u,v)\right]cd\right)$$

for (u,c), $(v,d)\in\mathbb{V}\times\mathbb{C}^*$, where $e[x]=\exp(2\pi ix)$. With this multiplication $\mathbb{V}\times\mathbb{C}^*$ is a connected real Lie group, called the Heisenberg group. The left translations are holomorphic with respect to the natural complex structure on $\mathbb{V}\times\mathbb{C}^*$.

Let ψ be a semi-character of L relative to A [p. 110, 9], namely, a map of L into the group $U(1)$ of complex numbers with modulus one satisfying the condition that

$$\psi(\ell+\ell') = \psi(\ell)\psi(\ell')e\left[\frac{1}{2}A(\ell,\ell')\right] \quad , \quad \ell,\ell'\in L \ .$$

Then, the subset L_ψ in the group $\mathbb{V}\times\mathbb{C}^*$ consisting of the elements of the form

$$\left(\ell,\psi(\ell)e\left[\frac{1}{4i}H(\ell,\ell)\right]\right) , \qquad \ell\in L \quad ,$$

is a discrete subgroup.

Let $\mathbb{P}$ be the quotient space $\mathbb{V}\times\mathbb{C}^*$ by the left action of the subgroup L_ψ. We obtain the following commutative diagram:

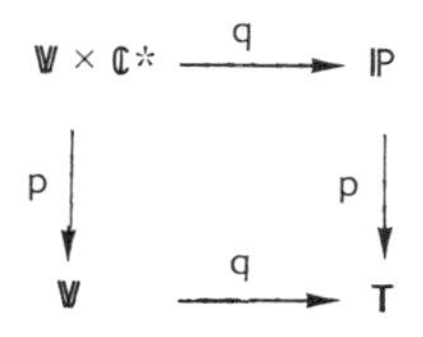

The row arrows are the quotient maps and the left hand side column arrow is the projection of $\mathbb{V}\times\mathbb{C}^*$ into its first component, which induces the projection p of $\mathbb{P}$ onto $\mathbb{T}$. The triple $(\mathbb{P},p,\mathbb{T})$ is a holomorphic principal bundle over the complex torus $\mathbb{T}$ with structure group $\mathbb{C}^*$, and $(\mathbb{V}\times\mathbb{C}^*,p,\mathbb{V})$ is the reciprocal image of $(\mathbb{P},p,\mathbb{T})$ under the quotient map $q:\mathbb{V}\to\mathbb{T}$.

The Chern class of the principal bundle $(\mathbb{P},p,\mathbb{T})$ is the integral cohomology class of $\mathbb{T}$ associated to the Riemann form A. Namely if (a_i) denotes a real cartesian coordinate system on V and if

$A = \Sigma A_{ij} a_i a_j$, the cohomology class is represented by the 2-form $\Sigma_{i<j} A_{ij} da_i \wedge da_j$ on $\mathbb{T}$.

Conversely, suppose a holomorphic principal bundle $\mathbb{P}$ over a complex torus $\mathbb{T}$ with structure group $\mathbb{C}^*$ is given. Its Chern class determines an alternating $\mathbb{R}$-bilinear form A on the vector space $\mathbb{V}$ which satisfies the above conditions (i) and (ii) (not necessarily (iii)). Then, $\mathbb{P}$ is obtained by the above construction determined by A and a suitable choice of a semi-character ψ.

We define a reduction of the structure group $\mathbb{C}^*$ to the subgroup $U(1)$ on the principal bundle $(\mathbb{P},p,\mathbb{T})$, which is canonically determined by the Riemann form A and the complex structure J. For this purpose let G be the subset in $\mathbb{V} \times \mathbb{C}$ consisting of the elements of the form

$$\left(u, e\left[\frac{1}{4i} H(u,u)\right] c\right) \qquad \text{with} \quad c \in U(1) \quad .$$

To be brief, such an element will be denoted by $\{u,c\}$.

Since

$$\{u,c\}^{-1} = \{-u,c^{-1}\} \quad ,$$

and

$$\{u,c\} \cdot \{v,d\} = \left\{u+v, e\left[\frac{1}{2} A(u,v)\right] cd\right\} \quad ,$$

G is a subgroup of $\mathbb{V} \times \mathbb{C}^*$. The discrete subgroup L_ψ defined above consists of the elements of the form $\{\ell,\psi(\ell)\}$, $\ell \in L$, and is a subgroup in G.

We regard $(G,p,\mathbb{V})$ as a sub-bundle of $(\mathbb{V} \times \mathbb{C}^*,p,V)$ whose structure group is $U(1)$. Thus we have a reduction of the structure group $\mathbb{C}^*$ to $U(1)$. Since the group L_ψ of bundle-automorphisms preserves the sub-bundle $(G,p,\mathbb{V})$, we obtain the following commutative diagram:

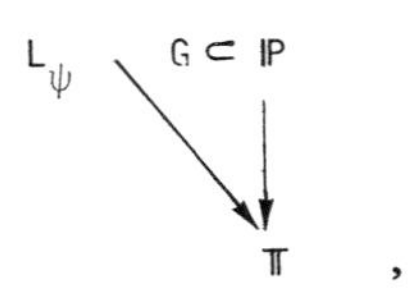

which defines a reduction of the structure group $\mathbb{C}^*$ to its subgroup $U(1)$ on the principal bundle $(\mathbb{P},p,\mathbb{T})$.

In an obvious way, $(\mathbf{V}\times\mathbb{C},p,\mathbb{V})$ is the complex line bundle associated to $(\mathbb{V}\times\mathbb{C}^*,p,\mathbf{V})$. The quotient $\mathbb{F}$ of $(\mathbb{V}\times\mathbb{C},p,\mathbb{V})$ by the left action of the group L_ψ is the complex line bundle associated to $(\mathbb{P},p,\mathbf{T})$.

Put

$$\tilde{a}(u,c) = e\left[-\frac{1}{2i}H(u,u)\right]|c|^2 \qquad \text{for} \quad (u,c)\in\mathbb{V}\times\mathbb{C}^* .$$

Then, $\tilde{a}$ is a hermitian inner product along fibers and the subset G in $\mathbf{V}\times\mathbb{C}^*$ is the subset of all points where $\tilde{a}$ assumes the value 1. Under the group G of bundle-automorphisms, $\tilde{a}$ is kept invariant. Indeed, if $\{u,c\}\in G$ and $(v,d)\in\mathbf{V}\times\mathbb{C}$

$$\tilde{a}(\{u,c\}(v,d)) = \tilde{a}(v,d) \qquad .$$

We take a hermitian metric on $\mathbf{V}$ determined by the hermitian form H in 1 and the above hermitian inner product along the fibers. Then, there corresponds the complex Laplace-Beltrami operator $\square_{\tilde{a}}$ acting on smooth sections of the line bundle $(\mathbf{V}\times\mathbb{C},p,\mathbb{V})$.

Since the group L_ψ is a subgroup in G, $\tilde{a}$ induces a hermitian inner product a along fibers on the complex line bundle $\mathbb{F}$ via the quotient map q. If we denote by $\square_a$ the complex Laplace-Beltrami operator on the line bundle $\mathbb{F}$ determined by the hermitian metric on $\mathbb{T}$ associated to H and the hermitian inner product a, then

$$\square_{\tilde{a}}\, q_* = q_*\square_a \qquad .$$

2. The operator $\square_{\tilde{a}}$ acts on smooth sections of the complex line bundle $(\mathbb{V}\times\mathbb{C},p,\mathbf{V})$. Given a section s, we denote by φ_s a $\mathbb{C}$-valued function on $\mathbb{V}$ determined by

$$s(u) = (u,\varphi_s(u)) \qquad , \qquad u\in\mathbf{V} \quad .$$

We define a differential operator acting on smooth functions on $\mathbb{V}$ by

$$(\square_{\tilde{a}}s)(u) = (u,(E\varphi_s)(u)) \ .$$

In terms of a complex coordinate system (u_α) on $\mathbf{V}$,

$$E = - \sum_{\alpha,\beta=1}^{n} h^{-1}_{\alpha\beta} \frac{\partial^2}{\partial\bar{u}_\alpha \partial u_\beta} + \pi \sum_{\alpha=1}^{n} \bar{u}_\alpha \frac{\partial}{\partial\bar{u}_\alpha} ,$$

where $(1/2)\,\Sigma\, h_{\alpha\beta}\, d\bar{u}_\alpha du_\beta$, $\bar{h}_{\alpha\beta} = h_{\beta\alpha}$, is the hermitian metric on $\mathbb{W}$ associated to the hermitian form H ([7]).

For later use, we consider another representation of a section s. We define a $\mathbb{C}$-valued function f_s on $\mathbb{W}$ by

$$s(u) = \left(u, e\left[\frac{1}{4i} H(u,u)\right] f_s(u)\right) . \tag{2.1}$$

If we put

$$(\square_{\tilde{a}} s)(u) = \left(u, e\left[\frac{1}{4i} H(u,u)\right] (Df_s)(u)\right)$$

then,

$$(Df_s)(u) = \exp\left(-\frac{\pi}{2} H(u,u)\right) E\left(\exp\left(\frac{\pi}{2} H(u,u)\right) f_s(u)\right) .$$

Since the left action of the group G on the complex line bundle $(\mathbb{W} \times \mathbb{C}, p, \mathbb{W})$ is a group of bundle-automorphisms, if we put

$$(\rho_{\{u,c\}} f_s)(v) = f_s(u+v)\, e\left[\frac{1}{2} A(u,v)\right] c^{-1}$$

for $\{u,c\} \in G$, then the map

$$\{u,c\} \to \rho_{\{u,c\}}$$

is a homomorphism of the group G into the group of linear operators. The fact that the Laplace-Beltrami operator $\square_{\tilde{a}}$ is invariant under the group G is written as

$$D\rho_{\{u,c\}} = \rho_{\{u,c\}} D, \qquad \{u,c\} \in G .$$

Let us denote by $L_2(\mathbf{T},\mathbf{F})$ the Hilbert space of square integrable sections of the line bundle $\mathbb{F}$ with respect to the volume element

$$du = (i/2)^n \det(h_{\alpha\beta})\, du_1 d\bar{u}_1 \ldots du_n d\bar{u}_n$$

on $\mathbb{T}$ and the hermitian inner product a along fibers. For the sake of practicality, we identify $L_2(\mathbb{T},\mathbb{F})$ with the Hilbert space of $\mathbb{C}$-valued functions f on $\mathbb{W}$ subject to the following conditions:

$$\text{i)} \quad f(u+\ell) = f(u)\psi(\ell)e\left[\frac{1}{2} A(\ell,u)\right] , \quad \ell \in L , \tag{2.3}$$

and

$$\text{ii)} \quad \int_F |f(u)|^2 \, du < \infty$$

where F denotes a fundamental domain of the lattice group L in $\mathbb{W}$. In this representation of $L_2(\mathbb{T},\mathbb{F})$, the complex Laplace-Beltrami operator is realized as the differential operator D above.

Here we collect some results on the differential operator

$$E = -\sum h^{-1}_{\alpha\beta} \frac{\partial^2}{\partial\bar{u}_\alpha \partial u_\beta} + \pi \sum \bar{u}_\alpha \frac{\partial}{\partial \bar{u}_\alpha} .$$

Suppose that a smooth function f on $\mathbb{W}$ is radial, that is, $f(u) = f(v)$ whenever $H(u,u) = \|u\| = \|v\|$. Put $f(u) = g(r)$ with $r = \|u\|$. Then

$$(Ef)(u) = -\frac{1}{4}\left(g'' + \left(\frac{2n-1}{r} - 2\pi r\right) g'\right)(r) .$$

Consider the 2nd order ordinary differential equation

$$g'' + \left(\frac{2n-1}{r} - 2\pi r\right) g' + 4\lambda g = 0 \tag{2.4}$$

on the interval $(0,\infty)$ with parameter λ. The point $r = 0$ is only one finite singular point of the equation and is a regular singular point. At $r = 0$, the roots of the indicial equation are 0 and $2(1-n) \leqslant 0$, where n is the complex dimension of the complex torus $\mathbb{T}$. For any positive integer n, the equation has two linearly independent solutions on the interval $(0,\infty)$, one of which is finite at $r = 0$ and the other infinite at $r = 0$. Thus, using the method of Frobenius, we obtain a unique real analytic solution $g(\lambda,r)$ subject to the initial condition $g(\lambda,0)=1$. The solution $g(\lambda,r)$ is given by

$$1+\sum_{k=1}^{\infty}\frac{(k-1-\lambda/\pi)(k-2-\lambda/\pi)\cdots(-\lambda/\pi)}{(k-1+n)(k-2+n)\cdots(n)}\frac{\pi^k}{k!}r^{2k}$$

whose radius of convergence is ∞. If we denote by $F(\mu,\nu,r)$ the confluent hypergeometric function of Kummer,

$$g(\lambda,r) = F(-\lambda/\pi,n,\pi r^2) .$$

We need the following estimate: If $n\geqslant 0$ and if $\lambda\in[0,\lambda_o]\,(\lambda_o>0)$, then

$$F(-\lambda/\pi,n,\pi r^2) < ce^{\pi r^2} , \qquad r\in(-\infty,\infty) , \tag{2.5}$$

where C is a constant depending on n and λ_o. This is easily checked. Write the series in the form

$$\sum_{k=o}^{\infty} B_k \frac{(\pi r^2)^k}{k!} .$$

Then

$$\left|\frac{B_k+1}{B_k}\right| = \left|\frac{k-\lambda/\pi}{k+n}\right| < 1 \qquad \text{for } k\geqslant k_o>\lambda_o/\pi ,$$

and

$$\sum_{k=k_o}^{\infty}|B_k|\frac{(\pi r^2)^k}{k!} < \left|B_{k_o}\right| e^{\pi r^2} .$$

Thus, we can choose a constant C so that the inequality in question is valid.

3. We introduce a class of linear operators of Hilbert-Schmidt type on the Hilbert space $L_2(\mathbf{T},\mathbf{F})$. Here the Hilbert space is realized as the space of functions on $\mathbf{V}$ subject to the conditions (2.3),(i) and (ii). Throughout this number, Φ denotes a function on $(-\infty,\infty)$ of the form

$$\Phi(t) = e^{-\pi t/2}s(t)$$

where S is a function such that $t \to S(t^2)$ is a rapidly decreasing function in the Schwartz' sense.

Put

$$k(u,v) = \Phi(\|u-v\|^2)e\left[-\frac{1}{2}A(u,v)\right]$$

for $(u,v) \in V \times V$, where $\|v\|^2 = H(v,v)$. Then k satisfies the following properties:

(i) $k(u+x,v+x) = k(u,v)e\left[\frac{1}{2}A(x,u-v)\right], \quad x \in \mathbb{V}$

(ii) $k(v,u) = \overline{k(u,v)}$.

Next, we put

$$K(u,v) = \sum_{\ell \in L} k(u,v+\ell)\psi(\ell)e\left[\frac{1}{2}A(\ell,v)\right].$$

Then, K satisfies the following properties:

(i) K is a smooth function on $V \times V$ and its partial derivative of any order is obtained by term by term differentiations.

(ii) $K(u+\ell,v+m) = K(u,v)\psi(\ell)\psi(m)^{-1}e\left[\frac{1}{2}A(\ell,u)\right]e\left[-\frac{1}{2}A(m,v)\right]$ for $\ell,m \in L$.

(iii) $K(v,u) = \overline{K(u,v)}$.

The last two properties of K above follows easily from the properties of k. Here we prove the property (i). Consider the function

$$(u,v,w) \mapsto k(u,v+w)e\left[\frac{1}{2}A(w,v)\right].$$

Because of the form of the function k, we see that this function, as well as its partial derivative with respect to u and v of any order is written as

$$h(u,v,w)e\left[\frac{1}{2}A(w,v)\right]$$

with a rapidly decreasing function h on $(u,v,w) \in \mathbb{R}^{6n}$. What we have to show is that the series

$$\sum_{\ell \in L} h(u,v,\ell) e\left[\frac{1}{2} A(\ell,v)\right] \psi(\ell)$$

converges absolutely and uniformly on (u,v) in a compact subset X in $V \times V$.

We count the number $N(r)$ of lattice points inside the sphere of radius r centered at the origin, which is $O(r^{2n})$ as $r \to \infty$. We can choose a constant C depending only on X such that

$$|h(u,v,\ell)| < C(1 + \|\ell\|)^{-(2n+1)}$$

for $(u,v) \in X$ and for all $\ell \in L$. Now

$$\sum_{\ell \in L} |h(u,v,\ell)| < C \int_0^\infty (1+r)^{-(2n+1)} \, dN(r)$$

$$< C' \int_0^\infty \frac{r^{2n-1}}{(1+r)^{-(2n+1)}} \, dr < \infty \quad .$$

We have finished the proof of (i).

Utilizing k, we define an integral operator M by

$$(Mf)(u) = \int_V k(v,u(f(v) \, dv$$

for a function f on V. If, especially, f represents a section in $L_2(\mathbb{T},\mathbb{F})$, in other words, if f satisfies the condition (2.3),

$$(Mf)(u) = \int_F K(v,u) f(v) \, dv \quad ,$$

where F is an arbitrary fundamental domain of the lattice group L. In this case, Mf also satisfies (3.3) and defines a section in $L_2(\mathbb{T},\mathbb{F})$. Thus M induces a bounded linear operator of $L_2(\mathbb{T},\mathbb{F})$ into itself.

4. Now we are ready to prove a trace formula. We begin with the following

Lemma. *Suppose that function* f *on* $\mathbb{W}$ *is associated to an eigen section of the complex Laplace-Beltrami operator* $\square_a$, *namely, suppose that* f *satisfies* (3.3) *and the equation*

$$Df = \lambda f \tag{4.1}$$

for some constant λ. *Then*

$$Mf = \Phi^*(\lambda) f ,$$

where

$$\Phi^*(\lambda) = \frac{2\pi}{\Gamma(n)} \int_0^\infty F\left(-\frac{\lambda}{\pi}, n, \pi r^2\right) e^{-\pi r^2/2} r^{2n-1} \Phi(r^2)\, dr .$$

Proof. The convergence of the above integral follows from the estimate (2.5) for the function F, and from the definition of Φ. A solution of the elliptic partial differential equation (4.1) is always smooth.

We remark that in order to prove the lemma, it suffices to show that if f is a solution of (4.1), then

$$(Mf)(0) = \Phi^*(\lambda) f(0) . \tag{4.2}$$

In virtue of the commutativity of D with $\rho_{\{u,1\}}$, f is an eigenfunction belonging to an eigenvalue λ if and only if $\rho_{\{u,1\}} f$ is an eigenfunction belonging to the same eigenvalue. The operator M also commutes with $\rho_{\{u,1\}}$.

By the property (i) of k,

$$\begin{aligned}(Mf)(u) &= \int_V k(v,u) f(v)\, dv \\ &= \int_V k(w,0)\, e\left[-\tfrac{1}{2} A(u,w)\right] f(u+w)\, dw \\ &= \int_V k(w,0)\, (\rho_{\{u,1\}} f)(w)\, dw \\ &= (M\rho_{\{u,1\}} f)(0) .\end{aligned}$$

Thus, our assumption that

$$(M\rho_{\{u,1\}} f)(0) = \Phi^*(\lambda)(\rho_{\{u,1\}} f)(0)$$

implies

$$(Mf)(u) = \Phi^*(\lambda)f(u) \quad .$$

In order to prove (6.2) we take in the group $U(H)$ of all real linear transformations on V which are commutative with J and leave A invariant. Obviously $U(H)$ is the unitary group of H on $\mathbb{V}$. If $\sigma \in U(H)$, the map $(u,c) \to (\sigma(u),c)$ is a holomorphic bundle-automorphism of $(\mathbb{V} \times \mathbb{C}, p, \mathbb{V})$ and preserves the hermitian inner product $\tilde{a}$. Thus commutes with $\Box_{\tilde{a}}$, and equivalently

$$D\sigma^* = \sigma^* D \quad .$$

Moreover, the function k is point-pair invariant under the group $U(H)$, that is,

$$K(\sigma(u),\sigma(v)) = k(u,v) \; .$$

Given a smooth function f on V, we put

$$g(r) = g(\|u\|) = \int_{U(H)} f(\sigma(u))\, d\sigma$$

where the integral in the right hand side of the equality is over the group $U(H)$ and is normalized so that the total volume of the group is 1. If f satisfies the equation (4.1), g also satisfies the equation and $g(0) = f(0)$.

Suppose that f is an eigenfunction of D belonging to an eigenvalue λ and that f satisfies (2.3). Then the radial function

$$e^{\pi r^2/2} g(r)$$

satisfies the ordinary differential equation (4.1) and is finite at $r = 0$. In virtue of the uniqueness of such solution subject to a given finite initial condition at $r = 0$,

$$g(r) = f(0)F\left(\frac{\pi}{\lambda}, n, \pi r^2\right) e^{-\pi r^2/2} \quad .$$

Now,

$$
\begin{aligned}
(Mf)(0) &= \int_V k(v,0)f(v)\ dv \\
&= \int_V \int_{U(H)} k(\sigma^{-1}(v),0)f(v)\ d\sigma dv \\
&= \int_V \int_{U(H)} k(v,0)f(\sigma(v))\ d\sigma dv \\
&= \int_V k(v,0)g(v)\ dv \\
&= f(0) \int_V k(v,0)F\left(\frac{\pi}{\lambda}, n, \pi r^2\right)e^{-\pi r^2/2}\ dv \\
&= f(0)\ \frac{2\pi}{\Gamma(n)} \int_V^{\infty} F\left(-\frac{\pi}{\lambda}, n, \pi r^2\right)e^{-\pi r^2/2} r^{2n-1}\Phi(r^2)\ dr ,
\end{aligned}
$$

which completes the proof.

Let

$$0 \leqslant \lambda_0 \leqslant \lambda_1 \leqslant \lambda_2 \leqslant \cdots$$

be the spectrum of the complex Laplace-Beltrami operator $\Box_a$ acting on $L_2(\mathbb{T},\mathbb{F})$, and

$$s_0, s_1, s_2, \cdots$$

an orthonormal basis of $L_2(\mathbb{T},\mathbb{F})$ consisting of eigensections corresponding to the above spectrum. We denote by f_ν the function on V associated to the section s_ν.

Lemma. *Let* $\delta(A)$ *be the Paffian of the Riemann form* A *with repsect to the lattice group* L. *Then*

$$\sum_{\nu=0}^{\infty} \Phi^*(\lambda_\nu) = \delta(A)\cdot\Phi(0) .$$

Proof. Regarding M as a bounded linear operator of $L_2(\mathbb{T},\mathbb{F})$ into itself, we have

$$(Mf_s)(u) = \int_F K(v,u)f_s(v)\ dv .$$

On account of the previous lemma, the spectrum of M is

$$\Phi^*(\lambda_o),\ \Phi^*(\lambda_1),\ldots \quad .$$

Thus, the kernel $K(u,v)\in L_2(\mathbb{T},\mathbb{F})\otimes\overline{L_2(\mathbb{T},\mathbb{F})}$ admits Fourier series expansion

$$K(u,v)\sim\sum_{\nu=o}^{\infty}\Phi^*(\lambda_\nu)f_\nu(u)\overline{f_\nu}(v) \quad .$$

As we have seen in 3, K is smooth $V\times V$ and $((D_u\cdot\bar{D}_v)^m K)(u,v)$ is given by the series

$$\sum_{\ell\in L}((D_u\cdot\bar{D}_v)^m k)(u,v+\ell)\psi(\ell)e\left[\frac{1}{2}A(\ell,v)\right] .$$

Thus, $(D_u\cdot\bar{D}_v)^m K$ satisfies the same conditions (i) and (ii) for K in 3, and is contained in $L_2(\mathbf{T},\mathbb{F})\otimes\overline{L_2(\mathbb{T},\mathbb{F})}$ for any positive integer m. This fact yields that the above Fourier series expansion of $K(u,v)$ converges absolutely and uniformly to $K(u,v)$ on $F\times F$ (Theorem (23.35.12), p. 268 [4]). From this it follows that

$$\operatorname{trace} M = \sum_{\nu=o}^{\infty}\Phi^*(\lambda_\nu) = \int_F K(u,u)\,du .$$

On the other hand,

$$\begin{aligned}\int_F K(u,u)\,du &= \sum_\ell \int_F k(u,u+\ell)\psi(\ell)e\left[\frac{1}{2}A(\ell,u)\right]du\\ &= \sum_\ell \int_F \Phi\left(\|\ell\|^2\right)e\left[-\frac{1}{2}A(u,\ell)\right]\psi(\ell)e\left[\frac{1}{2}A(\ell,u)\right]du\\ &= \sum_\ell \int_F \Phi(\|\ell\|^2)\psi(\ell)e[A(\ell,u)]\,du \quad .\end{aligned}$$

The map $u\mapsto e[A(\ell,u)]$ induces a character of the torus group T and

$$\int_F e[A(\ell,u)]\,du = \begin{cases}0 & \text{if } \ell\neq 0 ,\\ \operatorname{vol}\mathbb{T} & \text{if } \ell=0 .\end{cases}$$

As is known, the volume of the torus $\mathbb{T}$ with respect to the hermitian metric H is the Paffian $\delta(A)$. Thus we conclude that

$$\sum_{\nu=0}^{\infty} \Phi^*(\lambda_\nu) = \delta(A)\Phi(0) ,$$

completing the proof.

5. Applying the trace formula in the previous number for a particular function, we obtain

Theorem 1. *Let* $\delta(A)$ *be the Paffian of the Riemann form* A *with respect to* L. *The set of distinct eigenvalues of the complex Laplace-Beltrami operator* $\square_a$ *on the complex line bundle* $\mathbb{F}$ *over* $\mathbb{T}$ *is*

$$\{0,\pi,2\pi,\ldots\}$$

and the multiplicity of the eigenvalue $\nu\pi(\nu=0,1,2,\ldots)$ *is*

$$\delta(A)\cdot\binom{\nu+n-1}{n-1} .$$

We choose $\Phi(t)$ to be

$$(1-e^{-\pi T})^{-n}\exp\left\{\frac{\pi}{2}t\left(1+\frac{1}{e^{\pi T}-1}\right)\right\}$$

with a constant $T>0$, and show that

$$\Phi^*(\lambda) = \frac{2\pi^n}{\Gamma(n)}\int_0^\infty F\left(-\frac{\lambda}{\pi},n,\pi r^2\right)\Phi(r^2)e^{-\pi r^2/2}\,r^{2n-1}\,dr$$

$$= e^{-\lambda T} . \tag{5.1}$$

Then, by the lemma in the previous number,

$$\sum_{\nu=0}^{\infty} e^{-\lambda_\nu T} = \delta(A)\Phi(0) = (1-e^{-\pi T})^{-n}$$

$$= \sum_{\nu=0}^{\infty}\binom{\nu+n-1}{n-1}e^{-\pi\nu T}$$

for any $T>0$. From this immediately we obtain the above theorem.

The equality (5.1) is derived from a known Laplace integral involving the confluent hypergeometric function (Bateman Manuscript Project [1], vol. 1, p. 215), which reads that for

$$M_{\kappa\mu}(z) = z^{\frac{1}{2}+\mu} e^{-\frac{z}{2}} F\left(\frac{1}{2}+\mu-\kappa, 2\mu+1, z\right)$$

$$(\mathrm{Re}(2\mu+1) > 0, \quad -\pi < \mathrm{arc}(z) < \pi) \quad ,$$

$$\int_0^\infty e^{-pt}\, t^{\mu-\frac{1}{2}} M_{\kappa\mu}(at)\, dt = a^{\mu+\frac{1}{2}}\, \Gamma(2\mu+1) \frac{\left(p-\frac{1}{2}a\right)^{\kappa-\mu-\frac{1}{2}}}{\left(p+\frac{1}{2}a\right)^{\kappa+\mu+\frac{1}{2}}}$$

$$\left(\mathrm{Re}\ p > \frac{1}{2}\,|\mathrm{Re}\ a|\right) \quad ,$$

where $\Gamma(x)$ is the Gamma function.

Apply the following substitution:

$$\frac{1}{2} + \mu - \kappa = -\frac{\lambda}{\pi} \quad ,$$

$$2\mu - 1 = n = \text{complex}\dim \mathbb{T} \quad ,$$

and obtain

$$a^{\frac{n}{2}} \int_0^\infty e^{-\left\{\frac{a}{2}+p\right\}t} F\left(-\frac{\lambda}{\pi}, n, at\right) t^{n-1}\, dt = a^{\frac{n}{2}}\, \Gamma(n) \frac{\left(p-\frac{1}{a}a\right)^{\frac{\lambda}{\pi}}}{\left(p+\frac{1}{2}a\right)^{n+\frac{\lambda}{\pi}}} .$$

In the left hand side integral, we substitute at by πr^2. Then

$$\int_0^\infty F\left(-\frac{\lambda}{\pi}, n, \pi r^2\right) e^{-\left(\frac{a}{2}+p\right)\frac{\pi r^2}{a}}\, r^{2n-1}\, dr = \frac{a^n}{2} \frac{\Gamma(n)}{\pi^n} \frac{\left(p-\frac{1}{2}a\right)^{\frac{\lambda}{\pi}}}{\left(p+\frac{1}{2}a\right)^{n+\frac{\lambda}{\pi}}} .$$

Finally we put

$$p = \frac{1}{2}(1+e^{-\pi T})$$

$$a = 1-e^{-\pi T} \quad ,$$

with $T>0$, and obtain the desired equality

$$\frac{2\pi^n}{\Gamma(n)}(1-e^{-\pi T})^{-n}\int_0^\infty F\left(-\frac{\lambda}{\pi},n,\pi r^2\right)\exp\left(-\pi r^2\left(1+\frac{1}{e^{\pi T}-1}\right)\right)r^{2n-1}\,dr$$

$$= e^{-\lambda T} \quad .$$

2. Eigenspaces

6. Our problem in this section is to determine the eigenspaces of the complex Laplace-Beltrami operator acting on $L_2(\mathbb{T},\mathbb{F})$. We begin with another representation of $L_2(\mathbb{T},\mathbb{F})$.

Here, the Hilbert space $\mathfrak{H}$ consists of functions φ on $\mathbf{V}$ satisfying the following conditions:

$$\varphi(u+\ell) = \varphi(u)\psi(\ell)e\left[\frac{1}{2i}H\left(\ell,u+\frac{1}{2}\ell\right)\right] \quad , \qquad u\in\mathbb{W}, \quad \ell\in L \; ,$$

and

$$\|\varphi\|^2 = \int_F |\varphi(u)|^2 e\left[-\frac{1}{2i}H(u,u)\right]\,du < \infty \quad .$$

This representation is isometric to the previous one in (2.3) by the map

$$\varphi(u) \to f(u) = \varphi(u)e\left[\frac{-1}{4i}H(u,u)\right] \quad .$$

This Hilbert space being regarded as $L_2(\mathbb{T},\mathbb{F})$, the complex Laplace-Beltrami operator is realized by the differential operator E in 2. A function φ in $\mathfrak{H}$ is holomorphic if and only if φ is an eigenfunction of E whose eigenvalue is 0 and a holomorphic function φ in

$\mathfrak{H}$ is nothing but a holomorphic theta function of reduced type (H,ψ) ([9], VI, no. 3).

The Hilbert space $\mathfrak{H}$ is the representation module of a unitary representation of the Heisenberg group, called the lattice representation (P. Cartier [3]). We assign to each $\{u,c\} \in G$, a unitary operator $U_{\{u,c\}}$ given by

$$(U_{\{u,c\}}\varphi)(v) = \varphi(u+v)e\left[-\frac{1}{2i}H\left(v+\frac{1}{2}u,u\right)\right]c^{-1} .$$

Let $\mathfrak{g}$ be the Lie algebra of the group G. The underlying vector space of $\mathfrak{g}$ is $\mathbf{V}+\mathbb{R}$ and its Lie bracket is defined by

$$[\{u,r\},\{v,s\}] = \{0,A(u,v)\},\{u,n\},\{v,s\} \in \mathfrak{g} .$$

The exponential map is given by

$$\exp\{u,r\} = \{u,e[r]\} .$$

Let us take a $\mathbb{Z}$-basis $(e_1,\ldots,e_{2n})$ of the lattice L adapted to the Riemann form A so that

$$A\left(\sum_{i=1}^{2n} x_i e_i, \sum_{i=1}^{2n} y_i e_i\right) = \sum_{\alpha-1}^{n} d_\alpha(a_\alpha y_{n+\alpha} - x_{n+\alpha}y_\alpha) ,$$

where the integers $d_1,d_2,\ldots,d_n$ are the elementary divisors of A. Then the Paffian $\delta(A)$ of A is $|d_1d_2\ldots d_n|$. Put

$$e^*_\alpha = \frac{1}{d_\alpha}e_\alpha, \quad e^*_{n+\alpha} = e_{n+\alpha}, \quad \alpha = 1,\ldots,n .$$

In the complex vector space $\mathbb{W} = (V,J)$, $(e^*_1,\ldots,e^*_n)$ is a $\mathbb{C}$-basis. Put

$$e^*_{n+\beta} = \sum_{\alpha=1}^{n} Z_{\alpha\beta}e^*_\alpha .$$

Then, the matrix $Z = (Z_{\alpha\beta})$ is a point in the Siegel's generalized upper half plane ([10] VI, Proposition 6). Namely Z is a symmetric $n \times n$ matrix whose imaginary part $Y = (Y_{\alpha\beta})$ is positive definite. We denote by X the real part of Z. From the definitions of Z and the hermitian form H, we see easily

$$H\left(\sum_{\alpha=1}^{n} u_\alpha e_\alpha^*\ ,\ \sum_{\alpha=1}^{n} v_\alpha e_\alpha^*\right) = \sum_{\alpha,\beta=1}^{n} Y_{\alpha\beta}^{-1}\bar{u}_\alpha v_\beta\ .$$

We denote by (u_α) the complex cartesian coordinates on $\mathbf{V}$ with respect to the C-basis $(e_1^*,\ldots,e_n^*)$ and put $u_\alpha = x_\alpha + iy_\alpha$. By a direct computation, we obtain

$$\left(\dot{U}_{\{e_\alpha^*,0\}}\varphi\right)(u) = \frac{\partial\varphi}{\partial x_\alpha}(u) - \pi H(u,e_\alpha^*) \qquad ,$$

$$\left(\dot{U}_{\{Je^*,0\}}\varphi\right)(u) = \frac{\partial\varphi}{\partial y_\alpha}(u) - i\pi H(u,e_\alpha^*) \qquad ,$$

for a smooth function $\varphi \in \mathfrak{H}$ $(\alpha = 1,\ldots,n)$.

These operators are skew self-adjoint, namely, they satisfy the equality $\dot{U}^* + \dot{U} = 0$.

We put

$$\mathfrak{a}_\alpha = \frac{1}{2}\left(\dot{U}_{\{e_\alpha^*,0\}} - i\dot{U}_{\{Je_\alpha^*,0\}}\right)$$

and

$$\mathfrak{b}_\alpha = \frac{1}{2}\left(\dot{U}_{\{e_\alpha^*,0\}} + i\dot{U}_{\{Je_\alpha^*,0\}}\right) \qquad .$$

Then,

$$\mathfrak{a}_\alpha = \frac{\partial}{\partial u_\alpha} - \pi \sum_\beta Y_{\beta\alpha}^{-1}\bar{u}_\beta \quad , \tag{6.1}$$

$$\mathfrak{b}_\alpha = \frac{\partial}{\partial \bar{u}_\alpha} \qquad . \tag{6.2}$$

The operators $\mathfrak{a}_\alpha$, $\mathfrak{b}_\alpha$ $(1 \leqslant \alpha \leqslant n)$ form a C-basis of the complex vector space of operators spanned by $\dot{U}(\mathfrak{g})$.

Here, we list some equalities involving these operators, each of which is verified easily.

$$\mathfrak{a}_\alpha^* + \mathfrak{b}_\alpha = 0, \quad \mathfrak{a}_\alpha + \mathfrak{b}_\alpha^* = 0 \qquad , \tag{6.3}$$

$$[\mathfrak{a}_\alpha,\mathfrak{a}_\beta]\varphi = 0, \quad [\mathfrak{b}_\alpha,\mathfrak{b}_\beta]\varphi = 0 \qquad , \tag{6.4}$$

$$[\mathfrak{a}_\alpha, \mathfrak{b}_\beta]\varphi = \pi Y^{-1}_{\alpha\beta}\varphi \quad , \tag{6.5}$$

where φ is a smooth function in $\mathfrak{H}$.

Moreover, the differential operator E can be written in terms of $\mathfrak{a}_\alpha$, $\mathfrak{b}_\alpha$ $(\alpha = 1,\dots,n)$ as

$$E = -\Sigma\, Y_{\alpha\beta}\mathfrak{a}_\alpha \mathfrak{b}_\beta \quad . \tag{6.6}$$

From this, it follows easily that

$$[E, \mathfrak{a}_\alpha]\varphi = \pi\mathfrak{a}_\alpha\varphi \tag{6.7}$$

$$[E, \mathfrak{b}_\alpha]\varphi = -\pi\mathfrak{b}_\alpha\varphi \tag{6.8}$$

for a smooth function φ in $\mathfrak{H}$.

7. Given a non-negative integer j, we denote by $[\nu]$ an ordered set of n non-negative integers $\nu_1,\dots,\nu_n$ whose sum is j and by $p(j)$ the collection of all such ordered sets. The cardinality of $p(j)$ is $\binom{n+j-1}{n-1}$. For $[\nu] \in p(j)$, we put

$$P_{[\nu]} = \mathfrak{a}_1^{\nu_1}\mathfrak{a}_2^{\nu_2}\cdots\mathfrak{a}_n^{\nu_n} \quad .$$

<u>Theorem 2</u>. *Let* $\mathfrak{H}_\nu$ *be the eigenspace of the operator* E *in* $\mathfrak{H}$ *belonging to the eigenvalue* $\pi\nu$. *The subspace* $\mathfrak{H}_0$ *consists of holomorphic theta functions of reduced type* (H,ψ) *and the identically zero function. Let* $\{\varphi_k,\ k = 1,\dots,\delta(A)\}$ *be an orthonormal basis of* $\mathfrak{H}_0$.

Then, the following $\delta(A)\cdot\binom{n+j-1}{n-1}$ *functions*

$$P_{[\nu]}\varphi_k = \mathfrak{a}_1^{\nu_1}\cdots\mathfrak{a}_n^{\nu_n}\varphi_k \quad .$$

$$[\nu] = (\nu_1,\dots,\nu_n) \in p(j), \quad k = 1,\dots,\delta(A) \quad ,$$

form a $\mathbb{C}$*-basis of* $\mathfrak{H}_j$ $(j = 1,2,\dots)$.

<u>Proof</u>. The first half of the theorem is already explained in 6. In order to prove the second half, we use induction on j. If $j = 0$, the statement is trivially true. Suppose that the statement is true for $j - 1$.

First, a function of the form

$$P_{[\nu]}\varphi_k = \mathfrak{a}_\rho^{\nu_\rho}\cdots\mathfrak{a}_n^{\nu_n}\varphi_k ,$$

where $\nu_\rho > 0$, $1 \leqslant \rho \leqslant n$ and $[\nu] \in p(j)$, belongs to $\mathfrak{H}_j$. Indeed, from (6.7)

$$E\mathfrak{a}_\rho^{\nu_\rho}\cdots\mathfrak{a}_n^{\nu_n}\varphi_k = \mathfrak{a}_\rho E\mathfrak{a}_\rho^{\nu_\rho - 1}\cdots\mathfrak{a}_n^{\nu_n}\varphi_k + \pi\mathfrak{a}_\rho^{\nu_\rho}\cdots\mathfrak{a}_n^{\nu_n}\varphi_k .$$

By the induction hypothesis, the first term of the right hand side of the above equality is $\pi(j-1)\mathfrak{a}_\rho^{\nu_\rho - 1}\cdots\mathfrak{a}_n^{\nu_n}\varphi_k$. Thus,

$$EP_{[\nu]}\varphi_k = \pi j P_{[\nu]}\varphi_k .$$

Put

$$\mathfrak{b}'_\rho = -\sum_\alpha Y^{-1}_{\rho\alpha}\mathfrak{b}_\alpha .$$

From the equality (6.5), it follows that

$$\mathfrak{b}'_\rho\mathfrak{a}_1^{\nu_1}\cdots\mathfrak{a}_n^{\nu_n}\varphi = \pi\nu_\rho\mathfrak{a}_1^{\nu_1}\cdots\mathfrak{a}_\rho^{\nu_\rho - 1}\cdots\mathfrak{a}_n^{\nu_n}\varphi$$

provided that φ is holomorphic.

Suppose that

$$\Sigma_{[\nu]}\Sigma_k\, C_{\nu_1\cdots\nu_n k}\,\mathfrak{a}_1^{\nu_1}\cdots\mathfrak{a}_n^{\nu_n}\varphi_k = 0 .$$

Apply the operator $\mathfrak{b}'_\rho$ on both sides of the equality. Then

$$\Sigma\,\Sigma_k\,\pi\nu_\rho C_{\nu_1\cdots\nu_n k}\,\mathfrak{a}_1^{\nu'_1}\cdots\mathfrak{a}_\rho^{\nu_\rho - 1}\cdots\mathfrak{a}_n^{\nu_n}\varphi_k = 0 .$$

From the induction hypothesis, we see that

$$C_{\nu_1\cdots\nu_n k} = 0 , \qquad \text{if} \quad \nu_\rho > 0 .$$

Since ρ is arbitrary, all coefficients $C_{\nu_1\cdots\nu_n k}$ are zero. Thus we have seen that these $\delta(A)\binom{n+j-1}{n-1}$ functions in the theorem are linearly independent. By Theorem 1, the dimension of the eigenspace $\mathfrak{H}_j$

is $\delta(A)\binom{n+j-1}{n-1}$, and hence they form a $\mathbb{C}$-basis of $\mathfrak{H}_j$. We have finished the proof.

8. As in Theorem 2, we denote by $\{\varphi_k;\ k=1,\dots,\delta(A)\}$ an orthonormal basis of $\mathfrak{H}_0$. For each k, $1\leqslant k\leqslant\delta(A)$, we denote by $\mathfrak{H}^{(k)}$ the closed subspace in $\mathfrak{H}$ spanned by the functions

$$P_{[\nu]}\varphi_k \qquad ,$$

$[\nu]\in p(j)$, $j=0,1,2,\dots$. These functions form an orthonormal basis of $\mathfrak{H}^{(k)}$. From (6.4) and (6.5), it follows that $\mathfrak{H}^{(k)}$ is $\mathfrak{g}$-stable and hence that $\mathfrak{H}^{(k)}$ is G-invariant. We denote by $(\mathfrak{H}^{(k)},U^{(k)})$ the representation of G induced on $\mathfrak{H}^{(k)}$. The correspondence

$$P_{[\nu]}\varphi_k \to P_{[\nu]}\varphi_\ell$$

gives rise to an equivalence of the unitary representations $(\mathfrak{H}^{(k)},U^{(k)})$ and $(\mathfrak{H}^{(\ell)},U^{(\ell)})$.

Theorem 3. *Each representation* $(\mathfrak{H}^{(k)},U^{(k)})$ *is irreducible and the representation* $(\mathfrak{H},U)$ *is the direct sum of the* $\delta(A)$ *irreducible representations* $(\mathfrak{H}^{(k)},U^{(k)})$, $1\leqslant k\leqslant\delta(A)$.

Remark. The conclusion that $(\mathfrak{H},U)$ is the direct sum of mutually equivalent $\delta(A)$ irreducible representations is a theorem by P. Cartier ([3], Theorem 2).

Proof. Let us take a non-trivial closed irreducible G-invariant subspace $\mathfrak{H}'$ of $\mathfrak{H}^{(k)}$, and denote by $\mathfrak{H}''$ the orthogonal complement of $\mathfrak{H}'$ in $\mathfrak{H}^{(k)}$. The subspace $\mathfrak{H}''$ is also G-invariant. On account of (6.6), $\mathfrak{H}'$ and $\mathfrak{H}''$ are stable under the operator E. Let $\varphi_k=\varphi'+\varphi''$ be the decomposition of φ_k according to the direct sum $\mathfrak{H}^{(k)}=\mathfrak{H}'+\mathfrak{H}''$. If E is defined on φ' and φ'', then both are eigenfunctions of E belonging to the eigenvalue 0. Since $\dim(\mathfrak{H}^{(k)}\cap\mathfrak{H}_0)=1$, either φ' or φ'' must be zero. Observing that the non-trivial subspace $\mathfrak{H}'$ is spanned by the functions of the form $P_{[\nu]}\varphi'$, we conclude that φ' is not identically zero, and hence $\varphi''=0$. Since $\mathfrak{H}''$ is spanned by the functions of the form $P_{[\nu]}\varphi''$, $\mathfrak{H}''=\{0\}$.

What remains to be verified is that φ' and φ'' are both in the domain of E. Since $\mathfrak{H}^{(k)}$ is the direct sum of the G-invariant subspaces $\mathfrak{H}'$ and $\mathfrak{H}''$, the Garding space of $\mathfrak{H}$ is the direct sum of the Garding spaces of $\mathfrak{H}'$ and $\mathfrak{H}''$. As a holomorphic function, φ_k

belongs to the Garding space of $\mathfrak{H}$. Therefore, φ' and φ'' are in the Garding spaces of $\mathfrak{H}'$ and $\mathfrak{H}''$, respectively, and they are in the domain of E, which is written in the form (6.6). We have finished the proof of the irreducibility of $\mathfrak{H}^{(k)}$.

We show that if $k \neq \ell$, then $\mathfrak{H}^{(k)}$ and $\mathfrak{H}^{(\ell)}$ are mutually orthogonal. For this, it suffices to prove that $\mathfrak{H}^{(k)} \cap \mathfrak{H}_i$ and $\mathfrak{H}^{(\ell)} \cap \mathfrak{H}_j$ are mutually orthogonal for all i and j. We verify this fact by induction on the sum $i+j$. If $i+j=0$, the statement is trivially true. Suppose that our claim is true if the sum is $i+j-1$. Take $[\mu] \in p(i)$ and $[\nu] \in p(j)$ arbitrarily. By (6.3),

$$\left(\mathfrak{a}_\rho^{\mu_\rho} \cdots \mathfrak{a}_n^{\mu_n} \varphi_k,\ P_{[\nu]}\varphi_\ell\right) = -\left(\mathfrak{a}_\rho^{\mu_\rho - 1} \cdots \mathfrak{a}_n^{\mu_n} \varphi_k,\ \mathfrak{b}_\rho P_{[\nu]}\varphi_\ell\right).$$

Since $\mathfrak{b}_\rho P_{[\nu]}\varphi_\ell \in \mathfrak{H}^{(\ell)} \cap \mathfrak{H}_{i-1}$ by (6.5), the right hand side of the above equality is 0. Thus, $\mathfrak{H}^{(k)} \cap \mathfrak{H}_i$ and $\mathfrak{H}^{(\ell)} \cap \mathfrak{H}_j$ are mutually orthogonal. We have seen that $\mathfrak{H}$ is the direct sum of these $\delta(A)$ subspaces $\mathfrak{H}^{(k)}$.

<u>Remark</u>. We have derived the Cartier's theorem from Theorem 1 in this section. Here we show that conversely we can determine the spectrum of the differential operator E acting on $\mathfrak{H}$ by making use of the Cartier's theorem.

Indeed, independently from Theorem 1, we see that the representation $(\mathfrak{H}, U)$ contains the direct sum $(\mathfrak{H}', U')$ of the $\delta(A)$ representations $(\mathfrak{H}^{(k)}, U^{(k)})$, $k = 1, \ldots, \delta(A)$. By the theorem of Cartier, $(\mathfrak{H}, U) = (\mathfrak{H}', U')$.

As is shown in the proof of Theorem 2,

$$\mathfrak{H}^{(k)} = \sum_{i=0}^{\infty} \mathfrak{H}^{(k)} \cap \mathfrak{H}_i \quad \text{(direct sum)}$$

and

$$\dim(\mathfrak{H}^{(k)} \cap \mathfrak{H}_i) = \binom{n+i-1}{n-1} .$$

Thus, the set of all distinct eigenvalues of the operator E is $\{0, \pi, 2\pi, \ldots\}$ and the multiplicity of the eigenvalue $\nu\pi$ is $\binom{n+\ -1}{n-1}$.

3. The Partial Fourier Series Expansion

9. In this section, we consider a realization of $L_2(\mathbb{T},\mathbb{F})$ in which the eigenspace belonging to the eigenvalue 0 of the complex Laplace-Beltrami operator is the space of the theta series, borrowing the theta-factors theory in Weil [10], Chapter VI. The purpose is to exhibit an eigenfunction as a Fourier series.

As in 6, let $(e_1,\ldots,e_{2n})$ be a $\mathbf{Z}$-basis of the lattice L adapted to the Riemann form A. We put

$$e^*_\alpha = \frac{1}{d_\alpha} e_\alpha, \quad e^*_{n+\alpha} = e_{n+\alpha}, \quad (\alpha = 1,\ldots,n) ,$$

where $d_1,\ldots,d_n$ are the elementary divisors of A. In the complex vector space $\mathbf{V} = (V,J)$, $(e^*_1,\ldots,e^*_n)$ is a $\mathbb{C}$-basis. Put

$$e^*_{n+\beta} = \sum_\alpha Z_{\alpha\beta} e^*_\alpha \quad . \tag{9.1}$$

Then, the matrix $Z = (Z_{\alpha\beta})$ is symmetric. We denote by X and Y its real and imaginary parts respectively. The matrix Y is positive definite.

From now on, we use the matrix notations with respect to the $\mathbb{C}$-basis $(e^*_1,\ldots,e^*_n)$ of $\mathbb{V}$, identifying a vector $u = \Sigma\, u_\alpha e^*_\alpha$ with the column vector (u_α). Thus

$$H(u,u) = {}^t\bar{u}Y^{-1}u \quad .$$

A quadratic form $u \to {}^tuQu$ will be denoted by $Q[u]$.

Define a $\mathbb{C}$-bilinear form Φ on $\mathbb{V}\times\mathbb{V}$ by

$$\Phi(u,v) = -\,{}^tuY^{-1}v \quad ,$$

and put

$$F = \frac{H+\Phi}{2i} \quad .$$

Then, F is an $\mathbb{R}$-bilinear form on $V \times V$. If we denote by V' (resp. V'') the real subspace in V spanned by $e^*_1,\ldots,e^*_n$ (resp. $e^*_{n+1},\ldots,e^*_{2n}$),

$$F(V',V'') = 0 \quad .$$

Obviously from the definition of F, the map $V \to F(u,v)$ is $\mathbb{C}$-linear.

We quote the following two equalities from [10], p. 118 (10) and (11):

$$F\left(\sum_{i=1}^{2n} x_i e_i, \sum_{\alpha=1}^{n} u_\alpha e_\alpha^*\right) = -\sum_{\alpha=1}^{n} x_{n+\alpha} u_\alpha \qquad (9.2)$$

$$F\left(\sum_{i=1}^{2n} x_i e_i, \sum_{i=1}^{2n} y_i e_i\right) = -\sum_{\alpha=1}^{n} x_{n+\alpha} y_\alpha - \sum_{\alpha_1 \beta=1}^{n} z_{\alpha\beta} x_{n+2} y_{n+\beta} \qquad (9.3)$$

We define a satellite form B of A by

$$B\left(\sum_{i=1}^{2n} x_i e_i, \sum_{i=1}^{2n} y_i e_i\right) = \sum_{\alpha=1}^{n} d_\alpha x_\alpha y_{n+\alpha} \quad .$$

If ψ is a semi-character of L relative to A,

$$\chi(\ell) = \psi(\ell) e\left[\frac{1}{2} B(\ell,\ell)\right] , \qquad \ell \in L \quad ,$$

is a character of L. Put

$$\chi(e_2) = e[d_\alpha \lambda_\alpha] \quad , \qquad \lambda_\alpha \in \mathbb{R} \quad . \qquad (9.4)$$

$$\chi(e_{n+2}) = e[\lambda_{n+\alpha}] \; , \quad \lambda_{n+\alpha} \in \mathbb{R}, \quad (\alpha = 1,\ldots,n) \quad . \qquad (9.5)$$

and define a $\mathbb{C}$-linear form λ on $\mathbb{W}$ by

$$\lambda\left(\sum_{\alpha=1}^{n} u_\alpha e_\alpha^*\right) = \sum_{\alpha=1}^{n} \lambda_\alpha u_\alpha \; .$$

Putting

$$\tilde{\chi}(\ell) = \chi(\ell) e[-\lambda(\ell)] \quad , \qquad (9.6)$$

we obtain a homomorphism of L into $\mathbb{C}$ which assumes 1 on $L \cap V'$. Obviously

$$\chi(\ell) e[-\lambda(\ell)] = \tilde{\chi}(\ell) e\left[\frac{1}{2} B(\ell,\ell)\right] \quad .$$

10. We denote by $\mathfrak{K}$ the Hilbert space of $\mathbb{C}$-valued measurable functions θ on $\mathbb{W}$ satisfying the equation

$$\theta(u+\ell) = \theta(u)\psi(\ell)e\left[F(\ell,u) + \frac{F(\ell,\ell)}{2} - \lambda(\ell)\right] , \qquad (10.1)$$

for all $\ell \in L$, with the norm defined by

$$\|\theta\|^2 = \int_F |\theta(u)|^2 \exp(2\pi \ \mathrm{Im}\ F(u,u) - 4\pi \ \mathrm{Im}\ \lambda(u))\ du ,$$

where the F measure du and the fundamental domain of F of L are as before. The weight in the above norm has the expression

$$\exp - 2\pi(Y^{-1}[y] + 2\lambda^t y) .$$

An isomorphism of the Hilbert space $L_2(\mathbb{T},\mathbb{F})$ with $\mathfrak{K}$ is given by the map

$$f(u) \to \theta(u) = f(u)e\left[\frac{1}{2} F(u,u) - \lambda(u)\right] ,$$

and the map

$$\iota : \varphi(u) \to \theta(u) = \varphi(u)e\left[\frac{1}{4i} \Phi(u,u) - \lambda(u)\right] \qquad (10.2)$$

is an isomorphism of the Hilbert space $\mathfrak{H}$ onto $\mathfrak{K}$.

One advantage of this realization of $L_2(\mathbb{T},\mathbb{F})$ is that if $\ell \in L \cap V'$, the condition (10.1) reduces to the periodicity condition

$$\theta(u+\ell) = \theta(u) , \quad \ell \in L \cap V' . \qquad (10.3)$$

This fact is an immediate consequence of (9.2), (9.3) and the definition of λ. If $\ell \in L \cap V''$,

$$\ell = \sum_\alpha m_\alpha e^*_{n+\alpha} , \quad (m_\alpha) \in \mathbb{Z}^n ,$$

and $\ell = Zm$ in the matrix notations. The condition (10.1) is written as

$$\theta(u+Zm) = \theta(u)\left(\prod_{\alpha=1}^{n} \tilde{\chi}_\alpha^{m_\alpha}\right) e\left[-\frac{1}{2} Z[m] - m^t u\right] . \qquad (10.4)$$

To each $\ell = Zm \in L \cap V'$, we assign an operator σ_ℓ acting on functions θ on $\mathbf{V}$, defined by

$$(\sigma_\ell \theta)(u) = \theta(u+\ell)\psi(\ell)^{-1} e\left[-F\left(\ell, u + \frac{1}{2}\ell\right) + \lambda(\ell)\right]$$

$$= \theta(u + Zm) \prod_{\alpha=1}^{n} \tilde{\chi}_\alpha^{-m_\alpha} e\left[\frac{1}{2} Z[m] + m^t u\right] \quad . \qquad (10.5)$$

Obviously, the correspondence $\ell \to \sigma_\ell$ is a homomorphism. We put

$$w = \sum_{\ell \in L \cap V'} \sigma_\ell \quad , \qquad (10.6)$$

which is called the Weil-Brezin operator in [1], [9]. If a continuous function θ on $\mathbb{W}$ is periodic with respect to $L \cap V'$, and if the infinite series $w\theta$ converges to a continuous function, then $w\theta$ satisfies the condition (10.4) and (10.5) and hence belongs to $\mathbb{K}$.

We denote by $\mathfrak{B}$ the set of $\delta(A)$ vectors in $\mathbb{R}^n$ of the form

$$b = \left(\frac{a_\alpha}{d_\alpha}\right), \quad (a_\alpha) \in \mathbf{Z}^n \quad \text{and} \quad 0 \leq a_\alpha < |d_\alpha| \ ,$$

where $(d_1, \ldots, d_n)$ is the set of the elementary divisors of A relative to L. If we denote by L_1 the discrete subgroup generated by $\{e_1^*, \ldots, e_n^*\}$, the set $\mathfrak{B}$ is a complete representative of the quotient group of L_1 modulo $L \cap V$.

For each $b \in \mathfrak{B}$, we consider the exponential function $e[b^t u]$ and form the series

$$w e[b^t u] \quad ,$$

and obtain $\delta(A)$ linearly independent theta series ([10], VI).

11. Let $\tilde{L}_2(\mathbb{R}^n)$ be the Hilbert space of measurable functions $g(t)$ on $\mathbb{R}^n$ with the weight function

$$\exp - \frac{1}{2} E[t] \quad ,$$

where E is the $n \times n$ unit matrix. The norm is defined by

$$\|g\|^2 = \int_{\mathbb{R}^n} |g(t)|^2 \exp\left(-\frac{1}{2} E[t]\right) dt .$$

For a multi-index $[\nu] = (\nu_1, \dots, \nu_n)$, $\nu \in p(\nu_0)$, $(\nu_0 = \nu_1 + \dots + \nu_n)$, we denote by $H_{[\nu]}(t)$ the Hermite polynomial

$$(-1)^{\nu_0} \exp\left(\frac{1}{2} E[t]\right) \frac{\partial^{\nu_0}}{\partial t_1^{\nu_1} \partial t_n^{\nu_n}} \exp\left(-\frac{1}{2} E[t]\right) .$$

It is well-known that the set of all Hermite polynomials form an orthogonal basis of $\tilde{L}_2(\mathbb{R}^n)$ and

$$\|H_{[\nu]}\| = (\nu_1! \dots \nu_n!)^{1/2} (2\pi)^{n/4} .$$

Moreover, each Hermite polynomial $H_{[\nu]}$ is an eigenfunction of the differential operator

$$- \sum_{\alpha} \left(\frac{\partial}{\partial t_\alpha^2} - t_\alpha \frac{\partial}{\partial t_\alpha} \right) ,$$

and its eigenvalue is ν_0. The dimension of the eigensapce of the operator belonging to the eigenvalue ν_0 is

$$\binom{n + \nu_0 - 1}{n - 1} .$$

From the definition of $H_{[\nu]}$, we obtain easily the equality

$$\left(\frac{\partial}{\partial t_\alpha} - t_\alpha \right) H_{[\nu]} = \nu_\alpha H_{[\nu']} ,$$

where $[\nu'] = (\nu_1, \dots, \nu_\alpha + 1, \dots, \nu_n)$. Combining this and the fact that $H_{[\nu]}$ is an eigenfunction of the operator (11.1), we see that

$$\frac{\partial}{\partial t_\alpha} H_{[\nu]} = \nu_\alpha H_{[\nu'']} ,$$

where $[\nu''] = (\nu_1, \dots, \nu_\alpha - 1, \dots, \nu_n)$ and $\nu_\alpha \geqslant 1$.

To each $b \in \mathfrak{B}$, we will define an isomorphism of $\tilde{L}_2(\mathbb{R}^n)$ into the Hilbert space $\mathfrak{K}$ which carries the operator (11.1) to the one corresponding to the complex Laplace-Beltrami operator on $L_2(\mathbb{T},\mathbb{F})$.

12. Since Y is a real positive definite symmetric matrix, we choose an orthogonal matrix O whose determinant is 1 such that

$$O^tYO = P \quad , \tag{12.1}$$

where P is a diagonal matrix with positive $\rho_1,\ldots,\rho_n$ on the diagonal. We denote by $P^{1/2}$ the positive definite diagonal matrix whose square is P.

To each $b \in \mathfrak{B}$, we assign a linear map $\mathfrak{L}_b : \tilde{L}_2(\mathbb{R}^n) \to \mathfrak{K}$ defined by

$$\mathfrak{L}_b g = c_b w \cdot \theta \quad , \tag{12.2}$$

where w is the series forming operators (10.5 and 6),

$$\theta(u) = e[b^tu]g(2\pi^{1/2}P^{-1/2}O^t(y + Y(\lambda + b))) \tag{12.3}$$

and

$$c_b = 2^{n/2}\pi^{n/4}(\det Y)^{1/4} \exp(-\pi Y[b+\lambda])\delta(A)^{-1/2} \quad .$$

Lemma 1. *The linear map $\mathfrak{L}_b$ is an isomorphism of the Hilbert space $\tilde{L}_2(\mathbb{R}^n)$ into the Hilbert space $\mathfrak{K}$. Let $\mathfrak{K}_b$ be the image of $\tilde{L}_2(\mathbb{R}^n)$ under $\mathfrak{L}_b$. Then if b and b' are distinct vectors in $\mathfrak{B}$, $\mathfrak{K}_b$ and $\mathfrak{K}_{b'}$ are mutually orthogonal.*

Proof. In order to prove the first half of the lemma, it suffices to show that the map $\mathfrak{L}_b$ is norm-preserving.

Put

$$u = \sum_\alpha (x_\alpha + iy_\alpha)e^*_\alpha = \sum_\alpha \xi_\alpha e^* + \sum_\alpha \eta_\alpha e^*_{n+\alpha} ,$$

where (x_α), (y_α), (ξ_α) and (η_α) are all in $\mathbb{R}^n$. From the equality (9.1),

$$x = \xi + X\eta, \quad y = Y\eta \quad .$$

Thus, the integral of a function $f(x,y)$ over a fundamental domain F can be performed by the following iterated integral.

$$\int_D \left(\int_{D(y)} f(x,y)\,dx_1 \ldots dx_n \right) (\det Y)^{-1}\, dy_1 \ldots dy_n \qquad ,$$

where

$$D(y) = \left\{ x; (XY^{-1}y)_\alpha \leqslant x_\alpha \leqslant (XY^{-1}y)_\alpha + |d_\alpha| \right\} ,$$

$$D = \{ y = Y\eta;\ 0 \leqslant \eta_\alpha \leqslant 1 \}$$

By the definition (10.6) of the operator w,

$$(w\theta)(u) = \sum_{m \in Z^n} e[(b+m)^t x] A_m(y) \tag{12.4}$$

where

$$A_m(y) = g(t_m) e\left[b^t Xm + \frac{1}{2} X[m] \right] \left(\Pi\ X_\alpha^{-m_\alpha} \right) \cdot$$
$$\exp\left\{ -2\pi \left(\frac{1}{2} Y[m] + b^t Ym + (b+m)^t y \right) \right\}$$

and

$$t_m = 2\pi^{1/2} P^{-1/2} O^t (y + Y(\lambda + b + m)) .$$

Since

$$\int_{D(y)} e[(m-m')^t x]\, dx_1 \ldots dx_n = \begin{cases} \delta(A) & \text{when } m = m' , \\ 0 & \text{otherwise} , \end{cases}$$

for $m, m' \in \mathbf{Z}^n$,

$$\|w\theta\|^2 = \delta(A) \int_D \sum_{m \in \mathbf{Z}^n} |Am(y)|^2 \exp\{-2\pi(Y^{-1}[y] + 2\lambda^t y)\} (\det Y)^{-1} \cdot$$
$$dy_1 \ldots dy_n \qquad .$$

Making use of the equality

$$\left| \prod_{\alpha=1}^{n} \chi_\alpha^{m_\alpha} \right|^2 = \exp(4\pi\lambda^t Ym) \qquad ,$$

we have

$$\|w\theta\|^2 = \delta(A) \exp 2\pi Y[b+\lambda]\cdot\det Y^{-1}\cdot \sum_m \int_D |g(2\pi^{1/2}P^{-1/2}O^t(y+Y(m+\lambda+b)))|^2\cdot\exp(-2\pi Y^{-1}[y+Y(m+\lambda+b)]) .$$

The integral over D in the right hand side of the above equality is equal to

$$\int_{D-Ym} |g(2\pi^{1/2}P^{-1/2}O^t(y+Y(\lambda+b)))|^2 \exp(-2\pi Y^{-1}[y+Y(\lambda+b)]) .$$

Therefore

$$\|w\theta\|^2 = \delta(A) \exp 2\pi Y[b+\lambda]\cdot\det Y^{-1}\cdot \int_{\mathbb{R}^n} |g(t)|^2\exp(-2\pi Y^{-1}[y+Y(\lambda+b)]),$$

where

$$t = 2\pi^{1/2}P^{-1/2}O^t(y+Y(\lambda+b)) .$$

On the other hand, from

$$-\frac{1}{2} E[t] = -2\pi Y^{-1}[y+Y(\lambda+b)]$$

and

$$dt_1\ldots dt_n = 2^n\pi^{n/2}(\det Y)^{-1/2}dy_1\ldots dy_n ,$$

$$\|g\|^2 = \delta(A)2^n\pi^{n/2}(\det Y)^{-1/2}\int_{\mathbb{R}^n} |g(t)|^2\exp(-2\pi Y^{-1}[y+Y(\lambda+b)])dy_1\ldots dy_n.$$

Therefore,

$$\|g\|^2 = \delta(A)^{-1}2^n\pi^{n/2}(\det Y)^{1/2}\exp(-2\pi Y[b+\lambda])\|w\theta\|^2 ,$$

completing the proof of the first half of the theorem.

The fact that if $b \neq b'$, then $\mathbb{K}_b$ and $\mathbb{K}_{b'}$ are mutually orthogonal follows easily from the equality

$$\int_{D(y)} e[(b+m)^t x]e[-(b'+m')^t x]\, dx_1\ldots dx_n = 0$$

for any $m, m' \in \mathbb{Z}^n$.

Lemma 2. *If* g *is a polynomial function on* t, *then obviously* g *belongs to* $\tilde{L}_2(\mathbf{R}^n)$ *and the series*

$$(\mathfrak{L}_b g)(u) = c_b w\{e[b^t u] g(2\pi^{1/2} P^{-1/2} 0^t (y + Y(\lambda + b)))\}$$

converges uniformly on any compact subset in $\mathbb{V}$. Moreover, its partial differentiation of any order can be performed term by term.

Proof. Take the partial derivative of a certain order of the m-th term of the series (12.4). Its absolute value is the form

$$|k(y,m)| \exp\left\{-2\pi\left(\frac{1}{2} Y[m] + b^t Ym + (b+m)^t y\right)\right\}$$

with a polynomial k of $2n$ variables $y_1, \ldots, y_n$, $m_1, \ldots, m_n$.

The matrix Y is a real positive definite symmetric matrix. Let ε be the smallest eigenvalue of Y. Let u be in a compact subset K in $\mathbb{V}$. Then the series obtained from the series (12.4) by taking term by term differentiation of a certain order is majored term by term by a convergent series

$$C_K \sum_{m \in \mathbb{Z}} n \exp\left(-\frac{1}{2} \pi\varepsilon E[m]\right)$$

where C_K is a constant depending on the compact set K. Thus we see the validity of the lemma.

13. First, we transfer the unitary representation $(\mathfrak{H}, U)$ in 6 to $(\mathfrak{K}, V)$ by the isomorphism $\iota : \mathfrak{H} \to \mathfrak{K}$ (10.2). We put

$$V_g = \iota U_g \iota^{-1} , \qquad g \in G .$$

From the definition of $(\mathfrak{H}, U)$ and from (10.2), it follows that

$$(V_{\{u,c\}}\theta)(v) = \theta(u+v) e\left[-F\left(v + \frac{1}{2} u, u\right) + \lambda(u)\right]$$

for $\{u,c\} \in G$ and $\theta \in \mathfrak{K}$.

For later use, we remark that the operator σ_ℓ, $\ell \in L$, defined by (10.5) and the operator $V_{\{u,c\}}$ are commutative, i.e., for any continuous function θ defined on $\mathbf{V}$,

$$V_{\{u,c\}} \sigma_\ell \theta = \sigma_\ell V_{\{u,c\}} \theta ,$$

as is easily checked. Indeed, this is nothing but the commutativity of the right and left regular representations of the group G. Immediately, it follows that

$$V_{\{u,c\}}w\theta = w\cdot V_{\{u,c\}}\theta ,$$

if $w\theta$ and the right hand side of the above equality are well defined.

As for the Lie algebra representation induced from $(\mathfrak{K},V)$,

$$\dot{V}_{\{e^*_\alpha,0\}} = \frac{\partial}{\partial x_\alpha} + 2\pi i(\Sigma_\beta Y^{-1}_{\alpha\beta}y_\beta + \lambda_\alpha) ,$$

$$\dot{V}_{\{Je^*_\alpha,0\}} = \frac{\partial}{\partial y_\alpha} - 2\pi(\Sigma_\beta Y^{-1}_{\alpha\beta}y_\beta + \lambda_\alpha) ,$$

$\alpha = 1,\ldots,n$, where $u = x+iy$, $x,y \in \mathbb{R}^n$.

From these, it follows that

$$\iota\mathfrak{a}_\alpha\iota^{-1} = \frac{\partial}{\partial u_\alpha} + 2\pi i(\Sigma_\beta Y^{-1}_{\alpha\beta}y_\beta + \lambda_\alpha) ,$$

$$\iota\mathfrak{b}_\alpha\iota^{-1} = \frac{\partial}{\partial \bar{u}_\alpha} .$$

The operator $\tilde{E} = \iota E\iota^{-1}$ corresponds to the complex Laplace-Beltrami operator in $L_2(\mathbb{T},\mathbb{F})$.

Lemma. *Put*

$$\tilde{\mathfrak{a}}_\alpha = \Sigma_\beta\ \rho_\alpha^{1/2} 0_{\beta\alpha}(\iota\mathfrak{a}_\beta\iota^{-1}) , \qquad (13.1)$$

and

$$\tilde{\mathfrak{b}}_\alpha = \Sigma_\beta\ \rho_\alpha^{1/2} 0_{\beta\alpha}(\iota\mathfrak{b}_\beta\iota^{-1}) , \qquad \alpha = 1,\ldots,n, \qquad (13.2)$$

where $0^tY0 = P = \mathrm{diag}(\rho_1,\ldots,\rho_n)$ (12.1).

Then, for any polynomial function $g(t) \in \tilde{L}_2(\mathbb{R}^n)$,

$$\mathfrak{L}_b\left(\sqrt{\pi}\ i\left(t_\alpha - \frac{\partial}{\partial t_\alpha}\right)g\right) = \tilde{\mathfrak{a}}_\alpha\mathfrak{L}_b g ,$$

and

$$\mathfrak{L}_b\left(\sqrt{\pi}\ i\ \frac{\partial}{\partial t_\alpha} g\right) = \tilde{\mathfrak{b}}_\alpha \mathfrak{L}_b\ g \quad .$$

Proof. Given a polynomial function $g(t)$, we denote by $\theta(g,u)$ tentatively the function defined by (12.3). By straightforward computation, we see easily that

$$\theta\left(\sqrt{\pi}\ i\left(t_\alpha - \frac{\partial}{\partial t_\alpha}\right) g, u\right) = \tilde{\mathfrak{a}}_\alpha \theta(g,u) \quad ,$$

and

$$\theta\left(\sqrt{\pi}\ i\ \frac{\partial}{\partial t_\alpha} g, u\right) = \tilde{\mathfrak{b}}_\alpha \theta(g,u) \quad .$$

Now, the commutativity of w with the operator $V_{\{u,c\}}$, $\{u,c\} \in G$, mentioned above, yields that of w with the operator $\tilde{\mathfrak{a}}_\alpha$ and $\tilde{\mathfrak{b}}_\alpha$. Thus, we obtain the result in question. The convergence and the differentiability of series involved are guaranteed by the second lemma in 12.

On account of (13.2) and (13.3), the operator $\tilde{E}$ corresponding to the complex Laplace-Beltrami operator on $L_2(\mathbb{T},\mathbb{F})$ is expressed as

$$\tilde{E} = -\Sigma_\alpha\ \tilde{\mathfrak{a}}_\alpha \tilde{\mathfrak{b}}_\alpha \quad .$$

By the above lemma, for any polynomial function $g(t) \in \tilde{L}_2(\mathbb{R}^n)$,

$$\mathfrak{L}_b\left(-\pi \sum_\alpha \left(\frac{\partial^2}{\partial t_\alpha^2} - t_\alpha \frac{\partial}{\partial t_\alpha}\right) g\right) = \tilde{E}\mathfrak{L}_b g \quad .$$

14. Theorem 4. *Let* $\mathfrak{L}_b$ *be the isomorphism of* $\tilde{L}_2(\mathbb{R}^n)$ *into* $\mathfrak{K}$ *defined by* (12.2), *and let* $\mathfrak{K}^{(b)}$ *be the image of* $\tilde{L}_2(\mathbb{R}^n)$ *under* $\mathfrak{L}_b$, $b \in \mathfrak{B}$. *Then, the Hilbert space* $\mathfrak{K}$ *is the direct sum of the* $\delta(A)$ *subspaces* $\mathfrak{K}^{(b)}$, $b \in \mathfrak{B}$.

Proof. Let $\mathfrak{K}_{\nu_0}$ be the eigenspace of the operator $\tilde{E}$ acting on $\mathfrak{K}$ whose eigenvalue is ν_0. From a remark made in 11 and from (13.3)

$$\dim \mathfrak{K}_{\nu_0} \cap \mathfrak{K}^{(b)} = \binom{n+\nu_0-1}{n-1} \quad .$$

By Lemma in 12,

$$\dim \mathfrak{K}_{\nu_0} \cap \left(\sum_{b\in\mathfrak{B}} \mathfrak{K}^{(b)} \right) = \delta(A) \binom{n+\nu_0-1}{n-1} .$$

On the other hand, by Theorem 1,

$$\dim \mathfrak{K}_{\nu_0} = \delta(A) \binom{n+\nu_0-1}{n-1} .$$

Thus, $\mathfrak{K}$ is the direct sum of the subspaces $\mathfrak{K}^{(b)}$, $b \in \mathfrak{B}$. We have finished the proof.

The results of the section are summarized in the following

Theorem 5. *Let* $H_{[\nu]}(t)$ *be the* $[\nu]$*-th Hermite polynomial in* $t \in \mathbb{R}^n$*, and let* w *be the operator defined by* (10.6). *Put*

$$\theta(u;b,[\nu]) = (\nu_1!\ldots\nu_n!)^{-1/2}(2\pi)^{-n/4}\mathfrak{L}_b H_{[\nu]} .$$

Then the function $\theta(u;b,[\nu])$ *is an eigenfunction of the operator* $\tilde{E}$ *on* $\mathfrak{K}$ *belonging to the eigenvalue* $\nu_0 = \nu_1 + \cdots + \nu_n$*, and has the expression*

$$\lambda(b,[\nu]) \sum_{m\in\mathbb{Z}^n} H_{[\nu]}(2\pi^{1/2}P^{-1/2}{}_0{}^t(y + Y(\lambda+b+m)))$$
$$e\left[\frac{1}{2} Z[\lambda+b+m] + (m+b)^t(u-\lambda')\right] ,$$

where

$$\lambda(b,[\nu]) = (\nu_1!\ldots\nu_n!)^{-1/2}2^{n/4}e\left[-\frac{1}{2}X[b+\lambda] + b^t\lambda'\right]\delta(A)^{-1/2}(\det Y)^{1/4}$$

$$\lambda' = (\lambda_{n+2}) \qquad (9.5) .$$

The set of the functions $\theta(u;b,[\nu])$, $b\in\mathfrak{B}$, $[\nu]\in p(\nu_0)$, $\nu_0 = 0,1,\ldots$, *forms an orthonormal basis of the Hilbert space* $\mathfrak{K}$. *Particularly,*

$$\theta(u;b) = \theta(u;b,[0]), \qquad b\in\mathfrak{B} ,$$

is a holomorphic theta series and

$$\theta(u;b,[\nu]) = (-i\sqrt{\pi})^{\nu_0} \nu_1!\ldots\nu_n!\,\tilde{\mathfrak{a}}_1^{\nu_1}\ldots\tilde{\mathfrak{a}}_n^{\nu_n}\,\theta(u,b) .$$

4. Acknowledgment

Partially supported by N.S.F. Grant 7701841.

5. References

[1] Auslander, L. and Tolimieri, R., "Abelian Harmonic Analysis, Theta Function and Functional Algebra on a Nilmanifold," Lecture Notes in Math. 436 (1975), Springer Verlag, Berlin, Heidelberg, New York.

[2] Beteman, H., *Beteman Manuscript Project, Table of Integral Transforms I*, McGraw-Hill, New York, 1954.

[3] Cartier, P., "Quantum mechanical commutation relations and theta functions," *Proc. Symp. Pure Math. IX, A.M.S.* (1966), 361-383.

[4] Dieudonné, J., *Éléments d'Analyse VII*, Gauthier-Villars, Paris, 1978.

[5] Hejhal, D.A., "The Selberg Trace Formula for $PSL(2,\mathbb{R})$ I," Lecture Notes in Math. 548 (1976), Springer Verlag, Berlin, Heidelberg, New York.

[6] Igusa, J., *Theta Functions*, Springer Verlag, Berlin, Heidelberg, New York, 1972.

[7] Kodaira, K., "On cohomology groups of compact analytic varieties with coefficients in some analytic faisceaux," *Proc. Nat. Acad. Sci. U.S.A.* 39 (1953), 865-868.

[8] Shimura, G., "On the derivatives of theta functions and modular forms," *Duke Math. J.* 44 (1977), 365-387.

[9] Tolimieri, R., "Heisenberg manifolds and theta functions," *Trans. A.M.S.* 239 (1978), 293-320.

[10] Weil, A., *Introduction à l'Étude des Variétés Kählériennes*, Hermann, Paris, 1958.

Washington University
St. Louis, Mo. 63130
USA

(Received September 30, 1980)

ON THE ORDERS OF THE AUTOMORPHISM GROUPS OF CERTAIN PROJECTIVE MANIFOLDS

Alan Howard and Andrew John Sommese

1. Summary of Results. It is a well-known theorem of Hurwitz that the automorphism group of a compact Riemann surface of genus $g > 1$ has order not larger than $84(g-1)$. This was generalized by Bochner who proved that a compact Riemannian manifold with negative Ricci tensor has a finite automorphism group, and Kobayashi who derived the same conclusion for a compact complex manifold with negative first Chern class [K]. The group of birational transformations was studied by Matsumura [M_1] who proved that it contains no one-parameter subgroup, provided the manifold has ample canonical bundle.

The first attempt to specify an upper bound on the order of the group was made, as far as we know, by Andreotti [A] who proved that the order of the (biholomorphic) automorphism group of a complex projective surface of general type has an upper bound which depends only on the Chern numbers of the surface. Our first theorem extends Andreotti's theorem to all dimensions, although with somewhat stronger hypotheses.

Theorem 1. *Let* M *be a compact complex manifold with ample canonical bundle. The order of its automorphism group is bounded above by a number which depends only on the Chern numbers of* M.

Unfortunately the method of proof sheds no light on the nature of the dependence, and, as Andreotti pointed out, the estimates one obtains for algebraic surfaces are far from sharp. (Indeed if one applies his technique to a nonhyperelliptic curve of genus g one obtains the upper bound $(6g-6)^{g^2+1}$ which is certainly less than optimal.) It would thus seem interesting to specify the nature of the upper bound and to ask, for example, if it is a linear or polynomial function of the Chern numbers, at least if one restricts the class of manifolds under consideration. A striking example was pointed out to us by Professor A. Borel[1] who observed that the following result is an easy consequence of the Kazdan-Margolis theorem [R_1, p. 177].

For each n *there is a constant* k_n *such that if* M *is any* n-*dimensional compact complex manifold whose universal cover is a bounded symmetric domain, then*

$$|\mathrm{Aut}(M)| \leqslant k_n e(M) \quad .$$

(Here and throughout this paper $|S|$ denotes the order of the finite set S and $e(M)$ denotes the Euler characteristic of M.)

To be explicit let D be the universal cover of M. It follows from the Kazdan-Margolis theorem that there is a lower bound ε for the volume of a fundamental region of any discrete subgroup of $\mathrm{Aut}(D)$. Hence one obtains $|\mathrm{Aut}(M)| < \mathrm{vol}(M)/\varepsilon$, and since the volume and Gauss-Bonnet forms are invariant on D they differ by a constant factor.

Another class of manifolds for which the upper bound has a reasonable form is that of complex projective hypersurfaces, and we can prove the following.

Theorem 2. *For each* $n \geqslant 3$ *there is a constant* k_n *such that for every nonsingular hypersurface of degree* d *in complex projective* n-*space*

$$|\mathrm{Aut}(M)| \leqslant k_n d^n$$

with the exception of the cases $d = 1,2$ *and* $(n,d) = (3,4)$.

Since the Euler characteristic $e(M)$ of a hypersurface in $\mathbb{P}^n$ is an n-th order polynomial in the degree we obtain

Corollary. *For each* $n \geqslant 3$ *there is a constant* b_n *such that if* M *is a hypersurface in* $\mathbb{P}^n$ *other than a quadric, a hyperplane, or a quartic surface in* $\mathbb{P}^3$,

$$|\mathrm{Aut}(M)| \leqslant b_n e(M) \quad .$$

It should be mentioned that Bott and Tate proved some time ago the existence of an upper bound depending only on n and d for the order of the group $\mathrm{Lin}(M)$ of automorphisms of $\mathbb{P}^n$ which preserve M (see [OS]). Moreover, according to a theorem of Matsumura and Monsky [MM], $\mathrm{Lin}(M) = \mathrm{Aut}(M)$ provided $n \geqslant 3$, $d \neq 1,2$, and $(n,d) \neq (3,4)$. Thus our theorem strengthens the Bott-Tate result by giving a more explicit description of the upper bound. That the order of growth given by Theorem 2 is optimal can be seen by the example of the so-called Fermat hypersurface $M = \{z \mid z_0^d + \cdots + z_n^d = 0\}$. The group $\mathrm{Lin}(M)$ contains the symmetric group S_{n+1} of all permutations of the coordinates as well as the group of all diagonal matrices with d-th roots of unity along the diagonal (modulo the center). In fact an easy induction argument, using the Hessian hypersurface, shows that $\mathrm{Lin}(M)$ is the semi-direct product of these two subgroups, and thus has order $(n+1)!d^n$.

In the last section of this paper we consider smooth algebraic surfaces of general type and derive an estimate of the form $|\mathrm{Aut}(M)| \leqslant k \cdot e(M)^4$ with universal constant k, valid in the following cases: first, if the geometric genus is zero or if the canonical map is not a holomorphic immersion (Theorem 4.6), and, second, if the first Betti number is not zero and the Albanese map is not an immersion (Theorem 4.7).

2. Proof of Theorem 1. The proof is essentially that of Andreotti combined with the previously mentioned result of Kobayashi and the so-called "big theorem" of Matsusaka [see, e.g., LM].

We first recall the notion and some basic properties of the dual of a projective variety. If X is a subvariety of $\mathbb{P}^m$ we consider the set of all points $y \in \check{\mathbb{P}}^m$ of the dual projective space which contain the projective tangent space to X at some regular point. The Zariski closure of this set is the *dual variety* $\check{X}$. The following facts are well-known but we include a proof of (ii) for convenience. (For a different proof of (ii) see [GH, Section 3]. A proof of (i) is in [M_3].)

Lemma 1.1. (i) $\check{\check{X}} = X$.

(ii) *If* X *is nonsingular and has any plurigenus positive then* $\operatorname{codim} \check{X} = 1$.

Proof. Let $E = \{(x,y) \in X \times \check{X} : y$ contains the projective tangent space to X at $x\}$, and let π and $\check{\pi}$ be the projections of E to the first and second factors respectively. Then $\check{\pi}^{-1}(\check{X}_{reg})$ is a $\mathbf{P}^k$-bundle over the set of regular points of $\check{X}$, where $k+1$ is the codimension of $\check{X}$. Moreover, the fiber over any $y \in \check{X}_{reg}$ consists of all hyperplanes of $\check{\mathbb{P}}^m$ containing the projective tangent space to $\check{X}$ at y, and is imbedded into X by π. Letting $n = \dim X$, it follows that there is an $(n-k)$-polydisc Δ imbedded in $\check{X}_{reg}$ such that π maps $\check{\pi}^{-1}(\Delta)$ onto an open subset of X. Hence there must be a pluri-canonical section on $\check{\pi}^{-1}(\Delta)$, but this can only occur if $k = 0$. ∎

Given M with ample canonical bundle there is some p such that the pluricanonical bundle K^p imbeds M into $\mathbb{P}^m$ with $m+1 = \dim H^0(M,K^p)$. From Matsusaka's big theorem it follows that the smallest such p depends only on the Chern numbers of M, and by applying the Hirzebruch-Riemann-Roch and Kodaira vanishing theorems, one obtains the same conclusion about m. Since M is imbedded by pluricanonical sections, every automorphism extends to an M-preserving automorphism of $\mathbb{P}^m$, and hence to an $\check{M}$-preserving automorphism of $\check{\mathbb{P}}^m$. By (i) of Lemma 1.1 this homomorphism of $\mathrm{Aut}(M)$ into $\mathrm{Lin}(\check{M})$ is an isomorphism.

To put an upper bound on $|\mathrm{Lin}(\check{M})|$ we proceed as follows. Represent an element of $\mathrm{Lin}(\check{M})$ as an $(m+1)$ by $(m+1)$ matrix, and think of the entries of A as homogeneous coordinates of a point in $\mathbb{P}^r$ where $r = (m+1)^2 - 1$. Writing $\check{M}$ as $\{x \in \mathbb{P}^m | F(x) = 0\}$ for some homogeneous polynomial F, we see that for each fixed $x \in \check{M}$ the entries of A satisfy the equation $F(Ax) = 0$ which defines a hypersurface $H(x)$ in $\mathbb{P}^r$, and thus $\mathrm{Lin}(\check{M}) = \cap_x H(x)$. By Kobayashi's theorem, $\mathrm{Aut}(M)$, and hence $\mathrm{Lin}(\check{M})$ is finite. Thus we may choose $x_1,\ldots,x_r$ in M so that $H(x_k)$ does not contain any irreducible component of $H(x_1) \cap \ldots \cap H(x_{k-1})$, and $\mathrm{Lin}(\check{M}) = \cap_{i=1}^r H(x_i)$. Since $\deg H(x_i) = \deg \check{M}$, we obtain from Bezout's theorem

$$|\mathrm{Aut}(M)| = |\mathrm{Lin}(\check{M})| \leqslant (\deg \check{M})^{(m+1)^2-1} .$$

The proof is then concluded by expressing $\deg \check{M}$ in terms of the Chern numbers of M, which is achieved by applying the following lemma to the case $X = M$ and $L = K^p$.

Lemma 1.2. *If* X *is an* n*-dimensional projective manifold imbedded in* $\mathbb{P}^m$ *by the line bundle* L, *and if* $\operatorname{codim} \check{X} = 1$, *then*

$$\deg \check{X} = \sum_{j=0}^{n} (-1)^{n+j}(1+j)c(L)^j c_{n-j}(X) .$$

This formula is a special case of Corollaire 5.6, exposé XVII of [DK]. An elementary proof (over the complex field) can be given in at least three ways: by counting the number of double points of a Lefschetz pencil, or by Morse theory as in the proof of the second Lefschetz theorem, or by noting that $\deg \check{X}$ is the highest Chern number of the first holomorphic jet bundle of L.

3. Proof of Theorem 2. As we remarked in the last section, $\mathrm{Aut}(M)$ can be identified with the subgroup $\mathrm{Lin}(M)$ of projective linear transformations which preserve M, and this group is known to be finite [MM]. We will make use of the following well-known theorem of Jordan [J]: *for each* r *there is a number* c_r *with the property that any finite subgroup of* $GL(r,\mathbb{C})$ *has an abelian normal subgroup of index* $\leq c_r$. In our case we take the pull-back of $\mathrm{Aut}(M)$ to $SL(n+1,\mathbb{C})$, pass to an abelian subgroup $\tilde{G}$ of index $\leq c_{n+1}$, and then project to G. Thus we obtain an abelian subgroup G of $\mathrm{Aut}(M)$ of index $\leq c_{n+1}$, and it suffices to obtain an upper bound for $|G|$. Moreover, since $\tilde{G}$ can be simultaneously diagonalized, there is a set of points $p_i \in \mathbb{P}^n$, $i = 0,\dots,n$, in general position, each of which is a fixed point for G. We will write $\langle p_{i_0},\dots,p_{i_t}\rangle$ to denote the projective linear subspace spanned by $p_{i_0},\dots,p_{i_t}$, and refer to it as a t-simplex, using the terms vertex and edge for the cases $t=0$ and 1 respectively. We shall have repeated occasion to use the following elementary counting principle: *if a finite group* H *acts on a set* S *and* H_x *is the isotropy subgroup of* x, *then* $|H| = |H_x|\cdot|\mathrm{orbit}(x)|$.

Lemma 3.1. *Suppose there is an edge which does not lie in* M *and which meets* M *at some point which is not a vertex. Then* $|G| \leq d^n$.

Proof. We may assume that the edge in question is $\langle p_0,p_1\rangle$ and choose $q_1 \in M \cap \langle p_0,p_1\rangle$ with $q_1 \neq p_0,p_1$. Let K_1 be the subgroup of G which leaves $\langle p_0,p_1\rangle$ pointwise fixed. Since the quotient $H_1 = G/K_1$ acts effectively on $\langle p_0,p_1\rangle$ and fixes p_0 and p_1, it follows that the isotropy subgroup of q_1 in H_1 is $\{1\}$. But H_1 also acts effectively on the finite set $M \cap \langle p_0,p_1\rangle$, so by the counting principle we get

$$|H_1| \;=\; |M \cap \langle p_0,p_1\rangle| \leq d \qquad .$$

Now let K_2 be the subgroup of K_1 which leaves $\langle p_0,p_1,p_2\rangle$ pointwise fixed, and let $H_2 = K_1/K_2$. Since $\langle p_0,p_1\rangle$ does not lie in M we can find a point $z \in \langle p_0,p_1\rangle$, $z \neq p_0,p_1$, such that the line L through p_2 and z does not lie in M and meets M in some point $q_2 \neq p_2,z$. Since H_2 fixes p_2 and every point of $\langle p_0,p_1\rangle$, it follows that the isotropy subgroup of q_2 in H_2 is $\{1\}$, and since H_2 acts effectively on the finite set $M \cap L$, the counting principle gives

$|H_2| = |M \cap L| \leqslant d$, so that $|G/K_2| \leqslant d^2$. Proceeding inductively we arrive at the lemma.

We may thus assume that every edge $<p_i, p_j>$ either lies entirely in M or else meets M only in one or both of the vertices p_i and p_j. By a change of coordinates we may assume that the point p_k is given by $z_i = 0$ for $i \neq k$. Let $F(z_0, \ldots, z_n)$ be the homogeneous polynomial defining M.

Obviously some of the points p_j must lie on M. Choose one and assume that it is p_0. Since there are n distinct edges $<p_0, p_i>$ and M has dimension $n-1$, at least one such edge, say $<p_0, p_1>$, must be transverse to M at p_0. On the other hand it meets M at no points other than p_0 and p_1, and therefore must be tangent to M at p_1. It follows that $F(z_0, z_1, 0, \ldots, 0) = A_0 z_0^{d-1} z_1$. Now consider the edges $<p_1, p_i>$. As before one of them must be transverse to M at p_1 (from which we can see that $i \neq 0$), and we may assume $i = 2$. The edge $<p_1, p_2>$ is the tangent to M at p_2, so that $F(0, z_1, z_2, 0, \ldots, 0) = A_1 z_1^{d-1} z_2$. We continue in this manner until we have a chain $p_0, \ldots, p_k$ with $2 \leqslant k \leqslant n$ such that $<p_i, p_{i+1}>$ is transverse to M at p_i and tangent at p_{i+1} for $0 \leqslant i \leqslant k-1$, while $<p_k, p_0>$ is transverse to M at p_k and tangent at p_0. We may thus write

$$F(z_0, \ldots, z_n) = \sum_{i=0}^{k-1} A_i z_i^{d-1} z_{i+1} + A_k z_k^{d-1} z_0$$

$$+ \text{ other monomials with each } A_j \neq 0.$$

Any element of the group G can be represented by a matrix of the form

$$\begin{pmatrix} 1 & & & \\ & \rho_1 & & 0 \\ & & \ddots & \\ 0 & & & \rho_n \end{pmatrix},$$

and the relation $F(z_0, \rho_1 z_1, \ldots, \rho_n z_n) = cF(z_0, z_1, \ldots, z_n)$ for some $c \neq 0$ easily yields the set of equations

$$\rho_{m+1} = \rho_1^{\lambda_m} \qquad 1 \leqslant m \leqslant k-1$$

$$1 = \rho_1^{\lambda_k}$$

where

$$\lambda_m = \sum_{i=0}^{m} (1-d)^i .$$

It follows that there is a subgroup K_1 of G which leaves every point of the simplex $\langle p_0,\ldots,p_k\rangle$ fixed and whose index is at most $|\lambda_k|$.

From here the proof proceeds more or less as that of Lemma 3.1. Our assumption assures that there is a point $z \in \langle p_0,\ldots,p_k\rangle$, not lying in any (k-1)-simplex, such that the line L through p_{k+1} and z does not lie in M and meets M in some point $q \notin \langle p_0,\ldots,p_k\rangle$, $q \neq p_{k+1}$. If we let K_2 be the subgroup of K_1 leaving every point of $\langle p_0,\ldots,p_{k+1}\rangle$ fixed, then the quotient group $H_2 = K_1/K_2$ acts effectively on the finite set $M \cap L$, and the isotropy group of q in H_2 is $\{1\}$. Hence $|H_2| \leqslant d$. Continuing in this way we arrive at the estimate $|G| \leqslant d^{n-k}|\lambda_k| \leqslant d^n$, and the proof is complete. ■

4. Surfaces of General Type. Throughout this section M will denote a *smooth surface of general type*, and we use freely the standard results concerning such surfaces [see B or S]. We will need the following lemma which, although perhaps familiar to some experts, we cannot find a proof for in the literature. (A proof may also be constructed from the techniques of [T].)

Lemma 4.1. *Given any holomorphic map* $\pi : M \to \mathbb{P}^1$ *there must be at least three singular fibers.*

Proof. Since M is of general type, a generic fiber of π is neither rational nor elliptic. Suppose that there are less than three singular fibers. Letting W denote M minus the singular fibers and $D = \pi(W)$, it follows that $D = \mathbb{C}$ or $\mathbb{C}^*$, so that the period map is constant on D. By passing, if necessary, to a finite cover $\mathbb{C}^* \to \mathbb{C}^*$, we may assume that there is no monodromy and that a generic fiber is irreducible [G]. From the Leray spectral sequence for the sheaf $\mathcal{O}_W$ one easily sees that the restriction map $H^1(W,\mathcal{O}_W) \to H^1(F,\mathcal{O}_F)$ is surjective, where F is a general fiber of π. By a theorem of Deligne [D, pp. 39-40] it follows that $H^1(M,\mathcal{O}_M) \to H^1(F,\mathcal{O}_F)$ is surjective. A standard argument using the Ramanujam vanishing theorem $[R_2]$

shows that $H^1(M,\mathcal{O}_M) \to H^1(F,\mathcal{O}_F)$ is also injective.

We next show that $H^2(M,\mathcal{O}_M) = 0$. From the Leray spectral sequence $H^2(M,\mathcal{O}_M) \cong H^1(\mathbb{P}^1, p_{(1)}\mathcal{O}_M)$. On the other hand $p_{(1)}\mathcal{O}_M$ is coherent, and since it has no torsion elements, it is locally free. (Any torsion element would have isolated support in contradiction to the isomorphism $H^1(M,\mathcal{O}_M) \overset{\cong}{\to} H^1(F,\mathcal{O}_F)$ just established.) Hence $p_{(1)}\mathcal{O}_M$ is the direct sum of line bundles on $\mathbb{P}^1$ each of which is easily seen, again using the above isomorphism, to be trivial.

From $H^2(M,\mathcal{O}_M) = 0$ it follows that M is mapped, via the Albanese mapping to a (nonsingular) curve $C \subset \mathrm{Alb}(M)$, and since $\dim \mathrm{Alb}(F) = \dim \mathrm{Alb}(M)$ we see that $F \cong C$. By taking the product of π with the Albanese mapping on M one obtains a birational map of M onto $\mathbb{P}^1 \times F$, a contradiction which completes the proof.

In studying the order of $\mathrm{Aut}(M)$ our strategy is to find an abelian subgroup whose index is bounded by a power of the Euler number $e(M)$. For an abelian group of automorphisms we have the following theorem and corollary.

Theorem 4.2. *Let* G *be an abelian subgroup of* $\mathrm{Aut}(M)$, *and suppose that the action of* G *lifts to an action on a line bundle* L *over* M. *Assume further that* (i) $\dim H^0(L) \geq 2$, (ii) L^N *is spanned for some* $N > 0$, *and* (iii) $K \cdot D \geqslant 0$ *where* K *is the canonical bundle of* M *and* D *is the base curve of the divisors of* L. *Then the order of* G *satisfies*

$$|G| \leqslant 42(2K\cdot L + 3L\cdot L + 3e(M))(L\cdot L + K\cdot L) \quad .$$

Proof. We may assume that L has no base curve. Otherwise write $L = D + E$ where E is free of base curves. Since L^N is spanned we get

$$L\cdot L = L\cdot D + L\cdot E \geqslant L\cdot E = D\cdot E + E\cdot E \geqslant E\cdot E$$

and, since $K\cdot D \geqslant 0$, $K\cdot L \geqslant K\cdot E$. Hence we may replace L by E and obtain an estimate for $|G|$ at least as good as the one desired.

Since G is abelian we can find a G-invariant pencil of sections of L, and by blowing up we obtain an equivariant holomorphic map $\pi: \tilde{M} \to \mathbb{P}^1$ with $\tilde{M}$ birationally equivalent to M. Since L has no base curve, a standard argument yields $e(\tilde{M}) \leqslant e(M) + L\cdot L$.

Let r be the number of singular fibers of π. We obtain that

$$3 \leqslant r \leqslant 3L \cdot L + 2K \cdot L + 3e(M) \ .$$

The lower bound comes from Lemma 4.1, and to obtain the upper bound we let $F_1, \ldots, F_r$ be the singular fibers and F be a general fiber. Then

$$\begin{aligned} L \cdot L + e(M) &\geqslant e(\tilde{M}) \\ &= (2 - r)e(F) + \sum_{i=1}^{r} e(F_i) \\ &= e(F_1) + e(F_2) + \sum_{i=3}^{r} [e(F_1) - e(F)] \\ &\geqslant e(F_1) + e(F_2) + (r - 2) \end{aligned}$$

since $e(F_i) > e(F)$ [S, p. 60, Theorem 7]. Hence

$$\begin{aligned} r &\leqslant 2 + L \cdot L + e(M) - e(F_1) - e(F_2) \\ &\leqslant 2 + L \cdot L + e(M) - 2e(F) - 2 \\ &= L \cdot L + e(M) - 2(2 - 2g(F)) \\ &= 3L \cdot L + 2K \cdot L + e(M) \end{aligned}$$

from the adjunction formula.

To complete the proof we observe that G is a finite abelian group acting on $\mathbb{P}^1$, and thus has two fixed points which we may take to be 0 and ∞. Since $r \geqslant 3$ there is at least one singular fiber, say F_1, with $\pi(F_1) \neq 0, \infty$, and the subgroup of G which fixes the point $\pi(F_1)$ (and hence all of $\mathbb{P}^1$) has index $\leqslant r$. By the Hurwitz theorem the order of this subgroup can be at most $84(g(F) - 1) = 42(L \cdot L + K \cdot L)$ since no nontrivial automorphism can act as the identity on both the base and a nonsingular fiber. (We remark that $g(F) > 1$ since M is of general type.) Combining this estimate with the upper bound on r completes the proof. ∎

<u>Corollary 4.3</u>. *Assume that* M *is a minimal model and that* $p_g \geqslant 2$. *If* G *is an abelian subgroup of* $\mathrm{Aut}(M)$ *then*

$$|G| \leqslant c \cdot e(M)^2$$

where c *is a universal constant.*

Proof. Apply Theorem 4.2 to the special case $L = K$, and use Miyaoka's inequality $K \cdot K \leqslant 3e(M)$ [M_2]. ∎

We will need the following elementary lemma.

Lemma 4.4. *Let* G *be a finite group acting effectively on a complex* n*-dimensional manifold with a fixed point. Then* G *contains an abelian subgroup whose index is bounded by a number depending only on* n.

Proof. Since G is finite there exists a complete G-invariant hermitian metric on the manifold. Hence the isotropy representation at the fixed point is faithful, and we apply the previously cited theorem of Jordan. ∎

Next we consider the case when M is not minimal.

Theorem 4.5. *If* M *is of general type and is not a minimal model then*

$$|\mathrm{Aut}(M)| \leqslant c \cdot e(M)^3$$

where c *is a universal constant.*

Proof. We use the following known result whose proof is implicit in [H, Section 4.4]: *if a smooth algebraic surface* S *contains two exceptional curves (of the first kind)* E_1 *and* E_2 *with* $E_1 \cap E_2 \neq \emptyset$ *then all plurigenera of* S *vanish.* Since M is of general type we thus obtain

$$\begin{aligned}
&(\text{number of exceptional curves in } M) \leqslant h^{1,1}(M) \\
&\quad \leqslant e(M) + 4h^{1,0}(M) \\
&\quad \leqslant e(M) + 4p_g && (\text{since } \chi(\mathcal{O}_M) > 0) \\
&\quad \leqslant e(M) + 2K \cdot K + 8 && (\text{Noether's inequality}) \\
&\quad \leqslant 7e(M) + 8 && (\text{Miyaoka's inequality}) \\
&\quad \leqslant 15e(M) && (\text{since } e(M) > 0) \quad .
\end{aligned}$$

Letting E denote one of the exceptional curves and G the subgroup of $\mathrm{Aut}(M)$ stabilizing E, we see that the index of G is bounded by $15e(M)$. Furthermore, letting $\pi: M \to M'$ be the map onto the minimal model, the group G is mapped injectively into the group of automorphisms of M' fixing the point $\pi(E)$. For from the above-stated fact it follows that π is the composition of a sequence of maps $M = M_0 \to M_1 \to \cdots \to M_n = M'$ where each map $M_i \to M_{i+1}$ is achieved by blowing down all exceptional curves simultaneously. Thus every automorphism of M_i induces a unique automorphism of M_{i+1}. The theorem now follows from Corollary 4.3 and Lemma 4.4.

We now prove the first main result of this section.

Theorem 4.6. *Let* M *be a smooth complex projective surface of general type, and assume that the canonical map is not a holomorphic immersion. Then*

$$|\mathrm{Aut}(M)| \leqslant c \cdot e(M)^4$$

where c *is a universal constant.*

Proof. By Theorem 4.5 we may assume that M is minimal. We consider several cases.

Case 1. $p_g = 0$. In this case $c_1^2 + c_2 \leqslant 12$, and since c_1^2 and c_2 are both positive, there are only a finite number of possible pairs c_1^2, c_2. It follows from Theorem 1 that we can choose a single upper bound $c \geqslant |\mathrm{Aut}(M)|$ for all M in this class, and we trivially obtain $|\mathrm{Aut}(M)| \leqslant c \cdot e(M)^4$.

Case 2. $p_g \neq 0$ but K is not spanned. If K has no base curve then there are isolated base points $p_1, \ldots, p_t$ where $t \leqslant K \cdot K$. Since $\mathrm{Aut}(M)$ permutes the base points of K, the isotropy group of p_1 must have index at most t, and by Lemma 4.4 it has an abelian subgroup whose index is bounded by a universal constant. We thus obtain from Corollary 4.3 and Miyaoka's inequality

$$|\mathrm{Aut}(M)| \leqslant cte(M)^2 \leqslant 3ce(M)^3 \qquad .$$

If K has a base curve we write it in the form $\Sigma\, C_i$ where each C_i is irreducible. We claim that there are at most $18 \cdot e(M)$ such C_i. For since $\Sigma\, KC_i \leqslant K^2 \leqslant 3e(M)$ it follows that there are at most $3e(M)$ components satisfying $KC_i > 0$. On the other hand if $KC_i = 0$

then $C_i \cdot C_i = -2$, and the number of such components cannot exceed $h^{1,1} - 1$. As in the proof of Theorem 4.5 we obtain an upper bound of $15e(M)$ for this number, and the assertion follows by combining these two estimates. It follows that there is a subgroup $G \subset \mathrm{Aut}(M)$ of index at most $18e(M)$ which leaves C_1 invariant.

If the nonsingular model $\tilde{C}_1$ has genus > 1, then the kernel of the restriction $G|C_1$ has

$$\text{index} \leqslant 84(g(\tilde{C}_1) - 1) \leqslant 42(KC_1 + C_1^2) \leqslant 84K^2 \leqslant 252e(M) .$$

From Lemma 4.4 we see that this kernel has an abelian subgroup whose index is bounded by a universal constant. Combining these as before yields $|\mathrm{Aut}(M)| \leqslant c \cdot e(M)^4$.

If C_1 is rational then $G|C_1$ has a subgroup of universally bounded index which leaves a point fixed, and, therefore, so does G. Arguing as before we get $|\mathrm{Aut}(M)| \leqslant c \cdot e(M)^3$.

If C_1 is elliptic then $G|C_1$ has an abelian subgroup of index $\leqslant 6$ which acts on $H^0(C_1, K|C_1)$. By choosing a simultaneous eigenvector for this action we obtain a divisor on C_1 of degree $\leqslant KC_1 \leqslant K^2$. Thus we obtain a subgroup of G of index $\leqslant 6K^2$ which fixes a point of C_1. As before we thus get $|\mathrm{Aut}(M)| \leqslant c \cdot e(M)^4$.

Case 3. K is spanned. Since M is minimal, the holomorphic mapping associated to K is generically, but not everywhere, an immersion. Hence the first jet bundle $J_1(K)$ is generically, but not everywhere, spanned. Thus the homomorphism $\Lambda^3 H^0(J_1(K)) \to H^0(\Lambda^3 J_1(K)) = H^0(K^4)$ has as image a nontrivial $\mathrm{Aut}(M)$-invariant subspace with non-empty base locus. We may then apply the arguments of cases 1 and 2 to this base locus to obtain estimates of the same type differing only in the constant factor from those already obtained. So for all cases we have obtained $|\mathrm{Aut}(M)| \leqslant c \cdot e(M)^4$ with universal constant c, and the proof is complete.

Our second main result in this section is

Theorem 4.7. *Suppose* M *is a complex projective surface of general type with* $b_1(M) \neq 0$. *If the Albanese map* $\alpha: M \to \mathrm{Alb}(M)$ *is not an immersion, then* $|\mathrm{Aut}(M)| \leqslant c \cdot e(M)^4$ *where* c *is a universal constant.*

Proof. If α is generically an immersion, i.e., $\dim \alpha(M) = 2$, we consider the linear system of holomorphic 2-forms spanned by $\phi_i \wedge \phi_j$ where the ϕ_i's form a basis for the holomorphic 1-forms. Since α is not everywhere an immersion this system must have base points (or a base curve), and we can use the argument of Case 2 in the proof of Theorem 4.6 to obtain the desired bound.

Next suppose that $\dim \alpha(M) = 1$, so that $\alpha(M)$ is a nonsingular curve in $\mathrm{Alb}(M)$, and also suppose that the genus of $\alpha(M)$ is greater than one. Since M is of general type the genus of a generic fiber F is also greater than one, and we obtain

$$|\mathrm{Aut}(M)| \leqslant (42)^2 e(\alpha(M)) \cdot e(F) \leqslant (42)^2 e(M) \quad .$$

Finally if $\dim \alpha(M) = \text{genus of } \alpha(M) = 1$, i.e., $h^{1,0} = 1$, let ϕ be the unique (up to scalar multiple) holomorphic 1-form on M. The zero set of ϕ is non-empty, for otherwise M would be differentiably fibered over a torus in which case $e(M) = 0$ and M would not be of general type. If the zeroes of ϕ are isolated there can be at most $e(M)$ of them, so that there is a subgroup G of $\mathrm{Aut}(M)$ of index at most $e(M)$ which fixes one of these points. We then proceed as in the proof of Theorem 4.6. If, on the other hand, ϕ has a nonisolated zero set, let D be the divisor on which ϕ is zero (counted with multiplicities). Then by [M_2, Prop. 1] we see that $D \cdot K \leqslant e(M)$ so that, as in the proof of Theorem 4.6, there are at most $16e(M)$ components of D. (We again assume that M is minimal.) For any such component D_1 we have $D_1 \cdot D_1 \leqslant 0$ so the genus of the nonsingular model satisfies

$$g(\tilde{D}_1) \leqslant KD_1 \leqslant e(M) \quad .$$

The proof then proceeds exactly as that of Case 2 of Theorem 4.6.

5. References

[A] Andreotti, A., "Sopra le superficie che possegono trasformazioni birazionali in se," Univ. Roma Ist. Naz. Alta Math., Red. Mat. e Appl. 9 (1950), 255-279.

[B] Bombieri, E., "Canonical models of surfaces of general type," *Publ. Math. IHES* 42 (1973), 171-219.

[D] Deligne, P., "Théorie de Hodge II, *Publ. Math. IHES* 40 (1971), 5-57.

[DK] Deligne, P. and Katz, N., *Groupes de Monodromie en Géométrie Algébrique (SGA 7 II)*, Springer Lecture Notes 370, Springer Verlag, Heidelberg, 1973.

[G] Griffiths, P., "Periods of integrals on algebraic manifolds III," *Publ. Math. IHES* 38 (1970), 125-180.

[GH] Griffiths, P. and Harris, J., "Algebraic geometry and local differential geometry," *Ann. Sci. École Norm. Sup.* 4e série t. 12 (1979), 355-432.

[H] Hirzebruch, F., "Hilbert modular surfaces," *Enseignement Math.* 19 (1973), 183-281.

[J] Jordan, C., "Memoire sur les equations differentielles linéaires à integrales algebriques," *J. Reine Angew. Math.* 84 (1878), 89-215.

[K] Kobayashi, S., *Transformation Groups in Differential Geometry*, Ergebnisse 70, Springer Verlag, Heidelberg, 1972.

[LM] Lieberman, D. and Mumford, D., "Matsusaka's big theorem," *AMS Proceedings of Symposia in Pure Mathematics* 29 (1975), 513-530.

[M_1] Matsumura, H., "On algebraic groups of birational transformations," *Atti Accad. Naz. dei Lincei* 34 (1963), 151-155.

[M_2] Miyaoka, Y., "On the Chern numbers of surfaces of general type," *Inv. Math.* 42 (1977), 225-237.

[M_3] Moishezon, B.G., *Complex Surfaces and Connected Sums of Complex Projective Planes*, Springer Lecture Notes 603, Springer Verlag, Heidelberg, 1977.

[MM] Matsumura, H. and Monsky, P., "On the automorphisms of hypersurfaces," *J. Math. Kyoto Univ.* 3 (1964), 347-361.

[OS] Orlick, P. and Solomon, L., "Singularities II: Automorphisms of forms," *Math. Ann.* 231 (1978), 229-240.

[R_1] Ragunathan, M.S., *Discrete Subgroups of Lie Groups*, Ergebnisse 68, Springer Verlag, Heidelberg, 1972.

[R_2] Ramanujam, C.P., "Remarks on the Kodaira vanishing theorem, *J. Indian Math. Soc.* 39 (1972), 41-51.

[S] Shafarevich, I.R., *Algebraic Surfaces*, Proceedings of the Steklov Institute 75, American Mathematical Society, Providence, 1967.

[T] Tu, L.W., "Variation of Hodge structure and the local Torelli problem," Harvard University Thesis, 1979.

University of Notre Dame
Notre Dame, Indiana 46556, U.S.A.

6. Notes

1. Appropriately to this paper this occurred at a conference in honor of Professor Matsushima.

(Received February 3, 1981)

HOMOGENEOUS SPACES FROM A COMPLEX ANALYTIC VIEWPOINT

A.T. Huckleberry and E. Oeljeklaus

0. Manifolds having many automorphisms play a fundamental role in geometry. If X is a *compact* complex manifold, then the group $Aut(X)$ of holomorphic automorphisms of X is, when equipped with the compact-open topology, a complex Lie group. If $G = Aut(X)$, then an orbit $G(p)$, $p \in X$, may be holomorphically identified with the quotient manifold G/H, where $H := \{g \in G | g(p) = p\}$ is the *isotropy group* of the G-action at p. Thus, studying quotients G/H of a complex Lie group G by a closed subgroup H becomes relevant. A natural first step is to analyze the structure of compact *homogeneous* spaces $X = G/H$. In the early 1950's, with development of Lie theory along with the theory of algebraic groups, a number of very sharp results were obtained by algebraic methods. The works of Borel (e.g., [10]), Goto [22], Tits [64] and Wang [67] are typical of this direction. Later, but still in an algebraic geometry spirit, many general methods were developed (e.g., see the papers of Hochschild, Mostow, Rosenlicht *et al.*).

Beginning in the late 1950's, Matsushima was one of the first to blend the newly developed methods of several complex variables with the Lie theory. We find this "complex analytic viewpoint" appealing, and have been strongly influenced by Matsushima's work.[1] Our first objective here is to outline some of his contributions to this area, particularly those which emphasize the above-mentioned "viewpoint". We then summarize some recent developments along these lines. Finally, we prove three theorems about complex homogeneous spaces. These concern solv-manifolds with many holomorphic functions, rationality of almost homogeneous Kähler manifolds with $b_1 = 0$, and G-equivariant compactifications of $\mathbb{C}^n = X = G/H$.

1. Some Structure Theorems of Matsushima

Let G be a connected, complex Lie group, and H a closed complex subgroup. We consider the (not necessarily compact) homogeneous space $X = G/H$. Due to the complexity of the situation, it is reasonable to restrict our attention to particular classes of groups. The importance of the class of *reductive* groups is strongly underlined by Matsushima's work. From the complex analytic viewpoint, G is reductive if its maximal compact subgroup K (uniquely determined up to conjugacy) is a totally real submanifold of G, or equivalently $\mathfrak{k} \cap \sqrt{-1}\mathfrak{k} = \{0\}$, and there is no proper complex subgroup of G which contains K. Thus G is the complexification $K^{\mathbb{C}}$ of K. More generally we call H a reductive group if H^o is reductive and H/H^o is finite.

We begin by considering $X = G/H$ where G is reductive and X has many analytic functions, the latter being made precise by assuming that if $\dim_{\mathbb{C}} X = n$, then there are analytically independent functions $f_1, \ldots, f_n \in \mathcal{O}(X)$ (i.e., $df_1 \wedge \ldots \wedge df_n \not\equiv 0$). Since $\mathbb{C}^n \setminus \{0\}$ under the action of SL_n is an example of such an X, the best we could hope to prove is that such a manifold is "quasi-affine," equivariantly embedded as a Zariski open set in an affine variety Y. Under slightly stronger assumptions, results of this type can be found in [8], [30].

If G acts holomorphically on X, then one has the induced action on the Fréchet space $\mathcal{O}(X)$. If G is reductive, then a theorem of Harish-Chandra [27] guarantees the existence of a dense subset $A \subset \mathcal{O}(X)$ so that the orbit $G(f)$ in $\mathcal{O}(X)$ is *finite dimensional* for all $f \in A$. Simultaneously approximating $f_1, \ldots, f_n$ given above, we obtain a finite dimensional G-invariant vector space $V = \langle\langle g_1, \ldots, g_N \rangle\rangle_{\mathbb{C}} \subset \mathcal{O}(X)$ so that the map $F := (g_1, \ldots, g_N) X \to \mathbb{C}^N$ has rank n at some point, and is G-equivariant into GL_N. The equivariance and transitivity of G on X implies that $\operatorname{rank}_x F = n$ for all $x \in X$. Since G is reductive, the induced representation of G is algebraic [29], and the image $F(X) = Y$ is Zariski open in its G-stable, affine-algebraic closure $\tilde{Y}$. Now, since the isotropy group I is algebraic, the fibration $F: X = G/H \to Y = G/I$ has finite fiber. Stein [63] derived the existence of an irreducible complex space $\tilde{X}$ which contains X as the complement of a closed thin analytic set, and an extension $\tilde{F}: \tilde{X} \to \tilde{Y}$ of F as a (ramified) covering map. Therefore $\tilde{X}$ is Stein and X is holomorphically separable. Approximating an element $f \in \mathcal{O}(X)$ which separates the points in some fiber of F with an element of A, we may enlarge V so that the corresponding map F is injective. Thus, if $X = G/H$, G is reductive, then the existence of n analytically independent functions on X implies that X carries a quasi-affine structure. These structures are studied in [8], [30].

In the above setting, Matsushima [40] showed that X is Stein if and only if H is itself reductive. Furthermore, the Stein structures are affine algebraic. Since going up or down by finite covers preserves Steinness, and since H/H^o is finite under either assumption, in order to prove Matsushima's theorem, we may assume that H is connected. Assume first that G/H is Stein. If H were not reductive, then we could split off the *unipotent radical* U of $H = U \times R$ where R is a maximal reductive subgroup. Note, since we have already shown that H is algebraic, this does not require the analytic structure theorem for Lie groups which we will discuss later.

Following Matsushima, we consider the fibration $G/U \to G/H$. Since $U \lhd H$, this is an R-principal bundle. Since R is realizable as a linear algebraic group, if G/H is Stein, then we can apply Serre's theorem to show that G/U is Stein. Now, since U is contractible, the principal bundle $G \to G/U$ is topologically trivial. Applying Grauert's Oka principle, we see that G (as a complex manifold) splits, $G = U \times G/U$. But G is reductive, and thus has the same topological type as K. Hence G has nontrivial singular homology H_n, $n = \dim_{\mathbb{C}} G$. Since G/U is Stein, $H_j(G/U) = \{0\}$ for $j > \dim_{\mathbb{C}} G/U$. But U is contractible, and thus G/U has the same homology as G. Thus U is trivial, and H is therefore reductive.

Conversely, if H is reductive, $H = L^{\mathbb{C}}$, then one can define a "Reynolds Operator" $E : \mathcal{O}(G) \to \mathcal{O}(X)$, by $Ef(g) = 1/a \int_{h \in L} f(gh)\, d\mu(h)$, $a = \int_{h \in L} d\mu(h)$, where μ is the Haar measure on L. Since $L^{\mathbb{C}} = H$, Ef is an H-invariant function on G (i.e., $Ef \in \mathcal{O}(X)$). Since G is Stein, we may choose f appropriately in order to show that the axioms of Stein are fulfilled on X. For example, if $x_o, x_1 \in X$ and F_o and F_1 are corresponding fibers of the map $G \to G/H = X$, then there exists $f \in \mathcal{O}(G)$ with $f|F_\nu = \nu$. Thus $Ef(x_\nu) \equiv \nu$, and $\mathcal{O}(X)$ is seen to separate the points of X.

Although Matsushima's proof of the above theorem on "reductive pairs" is easily explained, it uses a number of deep complex analytic facts:

1) Principal bundles of algebraic groups over Stein manifolds are Stein; 2) the singular homology of a Stein manifold vanishes beyond its complex dimension; 3) a topologically trivial holomorphic fiber bundle over a Stein manifold is analytically trivial. Thus this

proof does not carry over to the algebraic case, which has been handled by totally different techniques [57].

Matsushima's structure theorem for complex Lie groups [40], and its application (with Morimoto [43]) to the solution of Serre's problem for *connected* complex Lie structure groups are further examples of a splendid mix of methods from complex analysis and Lie theory. In the former, Matsushima splits off the "compact" part: Let G be a connected complex Lie group and K a maximal compact subgroup. The subgroup $\tilde{K}$ associated to the smallest complex Lie subalgebra $\tilde{\mathfrak{k}}$ in $\mathfrak{g}$ which contains $\mathfrak{k}$ is closed (a lemma of Goto), and thus one may consider the principal bundle $G \to G/\tilde{K}$. It follows from the Iwasawa decomposition that G/K is a cell, and thus the homotopy sequence for $G/K \to G/\tilde{K}$ shows that the base $G/\tilde{K}$ is simply-connected. If R is a maximal connected normal solvable Lie subgroup of G (the "radical"), then the Levi-Malcev decomposition shows that $G = R\cdot\tilde{K}$, and thus R acts transitively on the base $G/\tilde{K}$.

We now consider the Steinizer G_o of the group G [45]. This is a connected central subgroup with $\mathcal{O}(G_o) = \mathbb{C}$, and G/G_o Stein. The fact that $\mathcal{O}(G_o) = \mathbb{C}$ immediately implies that G_o is the smallest complex group containing its maximal compact subgroup K_o. Thus, choosing $\tilde{K} \supset K_o$, we may assume that $G_o \subset \tilde{K}$. Considering the fibration $G/G_o \to G/\tilde{K} = X$ we may define a Reynolds operator as above by integrating over $\tilde{K}/K_o$. In exactly the same way as above, we see that X is Stein. Since X is simply-connected, it is realized as R/I where I is connected. Applying Theorem 1 below, we see that $X = \mathbb{C}^s$. An application of Grauert's Oka principle to the fibration $G \to G/\tilde{K}$ shows that $G = \mathbb{C}^s \times \tilde{K}$. This is of course not a group theoretic splitting.

The group $\tilde{K}$ is rather easily analyzed. First, its Lie algebra splits: $\tilde{\mathfrak{k}} = \tilde{\mathfrak{s}} \times \tilde{\mathfrak{z}}$, where $\tilde{\mathfrak{s}}$ is semi-simple and $\tilde{\mathfrak{z}}$ is central, and, second, the obvious map $\tilde{S} \times \tilde{Z} \to \tilde{K}$ is finite. Since semi-simple groups and their finite centers are well-understood, it remains to analyze the possibilities for Z (i.e., complex abelian groups). One shows that $\tilde{Z} = T/(\mathbb{C}^*)^r\mathbb{C}^s$, where T is a "reduced" group, $T = \mathbb{C}^n/\Gamma_{n+m}$, where Γ is a lattice of rank $n+m$, $0 < m \leqslant n$, and $\mathcal{O}(T) = \mathbb{C}$ (see [7], [46]). One can transfer the condition $\mathcal{O}(T) = \mathbb{C}$ to an irrationality condition on Γ (see [46]). However, the manifolds T are not well-understood at this point in time. For example, although every such T is *pseudo-convex* in an essentially canonical way, one can arrange T to have infinite dimensional $H^1(T,\mathcal{O})$, or a function field $\mathcal{M}(T)$ of infinite transcendence degree. The main complication seems to be that if T is non-compact, consideration of Θ-functions is not enough.

Even though the splitting $G = \mathbb{C}^s \times \tilde{K}$ is not group theoretic, and even though the complex analytic properties of $\tilde{K}$ are not fully understood, one can nevertheless make good use of this decomposition. A good example is the Matsushima-Morimoto solution of Serre's problem in the case of a connected complex structure group [43].

The setting of Serre's problem is the following: Let $E \to B$ be a holomorphic fiber bundle with Stein base B and Stein fiber F. Is E Stein? The answer in general is negative (see [61]), but known examples involve extremely complicated structure groups. The question is still open in the homogeneous case: Let G be a complex Lie group, H and J closed complex subgroups with the bundle $G/H \to G/J$ having the above properties. Is G/H Stein? Matsushima and Morimoto affirmatively answered Serre's question when the structure group S of the bundle $E \to B$ is a *connected* complex Lie group. The following is a proof of this extremely useful fact.

We may assume that S acts effectively on F, and since the Steinizer S_o must have trivial orbits on the Stein manifold F, $S_o = \{e\}$ and S is Stein. Thus $S = \mathbb{C}^s \times \tilde{K}$ where $\tilde{K}$ is Stein. Consequently $\tilde{K}$ is reductive and agrees with the complexification $K^{\mathbb{C}}$. Of course such groups carry linear algebraic structure. Grauert's Oka principle allows us to *holomorphically* reduce the structure group to $\tilde{K}$, and thus by the theorem of Serre [60] the principal bundle space $P = P(B,K)$ is Stein. The usual construction realizes E as the base of a K-principal bundle: $P \times F \xrightarrow[\pi]{} P \times_{\tilde{K}} F = E$. Since $P \times F$ is Stein and $\tilde{K} = K^{\mathbb{C}}$, one defines the Reynolds operation by integration over K and proves E is Stein as above. For example, if $\{x_\nu\} \subset E$ is a divergent sequence, then there exists $f \in \mathcal{O}(P \times F)$ with $f|\pi^{-1}(x_\nu) \equiv \nu$. Thus $Ef(x_\nu) = \nu$ and we see that E is holomorphically convex.

In a different direction, Matsushima [39] (see also Matsushima and Hano [26]) again emphasized the role of the compact group. In particular, he showed that a compact Kähler manifold X which is homogeneous under the action of a compact group $K \subset \mathrm{Aut}(X)$ is the product $Q \times T$ of a homogeneous rational manifold Q with a complex torus T. Later, Borel and Remmert [13] used different methods to prove this theorem without the compactness assumption on K.

It is perhaps interesting to note that without much difficulty the general theorem can be reduced to Matsushima's special case. For this purpose let X be a compact homogeneous Kähler manifold, $G = \mathrm{Aut}(X)^{o}$, $G = RS$ a Levi-Malcev decomposition of G, and write $X = G/N$. We consider the Tits fibration $G/H \to G/N$, where $N := N_G(H^o)$ is the normalizer in G of the identity component H^o. We recall that the base $G/N = Q$ is naturally G-equivariantly embedded in projective space, and a flag argument shows that R acts trivially on Q (see [22], [13]). Thus $Q = S/H$, and, since S is acting algebraically, $\pi_1(Q)$ is finite. Consequently a maximal compact group K of G acts transitively on Q [44].

The fiber $N/H = N/H^o/H/H^o$ is parallelizable. Thus, by a theorem of Wang [68], N/H is a torus. To reduce to Matsushima's case, it is enough to find a compact torus in G which acts transitively on the fiber N/H. Of course, using the existence of a Bruhat cell, one sees that Q is simply-connected, and consequently $N = N^o$. Thus $H \triangleleft N$, and the right action of N/H on the coset space G/H provides the desired torus. But we would like to prove this independent of the theory of roots, parabolic groups, etc.

Note that we have the diagram,

$$\begin{array}{rcccl} & X = & G/H & \longrightarrow & G/N \\ \text{finite} & & \uparrow & & \uparrow \ \text{finite} \\ & \tilde{X} := & G/\tilde{H} & \longrightarrow & G/N^o \end{array} \quad ,$$

where $\tilde{H} = N^o \cap H$. We see that $\tilde{X}$ is Kähler, and, since N^o/H^o is abelian, $\tilde{H} \triangleleft N^o$. Hence the torus $N^o/\tilde{H} =: \tilde{T}$ is embedded in $\mathrm{Aut}(\tilde{X})^o$ via action from the right. Thus any maximal compact subgroup of $\mathrm{Aut}(\tilde{X})^o$ acts transitively, and we may apply Matsushima's result, $\tilde{X} = \tilde{Q} \times \tilde{T}$. Unfortunately, this compact group may not be in G. However, one sees from the splitting $\tilde{X} = \tilde{Q} \times \tilde{T}$ that the radical of any effective transitive connected group on $\tilde{X}$ is in fact $\tilde{T}$ itself. Thus $\tilde{T} \subset G$, and the above diagram yields a compact subgroup of G acting transitively on X.

2. Recent Developments

In more recent times there has been some progress toward classification of non-compact homogeneous spaces. These results have either involved strong conditions on the group, such as Matsushima's theorem on reductive pairs, or conditions "near infinity" such as compactifiability or Levi curvature. We now summarize some of these contributions.

Remmert and van de Ven introduced the concept of an "almost homogeneous" complex space: An irreducible compact complex space X so that the group $\mathrm{Aut}(X)$ has an open orbit Ω. Elementary considerations show that there is exactly one such open orbit and that its complement $E := X\backslash\Omega$ is a closed proper analytic subspace. One can think of X as an equivariant compactification of Ω, and hence a condition at infinity is imposed. Of course from this point of view one should allow the group to be (possibly) smaller than $\mathrm{Aut}(X)$, and furthermore there are interesting situations where X is not compact. Thus we consider complex almost homogeneous pairs (X,G), where X is an irreducible complex space, and G is a complex Lie group acting holomorphically on X and having an open orbit Ω_G with $E_G := X\backslash\Omega_G$.

The basic facts for almost homogeneous spaces were set down by Remmert and van de Ven (see [51], [54], [56]). One should particularly emphasize the importance of the structure of the Albanese fibration, and equivariance theorems for proper maps [56]. The first classification theorem was that of Potters [54], where almost homogeneous compact surface were listed. Except for almost homogeneous Hopf surfaces (i.e., the Hopf surfaces S with $\pi_1(S)$ abelian), all such surfaces are Kähler: 1) Rational surfaces: $\mathbb{P}_2$, $\mathbb{P}_1\times\mathbb{P}_1$, and the Hirzebruch surfaces Σ_n with (possibly) certain points blown up; 2) topologically trivial $\mathbb{P}_1$-bundles over elliptic curves; 3) Tori.

Potters used a great deal of special information about surfaces which has not been proved in higher dimensions. For example, he makes essential use of the fact that $\mathbb{C}^2$ has only algebraic compactifications. Moreover, the "parallelisable" case (i.e., $\Omega = G/H$ with H discrete) is handleable, because there are only two 2-dimensional simply-connected complex Lie groups, and the non-abelian group is not complicated. Using this sort of special information, one can even classify the homogeneous surfaces $X = G/H$ where G is a complex Lie group [31], [62]. It is perhaps of interest that if G is the above-mentioned non-abelian (solvable) group and H is any non-central discrete subgroup, then G/H is not equivariantly compactifiable [62].

One might hope to carry out Potters' program in higher dimensions. This is impossible in general, but one might have a chance in dimension 3. The difficulties arise in the parallelisable case. However, except for $SL_2(\mathbb{C})$, the group G must be solvable. A complete list of discrete subgroups Γ of these solvable groups is known [62], and, unless G is abelian, we suspect that such G/Γ are almost never equivariantly compactifiable. Therefore, one could hope to reduce matters to discrete subgroups of SL_2. This is on the surface no great progress, but in our case we know that the open orbit $\Omega = SL_2/\Gamma$ has an equivariant compactification. We believe that this significantly limits the possibilities for Γ, and merits further study.

The class of non-compact homogeneous spaces is enormous, and it is consequently reasonable to try to cut it down to size by imposing Levi curvature conditions near infinity or conditions on the normal bundle of E_G in the case of pairs (X,G). One of the strongest possible such conditions is to assume that E_G or one of its components has a strongly pseudoconvex neighborhood, or equivalently that it can be blown down to a point. Since such modifications are equivariant, we might as well study the situation where E_G contains an isolated point. The first classification theorem in this context was proved in [52] under the additional assumption that X is a compact complex manifold. Ahiezer [2] handled the situation in the algebraic category. The final classification was proved in [32]:

Cone Theorem. *Let* X *be an irreducible complex space, and* G *a complex Lie group acting holomorphically on* X. *Assume that* G *has an open orbit* Ω *so that its complement* $E = X\setminus\Omega$ *contains an isolated point. Then, after a* 1-1 *normalisation,* X *is a cone (affine or projective rational, depending on compactness) over a homogeneous projective rational manifold. The isolated point is the vertex. If* X *is a manifold, then it is either* $\mathbb{C}^n$ *or* $\mathbb{P}^n$. *Furthermore, a list of all possible* G *can be determined.*

The classification of strongly pseudoconcave homogeneous spaces Ω is obviously related to the above, and in fact, if a complex Lie group acts transitively on Ω, then Ω has a 1-point equivariant compactification as such a rational cone [34]. On the other hand, there are homogeneous strongly pseudoconcave manifolds where *no* complex Lie group acts holomorphically and transitively (e.g., $\mathbb{P}_n \setminus \mathbb{B}_n$, where $\mathbb{B}_n$ is a Euclidean ball). However, if Ω is strongly pseudoconcave, then $\mathrm{Aut}(\Omega)$, equipped with the compact-open topology, is a Lie group. Suppose in addition that Ω is homogeneous and let $G := \mathrm{Aut}(\Omega)^o$. If

$\dim_{\mathbb{C}}\Omega > 2$, then Ω has a G-equivariant compactification X. We now outline the classification theorem for this setting [35].

The compactification X is unique if it is taken to be minimal and normal. If the difference $E_G := X\backslash\Omega$ is a point, then we are in the cone situation as above. If E_G has interior, then $X = \mathbb{P}_n$ and $E_G = \mathbb{B}_n$. In the remaining cases, E_G is a totally real submanifold of X, and in fact should be thought of as the "real points" of X. As a paradym the reader might consider the situation in $\mathbb{P}_n(\mathbb{C})$, where $E_G = \mathbb{P}_n(\mathbb{R})$. Let $G := SL_{n+1}(\mathbb{R})$ acting via the standard representation on $\mathbb{P}_n(C)$. Then G has two orbits: E_G and $\Omega = P_n(\mathbb{C})\backslash E_G$. The maximal compact group $K = SO_{n+1}(\mathbb{R})$ of G has three orbit types: E_G, a complex quadric Q, and the real hypersurfaces "around E_G" which yield a strongly pseudoconvex exhaustion of $P_n(\mathbb{C})\backslash Q$. The complex group $K^{\mathbb{C}} = SO_{n+1}(\mathbb{C})$ has two orbits: Q and the Stein manifold $\mathbb{P}_n(\mathbb{C})\backslash Q$. Finally G is a real form of the complex group $SL_{n+1}(\mathbb{C})$ which of course acts transitively on $\mathbb{P}_n(\mathbb{C})$.

The general situation where E_G is totally real can be described as follows. Let $\mathbb{A}$ be one of the division algebras $\mathbb{R}$, $\mathbb{C}$, $\mathbb{H}$ or $\mathfrak{C}$ and let $E_G := \mathbb{P}_n(\mathbb{A})$ be the n-dimensional projective space over $\mathbb{A}$. Of course, $\mathbb{P}_n(\mathbb{A})$ is not necessarily n-dimensional as a differentiable manifold, and, in the case of the Cayley numbers, we only have the Cayley plane $\mathbb{P}_2(\mathfrak{C})$. The tangent bundle $T(E_G)$ is equipped with a natural metric and complex structure so that the appropriate compact group $K|(K_{\mathbb{R}} = SO$, $K_{\mathbb{C}} = SU$, $K_{\mathbb{H}} = Sp$, and $K_{\mathfrak{C}} = F_4)$ acts as a group of holomorphic isometries. The bundle $T(E_G)$ has a natural K-equivariant compactification X by adding a K-orbit which is a compact complex divisor Q. The group $K^{\mathbb{C}}$ has two orbits: Q and the Stein manifold $X\backslash Q = T(E)$. For example, $X_{\mathbb{R}} = \mathbb{P}_n(\mathbb{C})$, $X_{\mathbb{C}} = \mathbb{P}_n(\mathbb{C})$, where $\mathbb{P}_n(\mathbb{C})$ is embedded in a totally real way by $p \to (p,\bar{p})$, and $X_{\mathbb{H}} = G_{2,2n}$, the Grassmann manifold of 2-planes in $\mathbb{C}^{2n}$.

The manifolds X are homogeneous and the stabilizer G of E_G in $\mathrm{Aut}(X)$ is a real form of $\mathrm{Aut}(X)^{o}$ which has two orbits: E_G, and the strongly pseudoconcave complement $\Omega = X\backslash E$. One can build another example, the complex n-dimensional quadric whose real points form the n-sphere, but this is just a 2-1 cover of $X_{\mathbb{R}} = \mathbb{P}_n(\mathbb{C})$ which is ramified over the quadric Q_{n-1} in $\mathbb{P}_n$. The manifolds Ω above (including cones and $\mathbb{P}_n\backslash\mathbb{B}_n$) form the list of *all* strongly pseudoconcave homogeneous manifolds. We should note that the generic K-orbits Σ in Ω form the list of all strictly pseudoconvex homogeneous hypersurfaces (see [47], [59], and [15]). The proof of the pseudoconcave classification does not seem to be reducible to that for the hypersurfaces. As observed in [3] and [36], the manifolds $X_{\mathbb{R}}$, $X_{\mathbb{C}}$, $X_{\mathbb{H}}$, and $X_{\mathfrak{C}}$, along with projective rational manifolds and tori, form the building blocks for a complete classification of Kähler almost homogeneous pairs (X,G), where E_G is a divisor orbit of the group G.

The above discussion arose in connection with almost homogeneous pairs (X,G) where E_G contains an isolated point. If X is a compact manifold and E_G is 1-dimensional, then there are again very few examples. If $\dim_{\mathbb{C}} X = 2$, then Potters' classification tells the full story. Otherwise we have the following [33]:

Curve Theorem. *Let* X *be a compact complex manifold of dimension greater than two. Let* (X,G) *be a complex almost homogeneous pair, and assume that* E_G *is* 1*-dimensional. Then* X *and* E_G *are as follows, and the possibilities for* G *are known:*

1) $X = \mathbb{P}_n$, *and* E_G *is a complex line;*
2) $X = \mathbb{P}_3$, *and* E_G *is the union of two disjoint complex lines;*
3) $X = \mathbb{P}_1 \times \mathbb{P}_{n-1}$, *and* $E_G = \mathbb{P}_1 \times \{0\}$;
4) $X = Q_3$, *the hyperquadric in* $\mathbb{P}_4$, *and* E_G *is a complex line in* Q_3;
5) X *is a desingularized Segre cone, and* E_G *is the rational curve lying over the vertex;*
6) X *is a homogeneous Hopf manifold, and* E_G *is a fiber (elliptic curve) of the canonical projection* $X \to \mathbb{P}_{n-1}$.

A Segre cone is formed by taking the projective cone over the Segre embedding of $\mathbb{P}_1 \times \mathbb{P}_{n-2}$. It is naturally desingularized by blowing up the vertex to a rational curve. The homogeneous Hopf manifolds are formed by taking the quotient of $\mathbb{C}^n \setminus \{0\}$ by a discrete cyclic group generated by a homothety $z \to \lambda \cdot z$, $|\lambda| < 1$. We should note that all examples except for the Hopf manifolds are rational, and, except for the Segre cone case, are all in reality homogeneous. That is, if we take G to be $\mathrm{Aut}(X)$, then the examples where E_G is 1-dimensional occur only in dimension two or in the case of the desingularized Segre cone. We should also note that it is possible to obtain a classification in the case $\dim_{\mathbb{C}} E_G = 1$ when X is an algebraic variety. The question for complex spaces remains open, but would be completely resolved if it were known which quotient SL_2/Γ were equivariantly compactifiable.

It might be possible to obtain a classification in the case when $\dim_{\mathbb{C}} E_G = 2$, but this would probably require determining a complete list of compact surfaces which have a positive dimensional automorphism group. As far as we know, this has only been carried out under additional assumptions [16].

We certainly do not suggest attempting to obtain a detailed classification even for compact almost homogeneous manifolds. But it would be useful and perhaps even possible to understand the algebraic or Kähler case when the components of E_G are orbits. If E_G has more than one component, then this has been done [2], [21]. Other than the above-mentioned divisor case, not much is known when E_G is connected. In this regard, Borel has shown that a symmetric space (i.e., a compact complex space X such that for all $x \in X$ there exists $s_x \in G \subset \mathrm{Aut}(X)$ with an isolated fixed point at x and $s_x^2 = \mathrm{id}$) is almost homogeneous, and any complex Lie group G which realizes X as a symmetric space has finitely many orbits [12]. Thus a better understanding of the two orbit situation mentioned above, should lead to theorems about symmetric spaces.

In a slightly different direction, but still related to how the complex homogeneous space $\Omega = G/H$ looks near infinity, there has been substantial recent progress in classification questions provided Ω has more than one end. For the definition of "ends" we refer the reader to the basic paper of Freudenthal [18]. If Ω is compactifiable as above, then the ends of Ω correspond to the connected components of E_G. Borel [11] noted that if H is connected then G/H has at most two ends, and, if it has two ends, it is a product $K/L \times \mathbb{R}$, where K/L is the orbit of a maximal compact subgroup K in Ω.

In the complex analytic category, one should attempt to split off a $\mathbb{C}^*$ to display the two ends. In many cases this is possible. For example, if (X,G) is a compact Kähler almost homogeneous pair where E_G has more than one component, then, by analyzing the Albanese fibration, it is possible to apply Borel's result to show that E_G has exactly two components [5]. In this case X is a homogeneous $\mathbb{C}^*$-bundle space over a homogeneous Kähler manifold $Q \times T$, where Q is a rational manifold and T is a torus. The space X is formed by (perhaps) blowing down either the ∞- or 0-section (or both) of the associated $\mathbb{P}_1$-bundle (see [36]).

A general structure theorem is possible when G/H has two ends and H/H^{o} is finite [21]: There is a fibration

$$G/H \to G/J \to G/P \qquad ,$$

where G/P is compact rational and, either $J/H = \mathbb{C}^*$ and P/J is a compact torus, or vice versa. The former situation (homogeneous $\mathbb{C}^*$-bundle over a generalized Hopf manifold) is basically understandable (see [1]), but the reverse case is seemingly quite complicated. For example, there are homogeneous bundles of elliptic curves over $\mathbb{C}^*$ which are not equivariantly compactifiable (see [62]). Furthermore, the above results are heavily dependent on the seemingly artificial assumption that H/H^{o} is finite. Without this assumption, one has quite complicated examples arising from arithmetic groups (e.g., $SL_2(\mathbb{C})/\Gamma_k$ with k-ends, $k = 1,2,\ldots$ (see [21]). However, in the presence of nonconstant holomorphic functions, Gilligan [19] has recently observed that the situation is really quite simple: Let $X = G/H$ have more than one end, and let $G/H \to G/J$ be the holomorphic separation fibration. (This fibration is defined by the equivalence relation $p \sim q$ iff $f(p) = f(q)$ for all $f \in \mathcal{O}(X)$). Assuming $\mathcal{O}(X) \neq \mathbb{C}$, then J/H is compact and G/J is an affine cone with vertex removed. This is therefore one case where the "envelope of holomorphy" of X is transparent. For general homogeneous $X = G/H$, we don't even know if the envelope of holomorphy exists.

It is obviously beyond the scope of this brief exposition to go into the details of the above results. However, the basic idea is to fiber the open orbit, extend the fibration (at least meromorphically) to X, and apply induction. The following is an example of an important tool: Let X be a compact complex manifold and (X,G) a compact almost homogeneous pair with $\dim_{\mathbb{C}} E_G = k$. Then there exists a fibration $\Omega = G/H \xrightarrow[\pi]{} G/P = Q$ of the open orbit onto a projective rational manifold Q with $\dim_{\mathbb{C}} P/H \leqslant k+1$. The map π is G-equivariantly meromorphically extendable to X.

Of course the Levi-Malcev decomposition $G = R \rtimes S$ (R = radical, S = maximal semi-simple) plays a fundamental role. One normally handles problems with S via algebraic group methods, and, although the system of R-orbits can in general be quite complicated, the analytic assumptions or flag arguments usually save the day. Moreover, if G is solvable and Γ is discrete and closed in G, then with the exception of simple tori, one always has a fibration $G/\Gamma \to G/J$ where J is positive dimensional.

At the beginning of this section we commented that there are at least two natural types of assumptions on non-compact homogeneous manifolds: Conditions near infinity, which we discussed above, and conditions on the group involved. Matsushima's theorem on reductive pairs was mentioned as typical of the latter. His study of complex nil-manifolds is another. For example, if G is nilpotent and H is connected, then $X = G/H$ is Stein if and only if $X = (\mathbb{C}^*)^s \times \mathbb{C}^t$ [42]. We prove a similar remark in the solvable situation in the present paper (see Theorem 1). The assumption H connected is not desirable. In the nilpotent case a structure theorem for arbitrary G/H has been proved. In this case G/H is realized as a fiber bundle tower of abelian groups: $G/H \to G/H_1 \to \cdots \to G/H_n$, where each extension is central. This is a result of Malcev in the compact case (see [37]), and Barth and Otte [6] and Mostow [50] have proved related results.

It is of course not true that a Stein nil-manifold is just $(\mathbb{C}^*)^s \times \mathbb{C}^t$. But at least there are no problems with envelopes of holomorphy: Let $X = G/H$ be an n-dimensional nil-manifold with n analytically independent functions. Then X is Stein. Moreover, if $G/H \to G/J$ is the holomorphic separation fibration, then $\mathcal{O}(J/H) = \mathbb{C}$ [20]. This is not true in the reductive setting (see [7]), but in that setting the envelope of holomorphy exists and is an affine algebraic variety. It would be very useful to know whether or not a holomorphically separable homogeneous space G/H has an envelope of holomorphy. The above indicate that if difficulties arise, they will be caused by solvable groups. This seems to be typical of the present status of the subject. For example, the above theorems for nil-manifolds are completely open in the solvable case. Furthermore, a fundamental tool, the positive answer to the Serre problem for homogeneous manifolds, is missing, i.e. if $G/H \to G/J$ is a fibration of complex Lie groups with J/H and G/J Stein, is G/H Stein? For G nilpotent or reductive, this is known to be true [20]. But there is no progress in the general or even solvable case.

3. On the Holomorphic Reduction of Solv-Manifolds

If X is a complex space, then one naturally has the relation $p \sim q$ whenever $f(p) = f(q)$ for all $f \in \mathcal{O}(X)$. In general the quotient space $X/\sim$ is not even Hausdorff [24]. However, if $X = G/H$ is the homogeneous space of a complex Lie group, then one has the "holomorphic separation fibration" $G/H \to G/I$ defined by this relation [45]. In this setting the following questions seem natural:

1) When is $\mathcal{O}(I/H) = \mathbb{C}$?
2) When is G/I Stein?
3) If G/I is not Stein, does it at least have an envelope of holomorphy?

Simple examples [7] show that I/H may even be Stein, and G/I may not be Stein (e.g., $\mathbb{C}^2 \setminus \{0\}$). If G is reductive, then G/I is quasi-affine, and if G is nilpotent, then $\mathcal{O}(I/H) \equiv \mathbb{C}$ and G/I is Stein [20]. Unfortunately not much more is known. Thus it still seems to make sense to place constraints on the group G.

If G is solvable, then, particularly in light of the methods of Mostow [49], it is reasonable to hope that $\mathcal{O}(I/H) \equiv \mathbb{C}$ and G/I is Stein. The purpose of this section is to make a first small step in this direction:

<u>Theorem 1</u>. *Let* $X = G/H$, *where* G *is a connected solvable complex Lie group, and* H *is a closed complex subgroup with finitely many connectivity components. Let* $G/H \to G/I$ *be the holomorphic separation fibration. Then* $\mathcal{O}(I/H) \equiv \mathbb{C}$, *and* G/I *is Stein. The fiber* I/H *is holomorphically equivalent to a reduced abelian group. Moreover* $H^0 \supset (I^0)'$, *and the holomorphic reduction* $G/H^0 \to G/I^0$ *is a trivial bundle with reduced abelian group as fiber and* $\mathbb{C}^s \times (\mathbb{C}^*)^t$ *as base.* (See Section I for some details on "reduced abelian groups.")

Since going down by finite maps preserves Steinness, it is enough to prove the theorem for $H = H^0$ (e.g., $I := I^0 H$). For this, the following is essential.

<u>Proposition</u>. *Let* $X = G/H$, *where* G *and* H *are connected solvable complex. Let* $n := \dim_{\mathbb{C}} X$, *and assume that there exist* $f_1, \ldots, f_n \in \mathcal{O}(X)$ *which are analytically independent (i.e.,* $df_1 \wedge \ldots \wedge df_n \not\equiv 0$). *Then* $X = \mathbb{C}^s \times (\mathbb{C}^*)^t$. *Moreover there is a connected closed complex subgroup* J *of* G *containing* H *so that* $G/J = \mathbb{C}^k$ *or* $G/J = \mathbb{C}^*$.

Proof. Assume we have such a subgroup J. Since the structure group is connected, the bundle $G/H \to G/J$ is trivial [23], and the proposition follows by induction. Thus it remains to construct J.

Let $N := N_G(H)$ be the normaliser of H in G, and consider the "Tits fibration" $G/H \to G/N$. If $N = G$, then $G/H =: \tilde{G}$ is itself a solvable complex Lie group. Since solvable compact connected Lie groups are necessarily abelian, the complex Lie group $\tilde{K}$ in the Matsushima decomposition $\tilde{G} = \mathbb{C}^k \times \tilde{K}$ is an abelian Stein group and therefore $\tilde{K} = \mathbb{C}^\ell \times (\mathbb{C}^*)^t$ [43]. Thus we may assume that $N \neq G$.

Let $\dim_{\mathbb{C}} \mathfrak{g} = m$, $\dim_{\mathbb{C}} \mathfrak{h} = k$, and consider the action of G on the Grassmann manifold of k-planes in m-space via $Ad(G)$ on $\mathfrak{g}$. The orbit of the "point" $\mathfrak{h}$ in this Grassmann manifold is just G/N. Thus the Plücker embedding yields a linear representation of G, where G/N is an orbit. We refer to this as the "Tits representation." The induced representation of the commutator group G' is algebraic [17], and thus the G'-orbits in G/N are closed. Hence we may consider the fibration $G/N \to G/G'N$. The algebraic lemma below shows that $G/G'N$ is a holomorphically separable abelian group, $G/G'N = \mathbb{C}^a \times (\mathbb{C}^*)^b$. If $a + b > 0$, the existence of $J \supset (G'N)^o$ is immediate. If $a + b = 0$, then G' acts transitively on G/N. But this unipotent group has only $\mathbb{C}^k$ as orbits. Hence $J := N$ does the job in this case.

Lemma. *Let* G *be complex Lie group in an affine algebraic group* A. *Let* I *be an algebraic subgroup of* A, *and* $N := G \cap I$. *Let* $\bar{G}$, $\bar{N}$ *be algebraic closures of* G, N *in* A. *The natural action of* G *on the affine algebraic abelian group* G/NG' *yields an embedding* $G/NG' \to \bar{G}/\bar{N}G'$. *In particular,* $\bar{G}/\bar{N}G' = \mathbb{C}^a \times (\mathbb{C}^*)^b$.

Proof. Let $g \in G$, and suppose $g\bar{N}G' = \bar{N}G'$. Since $G \cap \bar{N} = N$, it follows that $g \in NG'$. Since $\bar{G}/\bar{N}G'$ is affine algebraic, G/NG' is holomorphically separable, and thus $G/NG' = \mathbb{C}^a \times (\mathbb{C}^*)^b$ [43].

We now complete the proof of Theorem 1. As noted earlier, we may assume that $H = H^o$. Let $G/H \to G/I$ be the holomorphic separation fibration. It follows from the above proposition that $G/I^o = \mathbb{C}^s \times (\mathbb{C}^*)^t$. Since $H \subset I^o$, it follows that $I = I^o$. If $I/H \to I/L$ is the holomorphic separation fibration of the fiber, then the fibration $G/L \to G/I$ has Stein base and fiber, and the structure group is connected. Thus G/L is Stein [43]. Consequently, $I = L$ and $\mathcal{O}(I/H) = \mathbb{C}$.

Since G is solvable, it stabilizes a flag in the Tits representation, and thus the orbit G/N is contained in some $\mathbb{C}^n$. The holomorphic separability of G/N implies that $I \subset N$. Consequently, I/H is itself a Lie group. Such Lie groups (i.e., $\mathcal{O}(I/H) \equiv \mathbb{C}$) are necessarily abelian reduced groups [46]. Thus $H \supset I'$.

It remains to show that the bundle $G/H \to G/I$ is trivial. The proposition gives us a connected closed subgroup J of G containing I so that $G/J = \mathbb{C}^k$ or $G/J = \mathbb{C}^*$. Since the structure group of the fibration $G/H \to G/J$ is connected, this fibration is trivial [23]. Hence the desired result follows applying the induction assumption to the fiber J/H.

4. Remarks on $\mathbb{C}^n$

As a result of the Mostow fibration [48], an "aspherical" (i.e., all homotopy groups vanish) homogeneous space of a Lie group is a cell. This result immediately carries over to the complex analytic category:

Proposition. *Let* G *be a complex Lie group and* H *a closed complex subgroup. Assume that* $\Omega := G/H$ *is aspherical. Then* $\Omega = \mathbb{C}^n$.

Proof. We may assume that G and H are connected. Let L and K be maximal compact subgroups of H and G with $L \subset K$. Following Borel [11], we see that the topological splittings $G = K \times \mathbb{R}^s$, $H = L \times \mathbb{R}^t$, and the Künneth-formula for cohomology with compact supports (applied to the trivial fibering $G/L \to G/K = \mathbb{R}^s$) yield $H_c^i(G/L) \simeq H^{i-s}(K/L)$ for all $i \in \mathbb{N}$. Moreover $\pi_1(K/L) = \pi_1(G/L) = \pi_1(\Omega) = \{1\}$ and therefore K/L is orientable. A spectral-sequence argument (applied to the fibering $\mathbb{R}^t \to G/L \to G/H$) shows $H_c^i(G/L) \simeq H_c^{i-t}(G/H)$. Finally the Poincaré duality-theorem and the Hurewicz-theorem yield

$$H_i(\Omega,\mathbb{Z}) \simeq H_i(K/L,\mathbb{Z}) , \qquad 0 = \pi_i(\Omega) = H_i(\Omega,\mathbb{Z}), \qquad i \in \mathbb{N} .$$

Thus $K = L$ and H contains a maximal connected reductive complex subgroup $S := K^{\mathbb{C}}$ of G. Hence the radical R of G acts transitively: $\Omega = R/R \cap H$. Since $\pi_1(\Omega) = \{1\}$ the group $R \cap H$ is connected and Theorem 1 of the previous section shows that $\Omega = T \times \mathbb{C}^n \times (\mathbb{C}^*)^m$, where T is a reduced group. But $\pi_1(\Omega) = \{1\}$ implies that $\Omega = \mathbb{C}^n$. ∎

Obviously "aspherical" is a very strong topological assumption. For the results in this section we would weaken this somewhat, but that would be artificial, as we really don't know the "right" condition!

(e.g. Vanishing of cohomology with compact supports is probably enough to guarantee $\Omega = \mathbb{C}^n$.

In the category of complex spaces, there are a number of interesting results on compactifications of $\mathbb{C}^n$ (i.e., $\mathbb{C}^n$ is a Zariski open subset of the compact complex space X). We refer the reader to [55], [66], and [14]. For example, every nonsingular compactification of a complex homology 2-cell is algebraic [66]. We are interested in this question in the almost homogeneous setting: Let X be a compact almost homogeneous space under the action of the complex Lie group G. Assuming that the open orbit $\Omega = G/H$ is aspherical, what can be said about X? We refer to X as an "equivariant compactification of $\mathbb{C}^n$."

Let X be a Kähler manifold which is an equivariant compactification of $\mathbb{C}^n$. Since $b_1(X) = 0$, it follows that X is algebraic. The proof of the above proposition shows that R acts transitively on Ω. Hence it follows from the theorem of Rosenlicht-Matsumura (see Section 5) that X is rational. We would hope that the Kähler assumption could be removed, but in cases where $\mathfrak{M}(\mathbb{C}) \cong \mathbb{C}$ there does not seem to be any good way to proceed. In special cases, however, this is possible. For example, if X is itself homogeneous under a larger group $\tilde{G}$, then we consider the Tits fibration, $X = \tilde{G}/\tilde{H} \to \tilde{G}/\tilde{N}$. The fiber $\tilde{N}/\tilde{H}^o/\tilde{H}/\tilde{H}^o = M/\Gamma$ is parallelisable. Thus any Lie group which acts almost transitively on it must in fact act transitively. If $\Omega \cap \tilde{N}/\tilde{H}$ is compact in Ω, then, since $\Omega = \mathbb{C}^n$, it is finite. Since $G \cap \tilde{N}$ acts almost transitively on $\tilde{N}/\tilde{H}$, it therefore follows that $\tilde{H} = \tilde{N}$ and X is rational. We conclude this section with a comment on the two orbit situation:

Theorem 2. *Let* X *be a compact complex space which is almost homogeneous under the action of the complex Lie group* G. *Assume that* $\Omega = G/H$ *is aspherical and* $E := X \backslash \Omega$ *is a* G-*orbit. Then, after a* 1-1 *normalization,* $X = \mathbb{P}_n$, *and* E *is a hyperplane.*

Proof. We may assume that G is connected and acts almost effectively on X. By the above propostion $\Omega = \mathbb{C}^n$, and therefore E is a divisor. Let $\nu: \tilde{X} \to X$ be the normalization. Since ν is bimeromorphic and has non-trivial fibers only over a set of codimension two, and since the only singularities of X lie in E which is a divisor, ν is 1-1. Further, since ν is equivariant and the singularity set of $\tilde{X}$ is at most 2-codimensional, it follows that $\tilde{X}$ is in fact a manifold. Thus we may assume that X is a complex manifold.

Let $\Omega = G/H \to G/N =: \Omega'$ be the Tits fibration of the open orbit. Since this map is given by the sections of the anti-canonical bundle of X which are induced from $\mathfrak{g}$, it is meromorphically and G-equivariantly extendable to X. Since E is a divisor and its indeterminacy set is G-stable, it follows that the Tits fibration is extendable to a holomorphic map $\tau: X \to X'$, where X' is G-equivariantly embedded in $\mathbb{P}_n$. If X' is a point, then the almost effectivity of the G-action implies that $H^o = \{e\}$. But Ω is aspherical. Therefore $H = H^o$, and $\Omega = G$. A further application of this topological assumption shows that G is solvable. Thus E is a compact solv-manifold, and consequently $b_1(E) \neq 0$. But Ω being a cell implies $H_1(E,\mathbb{Z}) = \{0\}$ [55]! Thus the space X' is positive dimensional.

Let $x' \in \Omega'$. If $\tau^{-1}(x') \cap E \neq \emptyset$, then $X' = \tau(X) = \tau(E) = E'$. Now E' is a compact homogeneous space equivariantly embedded in $\mathbb{P}_n$. Thus a maximal reductive part S of G acts transitively on E'. On the other hand, as in the proof of the proposition, S has a fixed point in Ω. Thus $\tau^{-1}(x') \cap E = \emptyset$. In other words $\tau^{-1}(x')$ is compact, and therefore finite, in Ω. Since the radical R of G acts transitively on Ω' and furthermore stabilizes a flag in $\mathbb{P}_n$, it follows that E' is a hyperplane section. Thus S also acts transitively on E.

Our situation is now the following: A maximal reductive part S of G has a fixed point $x_o \in \Omega$ and acts transitively on the divisor E. Since Ω is holomorphically separable, the generic S-orbit in Ω is open and all other orbits in Ω are isolated points. Thus it follows that x_o is the only fixed point of S. Then $X = \mathbb{P}_n$ and E is a hyperplane section. ■

5. Rationality of Certain Almost Homogeneous Manifolds

If X is a compact homogeneous Kähler manifold with $b_1(X) = 0$, then $X = S/P$ where S is semi-simple and P contains a maximal connected solvable ("Borel") subgroup B of S. The top dimensional B-orbit in X is a Zariski open copy of $\mathbb{C}^n$ (a "Bruhat cell"). The injection of this cell in X can be extended to a bimeromorphic map $\mathbb{P}_n \overset{\sim}{\to} X$ (i.e., X is "rational," see Goto [22]). This characterization of the function field of X has a more general version (see Grauert and Remmert [25]): Let $X = G/H$ be a compact homogeneous space with $b_1(X) = 0$, and let N be the normalizer of H^o in G. Then the Tits fibration $\tau: G/H \to G/N =: Y$ yields an isomorphism of function fields

$\tau^*: \mathfrak{M}(Y) \xrightarrow{\sim} \mathfrak{M}(X)$, and $Y \leftrightarrow \mathbb{P}_k$ as above. Even when $b_1(X) \neq 0$, the function field $\mathfrak{M}(X)$ is equivalent to that of the product of an abelian variety and a rational manifold.

The problem of characterizing the function fields of almost homogeneous manifolds seems to be substantially more difficult than the homogeneous case. In the Kähler setting, one at least sees where to begin: Let X be an almost homogeneous Kähler manifold with open orbit $\Omega = G/H$. Then the Albanese fibration $F \to X \to \mathrm{Alb}(X) = G/J$, although not necessarily trivial, has two useful properties: 1) $\dim_{\mathbb{C}} \mathrm{Alb}(X) = b_1(X)$; 2) $b_1(F) = 0$, and there is a J-equivariant embedding of F in projective space [53]. Thus a good starting point in the study of these fields is the special case of X projective algebraic with $b_1(X) = 0$. By Blanchards Theorem [9], there exists an $\mathrm{Aut}(X)^{\circ}$-equivariant embedding $X \hookrightarrow \mathbb{P}_n$. In this way $\mathrm{Aut}(X)^{\circ}$ is represented as the *linear algebraic group* in $\mathrm{Aut}(\mathbb{P}_N)$ which stabilizes X. Conversely, let G be a linear algebraic group and $H \subset G$ an algebraic subgroup. Then there is an algebraic representation of G in $\mathrm{Aut}(\mathbb{P}_n)$ and a point $p \in \mathbb{P}_n$ so that $G(p) = G/H = \Omega$. The orbit Ω is Zariski open in its closure X. Of course G acts on X, and, since H/H° is finite, $b_1(X) = 0$. On one hand, this reduces the matter to algebraic groups, but on the other hand the situation is still very general!

By analogy with the homogeneous case it is natural to ask if an almost homogeneous Kähler manifold X with $b_1(X) = 0$ is in fact rational. Primarily due to difficulties caused by finite groups, we rather doubt this, but this "Remmert-van de Ven question" is nevertheless an excellent focus point. So far it has been proved that there is always a finite meromorphic map $\mathbb{P}_n \rightharpoonup X$ (i.e., X is "unirational"), and if $\dim_{\mathbb{C}} X \leqslant 3$ then X is in fact rational (see [4], [54]). Our purpose here is to extend the rationality result to $\dim_{\mathbb{C}} X \leqslant 7$. We believe that it is quite possible that there are non-rational 8-dimensional examples.

As a result of the above remarks, we now consider a connected affine algebraic group G acting linearly on a projective algebraic manifold X. We assume that some G-orbit, $G(p) = G/I =: \Omega$, is open. Our goal is to prove the following:

<u>Proposition</u>. *Let* B *be a Borel subgroup of* G. *If* $\dim_{\mathbb{C}} X \leqslant 7$, *then there exists* $p \in \Omega$ *with* $5 \cdot \dim_{\mathbb{C}} B(p) \geqslant 3 \cdot \dim_{\mathbb{C}} X$. *In particular*, $\mathrm{codim}_{\mathbb{C}} B(p) \leqslant 2$.

As we shall now show, methods of Rosenlicht [58], Matsumura [38], and the Castelnuovo-Enriquez criterion yield the rationality result as a corollary.

By first splitting off orbits of 1-dimensional normal subgroups and then applying induction, Matsumura [38] proved the following extremely useful theorem: Let the solvable linear algebraic group G algebraically act on the projective variety X_n. If the generic G-orbit is m-dimensional, then X is bimeromorphically equivalent to the product $\mathbb{P}_m \times Y_{n-m}$, where Y corresponds to the orbit space of the G-action, having as function field the G-invariant functions on X.

Assuming the above proposition, this theorem shows that $X \underset{\text{bimer}}{\longleftrightarrow} \mathbb{P}_m \times Y$, where Y is an algebraic manifold and $\dim_{\mathbb{C}} Y \leqslant 2$. We note that for any algebraic variety Z the k-th irregularity $\dim_{\mathbb{C}} H^k(Z,\mathcal{O})$ and the plurigenus $P_k(Z)$ are bimeromorphic invariants. Moreover if Z is unirational, then $P_k(Z) = 0$ for all $k \geqslant 1$ (see [65]). We have $H^k(X,\mathcal{O}) = (0)$ for $k \geqslant 1$. Thus by applying the Künneth formula to $\mathbb{P}_m \times Y$, we see that $H^1(Y,\mathcal{O}) = H^2(Y,\mathcal{O}) = (0)$. The unirationality of X implies the unirationality of Y, which gives $P_2(Y) = 0$. Therefore by the Castelnuovo-Enriquez-theorem we know that Y is rational.

Hence the following is an immediate consequence of the Castelnuovo-Enriquez criterion:

Theorem 3. *Let* X *be an almost homogeneous compact Kähler manifold with* $b_1(X) = 0$. *If* $\dim_{\mathbb{C}} X \leqslant 7$, *then* X *is rational.*

It remains to prove the above proposition. To this end, since we are only interested in orbit dimensions, and since any such algebraic G/H has such an X as an equivariant compactification, we may always assume that H is connected and G is almost effective. Let R be the radical of G. Since R is algebraic, its orbits in Ω are closed, and we may consider the fibration

$$\Omega = G/H \to G/RH = S/S \cap H ,$$

where S is a maximal semi-simple part of $G = R \cdot S$. Let B_S be a Borel subgroup of S. Then $B = R \cdot B_S$ is a Borel subgroup of G. Thus it is enough to prove the proposition for G semi-simple (i.e., $R = (e)$), and since we may pass to a finite cover, we may also assume that $G = S_1 \times \ldots \times S_k$ is a product of its simple factors.

We prove the proposition by induction on $n = \dim_{\mathbb{C}} X$. It is obvious for $n = 1$, and if G is not simple, then the proposition is stated so that application of induction to the fiber and base of $G/H \to G/S_1H$ yields the result (e.g., if the fiber were 3-dim. and the base were 4-dim., then they would both have 1-codimensional B-orbits). Thus, we may assume that G is simple.

If X is G-homogeneous, then a Bruhat cell, as mentioned at the beginning, is an open B-orbit. If X is not G-homogeneous, then the minimal G-orbit in $E := X \backslash \Omega$ is compact. The following asserts that this orbit is non-trivial.

Lemma 1. *Let* X *be a connected, compact algebraic variety, and let* G *be semi-simple group acting linearly on* X. *Assuming the* G*-action is non-trivial, there exists* $p \in X$ *with* $G(p)$ *closed and positive dimensional.*

Proof. We may assume that G has a fixed point p_o in X, as otherwise a minimal G-orbit does the job. Since the action is linear, G stabilizes a hyperplane section $H \not\ni p_o$. Applying induction to the G-action on the components of H, we see that, unless G fixes H pointwise, we are finished. If G fixed H pointwise, linearization at points of H would show that the generic G-orbit is at most 1-dimensional. This is contrary to G being semi-simple. ∎

As a consequence of this lemma, we can quickly eliminate all but a finite number of possibilities.

Lemma 2. *Under the further assumption that* G *is simple, the proposition is true if any one of the following holds:*

1) $n \leq 2$
2) $n = 3$ *and* $\dim_{\mathbb{C}} G \geq 4$
3) $n \leq 5$ *and* $\dim_{\mathbb{C}} G \geq 9$
4) $n = 6$ *and* $\dim_{\mathbb{C}} G \geq 16$
5) $n = 7$ *and* $\dim_{\mathbb{C}} G \geq 25$
6) $n \leq 7$ *and* $G = G_2, B_3,$ *or* C_3.

Proof. Let $G(p)$ be a non-trivial closed orbit on X. Since G is simple, its action on $G(p)$ is almost effective. Thus

$$\dim_{\mathbb{C}} G \leq \dim_{\mathbb{C}} \mathrm{Aut}(X) \leq \dim_{\mathbb{C}} X(\dim_{\mathbb{C}} X + 2) \quad . \qquad (*)$$

This finishes the proof if any of (2)-(5) are satisfied. If $n=2$ and $\dim_{\mathbb{C}} G(p)=1$, then $G=A_1$, and the universal cover of the open orbit $\Omega=G/H$ is either $\mathbb{C}^2\setminus\{0\}$ or the affine quadric. An explicit check shows that B has an open orbit in Ω. Thus it remains to prove the proposition under (6). We do this by showing that if $G=G_2$, B_3, or C_3 and $G/P=:Q$ is compact, then $\dim_{\mathbb{C}} Q\geqslant 5$.

If P is as above, then $P=R\times S_P$, where the semi-simple part S_P has rank at most that of G. If $G=G_2$, then, using (*) one sees that the only possibilities for S_P are A_1, $A_1\times A_1$, and A_2. The latter two can be eliminated by looking at the root diagram for G. Thus, since we pick up only one negative root from A_1, it follows that $\dim_{\mathbb{C}} P\leqslant \dim_{\mathbb{C}} B_{G_2}+1=9$. Consequently, any closed $G_2(p)$ is at least 5-dimensional.

If $G=B_3$ or C_3, then $\dim_{\mathbb{C}} G=21$ and $\dim_{\mathbb{C}} B_G=12$. Using the bound on rank and (*), the only possibility which might yield $\dim_{\mathbb{C}} Q\leqslant 4$ is $S_P=B_2\times A_1=C_2\times A_1$ in which case $\dim_{\mathbb{C}} Q=4$. In this case we consider the action on the tangent space T_p, $P\hookrightarrow GL(T_p)=\mathbb{C}^*\times A_3$. This yields a faithful representation of S_P in A_3. Then $A_3/S_p:=\Omega$ is 2-dimensional, and a nontrivial closed A_3-orbit on an equivariant compactification of Ω is 1-dimensional, contrary to (*). ■

We now complete the proof of the proposition. For $n=\dim X=3$ it is enough to note that Lemma 2 takes care of all possibilities except for $G=A_1$. In this case H is discrete and all B-orbits in Ω are 1-dimensional. Unfortunately for $3<n\leqslant 7$ it seems necessary to consider a number of cases. However, with few exceptions we only need the methods listed below. Thus, instead of going through numerical calculations, we only indicate which methods are used and how.

Method I. One can always choose a Borel group B in G so that if B_H is a Borel group in H, then

$$\dim B\cap H\leqslant \dim B_H-1 \quad . \tag{*}$$

That is, the B-orbit $B/B\cap H$ has dimension at least $\dim B-\dim B_H+1$.

Proof. If (*) were violated for all such B, then the H-orbits on the flag manifold G/B are all compact and equivalent to the rational manifold $H/B_H\subsetneqq G/B$. Since the leaves of this foliation are simply-connected there is no holonomy and an orbit space Y exists,

$\dim Y \geq 1$. The projection $G/B \to Y$ is G-equivariant, and H acts trivially on Y. Thus $H \triangleright G$, contradicting the fact that G is simple.

Method II. Let B_H be a Borel group in H, and let B^+ be a Borel group in G with $B_H \subset B^+$. Let B^- be a Borel group in G which is obtained from B^+ by replacing positive root spaces with their negative counterparts. Let $T = B^+ \cap B^-$. Then

$$\dim(B^- \cap H) \leq \dim H + \dim(B_H \cap T) - \dim B_H \quad .$$

Proof. This follows immediately from the vector space decomposition $\mathfrak{g} = \mathfrak{b}^+ \oplus \mathfrak{u}^-$, where $\mathfrak{u}^-$ is the Lie algebra of the unipotent part of $\mathfrak{b}^-$.

In the following $H = R \cdot S$ will always denote a Levi-Malcev decomposition of H (R is the radical, and S is a maximal connected semisimple subgroup of H.

$\dim X = n \leq 5$: By Lemma 2, we only need consider the case when $G = A_2$. If $n = 4$ and $H = R \subset B^+$, then $\dim(H \cap B^-) \leq \dim T = 2$, and $B^-/B^- \cap H$ is at least 3-dimensional. Since one only needs a 2-codimensional B-orbit, the proof for $n = 5$ is the same.

$\dim X = n = 6$: By Lemma 2, we only need to consider $G = A_2$, $B_2 (= C_2)$, and A_3. If $G = A_2$, then H is solvable and 2-dimensional. Method I therefore yields a Borel group which does the job. If $G = B_2$, then $\dim H = 4$ and $\dim B_G = 6$. If $H = R \subset B^+$, then $\dim(B^- \cap H) \leq \dim T = 2$, and the orbit $B^-/B^- \cap H$ is at most 2-dimensional. Otherwise $S = A_1$, $\dim B_H = 3$, and Method I yields the desired result. It remains to consider $G = A_3$ where $\dim B_G = \dim H = 9$. If $H = R = B^+$, then $B^-/B^- \cap H$ is open. If $S = A_1$ or $A_1 \times A_1$ then $\dim B_H = 8$ and 7, respectively. In each of these cases Method I yields $\dim(B^- \cap H) \leq 5$. If $S = A_2$, then $\dim B_H = 6$, and Method I yields the desired result. All other possible S are too big. Thus the case $n = 6$ is finished.

$\dim X = n = 7$: Since $\dim G \geq 25$ and $G = G_2$, B_3 or C_3 have been handled by Lemma 2, the only possibilities are $G = A_2$, $B_2 (= C_2)$, A_3, and A_4. If $G = A_2$, then $H = R = B_H$ is 1-dimensional and Method I yields B with a 5-dimensional orbit. If $G = B_2$, then either $H = R$ or $H = A_1$. An application of Method I handles the latter case. If $H \subset B^+$ and $H \not\supset T$, then Method II is enough. If $H \supset T$, we note that since the generic T-orbit in G/B is 2-dimensional, $\dim(B \cap H) \leq 1$ for a generic

B. Since $\dim B = 6$, this finishes the case $G = B_2$. If $G = A_3$, then $\dim H = 8$ and $\dim B = 9$. If $H = R \subset B^+$, then $\dim(B^- \cap H) \leqslant 3$ and B^- does the job. If $S = A$ or $A_1 \times A_1$, then $\dim B_H = 7$ and $\dim B_H = 6$, respectively. Since $\dim B = 9$, Method II yields the result unless $S = A_1 \times A_1$ and $H \supset T$. In this case, the reductive part of H is $\mathbb{C}^* \times A_1 \times A_1$ and since H is not reductive, it is contained in a maximal parabolic subgroup of A_3 which must in this case be an isotropy group for the usual action of A_3 on $\mathbb{P}_3$. Thus H must consist of this reductive part plus a positive root space. An easy check shows that no such group exists. Thus, when $G = A_3$, it remains to consider the case $S = A_2$, where $\dim B_H = 5$. Thus Method I yields the result.

It remains to consider the case $G = A_4$, where $\dim B_G = 14$ and $\dim H = 17$. Routine application of Method II handles all cases except for $S = A_3$ and $A_2 \times A_2$. Since the only 5-dimensional representation of A_2 are trivial, $A_2 \times A_2$ cannot be realized in A_4. Thus it remains to consider $S = A_3$. Unless $H \supset T$, it is enough to apply Method II. If $H \supset T$, then the reductive part of H is $\mathbb{C}^* \times A_3$, and, as we argued in the case $n = 6$, H must be the product of this group with a 1-dimensional positive root space. Since H can be conjugated into the standard isotropy group of the A_4-action on $\mathbb{P}_4$, one can easily check that no such group exists. ∎

We realize that the above arguments are too tedious, but we have attempted to hold the group theoretic prerequisites to a minimum.

6. Notes

1. The first author had the additional pleasure of being Matsushima's colleague for almost a decade at the University of Notre Dame.

7. References

[1] Ahiezer, D.N., "Cohomology of compact complex homogeneous spaces," *Math. USSR*, Sbornik 13, 285-296 (1971).

[2] Ahiezer, D.N., "Dense orbits with two ends," *Math. USSR*, Izvestija 11, 293-307 (1977).

[3] Ahiezer, D.N., "Algebraic groups acting transitively in the complement of a homogeneous hypersurface," *Soviet Math. Dokl.* 20, 278-281 (1979).

[4] Akao, K., "On prehomogeneous compact Kähler manifolds," Proc. Jap. Acad. 49, 483-485 (1973).

[5] Barth, W., Oeljeklaus, E., "Über die Albanese-Abbildung einer fast homogenen Kähler-Mannigfaltigkeit," *Math. Ann.* 211, 47-62 (1974).

[6] Barth, W., Otte, M., "Über fast-uniforme Untergruppen komplexer Liegruppen and auflösbare komplexe Mannigfaltigkeiten," *Comm. Math. Helv.* 44, 269-281 (1969).

[7] Barth, W., Otte, M., "Invariante holomorphe Funktionen auf reduktiven Liegruppen," *Math. Ann.* 201, 97-112 (1973).

[8] Bialynicki-Birula, A., Hochschild, G., Mostow, G.D., "Extension of representations of algebraic linear groups," *Am. J. Math.* 85, 131-144 (1963).

[9] Blanchard, A., "Sur les variétés analytiques complexes," *Ann. Sci. Ec. Norm. Sup.* 73, 157-202 (1956).

[10] Borel, A., "Kählerian coset spaces of semisimple Lie groups," Proc. Nat. Acad. Sci. USA 40, 1147-1151 (1954).

[11] Borel, A., "Les bouts des espaces homogènes de groupes de Lie," *Ann. of Math.* 58, 443-457 (1953).

[12] Borel, A., "Symmetric compact complex spaces," *Arch. Math.* 33, 49-56 (1979).

[13] Borel, A., Remmert, R., "Über kompakte homogene Kählersche Mannigfaltigkeiten," *Math. Ann.* 145, 429-439 (1962).

[14] Brenton, L., Morrow, J., "Compactifications of $\mathbb{C}^n$," *Trans. Am. Math. Soc.* 246, 139-153 (1978).

[15] Burns, D., Shnider, S., "Spherical hypersurfaces in complex manifolds," *Inv. Math.* 33, 223-246 (1976).

[16] Carrell, J., Howard, A., Kosniowski, C., "Holomorphic vector fields on complex surfaces," *Math. Ann.* 204, 73-81 (1973).

[17] Chevalley, C., *Théorie des groupes de Lie*, Hermann, Paris (1968).

[18] Freudenthal, H., "Über die Enden topologischer Räume und Gruppen," *Math. Z.* 33, 692-713 (1931).

[19] Gilligan, B., "Ends of complex homogeneous manifolds having non-constant holomorphic functions" (to appear).

[20] Gilligan, B., Huckleberry, A., "On non-compact complex nil-manifolds," *Math. Ann.* 238, 39-49 (1978).

[21] Gilligan, B., Huckleberry, A., "Complex homogeneous manifolds with two ends," *Mich. J. Math.* (to appear).

[22] Goto, M., "On algebraic homogeneous spaces," *Am. J. Math.* 76, 811-818 (1954).

[23] Grauert, H., "Analytische Faserungen über holomorph-vollständigen Räumen," *Math. Ann.* 135, 263-273 (1958).

[24] Grauert, H., "Bemerkenswerte pseudokonvexe Mannigfaltigkeiten," *Math. Z.* 81, 377-391 (1963).

[25] Grauert, H., Remmert, R., "Über kompakte homogene komplexe Mannigfaltigkeiten," *Arch. Math.* 13, 498-507 (1962).

[26] Hano, J., Matsushima, Y., "Some studies on Kählerian homogeneous spaces," *Nagoya Math. J.* 11, 77-92 (1957).

[27] Harish-Chandra, "Representations of a semi-simple Lie group on a Banach space," *Trans. AMS* 70, 28-96 (1953).

[28] Hochschild, G., Mostow, G.D., "Representations and representative functions of Lie groups," *Ann. of Math.* 66, 495-542 (1957).

[29] Hochschild, G., Mostow, G.D., "Representations and representative functions of Lie groups, III," *Ann. of Math.* 70, 85-100 (1959).

[30] Hochschild, G., Mostow, G.D., "Affine embeddings of complex analytic homogeneous spaces," *Am. J. Math.* 87, 807-839 (1965).

[31] Huckleberry, A., Livorni, L., "A classification of complex homogeneous surfaces," *Can. J. Math.* (to appear).

[32] Huckleberry, A., Oeljeklaus, E., "A characterization of complex homogeneous cones," *Math. Z.* 170, 181-194 (1980).

[33] Huckleberry, A., Oeljeklaus, E., "Sur les espaces analytiques complexes presque homogènes," C.R. Acad. Sc. Paris, Serie A, 447-448 (1980).

[34] Huckleberry, A., Snow, D., "Pseudoconcave homogeneous manifolds," *Ann. Scuola Norm. Sup.* Pisa, Serie IV, Vol. VII, 29-54 (1980).

[35] Huckleberry, A., Snow, D., "A classification of strictly pseudoconcave homogeneous manifolds," *Ann. Scuola Norm. Sup.* Pisa (to appear).

[36] Huckleberry, A., Snow, D., "Almost-homogeneous Kähler manifolds with hypersurface orbits" (to appear).

[37] Malcev, A.J., "On a class of homogeneous spaces," *AMS Trans.* 39 (1951).

[38] Matsumura, H., "On algebraic groups of birational transformation," *Rend. Acad. Naz. Lincei*, Ser. VII, 34, 151-155 (1963).

[39] Matsushima, Y., "Sur les espaces homogènes kähleriens d'un groupe de Lie reductif," *Nagoya Math. J.* 11, 53-60 (1957).

[40] Matsushima, Y., "Espaces homogènes de Stein des groupes de Lie complexes I," *Nagoya Math. J.* 16, 205-218 (1960).

[41] Matsushima, Y., "Sur certains variétés homogènes complexes," *Nagoya Math. J.* 18, 1-12 (1961).

[42] Matsushima, Y., "Espaces homogènes de Stein des groupes de Lie complexes II," *Nagoya Math. J.* 18, 153-164 (1961).

[43] Matsushima, Y., Morimoto, A., "Sur certains espaces fibrés holomorphes sur une variété de Stein," *Bull. Soc. Math. France* 88, 137-155 (1960).

[44] Montgomery, D., "Simply connected homogeneous spaces," Proc. Am. Math. Soc. 1, 467-469 (1950).

[45] Morimoto, A., "Non-compact complex Lie groups without non-constant holomorphic functions," Proc. of the Conf. on Complex Analysis, Minneapolis 1964, 256-272.

[46] Morimoto, A., "On the classification of non-compact abelian Lie groups," *Trans. AMS* 123, 200-228 (1966).

[47] Morimoto, Y., Nagano, T., "On pseudo-conformal transformations of hypersurfaces," *J. Math. Soc. Japan* 14, 289-300 (1963).

[48] Mostow, G.D., "On covariant fiberings of Klein spaces," *Am. J. Math.* 77, 247-278 (1955).

[49] Mostow, G.D., "Some applications of representative functions to solv-manifolds," *Am. J. Math.* 93, 11-32 (1971).

[50] Mostow, G.D., "Factor spaces of solvable groups," *Ann. of Math.* 60, 1-27 (1954).

[51] Oeljeklaus, E.,"Über fast homogene kompakte komplexe Mannigfaltigkeiten," *Schritenr. Math. Inst. Univ. Münster*, 2. Serie, Bd. 1 (1970).

[52] Oeljeklaus, E., "Ein Hebbarkeitssatz für Automorphismengruppen kompakter komplexer Mannigfaltigkeiten," *Math. Ann.* 190, 154-166 (1970).

[53] Oeljeklaus, E., "Fast homogene Kählermannigfaltigkeiten mit verschwindender erster Bettizahl," *Manuskr. Math.* 7, 175-183 (1972).

[54] Potters, J., "On almost homogeneous compact complex surfaces," *Inv. Math.* 8, 244-266 (1969).

[55] Remmert, R., van de Ven, T., "Zwei Sätze über die komplex-projektive Ebene," *Nieuw. Arch. Wisk.* (3), 8, 147-157 (1960).

[56] Remmert, R., van de Ven, T., "Zur Funktionentheorie homogener komplexer Mannigfaltigkeiten," *Topologie* 2, 137-157 (1963).

[57] Richardson, R.W., "Affine coset spaces of reductive algebraic groups," *Bull. London Math. Soc.* 9, 38-41 (1977).

[58] Rosenlicht, M., "Some basic theorems on algebraic groups," *Am. J. Math.* 78, 401-443 (1956).

[59] Rossi, H., "Homogeneous strongly pseudoconvex hyperfaces," *Rice Studies* 59(3), 131-145 (1973).

[60] Serre, J.-P., "Quelques problèmes globaux relatifs aux variétés de Stein," Colloque sur les fonctions de plusieurs var. Bruxelles 1953, 57-68.

[61] Skoda, H., "Fibrés holomorphes à base et à fibre de Stein," *Inv. Math.* 43, 97-107 (1977).

[62] Snow, J., "Complex solv-manifolds of dimension two and three," Thesis, Notre Dame University (1979).

[63] Stein, K., "Analytische Zerlegungen komplexer Räume," *Math. Ann.* 132, 63-93 (1956).

[64] Tits, J., "Espaces homogènes complexes compacts," *Comm. Math. Helv.* 37, 111-120 (1962).

[65] Ueno, K., "Classification theory of algebraic varieties and compact complex spaces," Lecture Notes in Mathematics, 439, Berlin-Heidelberg-New York: Springer 1975.

[66] van de Ven, T., "Analytic compactifications of complex homology cells," *Math. Ann.* 147, 189-204 (1962).

[67] Wang, H.C., "Closed manifolds with homogeneous complex structure," *Am. J. Math.* 76, 1-32 (1954).

[68] Wang, H.C., "Complex parallelisable manifolds," Proc. Am. Math. Soc. 5, 771-776 (1954).

A. Huckleberry
Fachbereich Mathematik
Universität Bochum
D-4630 Bochum, FRG

E. Oeljeklaus
Fachbereich Mathematik
Universität Bremen
D-2800 Bremen 33, FRG

(Received February 15, 1981)

ON LIE ALGEBRAS GENERATED BY TWO DIFFERENTIAL OPERATORS[1]

Jun-ichi Igusa

0. Introduction. We shall denote by K a field of characteristic 0, by $K[x]$ the ring of polynomials in r variables $x_1,\ldots,x_r$ with coefficients in K, and by D_i the K-derivation in $K[x]$ defined by $D_i x_j = \delta_{ij}$ for $1 \leqq i,\ j \leqq r$; then the multiplications by $x_1,\ldots,x_r$ in $K[x]$ and $D_1,\ldots,D_r$ generate a subalgebra A of the associative K-algebra of all K-linear transformations in $K[x]$. An element X of A can be written uniquely in the form

$$X = \sum a_{i_1\ldots i_r j_1\ldots j_r} x_1^{i_1}\ldots x_r^{i_r} D_1^{j_1}\ldots D_r^{j_r}$$

with $a_{i_1\ldots i_r j_1\ldots j_r}$ in K; it is a linear differential operator with polynomial coefficients. In this paper we shall prove a criterion for two such elements X, Y to generate an infinite dimensional Lie algebra over K; and as its application we shall prove the following remarkable fact:

If $f(x)$ is a homogeneous polynomial in $x_1,\ldots,x_r$ of degree m with coefficients in $\mathbb{R}$ and if we denote by U_t for every t in $\mathbb{R}$ the multiplication by $\exp(2\pi i t\cdot f(x))$ in $L^2(\mathbb{R}^r)$ and by V_t the conjugate of U_t under the Fourier transformation, then U_t, V_t always generate an infinite dimensional group for $m > 2$ in contrast to the classical theorem stating that they generate three-dimensional groups, a Heisenberg group for $m = 1$ and a metaplectic group for $m = 2$.

1. We shall denote elements of K^{2r} by column vectors ξ,η, etc. and convert K^{2r} into a symplectic space via the standard alternating form

$$B(\xi,\eta) = {}^t\xi E\eta = {}^t\xi \begin{pmatrix} 0 & 1_r \\ -1_r & 0 \end{pmatrix} \eta .$$

If X, Y are elements of any associative algebra, we put $[X,Y] = YX - XY$. The K-algebra A in the Introduction is the unique associative algebra generated over K by $2r$ elements $x_1,\dots,x_r$, $D_1,\dots,D_r$ with the Heisenberg commutation relations

$$[x_i,x_j] = [D_i,D_j] = 0, \quad [x_i,D_j] = \delta_{ij}$$

as its defining relations. Therefore if $\xi_1,\dots,\xi_r$, $\eta_1,\dots,\eta_r$ are $2r$ elements of K^{2r} and if, e.g., ξ_{ij} denotes the i-th coefficient of ξ_j for $1 \leqq i \leqq 2r$, then the K-linear change of generators

$$x_i' = \sum_{j=1}^{r} (\xi_{ij}x_j + \eta_{ij}D_j) \quad , \quad D_i' = \sum_{j=1}^{r} (\xi_{r+i,j}x_j + \eta_{r+i,j}D_j)$$

gives rise to an automorphism of A if and only if the matrix

$$\sigma = {}^t(\xi_1 \dots \xi_r \eta_1 \dots \eta_r)$$

has the property ${}^t\sigma E \sigma = E$. This can be rewritten as $\sigma E {}^t\sigma = E$ or, equivalently, as

$$B(\xi_i,\xi_j) = B(\eta_i,\eta_j) = 0, \quad B(\xi_i,\eta_j) = \delta_{ij} \quad ;$$

matrices such as σ form the symplectic group $Sp_{2r}(K)$. For the sake of clarity we mention that σX for any X in A is the element of A obtained from X under the substitution $x_i \to x_i'$, $D_i \to D_i'$ for $1 \leqq i \leqq r$.

If we replace $D_1,\dots,D_r$ by $y_1,\dots,y_r$ subject to $[x_i,y_j] = 0$ instead of $[x_i,D_j] = \delta_{ij}$, we get a ring of polynomials $K[x,y]$. If X is an element of A, we shall denote by P_X the highest homogeneous part of an element of $K[x,y]$ obtained from X under the substitution $D_i \to y_i$ for $1 \leqq i \leqq r$. We observe that P_X is uniquely determined by X. We also mention that σF for any F in $K[x,y]$ is the element of $K[x,y]$ defined by

$$(\sigma F)\binom{x}{y} = F({}^t\sigma\binom{x}{y}) \quad ,$$

in which, e.g., x denotes the column vector with x_i as its i-th coefficient for $1 \leqq i \leqq r$. In the following lemmas ∇ stands for the taking of a gradient:

Lemma 1. *We have* $\sigma P_X = P_{\sigma X}$ *and* $\sigma B(\nabla P_X, \nabla P_Y) = B(\nabla P_{\sigma X}, \nabla P_{\sigma Y})$ *for every* X, Y *in* A *and* σ *in* $Sp_{2r}(K)$. *Furthermore if*

$B(\nabla P_X, \nabla P_Y) \neq 0$, *then* $P_{[X,Y]} = B(\nabla P_X, \nabla P_Y)$.

Lemma 2. *If we put* ${}^t\sigma = (\xi_1 \dots \xi_r \eta_1 \dots \eta_r)$, *then for every* F *in* $K[x,y]$ *of degree* m *we have*

$$\sigma F = F(\eta_1) y_1^m + {}^t(\nabla F)(\eta_1) \left(\sum_{i=1}^{r} x_i \xi_i + \sum_{i=2}^{r} y_i \eta_i \right) y_1^{m-1} + \cdots ,$$

in which the unwritten part has a smaller degree in y_1 *than* $m-1$.

The verifications of the above lemmas are straightforward; hence they will be omitted.

2. We shall discuss a certain normalization of a pair of elements of A in the following lemma:

Lemma 3. *Let* X, Y *denote elements of* A *and put* $m = \deg(P_X)$, $n = \deg(P_Y)$; *assume that for some* ξ *in* K^{2r} (i) $P_X(\xi) P_Y(\xi) \neq 0$, (ii) $B(\nabla P_X, \nabla P_Y)(\xi) \neq 0$, (iii) $(\nabla P_X)(\xi)$, $(\nabla P_Y)(\xi)$, $E\xi$ *are linearly dependent. Then there exists an element* σ *of* $Sp_{2r}(K)$ *such that*

$$\sigma P_X = a_0 y_1^m + a_1 x_1 y_1^{m-1} + \cdots, \quad \sigma P_Y = b_0 y_1^n + b_1 x_1 y_1^{n-1} + \cdots,$$

in which $a_0 b_0 \neq 0$, $na_1 b_0 - ma_0 b_1 \neq 0$, *and the unwritten parts have smaller degrees in* y_1.

Proof. We put

$$\xi_1 = (mP_X(\xi))^{-1} E(\nabla P_X)(\xi), \qquad \eta_1 = \xi ;$$

the definition is legitimate because $P_X(\xi) \neq 0$ by (i). If we use Euler's identity, we get $B(\xi_1, \eta_1) = 1$. Therefore we can find an element σ of $Sp_{2r}(K)$ such that ${}^t\sigma = (\xi_1 \dots \xi_r \eta_1 \dots \eta_r)$ for some $\xi_2, \dots, \xi_r$, $\eta_2, \dots, \eta_r$ in K^{2r}. Since ξ_1, $E(\nabla P_Y)(\xi)$, η_1 are linearly dependent by (iii), we can write

$$E(\nabla P_Y)(\xi) = \alpha \xi_1 + \beta \eta_1$$

with some α, β in K. If we use Euler's identity again, we get

$$\alpha = nP_Y(\xi) \quad , \qquad \beta = (mP_X(\xi))^{-1} B(\nabla P_X, \nabla P_Y)(\xi) \quad .$$

If we apply Lemma 2 to $F = P_X$, we get

$$\sigma P_X = P_X(\xi)y_1^m + mP_X(\xi)\cdot B\left(\xi_1, \sum_{i=1}^r x_i\xi_i + \sum_{i=2}^r y_i\eta_i\right) y_1^{m-1} + \cdots$$

$$= a_0y_1^m + a_1x_1y_1^{m-1} + \cdots ,$$

in which $a_0 = P_X(\xi) \neq 0$ and $a_1 = 0$. In the same way we get

$$\sigma P_Y = b_0y_1^n + b_1x_1y_1^{n-1} + \cdots ,$$

in which $b_0 = P_Y(\xi) \neq 0$, $b_1 = -\beta$, and

$$na_1b_o - ma_0b_1 = B(\nabla P_X, \nabla P_Y)(\xi) \neq 0 .$$

■

3. We shall prove our criterion in a preliminary form after the following lemma:

Lemma 4. *Let* X,Y *denote elements of* A *such that*

$$\begin{cases} P_X = a_0y_1^m + \left(a_1x_1 + \sum_{i=2}^r (a_i'x_i + a_i''y_i)\right) y_1^{m-1} + \cdots \\ P_Y = b_0y_1^n + \left(b_1x_1 + \sum_{i=2}^r (b_i'x_i + b_i''y_i)\right) y_1^{n-1} + \cdots ; \end{cases}$$

put

$$c(X,Y) = (na_1b_0 - ma_0b_1) + \sum_{i=2}^r (a_i'b_i'' - a_i''b_i') ;$$

then we have $[X,Y] = c(X,Y)D_1^{m+n-2} + \cdots$.

This follows from the formula $P_{[X,Y]} = B(\nabla P_X, \nabla P_Y)$ in Lemma 1 or rather from the calculation behind that formula.

Lemma 5. *Let* X,Y *denote elements of* A *such that*

$$P_X = a_0y_1^m + a_1x_1y_1^{m-1} + \cdots, \quad P_Y = b_0y_1^n + b_1x_1y_1^{n-1} + \cdots ,$$

in which $a_0b_0 \neq 0$; *suppose that* $m, n > 2$ *and* $c(X,Y) \neq 0$. *Then* X,Y *generate an infinite dimensional Lie algebra over* K.

Proof. If we put $Z_1 = [X,Y]$, by Lemma 4 we get

$$P_{Z_1} = c_0 y_1^{m+n-2} + \left(c_1 x_1 + \sum_{i=2}^{r} (c_i' x_i + c_i'' y_i)\right) y_1^{m+n-3} + \cdots ,$$

in which $c_0 = c(X,Y) \neq 0$. Suppose that $c(X,Z_1) = c(Y,Z_1) = 0$; then both (ma_0, a_1) and (nb_0, b_1) are linearly dependent on $((m+n-2)c_0, c_1) \neq (0,0)$, hence $(nb_0)a_1 - (ma_0)b_1 = c(X,Y) = 0$, a contradiction. If $c(X,Z_1) \neq 0$, we put $Z_2 = [X,Z_1]$; then we get

$$P_{Z_2} = c(X,Z_1) y_1^{2m+n-4} + \cdots .$$

If $C(X,Z_1) = 0$, hence $c(Y,Z_1) \neq 0$, we put $Z_2 = [Y,Z_1]$; then we get

$$P_{Z_2} = c(Y,Z_1) y_1^{m+2n-4} + \cdots .$$

We can then process Z_2 just as we have processed Z_1. In that way we will get a sequence $Z_1, Z_2, Z_3, \ldots$ in the Lie algebra generated by X, Y such that

$$P_{Z_k} = c^{(k)} y_1^{im+jn-2k} + \cdots ,$$

in which $c^{(k)} \neq 0$, $i+j = k+1$, $i, j \geqq 1$, for $k = 1,2,3,\ldots$. Since $m, n > 2$ by assumption, we have

$$\deg\left(P_{Z_k}\right) = im + jn - 2k \geqq 3(i+j) - 2k = k+3 .$$

Therefore the Lie algebra generated by X,Y is infinite dimensional. ■

4. If we put Lemmas 3, 5 together and use the obvious fact that for any σ in $Sp_{2r}(K)$ the Lie algebra generated by X, Y is isomorphic to the Lie algebra generated by σX, σY, we get the following theorem:

Theorem 1. *Let* X,Y *denote linear differential operators in* r *variables* $x_1,\ldots,x_r$ *with polynomial coefficients over any field* K *of characteristic* 0; *let* P_X, P_Y *denote the highest homogeneous parts of the polynomials in* $x_1,\ldots,x_r$, $y_1,\ldots,y_r$ *obtained from* X,Y, *respectively, under the substitution* $D_i = \partial/\partial x_i \to y_i$ *for* $1 \leqq i \leqq r$; *let* $B(\xi,\eta) = {}^t\xi E \eta$ *denote the standard alternating form on* K^{2r}. *Suppose that* $\deg(P_X)$, $\deg(P_Y) > 2$ *and that*

(i) $P_X(\xi)P_Y(\xi) \neq 0$,

(ii) $B(\nabla P_X, \nabla P_Y)(\xi) \neq 0$,

(iii) $(\nabla P_X)(\xi), (\nabla P_Y)(\xi)$, $E\xi$ *are linearly dependent*

for some ξ *in* K^{2r}. *Then* X,Y *generate an infinite dimensional Lie algebra over* K.

We observe that if X, Y satisfy the second condition in the theorem, i.e., (i)-(iii), for ξ, then σX, σY for any σ in $Sp_{2r}(K)$ satisfy the same condition for $\eta = {}^t\sigma^{-1}\xi$. This follows from Lemma 1 and from the fact that $(\nabla P_{\sigma X})(\eta)$, $(\nabla P_{\sigma Y})(\eta)$, $E\eta$ are respectively equal to $(\nabla P_X)(\xi)$, $(\nabla P_Y)(\xi)$, $E\xi$ all multiplied by σ. In this sense the conditions in the theorem are $Sp_{2r}(K)$-invariant.

Corollary. *In the special case where* X, Y *are ordinary differential operators, if we put* $m = \deg(P_X)$, $n = \deg(P_Y)$, *the conditions in Theorem* 1 *simply become* $m, n > 2$ *and* $(P_X)^n \neq \mathit{const.}(P_Y)^m$.

In fact in the case where $r = 1$, the second condition is not satisfied if and only if

$$B(\nabla P_X, \nabla P_Y) = \partial(P_X, P_Y)/\partial(x,y) = 0 \quad ;$$

we have tacitly used the consequence $\mathrm{card}(K) = \infty$ of $\mathrm{char}(K) = 0$. Since P_X, P_Y are homogeneous of degrees m, n, respectively, this is equivalent to

$$n(\partial P_X/\partial x)P_Y - m(\partial P_Y/\partial x)P_X = 0 \quad ,$$

hence to $\partial((P_X)^n/(P_Y)^m)/\partial x = 0$. Therefore $(P_X)^n/(P_Y)^m$ is a homogeneous element of $K(x,y)$ of degree 0 contained in $K(y)$, hence it is in K.

5. We shall make a digression: for any r if $\deg(P_X)$, $\deg(P_Y) \leqq 2$, clearly X,Y generate a finite dimensional Lie algebra over K. A less obvious fact is that *the class of Lie algebras generated by* X,Y *for which* $\deg(P_X) = \deg(P_Y) = 2$ *contains all* K*-split semisimple Lie algebras;* the proof is as follows:

If $\mathfrak{g}$ is any finite dimensional Lie algebra over K and if $X_1, \ldots, X_r$ form a basis for $\mathfrak{g}$, then we have

$$[X_i, X_j] = \sum_{k=1}^{r} c_{ijk} X_k$$

with c_{ijk} in K; and the linear differential operators

$$Y_i = \sum_{j,k=1}^{r} c_{ijk} x_j D_k$$

generate a Lie algebra over K homomorphic to $\mathfrak{g}$ with the center of $\mathfrak{g}$ as its kernel under $X_i \to Y_i$ for $1 \leqq i \leqq r$. On the other hand if $\mathfrak{g}$ is a K-split semisimple Lie algebra and if

$$\mathfrak{g} = \mathfrak{h} + \sum_{\alpha \in \Phi} K X_\alpha$$

is its root space decomposition or Cartan decomposition in the standard notation, then $\mathfrak{g}$ is generated by a general element X of $\mathfrak{h}$ and the sum Y of all X_α; the condition on X is that it is not on any one of the hyperplanes $(\alpha - \beta)(h) = 0$ for $\alpha \neq \beta$ in Φ. In fact if we put $\ell = \dim(\mathfrak{h})$, then the system of affine linear equations

$$\sum_{\alpha \in \Phi} \alpha(X)^i X_\alpha = (\mathrm{ad}(X))^i Y$$

for $0 \leqq i < r - \ell$ can be solved in the X_α's. Therefore the Lie algebra generated by X, Y contains all X_α's, hence all $[X_\alpha, X_{-\alpha}]$'s, and hence it coincides with $\mathfrak{g}$.

6. We shall return from the digression. We shall consider the special case where $K = \mathbb{R}$ and derive the following theorem from Theorem 1.

Theorem 2. *Let* $f(x)$ *denote an arbitrary homogeneous polynomial of degree* m *in* r *variables* $x_1, \ldots, x_r$ *with coefficients in* $\mathbb{R}$; *let* U_t *for any* t *in* $\mathbb{R}$ *denote the multiplication by* $\exp(2\pi i t \cdot f(x))$ *in* $L^2(\mathbb{R}^r)$ *and* V_t *the conjugate of* U_t *under the Fourier transformation. Then* U_t, V_t *for all* t *generate a finite dimensional subgroup of the unitary group of* $L^2(\mathbb{R}^r)$ *if and only if* $m \leqq 2$.

Proof. We put $X = f(x)$ and $Y = f(D)$; then we have

$$(dU_t/dt)_{t=0} = 2\pi i X, \qquad (dV_t/dt)_{t=0} = \text{const.} Y,$$

in which the "const." is $(2\pi i)^{1-m}$ if the Fourier transformation is defined with respect to $\exp(2\pi i \cdot {}^t xy)$. Therefore we have only to show that the above X, Y generate an infinite dimensional Lie algebra over $\mathbb{R}$ if $m > 2$. Since the first condition in Theorem 1 is satisfies by assumption, we shall show that the second condition is also satisfied.

We restrict x to the unit sphere $\|x\| = ({}^t xx)^{1/2} = 1$ and choose a point a where $|f(x)|$ attains its maximum; then $f(a) \neq 0$ and $(\nabla f)(a) = \lambda a$ for some λ in $\mathbb{R}$. In view of ${}^t a(\nabla f)(a) = mf(a) \neq 0$ we then get $(\nabla f)(a) \neq 0$. We shall denote by ξ the element of $\mathbb{R}^{2r}$ with a as its first and the second entry vectors; then we have $P_X(\xi)P_Y(\xi) = f(a)^2 > 0$,

$$B(\nabla P_X, \nabla P_Y)(\xi) = \|(\nabla f)(a)\|^2 > 0$$

in view of $f(a) \neq 0$, $(\nabla f)(a) \neq 0$, and $(\nabla P_X)(\xi) - (\nabla P_Y)(\xi) - \lambda E\xi = 0$. These are respectively (i), (ii), (iii) in Theorem 1. ■

We might as well explain the well-known fact that the Lie algebra generated by X, Y is isomorphic to the Heisenberg Lie algebra $\mathbb{R}P + \mathbb{R}Q + \mathbb{R}1$, where $[P,Q] = 1$, if $m = 1$ and to the Lie algebra of $SL_2(\mathbb{R})$ if $m = 2$: in the first case by a change of coordinates in $\mathbb{R}^r$ we may assume that $f(x) = x_1$; then $\theta(P) = X$, $\theta(Q) = Y$ gives such an isomorphism. In the second case, again by a change of coordinates, we may assume that $f(x) = 1/2(a_1x_1^2 + \cdots + a_rx_r^2)$ with $a_i^3 = a_i$ for every i and $a_i \neq 0$ for at least one i; then

$$\theta\begin{pmatrix} 0 & 1 \\ 0 & 0 \end{pmatrix} = X, \quad \theta\begin{pmatrix} 0 & 0 \\ -1 & 0 \end{pmatrix} = Y$$

gives rise to such an isomorphism.

7. We shall briefly explain the background of this paper: in Vol. 111 and Vol. 113 of the *Acta mathematica* Weil clarified the roles of the metaplectic groups and the Siegel formula in the theory of quadratic forms. In his thesis of 1965, which appeared in Vol. 94 of the *Bulletin de la Société mathématique de France*, Mars proved a Siegel formula for the norm form of an exceptional simple Jordan algebra, which is a cubic form. Later in Vol. 47 of the *Nagoya Mathematical Journal* we proved a Siegel formula for the norm form of the simple Jordan algebra of quaternionic hermitian matrices of degree m, which is an m-form. If k denotes an algebraic number field which contains the coefficients of such a form, the Siegel formula gives rise to a

distribution which is invariant under the group of k-rational points of a "hypermetaplectic group." And Theorem 2 shows that the local hypermetaplectic group over $\mathbb{R}$ is infinite dimensional. Although it may take many years for us to be able to "use" such a group, our view is that it should not be discarded as pathological. In fact through the investigation of such groups we may be able to really understand the circle method in the theory of Diophantine equations.

Finally, as far as we know, our theorems are still new; the Corollary of Theorem 1 has been known to us for more than five years. At any rate we have learned from Guillemin the following result by Spafford which may have some relevance to this paper: the associative algebra A is simple and every left ideal of A is generated by two elements.

8. Notes

1. This work was partially supported by the National Science Foundation.

The Johns Hopkins University
Baltimore, Maryland 21218 USA

(Received December 19, 1980)

CONFORMALLY-FLATNESS AND STATIC SPACE-TIME

Osamu Kobayashi and Morio Obata

0. Introduction

A Lorentzian (n+1)-manifold $(\tilde{M}^{n+1},\tilde{g})$ is called (globally) *static* [1], [2] if $\tilde{M}$ is a product space $\mathbb{R}\times M$ of $\mathbb{R}$ with an n-manifold M^n and the metric $\tilde{g}$ has the form

$$\tilde{g} = -f(x)^2 dt^2 + x^*g \quad , \tag{0.1}$$

where $t:\tilde{M}\to\mathbb{R}$ and $x:\tilde{M}\to M$ are the natural projections, g a Riemannian metric on M, and f a positive function on M. We consider Einstein's equation on $(\tilde{M},\tilde{g})$ with perfect fluid as a matter field, i.e.,

$$\widetilde{Ric} - \frac{1}{2}\tilde{R}\tilde{g} = (\mu + p)\eta\otimes\eta + p\tilde{g} \quad , \tag{0.2}$$

where η is a 1-form with $\tilde{g}(\eta,\eta) = -1$, whose associated vector field represents the flux of the fluid, and μ and p are functions on $\tilde{M}$ which are called the *energy density* and the *pressure*, respectively [1]. In other words, (0.2) says that, at each point of $\tilde{M}$, the Ricci tensor $\widetilde{Ric}$ has at most two distinct eigenvalues with multiplicities 1 and n. It is known [2] that under the condition (0.1), (0.2) is equivalent to the following equation on (M,g);

$$Ric - \frac{\text{Hess } f}{f} = \frac{1}{n}\left(R - \frac{\Delta f}{f}\right) g \quad , \tag{0.3}$$

and then, there are relations;

$$\mu = \frac{R}{2} \quad ,$$

$$p = \frac{n-1}{n}\left(\frac{\Delta f}{f} - \frac{n-2}{2(n-1)} R\right). \tag{0.4}$$

In this paper we shall give some results on the equation (0.3) in the case when M or $\tilde{M}$ is conformally flat.

1. Preliminaries

Let (M,g) be a Riemannian n-manifold and f a positive function on M. We concentrate our attention mainly on the equation mentioned above:

$$\mathrm{Ric} - \frac{\mathrm{Hess}\ f}{f} = \frac{1}{n}\left(R - \frac{\Delta f}{f}\right) g \quad , \tag{1.1}$$

where Ric and R are the Ricci tensor and the scalar curvature respectively and Hess denotes the Hessian, i.e., $\mathrm{Hess}\ f = \nabla^2 f$.

Taking the divergence of (1.1) we have

$$d(Rf + (n-1)\Delta f) = \frac{n}{2} f dR \ , \tag{1.2}$$

or equivalently

$$d\left(\frac{n-2}{2(n-1)} R - \frac{\Delta f}{f}\right) = \frac{1}{f^2}\left(\frac{Rf}{n-1} + \Delta f\right) df \quad . \tag{1.3}$$

From (1.2) we see easily

$$df \wedge dR = dR \wedge d\Delta f = d\Delta f \wedge df = 0 \quad . \tag{1.4}$$

Define a symmetric 2-tensor L by

$$L = \mathrm{Ric} - \frac{R}{2(n-1)} g \quad . \tag{1.5}$$

Then (1.1) implies

$$L = \frac{\mathrm{Hess}\ f}{f} + \frac{1}{n}\left(\frac{n-2}{2(n-1)} R - \frac{\Delta f}{f}\right) g \quad . \tag{1.6}$$

From (1.3) and (1.6), the conformal curvature tensor C of Weyl and Schouten of type $(0,3)$ can be written as follows:

$$C(U,X,Y) = \underset{X,Y}{\mathrm{Alt}}\ (\nabla_X L)(Y,U)$$

$$= \underset{X,Y}{\mathrm{Alt}} \left\{ \frac{1}{f} (\nabla_X \text{ Hess } f)(Y,U) - \frac{X\cdot f}{f^2} \text{Hess } f(Y,U) + \frac{X\cdot f}{nf^2} \left(\frac{Rf}{n-1} + \Delta f \right) g(Y,U) \right\}$$

$$= \frac{1}{f} g(R(X,Y)\text{grad } f,U) - \underset{X,Y}{\mathrm{Alt}} \left[\frac{X\cdot f}{f^2} \left\{ \text{Hess } f(Y,U) - \frac{1}{n} \left(\frac{Rf}{n-1} + \Delta f \right) g(Y,U) \right\} \right]$$
$$U,X,Y \in T_xM \quad , \tag{1.7}$$

where Alt denotes the alternation, e.g., $\mathrm{Alt}_{X,Y}(\ldots X\ldots Y\ldots) = (\ldots X\ldots Y\ldots) - (\ldots Y\ldots X\ldots)$, and $R(\cdot,\cdot)$ is the curvature tensor for the Riemannian connection i.e., $R(X,Y) = [\nabla_X,\nabla_Y] - \nabla_{[X,Y]}$. Especially, if $U = \text{grad } f$ in (1.7), we have

$$C(\text{grad } f,X,Y) = \underset{X,Y}{\mathrm{Alt}} \left\{ -\frac{X\cdot f}{f^2} \text{Hess } f(Y,\text{grad } f) \right\}$$

$$= -\underset{X,Y}{\mathrm{Alt}} \left\{ \frac{X\cdot f}{f^2} \left(Y\cdot \frac{|df|^2}{2} \right) \right\}$$

$$= -\frac{1}{f^2} (df\wedge d|df|^2)(X,Y), \quad X,Y \in T_xM \quad . \tag{1.8}$$

On the other hand, Weyl's conformal curvature tensor W of type (1.3) is defined by

$$W(X,Y)Z = R(X,Y)Z - \frac{1}{n-2} \underset{X,Y}{\mathrm{Alt}} \left\{ L(Z,Y)X + g(Z,Y)L^{\#}(X) \right\}, \quad X,Y,Z \in T_xM, \tag{1.9}$$

where $L^{\#}$ is a (1,1)-tensor associated with L, i.e., $g(L^{\#}(X),Y) = L(X,Y)$ for $X,Y \in T_xM$. Putting $Z = \text{grad } f$ in (1.9) and contracting with U, we obtain from (1.6)

$$g(R(X,Y)\text{grad } f,U) - g(W(X,Y)\text{grad } f,U)$$

$$= \frac{1}{n-2} \underset{X,Y}{\mathrm{Alt}} \left\{ \frac{Y\cdot f}{f} \text{ Hess } f(X,U) + \frac{Y\cdot |df|^2}{2f} g(X,U) - \frac{Y\cdot f}{n} \left(\frac{n-2}{n-1} R - \frac{2\Delta f}{f} \right) g(X,U) \right\} \quad , \quad U,X,Y \in T_xM \quad . \tag{1.10}$$

Thus, substituting (1.10) to (1.7), we have

$$(n-2)f^2C(U,X,Y) - (n-2)fg(W(X,Y)\operatorname{grad} f,U)$$

$$= \operatorname*{Alt}_{X,Y}\left[(n-1)(Y\cdot f)\operatorname{Hess} f(X,U) + \left\{\frac{Y\cdot|df|^2}{2}(\Delta f)(Y\cdot f)\right\} g(X,U)\right], \quad U,X,Y \in T_xM. \tag{1.11}$$

These formulas will be used in Section 3.

2. A Generalization of the Schwarzschild Solution

Before entering into the case of a conformally flat M, we consider a function f in Section 1 on some special Riemannian manifold, which will be a model of the manifold considered in the next section.

Let I be an open interval, $(N,\bar{g})$ a Riemannian $(n-1)$-manifold and $r(s)$ a positive function on I. In this section, we assume $M = I\times_r N$, i.e., M is a product space $I\times M$ with projections $s:M\to I$ and $\pi:M\to N$ and the metric g of M has the form

$$g = ds^2 + r(s)^2\pi^*\bar{g} \quad . \tag{2.1}$$

Throughout this section ξ denotes the natural lift to M of the canonical vector field d/ds on I and A,B denote vector fields (or vectors) orthogonal to ξ. A tangent vector orthogonal to ξ and its projection to N will be denoted by the same letters.

From (2.1), a straightforward calculation shows

$$\nabla_\xi\xi = 0, \qquad \nabla_A\xi = \frac{\dot{r}}{r}A \quad ,$$

$$\nabla_AB = \bar{\nabla}_AB - r\dot{r}\bar{g}(A,B) \quad , \tag{2.2}$$

where ∇ and $\bar{\nabla}$ are the covariant differentiation of (M,g) and $(N,\bar{g})$ respectively, and $\dot{r} = dr/ds$. Hence,

$$\operatorname{Ric}(\xi,\xi) = -(n-1)\ddot{r}/r, \qquad \operatorname{Ric}(\xi,A) = 0 \quad ,$$

$$\operatorname{Ric}(A,B) = \overline{\operatorname{Ric}}(A,B) - (r\ddot{r} + (n-2)\dot{r}^2)\bar{g}(A,B) \quad , \tag{2.3}$$

$$R = -2(n-1)\frac{\ddot{r}}{r} - (n-1)(n-2)\left(\frac{\dot{r}}{r}\right)^2 + \frac{\bar{R}}{r^2} \quad ,$$

where $\ddot{r} = d^2r/ds^2$.

Now, we consider a positive function f on M like in Section 1. To do this, we assume here that f depends only on the parameter $s \in I$. Then, from (2.2), we have

$$\begin{aligned} \text{Hess } f(\xi,\xi) &= \ddot{f} , \qquad \text{Hess } f(\xi,A) = 0 , \\ \text{Hess } f(A,B) &= r\dot{r}\dot{f}\bar{g}(A,B) , \\ \Delta f &= \ddot{f} + (n-1)\dot{f}\frac{\dot{r}}{r} . \end{aligned} \tag{2.4}$$

Combining (2.3) and (2.4) we obtain immediately

Proposition 2.1. *Let* $M = I \times_r N$ *be as above and* f *a positive function on* M *depending only on the parameter* $s \in I$. *Then the metric* g *and the function* f *satisfy*

$$\mathrm{Ric} - \frac{\text{Hess } f}{f} = \frac{1}{n}\left(R - \frac{\Delta f}{f}\right) g , \tag{2.5}$$

if and only if $\bar{g}$ *is an Einstein metric on* N, *and* f *and* r *satisfy*

$$\frac{\bar{R}}{r^2} - (n-1)(n-2)\left\{\left(\frac{\dot{r}}{r}\right)^2 - \frac{\ddot{r}}{r}\right\} - (n-1)\left(\frac{\dot{f}}{f}\frac{\dot{r}}{r} - \frac{\ddot{f}}{f}\right) = 0 . \tag{2.6}$$

Corollary 2.2. *Under the same assumptions as in Proposition* 2.1, *assume further that* $\Delta f = R = 0$ *and* $n > 2$. *Then* g *and* f *satisfy* (2.5) *on* $M = I \times_r N$ *if and only if*

$$g = \frac{1}{a^2 f^2}\, dr^2 + r^2 \bar{g} , \tag{2.7}$$

$$a^2 f^2 = \frac{\bar{R}}{(n-1)(n-2)} - \frac{2am}{(n-2)\, r^{n-2}} , \tag{2.8}$$

where a, m *are constants and* $a \neq 0$, *and* $\bar{g}$ *is an Einstein metric on* N.

Proof. From $\Delta f = 0$, (2.4) gives

$$\ddot{f} + (n-1)\dot{f}\frac{\dot{r}}{r} = 0 , \tag{2.9}$$

which implies

$$\dot{f} r^{n-1} = m = \text{const.} . \tag{2.10}$$

From $R=0$, (2.3) gives

$$\frac{\bar{R}}{r^2} = 2(n-1)\frac{\ddot{r}}{r} + (n-1)(n-2)\left(\frac{\dot{r}}{r}\right)^2 , \tag{2.11}$$

which, together with (2.6), implies

$$n\frac{\ddot{r}}{r} + \frac{\ddot{f}}{f} - \frac{\dot{f}}{f}\frac{\dot{r}}{r} = 0 . \tag{2.12}$$

Combining (2.9) and (2.12), we obtain

$$\ddot{r}f - \dot{f}\dot{r} = 0 . \tag{2.13}$$

Hence we have

$$\dot{r} = af \qquad \text{or} \qquad ds = \frac{1}{af}\,dr , \tag{2.14}$$

for some nonzero constant a. From (2.10), (2.11) and (2.14) we obtain

$$a^2f^2 = \frac{\bar{R}}{(n-1)(n-2)} - \frac{2amr^{2-n}}{(n-2)} .$$

Remark 1. If $\Delta f = R = 0$, then $\bar{R}$ must be constant. In fact, since $\dot{f}=(df/dr)\dot{r}$, (2.10) and (2.14) give

$$af\frac{df}{dr} = mr^{1-n} ,$$

which implies

$$a^2f^2 + \frac{2am}{n-2}r^{2-n} = \text{const.} = \frac{\bar{R}}{(n-1)(n-2)} .$$

Remark 2. In view of (0.2) and (0.4), the condition $\Delta f = R = 0$ implies that the corresponding static model $(\tilde{M},\tilde{g})$ has zero Ricci curvature. In particular, when $a=1$, $n=3$ and $(N,\bar{g}) = (S^2,g_0)$, the solution in Corollary 2.2 is nothing but the Schwarzschild solution in general relativity.

3. Conformally Flat Space Section

We are now going into the case M is conformally flat. As is shown in [2], M admits a constant f as a solution to (0.3), if and only if M is an Einstein space, hence it is of constant curvature when it is conformally flat. So in the following only nonconstant f will be considered.

Theorem 3.1. *Let* (M,g) *be a conformally flat Riemannian* n*-manifold and* f *be a positive function on* M *which satisfies*

$$\mathrm{Ric} - \frac{\mathrm{Hess}\ f}{f} = \frac{1}{n}\left(R - \frac{\Delta f}{f}\right) g \quad . \tag{3.1}$$

Then, for a regular value c *of* f*, the hypersurface* $f^{-1}(c)$ *is of constant curvature. Moreover, around* $f^{-1}(c)$*, the metric* g *splits locally as follows*

$$g = ds^2 + r(s)^2 \bar{g} \quad , \tag{3.2}$$

where $ds = df/|df|$, $r(s)^2\bar{g} = g|f^{-1}(c)$ *(*s *can be considered as a function of* c*) and* $\bar{g}$ *is independent of* s.

Thus, f in the above theorem satisfies the equation (2.6) in Proposition 2.1. On the other hand, it is well-known that a metric g given by (3.2) with $\bar{g}$ of constant curvature is conformally flat. Hence, the above theorem together with Proposition 2.1 gives a complete solution to the equation (0.3) for a conformally flat M except the critical set of f.

Proof. Since M is conformally flat, the conformal curvature tensor C of type $(0,3)$ defined in Section 1 vanishes identically. Hence, from (1.8), we have

$$df \wedge d|df|^2 = 0 \quad , \tag{3.3}$$

which implies

$$d\,\frac{|df|^2}{2} = \mathrm{Hess}\ f(\xi,\xi)\,df \quad , \tag{3.4}$$

where

$$\xi = \mathrm{grad}\ f/|\mathrm{grad}\ f| \quad .$$

From (1.11) and (3.4), we get

$$\underset{X,Y}{\mathrm{Alt}}\left\{(n-1)\mathrm{Hess}\ f(X,U) + (\mathrm{Hess}\ f(\xi,\xi) - \Delta f)g(X,U)\right\}(Y\cdot f) = 0 \ , \quad U,X,Y \in T_xM \quad . \tag{3.5}$$

In particular, for vectors A and B orthogonal to ξ we have

$$(n-1)\text{Hess } f(A,B) + (\text{Hess } f(\xi,\xi) - \Delta f)g(A,B) = 0 . \tag{3.6}$$

Let c be a regular value of f and h the second fundamental form of $f^{-1}(c)$. Then it is given by

$$h(A,B) = -\frac{\text{Hess } f(A,B)}{|df|} , \qquad A,B \in T_x f^{-1}(c) . \tag{3.7}$$

Hence, from (3.6) we have

$$h(A,B) = \lambda g(A,B) , \qquad A,B \in T_x f^{-1}(c) , \tag{3.8}$$

where

$$\lambda = \frac{\Delta f - \text{Hess } f(\xi,\xi)}{(n-1)|df|} . \tag{3.9}$$

Thus $f^{-1}(c)$ is totally umbilical, hence it is conformally flat (cf. [4]). Furthermore, by (3.8), the contracted Gauss equation for $f^{-1}(c)$ is written as follows:

$$\overline{\text{Ric}}(A,B) - g(R(\xi,A)B,\xi) = \text{Ric}(A,B) - (n-2)\lambda^2 g(A,B) \tag{3.10}$$

where $\overline{\text{Ric}}$ is the Ricci tensor of $(f^{-1}(c), g|f^{-1}(c))$. On the other hand, from (3.1) and (3.6),

$$\text{Ric}(A,B) = \left\{ -\frac{\text{Hess } f(\xi,\xi) - \Delta f}{(n-1)f} + \frac{1}{n}\left(R - \frac{\Delta f}{f}\right)\right\} g(A,B) . \tag{3.11}$$

From (1.10), (3.4) and (3.6),

$$g(R(\xi,A)B,\xi) = \frac{1}{n(n-1)}\left(\frac{n \text{ Hess } f(\xi,\xi) - \Delta f}{f} + R\right) g(A,B) . \tag{3.12}$$

(3.10), (3.11) and (3.12) show that $f^{-1}(c)$ is an Einstein space. Hence it is of constant curvature.

The latter part of the theorem is an easy consequence from (3.3) and (3.6).

Remark. A similar result for $n = 3$ is found in [3].

4. Conformally Flat Static Space-Time

In this last section, we consider how conformal flatness of $(\tilde{M}^{n+1},\tilde{g})$ reflects the geometry of (M^n,g). Weyl's conformal curvature tensor for $(\tilde{M},\tilde{g})$ is defined in a way similar to (M,g):

$$\tilde{W}(\tilde{X},\tilde{Y})\tilde{Z} = \tilde{R}(\tilde{X},\tilde{Y})\tilde{Z} - \frac{1}{n-1} \operatorname*{Alt}_{\tilde{X},\tilde{Y}} \left\{ \tilde{L}(\tilde{Z},\tilde{Y})\tilde{X} + \tilde{g}(\tilde{Z},\tilde{Y})\tilde{L}^{\#}(\tilde{X}) \right\} ,$$

$$\tilde{X},\tilde{Y},\tilde{Z} \in T_{(t,x)}\tilde{M} , \qquad (4.1)$$

where $\tilde{R}(\cdot,\cdot)$ is the curvature tensor of $(\tilde{M},\tilde{g})$ and $\tilde{L}$, $\tilde{L}^{\#}$ are defined by

$$\tilde{L} = \widetilde{Ric} - \frac{\tilde{R}}{2n}\tilde{g} ,$$

$$\tilde{g}(\tilde{L}^{\#}(\tilde{X}),\tilde{Y}) = \tilde{L}(\tilde{X},\tilde{Y}) \quad \text{for} \quad \tilde{X},\tilde{Y} \in T_{(t,x)}\tilde{M} . \qquad (4.2)$$

<u>Theorem 4.1</u>. *Let* $(\tilde{M}^{n+1},\tilde{g})$ *and* (M^n,g) $(n > 2)$ *be as above and assume that*

$$Ric - \frac{\text{Hess } f}{f} = \frac{1}{n}\left(R - \frac{\Delta f}{f}\right) g . \qquad (4.3)$$

Then $\tilde{M}$ *is conformally flat if and only if* M *is of constant curvature.*

<u>Remark</u>. If M is of constant curvature, the solutions to (4.3) can be given explicitly [2].

<u>Proof</u>. We remark that if M is of constant curvature, $\tilde{M}$ is conformally flat, which can be shown by (0.1). So we prove the "only if" part of the statement.

By X, Y, Z we denote tangent vectors to $\tilde{M}$ such that $X\cdot t = Y\cdot t = Z\cdot t = 0$. Thus they are naturally identified with vectors tangent to M. From (0.1) we can show

$$\tilde{R}(X,Y)Z = R(X,Y)Z , \qquad (4.4)$$

$$\widetilde{Ric}(X,Y) = \left(Ric - \frac{\text{Hess } f}{f}\right)(X,Y) , \qquad (4.5)$$

$$\tilde{R} = R - 2\frac{\Delta f}{f} . \qquad (4.6)$$

Hence, from (4.2), (4.3), (4.5) and (4.6),

$$\tilde{L}(X,Y) = \frac{1}{2n} Rg(X,Y) \quad . \tag{4.7}$$

Then, from (4.1), (4.4) and (4.7), we have

$$\tilde{W}(X,Y)Z = R(X,Y)Z - \frac{R}{n(n-1)} \{g(Z,Y)X - g(Y,Z)X\} \ . \tag{4.8}$$

Thus, conformal flatness of $\tilde{M}$ implies that M is of constant curvature.

5. References

[1] S.W. Hawking and G.F.R. Ellis, *The Large Scale Structure of Space-Time*, Cambridge Univ. Press, 1973.

[2] O. Kobayashi and M. Obata, "Certain mathematical problems on static models in general relativity, to appear in Proc. Symp. Diff. Geom. and Partial Diff. Equ., Beijing, 1980.

[3] L. Lindblom, "Some properties of static general relativistic stellar models," *J. Math. Phys.* 21 (1980), 1455-1459.

[4] S. Nishikawa and Y. Maeda, "Conformally flat hypersurfaces in a conformally flat Riemannian manifold, *Tôhoku Math. J.* 26 (1974), 159-168.

Tokyo Metropolitan University
Tokyo 158, Japan

Faculty of Science and Technology
Keio University
Yokohama 223, Japan

(Received January 13, 1981)

HOLOMORPHIC STRUCTURES MODELED AFTER COMPACT HERMITIAN SYMMETRIC SPACES

Shoshichi Kobayashi and Takushiro Ochiai[1]

1. Introduction

Let M be a compact m-dimensional complex manifold and $F(M)$ the holomorphic frame bundle over M. Then $\pi: F(M) \to M$ is a holomorphic principal $GL(m;\mathbb{C})$-bundle over M. Let G be a complex Lie subgroup of $GL(m;\mathbb{C})$. A holomorphic principal G-subbundle $\pi: P \to M$ of $F(M)$ is called a *holomorphic G-structure* on M.

Let $\pi: P \to M$ be a holomorphic G-structure on M. We call P *integrable* if for each $z \in M$ there exists a holomorphic local coordinate system $(z^1, \ldots, z^m)$ defined in a neighborhood U of z such that for each $w \in U$ the frame $((\partial/\partial z^1)_w, \ldots, (\partial/\partial z^m)_w)$ belongs to P. We call P *semi-integrable* if for every $z \in M$ there exists an open neighborhood U of z such that the restriction $P|U$ admits a torsion-free holomorphic G-connection. An integrable holomorphic G-structure is automatically semi-integrable. (Since we do not have a "partition of unity" in the holomorphic category, a semi-integrable holomorphic G-structure may not admit a *global* holomorphic G-connection).

Now let S be an m-dimensional irreducible hermitian symmetric space of compact type. We fix a point o in S. Let $L(S)$ be the identity component of the group of all biholomorphic transformations of S, and $L_o(S)$ the isotropy subgroup of $L(S)$ at o so that $S = L(S)/L_o(S)$. Let $\mathbb{T}_o(S)$ denote the tangent space to S at o, and $GL(\mathbb{T}_o(S))$ the general linear group acting on $\mathbb{T}_o(S)$. Let $G(S) \subset GL(\mathbb{T}_o(S)) \approx GL(m;\mathbb{C})$ be the linear isotropy group of $L_o(S)$, i.e., the image of the linear isotropy representation of $L_o(S)$. For example, if S is a complex projective space $P^m(\mathbb{C})$, then $G(P^m(\mathbb{C})) = GL(m;\mathbb{C})$. If S is a (non-singular) hyperquadric $Q^m(\mathbb{C})$ in $P^{m+1}(\mathbb{C})$, then $G(Q^m(\mathbb{C})) = CO(m;\mathbb{C})$, where

$$CO(m;\mathbb{C}) = \{aA \in GL(m;\mathbb{C});\ {}^tAA = I_m \text{ and } a \in \mathbb{C} - \{0\}\} .$$

In this paper we shall prove the following three theorems.

(1.1) <u>Theorem</u>. *Let* S *be an* m*-dimensional irreducible compact hermitian symmetric space, not isomorphic to the projective space* $P^m(\mathbb{C})$, *and let* $G(S) \subset GL(m;\mathbb{C})$ *be as above. Let* M *be an* m*-dimensional compact complex manifold admitting a semi-integrable holomorphic* $G(S)$*-structure. If* A *is a weighted homogeneous polynomial of degree* m *such that the Chern number* $A(c_1(S),\ldots,c_m(S))$ *of* S *vanishes, then the corresponding Chern number* $A(c_1(M),\ldots,c_m(M))$ *of* M *vanishes. (Here,* $c_i(\cdot)$ *denotes the* i*-th Chern class.)*

(1.2) <u>Theorem</u>. *Let* S *and* $G(S)$ *be as in* (1.1). *Let* M *be an* m*-dimensional complex manifold satisfying one of the following two conditions:*

(1) M *is a compact Kähler manifold admitting a semi-integrable holomorphic* $G(S)$*-structure;*

(2) M *admits an integrable holomorphic* $G(S)$*-structure.*

Then the mapping $c_i(S) \mapsto c_i(M)$, $(1 = 1,\ldots,m)$, *induces a ring homomorphism of characteristic ring of* S *onto that of* M.

(1.3) <u>Theorem</u>. *Let* S *and* $G(S)$ *be as in* (1.1). *Let* M *be an* m*-dimensional compact Einstein-Kähler manifold admitting a holomorphic* $G(S)$*-structure. Then* M *is either isomorphic to* S, *or flat, or covered by the noncompact dual of* S *according as the Ricci tensor of* M *is positive,* 0 *or negative.*

From the existence of the "holomorphic Levi-Civita connection" for any holomorphic $O(m;\mathbb{C})$-structure, it follows that every holomorphic $CO(m;\mathbb{C})$-structure is semi-integrable. For holomorphic $CO(m;\mathbb{C})$-structures, the three theorems above have been proved in [5], whose first four sections may be read as an introduction to the present paper.

Generalizing the concepts of affine, projective and quadric structures, we can define that of holomorphic S-structure (see §4) and show that it is equivalent to the concept of integrable holomorphic $G(S)$-structure.

The case where $S = P^m(\mathbb{C})$ is excluded for the technical reason that a projective structure cannot be defined as the prolongation of a $G(S)$-structure but must be defined as a second order structure directly.

2. Irreducible Hermitian Symmetric Spaces of Compact Type

Let $S = L(S)/L_o(S)$ be as in Section 1. We write $\mathfrak{l}(S)$ (resp., $\mathfrak{l}_o(S)$) for the Lie algebra of $L(S)$ (resp., $L_o(S)$). We know that $L(S)$ is a complex Lie group and that $\mathfrak{l}(S)$ can be identified with the complex Lie algebra of all holomorphic vector fields on S. Then $\mathfrak{l}_o(S)$ can be identified with the complex subalgebra of holomorphic vector fields vanishing at $o \in S$.

Let $I(S)$ be the identity component of the group of isometries of S, and K the isotropy subgroup of $I(S)$ at o. Let $\mathfrak{u}$ (resp., $\mathfrak{k}$) be the Lie algebra of $I(S)$ (resp., K). Then $\mathfrak{u}$ can be identified with the real Lie algebra of all Killing vector fields on S, and $\mathfrak{k}$ with the subalgebra of Killing vector fields vanishing at $o \in S$.

Let J denote the (almost) complex structure of S. Then

$$\mathfrak{l}(S) = \mathfrak{u} \oplus J\mathfrak{u} \qquad \text{(vector space direct sum)}. \tag{2.1}$$

Let $\mathfrak{u} = \mathfrak{k} \oplus \mathfrak{p}$ be the usual Cartan decomposition of $\mathfrak{u}$ so that

$$[\mathfrak{k},\mathfrak{p}] \subset \mathfrak{p} \qquad \text{and} \qquad [\mathfrak{p},\mathfrak{p}] \subset \mathfrak{k} \quad . \tag{2.2}$$

Write $T_o(S)$ (resp., $\mathbf{T}_o(S)$) for the real (resp., holomorphic) tangent space of S at o. Then we have

$$T_o(S) \otimes \mathbb{C} = \mathbf{T}_o(S) \oplus \bar{\mathbf{T}}_o(S) \quad . \tag{2.3}$$

We have the natural identification

$$T_o(S) = \mathfrak{p} \tag{2.4}$$

so that for every $\sigma \in K$, $Ad(\sigma); \mathfrak{p} \to \mathfrak{p}$ is the linear isotropy representation of σ. From (2.3) and (2.4) we have the natural identification

$$\mathfrak{p} \otimes_{\mathbb{R}} \mathbb{C} = \mathbf{T}_o(S) \oplus \mathbf{T}_0(S) . \tag{2.5}$$

We write $\mathfrak{g}_{-1}(S)$ (resp., $\mathfrak{g}_1(S)$) for the complex subspace of $\mathfrak{p} \otimes_{\mathbb{R}} \mathbb{C}$ corresponding to $\mathbf{T}_o(S)$ (resp., $\bar{\mathbf{T}}_o(S)$). From (2.1) we have the natural identification

$$\mathfrak{l}(S) = \mathfrak{u} \otimes_{\mathbb{R}} \mathbb{C} \quad . \tag{2.6}$$

Write $\mathfrak{g}_0$ for $\mathfrak{k}\otimes_{\mathbb{R}}\mathbb{C}$. Then combining (2.5) and (2.6) we obtain

$$\mathfrak{l}(S) = \mathfrak{g}_{-1}(S) \oplus \mathfrak{g}_0(S) \oplus \mathfrak{g}_1(S) \qquad \text{(direct sum)} \quad . \tag{2.7}$$

We write simply $\mathfrak{g}_j$ instead of $\mathfrak{g}_j(S)$ when there is no danger of confusion. Well-known basic facts on $S = L(S)/L_o(S)$ can be summarized as follows (cf. [7],[3] or [6]).

(2.8) Facts. (a) $\mathfrak{l}(S) = \mathfrak{g}_{-1} \oplus \mathfrak{g}_0 \oplus \mathfrak{g}_1$ *is a graded Lie algebra (i.e.,* $[\mathfrak{g}_i,\mathfrak{g}_j] \subset \mathfrak{g}_{i+j}$*),* $\mathfrak{l}_o(S) = \mathfrak{g}_0 \oplus \mathfrak{g}_1$ *and* $[\mathfrak{g}_{-1},\mathfrak{g}_1] = \mathfrak{g}_0$.

(b) *If* $B:\mathfrak{l}(S)\times\mathfrak{l}(S)\to\mathbb{C}$ *is the Killing-Cartan form of* $\mathfrak{l}(S)$*, then its restriction* $B:\mathfrak{g}_{-1}\times\mathfrak{g}_1\to\mathbb{C}$ *is nondegenerate and identifies* $\mathfrak{g}_1$ *with the dual of* $\mathfrak{g}_{-1}$*. Then the adjoint representation* $\{\mathfrak{g}_0,\mathrm{ad},\mathfrak{g}_1\}$ *of* $\mathfrak{g}_0$ *on* $\mathfrak{g}_1$ *is the dual representation of* $\{\mathfrak{g}_0,\mathrm{ad},\mathfrak{g}_{-1}\}$.

(c) *Let* $G_0(S)$ *(resp.,* $G_{\pm 1}(S)$*) be the connected Lie subgroup of* $L(S)$ *generated by* $\mathfrak{g}_0$ *(resp.,* $\mathfrak{g}_{\pm 1}$*). Then* $G_1(S)$ *is a normal subgroup of* $L_o(S)$*, and we have* $L_o(S) = G_0(S)G_1(S)$ *(semi-direct product). Moreover,* $\mathrm{esp}:\mathfrak{g}_{\pm 1}\to G_{\pm 1}(S)$ *are isomorphisms.*

(d) *Let* $\rho:L_o(S)\to GL(\mathbf{T}_o(S))$ *be the linear isotropy representation of* $L_o(S)$*. Then,* $\ker\rho = G_1(S)$*. In particular,* $G_0(S)$ *is naturally isomorphic to the linear isotropy subgroup* $G(S) = \rho(L_o(S))$*. More precisely, for every* $\sigma\in G_0$*, we have the following commutative diagram:*

$$\begin{array}{ccc} \mathbf{T}_o(S) & \xrightarrow{\rho(\sigma)} & \mathbf{T}_o(S) \\ \Big\| \mathrm{id} & & \Big\| \mathrm{id} \\ \mathfrak{g}_{-1} & \xrightarrow{\mathrm{Ad}(\sigma)} & \mathfrak{g}_{-1} \end{array} .$$

Therefore, considering $G(S)$ *as a subgroup of* $GL(\mathfrak{g}_{-1})$ *under the identification* $\mathbf{T}_o(S) = \mathfrak{g}_{-1}$ *and* $G_0(S)$ *as a subgroup of* $GL(\mathfrak{g}_{-1})$ *via the adjoint representation, we have* $G_0(S) = G(S)$.

(e) *The linear isotropy group* $G_0(S) = G(S)$ *is irreducible and contains the group of scalar multiplications as its center.*

(f) *The mapping*

$$\tau : X\in\mathfrak{g}_{-1} \to (\exp X)o \in S$$

is a holomorphic imbedding of $\mathfrak{g}_{-1}$ *onto an open neighborhood of* o *in* S.

Now we fix an identification of $\mathfrak{g}_{-1}$ with $\mathbb{C}^m$. Define a tensor $\Phi_S \in \mathfrak{g}_0 \otimes (\mathfrak{g}_{-1})^* \otimes (\mathfrak{g}_1)^* \subset \mathbb{C}^m \otimes (\mathbb{C}^m)^* \otimes (\mathbb{C}^m)^* \otimes \mathbb{C}^m$ by

$$\Phi_S(v,\alpha) = [v,\alpha] \qquad \text{for } v \in \mathbb{C}^m \text{ and } \alpha \in (\mathbb{C}^m)^* .$$

(2.9) Lemma. $G(S)$ *is the identity component of the group*

$$G^* = \{\sigma \in GL(m;C)\,;\ \sigma(\Phi_S) = \Phi_S\} .$$

Proof. Let $\mathfrak{g}^*$ be the Lie algebra of G^*. For any $\sigma \in G(S)$, we have

$$\Phi_S(\sigma v, {}^t\sigma^{-1}\alpha) = [\mathrm{Ad}(\sigma)v, \mathrm{Ad}(\sigma)\alpha] = \sigma[v,\alpha]\sigma^{-1} = \sigma\Phi_S(v,\alpha)\sigma^{-1} .$$

So we have $G(S) \subset G^*$. Conversely, if $\sigma \in G^*$, then

$$\sigma[v,\alpha]\sigma^{-1} = [\sigma v, {}^t\sigma^{-1}\alpha] \in \mathfrak{g}_0 \qquad \text{for any } v \in \mathfrak{g}_{-1},\ \alpha \in \mathfrak{g}_1 .$$

From Fact (a) of (2.8), we see that $\sigma\mathfrak{g}_0\sigma^{-1} \subset \mathfrak{g}_0$. Thus $\mathfrak{g}_0$ is an ideal of $\mathfrak{g}^*$. Since $\mathfrak{g}_0$ is irreducible, so is $\mathfrak{g}^*$. In particular, $[\mathfrak{g}^*,\mathfrak{g}^*]$ is simple. From Fact (e) of (2.8), we have

$$\mathfrak{g}_0 = [\mathfrak{g}_0,\mathfrak{g}_0] + \mathbb{C}I_m , \qquad \mathfrak{g}^* = [\mathfrak{g}^*,\mathfrak{g}^*] + \mathbb{C}I_m .$$

Since $[\mathfrak{g}_0,\mathfrak{g}_0]$ is an ideal of $[\mathfrak{g}^*,\mathfrak{g}^*]$, we must have $\mathfrak{g}^* = \mathfrak{g}_0$. ∎

In general, let $\mathfrak{g}$ be any Lie subalgebra of $\mathfrak{gl}(m;\mathbb{C})$. For $k = 1,2,\ldots,$ let $\mathfrak{g}^{(k)}$ be the space of symmetric multilinear mappings

$$t : \underbrace{\mathbb{C}^m \times \cdots \times \mathbb{C}^m}_{(k+1)\text{-times}} \to \mathbb{C}^m$$

such that, for each fixed $v_1,\ldots,v_k \in \mathbb{C}^m$, the linear transformation

$$u \in \mathbb{C}^m \to t(u,v_1,\ldots,v_k) \in \mathbb{C}^m$$

belongs to $\mathfrak{g}$. The space $\mathfrak{g}^{(k)}$ is called the k-th prolongation of $\mathfrak{g}$. For each $\alpha \in \mathfrak{g}_1$, define $\hat{\alpha} : \mathbb{C}^m \times \mathbb{C}^m \to \mathbb{C}^m$ by

$$\hat{\alpha}(u,v) = [[\alpha,u],v] \in \mathfrak{g}_{-1} = \mathbb{C}^m .$$

From the Jacobi identity we see that $\hat{\alpha} \in (\mathfrak{g}_0)^{(1)}$. It is easy to see that the mapping $\alpha \in \mathfrak{g}_1 \mapsto \hat{\alpha} \in (\mathfrak{g}_0)^{(1)}$ is injective. From now on we identify α with $\hat{\alpha}$ so that $\mathfrak{g}_1 \subset (\mathfrak{g}_0)^{(1)}$.

(2.10) Lemma. *If* S *is not isomorphic to* $P^m(\mathbb{C})$, *then*

$$(\mathfrak{g}_0)^{(1)} = \mathfrak{g}_1 \quad \textit{and} \quad (\mathfrak{g}_0)^{(k)} = 0 \quad \textit{for} \quad k \geqq 2 .$$

Proof. This has been proved in [7]. However, for the sake of convenience, we shall give a proof. If $\mathfrak{g}_0$ is of infinite type (i.e., $(\mathfrak{g}_0)^{(k)} \neq 0$ for all k), then the irreducibility of $\mathfrak{g}_0$ with non-trivial center implies that $\mathfrak{g}_0$ is either $\mathfrak{gl}(m;\mathbb{C})$ or $csp(m/2;\mathbb{C})$, [3]. Since the natural representation of $csp(m/2;\mathbb{C})$ on $(csp(m/2;\mathbb{C}))^{(1)}$ is irreducible (cf. [3]), $(csp(m/2;\mathbb{C}))^{(1)}$ cannot admit an m-dimensional invariant subspace $\mathfrak{g}_1$. Hence, we must have $\mathfrak{g}_0 = \mathfrak{gl}(m;\mathbb{C})$. This is a contradiction since S is not isomorphic to $P^m(\mathbb{C})$. If $\mathfrak{g}_0$ is of finite type, then $\dim(\mathfrak{g}_0)^{(1)} = m$ and $(\mathfrak{g}_0)^{(k)} = 0$ for $k \geqq 2$. (This is true for any irreducible linear Lie algebra $\mathfrak{g}_0$ of finite type with $(\mathfrak{g}_0)^{(1)} \neq 0$, (cf. [3])). Hence, $(\mathfrak{g}_0)^{(1)} = \mathfrak{g}_1$ and $(\mathfrak{g}_0)^{(k)} = 0$ for $k \geqq 2$. ∎

Let R be the curvature tensor of the hermitian symmetric space S. Then under the identifications $\mathbf{T}_o(S) = \mathfrak{g}_{-1}$ and $\overline{\mathbf{T}}_o(S) = \mathfrak{g}_1$, we have

$$R(u,v) = [u,v] \qquad u \in \mathfrak{g}_{-1}, \quad v \in \mathfrak{g}_1 . \tag{2.11}$$

3. Proof of Theorem (1.1)

Let S be an m-dimensional irreducible hermitian symmetric space of compact type, not isomorphic to $P^m(\mathbb{C})$. Let $\pi : P \to M$ be a semi-integrable holomorphic G(S)-structure on M. There exists a simple open covering $\{U_\alpha\}_{\alpha \in A}$ of M with torsion-free holomorphic G(S)-connections ω_α in $P|U_\alpha$. In the intersection $U_\alpha \cap U_\beta$, the difference of two connections ω_α and ω_β is a holomorphic tensor field. More precisely, since both ω_α and ω_β are torsion-free, there exists a holomorphic mapping

$$\hat{\xi}_{\alpha\beta} : P|U_\alpha \cap U_\beta \to (\mathfrak{g}_0)^{(1)} \subset \mathfrak{g}_0 \otimes (\mathbb{C}^m)^*$$

such that

$$\omega_\alpha(u) - \omega_\beta(u) = \hat{\xi}_{\alpha\beta}(p)(\theta(u)) \in \mathfrak{g}_0 \qquad \text{for} \quad u \in \mathbb{T}_p(P) ,$$

where θ denotes the canonical $\mathbb{C}^m$-valued 1-form on P. From (2.10) there exists a holomorphic mapping

$$\xi_{\alpha\beta} : P|U_\alpha \cap U_\beta \to \mathfrak{g}_1$$

such that

$$\omega_\alpha - \omega_\beta = [\theta, \xi_{\alpha\beta}] \qquad \text{on} \quad P|U_\alpha \cap U_\beta . \tag{3.1}$$

We have also

$$\xi_{\alpha\beta}(p\sigma) = \mathrm{Ad}(\sigma^{-1})\xi_{\alpha\beta}(p) \qquad \text{for} \quad \sigma \in G(S), \quad p \in P|U_\alpha \cap U_\beta .$$

(This expresses the fact that $\xi_{\alpha\beta}$ corresponds to a tensor field on $U_\alpha \cap U_\beta$.) From Fact (b) of (2.8) we can consider $\xi_{\alpha\beta}$ as a holomorphic 1-form defined on $U_\alpha \cap U_\beta$. Clearly $\{\xi_{\alpha\beta}\}$ is a 1-cocycle (with coefficients in the sheaf of germs of holomorphic 1-forms). As a 1-cocycle with coefficients in the sheaf of germs of C^∞ 1-forms, $\{\xi_{\alpha\beta}\}$ is the coboundary of 1-cochain $\{\eta_\alpha\}$, $\eta_\alpha : P|U_\alpha \to \mathfrak{g}_1$:

$$\xi_{\alpha\beta} = \eta_\alpha - \eta_\beta \qquad \text{on} \quad P|U_\alpha \cap U_\beta , \tag{3.2}$$

$$\eta_\alpha(p\sigma) = \mathrm{Ad}(\sigma^{-1})\eta_\alpha(p) \qquad p \in P|U_\alpha, \ \sigma \in G(S) . \tag{3.3}$$

Now, set

$$\omega_\alpha^* = \omega_\alpha - [\theta, \eta_\alpha] \qquad \text{on} \quad P|U_\alpha .$$

Then from (3.1) and (3.2) we have $\omega_\alpha^* = \omega_\beta^*$ on $P|(U_\alpha \cap U_\beta)$. Hence $\{\omega_\alpha^*\}$ defines a global C^∞ 1-form ω^* with values in $\mathfrak{g}_0$ on P. From (3.3) we know that ω^* is a $G(S)$-connection in P. Now we compute the curvature form Ω^* of ω^*. On $P|U_\alpha$, we have

$$\Omega^* = d\omega^* + [\omega^*, \omega^*] = d(\omega_\alpha - [\theta, \eta_\alpha]) + [\omega_\alpha - [\theta, \eta_\alpha], \omega_\alpha - [\theta, \eta_\alpha]]$$

$$= -[\theta, \bar{\partial}\eta_\alpha] + \Psi_\alpha , \tag{3.4}$$

where Ψ_α is a C^∞ 2-form of degree (2,0) on $P|U_\alpha$. Let γ_i be the i-th Chern form of ω^*, i.e., γ_i is the closed 2i-form on M such that

$$\det\left(I + \frac{\sqrt{-1}}{2\pi}\ \Omega^*\right) = 1 + \pi^*\gamma_1 + \cdots + \pi^*\gamma_m \qquad , \tag{3.5}$$

where $\pi: P \to M$ is the projection. Let $\gamma_i^{(i,i)}$ denote the (i,i)-component of γ_i. Then from (3.4) and (3.5), we have

$$\det\left(I - \frac{\sqrt{-1}}{2}\ [\theta, \bar\partial\eta_\alpha]\right) = 1 + \pi^*\gamma_1^{(1,1)} + \cdots + \pi^*\gamma_m^{(m,m)} \tag{3.6}$$

on $P|U_\alpha$.

Let $\{e_1,\ldots,e_m\}$ (resp., $\{f_1,\ldots,f_m\}$) be a basis for $\mathfrak{g}_{-1}$ (resp., $\mathfrak{g}_1$). Let $c_{ij,ab}$ denote the (a,b)-entry of the matrix $[e_i, f_j] \in \mathfrak{g}_0 \subset \mathfrak{gl}(m;\mathbb{C})$. Set

$$\theta = \sum \theta^i e_i \qquad \text{and} \qquad \eta_\alpha = \sum \eta_\alpha^j f_j \quad .$$

Then

$$\text{the (a,b)-entry of the matrix } [\theta, \bar\partial\eta_\alpha] = \sum_{i,j} c_{ij,ab}\,\theta^i \wedge \bar\partial\eta_\alpha^j \ . \tag{3.7}$$

Let $\{\varphi_1,\ldots,\varphi_m\}$ (resp. $\{\psi_1,\ldots,\psi_m\}$) be the dual basis of $\{e_1,\ldots,e_m\}$ (resp. $\{f_1,\ldots,f_m\}$). Since $\mathfrak{p} \otimes_{\mathbb{R}} \mathbb{C} = \mathfrak{g}_{-1} + \mathfrak{g}_1$, by setting

$$h_\alpha(\varphi_i) = \theta^i \qquad \text{and} \qquad h_\alpha(\psi_j) = -\bar\partial\eta_\alpha^j \ , \quad 1 \leq i,j \leq m \quad ,$$

we obtain an algebra homomorphism

$$h_\alpha : \Lambda(\mathfrak{p}^* \otimes_{\mathbb{R}} \mathbb{C}) \to \Lambda(T^*(P|U_\alpha) \otimes_{\mathbb{R}} \mathbb{C}) \ .$$

From (2.11) we know

$$\text{the (a,b)-entry of the matrix } R = \sum_{i,j} c_{ij,ab}\ \varphi^i \wedge \psi^j \ . \tag{3.8}$$

Therefore from (3.7) and (3.8) we obtain

$$\text{the (a,b)-entry of the matrix } -[\theta, \bar\partial\eta_\alpha] = h_\alpha \text{ (the (a,b)-entry of R)}. \tag{3.9}$$

Let δ_i be the i-th Chern form of the hermitian symmetric space S. Then at the origin $o \in S$, we have

$$\det\left(I + \frac{\sqrt{-1}}{2} R\right) = 1 + \delta_1 + \cdots + \delta_m \quad . \tag{3.10}$$

From (3.6), (3.9) and (3.10) we have

$$h_\alpha((\delta_i)_o) = \pi^*\gamma_i^{(i,i)} \quad . \tag{3.11}$$

Let $A(c_1,\ldots,c_m)$ be the weighted homogeneous polynomial of degree m. Since $A(\delta_1,\ldots,\delta_m)$ is a 2m-form on S invariant under the group I(S) of isometries of S, it is the harmonic 2m-form representing the cohomology class $A(c_1(S),\ldots,c_m(S))$. Hence, we have

$$A(c_1(S),\ldots,c_m(S)) = 0 \quad \text{if and only if} \quad A((\delta_1)_o,\ldots,(\delta_m)_o) = 0 \; . \tag{3.12}$$

From (3.11) and (3.12) we have

$$A(c_1(S),\ldots,c_m(S)) = 0 \Rightarrow A(\pi^*\gamma_1^{(1,1)},\ldots,\pi^*\gamma_m^{(m,m)}) = 0 \quad \text{on} \quad P|U_\alpha$$

$$\Rightarrow \pi^* A(\gamma_1^{(1,1)},\ldots,\gamma_m^{(m,m)}) = 0 \quad \text{on} \quad P|U_\alpha$$

$$\Rightarrow A(\gamma_1^{(1,1)},\ldots,\gamma_m^{(m,m)}) = 0 \quad \text{on} \quad U_\alpha$$

(from the degree consideration)

$$\Rightarrow A(\gamma_1,\ldots,\gamma_m) = 0 \quad \text{on} \quad U_\alpha \; .$$

Therefore, $A(\gamma_1,\ldots,\gamma_m) = 0$ on M and hence, $A(c_1(M),\ldots,c_m(M)) = 0$. ■

(2.13) Remark. Suppose that M is moreover a compact Kähler manifold. Let $A(c_1,\ldots,c_m)$ be an weighted homogeneous polynomial of any degree, say ℓ. Then the argument following (3.11) can be strenghtened as follows:

$$A(c_1(S),\ldots,c_m(S)) = 0 \iff A((\delta_1)_o,\ldots,(\delta_m)_o) = 0$$

$$\Rightarrow A(\gamma_1^{(1,1)},\ldots,\gamma_m^{(m,m)}) = 0$$

$$\Rightarrow \text{ the } (\ell,\ell)\text{-component of } A(\gamma_1,\ldots,\gamma_m) = 0$$

$$\Rightarrow A(c_1(M),\ldots,c_m(M)) = 0 \quad .$$

The last implication follows from the assumption that M is compact Kähler.

4. Proof of Theorem (1.2)

We fix an identification of $\mathfrak{g}_{-1} = \mathfrak{g}_{-1}(S)$ with $\mathbb{C}^m$, and consider the holomorphic imbedding $\tau:\mathbb{C}^m \to S$ described in Fact (f) of (2.8). Let p_o be the holomorphic frame of S at the origin o given by

$$p_o = \tau_* : \mathbb{C}^m \cong \mathbf{T}_o(S) \quad .$$

Set

$$P_S = \{g_* \circ p_o ;\ g \in L(S)\} \quad .$$

Then from Facts (c) and (d) of (2.8), we see easily that $P_S \to S$ is a holomorphic $G(S)$-structure on S. Since the action of $G_1(S)$ on $\mathbb{C}^m \cong \tau(\mathbb{C}^m) \subset S$ is by translation, we see that $P_S \to S$ is an integrable holomorphic $G(S)$-structure. Using (2.10) we can prove the following (cf. [6])

(4.1) Fact. *Let* $f:U \cong V$ *be a local holomorphic automorphism of* S. *Then there is an element* $\tilde{f}$ *of* $L(S)$ *such that* $\tilde{f}|U = f$ *if and only if* f *leaves* P_S *invariant.*

By an S-*structure* on M, we mean an atlas of holomorphic coordinate charts $\{(U_\alpha, \varphi_\alpha, D_\alpha)\}_{\alpha \in A}$ of M such that

(i) φ_α maps an open set U_α of M biholomorphically onto an open set D_α of S;

(ii) for every pair (α,β) with $U_\alpha \cap U_\beta \neq \emptyset$, the coordinate change

$$\varphi_\alpha \circ \varphi_\beta^{-1} : \varphi_\beta(U_\alpha \cap U_\beta) \to \varphi_\alpha(U_\alpha \cap U_\beta)$$

is given by the restriction of some element of $L(S)$ acting on S.

(4.2) Lemma. *A complex manifold* M *admits an integrable holomorphic* $G(S)$*-structure if and only if it admits an* S*-structure.*

Proof. Assume that M admits an integrable holomorphic $G(S)$-structure P. Since $P_S \to S$ is an integrable holomorphic $G(S)$-structure on S, there exists an atlas of holomorphic coordinate charts $\{(U_\alpha, \varphi_\alpha, D_\alpha)\}$ such that (1) $M = \cup U_\alpha$, (2) $D_\alpha \subset \mathbb{C}^m$ $(\subset S)$, and (3) φ_α is a $G(S)$-isomorphism from $P|U_\alpha$ onto $P_S|D_\alpha$. For any pair (α,β) with $U_\alpha \cap U_\beta \neq \emptyset$, the coordinate change

$$\varphi_\alpha \circ \varphi_\beta^{-1} : \varphi_\beta(U_\alpha \cap U_\beta) \to \varphi_\alpha(U_\alpha \cap U_\beta)$$

leaves P_S invariant. From Fact (4.1), there exists $f_{\alpha\beta} \in L(S)$ such that $f_{\alpha\beta}|\varphi_\beta(U_\alpha \cap U_\beta) = \varphi_\alpha \circ \varphi_\beta^{-1}$. Hence, $\{(U_\alpha, \varphi_\alpha, D_\alpha)\}$ is an S-structure. The converse can be proved also from Fact (4.1). ∎

For a complex manifold N of dimension n, we denote by $F^{(2)}(N)$ the bundle of holomorphic frames of order 2. It is a holomorphic principal bundle with structure group $G^{(2)}(n;\mathbb{C})$, (cf. [6]).

(4.3) Facts. (cf. [6]). *Let* $\tau : \mathbb{C}^m \to S$ *be the imbedding described in Fact* (f) *of* (2.8). *Then*

(a) *The correspondence* $\sigma \in L_o(S) \mapsto j_o^2(\tau^{-1} \circ \sigma \circ \tau) \in G^{(2)}(m;\mathbb{C})$ *is a holomorphic isomorphism. Hence,* $L_o(S)$ *can be considered as a complex Lie subgroup of* $G^{(2)}(m;\mathbb{C})$.

(b) *The correspondence* $\sigma \in L(S) \mapsto j_o^2(\sigma \circ \tau) \in F^{(2)}(S)$ *is a holomorphic imbedding. Hence, the principal bundle* $L(S) \to S = L(S)/L_o(S)$ *can be considered as a holomorphic* $L_o(S)$*-subbundle of* $F^{(2)}(S) \to S$.

Now we shall prove Theorem (1.2). In the case where M satisfies Condition (1) of (1.2), our assertion follows from Remark (3.13). So assuming Condition (2) of (1.2), let $P \to M$ be an integrable holomorphic $G(S)$-structure on M. From Lemma (4.2) and Facts (4.3) we see easily that there exists a holomorphic $L_o(S)$-subbundle $Q \to M$ of $F^{(2)}(M)$ and a holomorphic S-Cartan connection ω on Q such that

$$d\omega + \frac{1}{2}[\omega,\omega] = 0 \quad , \tag{4.4}$$

$$\alpha(Q) = P \quad , \tag{4.5}$$

where $\alpha: F^{(2)}(M) \to F^{(1)}(M)$ is the natural projection. (In fact, using the corresponding S-structure $\{(U_\alpha, \varphi_\alpha, D_\alpha)\}$ on M, we pull back the $L_o(S)$-subbundle $L(S) \to S$ of $F^{(2)}(S)$ to an $L_o(S)$-subbundle $Q \to M$ of $F^{(2)}(M)$. Pulling back the holomorphic Maurer-Cartan form ω_S of $L(S)$ to Q, we obtain a holomorphic S-Cartan connection ω on Q.)

Since ω is $\mathfrak{l}(S)$-valued, we set

$$\omega = \omega_{-1} + \omega_0 + \omega_1$$

according to the decomposition $\mathfrak{l}(S) = \mathfrak{g}_{-1} + \mathfrak{g}_0 + \mathfrak{g}_1$. Then from (4.4) we have

$$d\omega_0 + \frac{1}{2}[\omega_0, \omega_0] = -[\omega_{-1}, \omega_1] \quad . \tag{4.6}$$

Let $s: P \to Q$ be any $G(S)$-invariant C^∞ cross-section of the fibering $Q \to P$. Put

$$\omega^* = s^*\omega_0 \quad .$$

Then ω^* is a C^∞ $G(S)$-connection in P (while $s^*\omega_{-1}$ is the canonical form θ of P). Let Ω^* denote the curvature form of ω^*. Then (4.6) implies

$$\Omega^* = -[\theta, s^*\omega_1] \quad . \tag{4.7}$$

Let γ_i be the i-th Chern form of ω^*, i.e., the closed 2i-form on M such that

$$\det\left(I - \frac{\sqrt{-1}}{2\pi}[\theta, s^*\omega_1]\right) = 1 + \pi^*\gamma_1 + \cdots + \pi^*\gamma_m \quad , \tag{4.8}$$

where $\pi: P \to M$ is the projection. Now, following the argument given after (3.5), we obtain Theorem (1.2). ∎

5. Proof of Theorem (1.3)

Let N be a connected irreducible complete Kähler manifold of dimension n. (By "irreducible" we mean "with irreducible restricted linear holonomy group" so that the universal covering space of N is not a product of lower dimensional Kähler manifolds.) Let H be the restricted linear holonomy group; it is a connected Lie subgroup of $U(n) \subset GL(n;\mathbb{C})$. Let $H_{\mathbb{C}}$ denote the connected complex Lie subgroup of $GL(n;\mathbb{C})$ generated by H.

(5.1) Lemma. *If* N *is not a hermitian symmetric space, then the complex linear group* $H_{\mathbb{C}}$ *is one of the following:*

$$GL(n;\mathbb{C}), \quad SL(n;\mathbb{C}), \quad Sp(n/2;\mathbb{C}), \quad CSp(n/2;\mathbb{C}),$$

and hence it is of infinite type.

Proof. This follows from Berger's holonomy classification theorem which says H is one of the following:

$$U(n), \quad SU(n), \quad Sp(n/2), \quad Sp(n/2)\times U(1) \quad .$$

For the fact that any of the four complex linear Lie group is of infinite type, see [3]. ∎

Let $P \to M$ be as in the proof of Theorem (1.2). From Lemma (2.9) there exists a nonzero holomorphic tensor field Φ of covariant degree 2 and contravariant degree 2. (The algebraic tensor Φ_S in (2.9) induces a tensor field Φ on M since Φ_S is invariant by the structure group $G(S)$ of P.) On a compact Einstein-Kähler manifold such a holomorphic tensor field (of equal covariant and contravariant degrees) is parallel (by Theorem 1 in [2]). We lift this parallel tensor field to the universal covering manifold $\tilde{M}$ of M. We shall show that $\tilde{M}$ is either S, or $\mathbb{C}^m$, or the noncompact dual of S according as the Ricci tensor is positive, 0 or negative. Let $\tilde{\Phi}$ denote the lift of Φ to $\tilde{M}$. Let

$$\tilde{M} = M_0 \times M_1 \times \dots \times M_r$$

be the de Rham decomposition of M, where M_0 is a complex Euclidean space and $M_1,\dots,M_r$ are all simply connected, complete, and irreducible Kähler manifolds. Fix a point $z_j \in M_j$, $(j = 0,1,\dots,r)$, and write T_j for $\mathbb{T}_{z_j}(M_j)$. Set

$$T = T_0 \oplus T_1 \oplus \cdots \oplus T_r \quad .$$

Let $H, H_1,\dots,H_r$ be the holonomy groups of $\tilde{M}, M_1,\dots,M_r$. Then

$$H = \{e\} \times H_1 \times \dots \times H_r \subset GL(T_0) \times GL(T_1) \times \dots \times GL(T_r) \subset GL(T) .$$

Since H is connected and leaves $\tilde{\Phi}$ invariant, Lemma (2.9) implies $H \subset G(S)$. In particular, we have

$$H_{\mathbb{C}} = \{e\} \times (H_1)_{\mathbb{C}} \times \dots \times (H_r)_{\mathbb{C}} \subset G(S) . \tag{5.2}$$

From our assumption that $S \neq P^m(\mathbb{C})$ we know that $G(S)$ is of finite type of order 2 (cf., Lemma (2.10)). Hence, each $(H_j)_{\mathbb{C}}$ is of finite order (cf., (4.2)). Then Lemma (4.1) implies that each M_j, $(j = 1,\dots,r)$, is a hermitian symmetric space. In particular, each M_j has nonzero Ricci tensor.

Hence, if M is Ricci-flat, then $\tilde{M} = M_0$, i.e., M is flat. We shall therefore assume that the Ricci tensor of M is either positive or negative definite. Then in the de Rham decomposition of $\tilde{M}$, the Euclidean factor M_0 does not appear and the irreducible factors $M_1,\dots,M_r$ have either all positive or all negative Ricci tensors. We denote by S_j the irreducible hermitian symmetric space M_j itself if M_j is compact and the compact dual of M_j if M_j is noncompact. Then we know (cf., (2.7))

$$(H_j)_{\mathbb{C}} = G_0(S_j) \qquad j = 1,\dots,r . \tag{5.3}$$

Since $(H_j)_{\mathbb{C}}$ is of finite type, none of S_j's is isomorphic to a projective space. From Lemma (2.10) we obtain

$$\mathfrak{l}(S) = T \oplus \mathfrak{g}_0(S) \oplus (\mathfrak{g}_0(S))^{(1)} ,$$

$$\mathfrak{l}(S_j) = T_j \oplus \mathfrak{g}_0(S_j) \oplus (\mathfrak{g}_0(S_j))^{(1)} , \qquad j = 1,\dots,r .$$

From (5.2) and (5.3) we have

$$(\mathfrak{g}_0(S_1))^{(1)} \oplus \dots \oplus (\mathfrak{g}_0(S_r))^{(1)} \subset (\mathfrak{g}_0(S))^{(1)} . \tag{5.4}$$

From Fact (b) of (2.8) and Lemma (2.10), we have

$$\dim \mathbf{T} = \dim(\mathfrak{g}_0(S))^{(1)} \geqq \sum_j \dim \mathfrak{g}_0(S_j)^{(1)} = \sum_j \dim \mathbf{T}_j = \dim \mathbf{T}. \tag{5.5}$$

Hence, we have

$$(\mathfrak{g}_0(S))^{(1)} = (\mathfrak{g}_0(S_1))^{(1)} + \cdots + (\mathfrak{g}_0(S_r))^{(1)} \quad . \tag{5.6}$$

From Fact (a) of (2.8) and Lemma (2.10), we have

$$\mathfrak{g}_0(S) = [\mathfrak{T},(\mathfrak{g}_0(S))^{(1)}] \subset \sum_j [\mathfrak{T}_j,(\mathfrak{g}_0(S_j))^{(1)}] = \sum_j \mathfrak{g}_0(S_j) \ . \tag{5.7}$$

Therefore, we have

$$\mathfrak{g}_0(S) = \mathfrak{g}_0(S_1) + \cdots + \mathfrak{g}_0(S_r) \quad . \tag{5.8}$$

Since $\mathfrak{g}_0(S)$ is irreducible, we have $r = 1$ and

$$\mathfrak{g}_0(S) = \mathfrak{g}_0(S_1) \quad . \tag{5.9}$$

Hence, $S = S_1$ and M is either S or the noncompact dual of S according as the Ricci tensor of M is positive or negative. ■

6. Notes

1. Partially supported by NSF Grant 79-02552.

7. References

1. M. Inoue, S. Kobayashi and T. Ochiai, "Holomorphic affine connections on compact complex surfaces," *J. Fac. Sci. Univ. Tokyo* 27 (1980), 247-264.
2. S. Kobayashi, "First Chern class and holomorphic tensor fields," *Nagoya Math. J.* 77 (1980), 5-11.
3. S. Kobayashi and T. Nagano, "On filtered Lie algebras and geometric structures," *I. J. Math. Mech.* 13 (1964), 873-908; II. 14 (1965), 513-522; III. 14 (1965), 679-706.
4. S. Kobayashi and T. Ochiai, "Holomorphic projective structures on compact complex surfaces," *Math. Ann.* 249 (1980), 75-94.
5. S. Kobayashi and T. Ochiai, "Holomorphic structures modeled after hyperquadrics," to appear.
6. T. Ochiai, "Geometry associated with semi-simple flat homogeneous spaces," *Trans. Amer. Math. Soc.* 152 (1970), 159-193.
7. N. Tanaka, "On the equivalence problems associated with a certain class of homogeneous spaces," *J. Math. Sco. Japan* 17 (1965), 103-139.

University of California
Berkeley, California 94720

University of Tokyo
Bunkyo-ku, Tokyo 113
Japan

(Received February 20, 1981)

GROUP COHOMOLOGY AND HECKE OPERATORS

Michio Kuga, Walter Parry, and Chih-Han Sah

Consider the Taylor expansion of the infinite product:

$$t \prod_{n\geq 1} (1-t^n)^2(1-t^{11n})^2 = \sum_{n\geq 1} a_n t^n \quad .$$

It is known that for a prime $p \neq 11$, we have:

$$1 - a_p + p \equiv 0 \mod 5 \quad . \tag{1}$$

On the other hand, for the Taylor expansion of the infinite product:

$$t \prod_{n\geq 1} (1-t^n)^{24} = \sum_{n\geq 1} \tau(n) t^n$$

it is known that for each prime p, we have:

$$1 - \tau(p) + p^{11} \equiv 0 \mod 691 \quad . \tag{2}$$

Among these coefficients, we can use the little Fermat theorem and get:

$$\tau(n) \equiv a_n \mod 11 \quad . \tag{3}$$

These are some of the many congruence relations of similar type discovered by old masters such as Ramanujan. They have been extended as congruence relations among eigenvalues of Hecke operators acting on automorphic forms in more recent years by many people, for example: Deligne, Doi, Koike, Serre, Swinnerton-Dyer among others; see [3,5,6,11,12,19,20,26].

The discovery of a "non-abelian class field theory" by Eichler [7] leads us to a connection between these congruence relations for automorphic forms of weight 2 and the arithmetic of division points of the Jacobian variety of the Shimura curve--a model of $\Gamma\backslash H$, where Γ is an arithmetic fuchsian group acting on the upper half plane H; see also [10] and Shimura [23]. For automorphic forms of higher weight m, the congruence relations mod ℓ can also be related to the arithmetic of ℓ-th division points of the Jacobian variety of $\Gamma(\ell)\backslash H$; see [13]. This philosophy of obtaining congruence relations from the arithmetic of division points has been developed extensively by many people.

In the present note, we introduce another approach, not as deep as the principle just mentioned, but perhaps broader. More precisely, a small number of these congruence relations can be proved through a more elementary principle--the functorial behavior of cohomology groups as Hecke ring modules. For example, (1) and (3) can be explained in this manner but many others similar to (2) found by Ramanujan, Serre, Doi, Koike, elude the elementary principle described in this note.

We restrict ourselves in the present work to fuchsian groups. The method can be extended to other groups Γ defining algebraic surfaces $\Gamma\backslash(H \times H)$, in fact even to groups quite unrelated to automorphic function theory in any direct manner. We can even extend ourselves out of our category of groups to a more general category with direct sum, fiber product (distributive with respect to direct sum), thereby dissolving the whole idea into the realm of "generalized nonsense" with even wider possible applications. As one example of suggestive works, see Dodson [4].

Eichler discovered that the space $S_{m+2}(\Gamma)$ of cusp forms of weight $m+2$ for a fuchsian group Γ is isomorphic to the cohomology group $H^{(1,0)}(\Gamma,\rho_m)_{par}$ where ρ_m denotes the m-th symmetric tensor representation of Γ; see [8]. As a consequence, in some part of the theory of automorphic forms, we can replace automorphic forms by cohomology groups. This loses much information, for example: that related to complex structures; the ring structure of automorphic forms; etc. Nevertheless, we can still obtain some clear interpretations of the classical results and even some new results through such abstractions.

The action of the Hecke ring on group cohomology was first defined by Rhie and Whaples [16], although there had been some suggestive statements made earlier. This interesting idea had been left undeveloped for many years while the theory of automorphic forms exploded in many different directions. Motivated by various questions in group theory, some computations of the cohomology groups of certain finite Chevalley groups with modular coefficient modules were made; see [17] and [18]. Such computations had been extensively developed by many other people; see

Cline, Parshall, Scott and van der Kallen [2]. On the other hand, the cohomology of discontinuous groups was investigated extensively by Matsushima, Murakami, Weil, Shimura, Borel, Wallach, etc. The time appears to be ripe to reconsider the original idea of Eichler--cohomology groups can be viewed as Hecke ring modules.

1. General Theory

1.1 Hecke Ring à la Shimura

We recall the definition and the fundamental properties of Hecke ring as defined by Shimura in his book [24].

In a group G, two subgroups Γ and Γ' are said to be commensurable, written $\Gamma \sim \Gamma'$, if $\Gamma \cap \Gamma'$ has finite index in both Γ and Γ'. For a fixed subgroup Γ of G, the commensurator of Γ in G is the subgroup $\tilde{\Gamma}$ defined by:

$$\tilde{\Gamma} = \{\alpha \in G \mid \alpha\Gamma\alpha^{-1} \sim \Gamma\} \quad .$$

Evidently, $\Gamma \subset \tilde{\Gamma} \subset G$. For $\alpha \in \tilde{\Gamma}$, $\Gamma\alpha\Gamma$ is a finite disjoint union of Γ-left as well as Γ-right cosets:

$$\Gamma\alpha\Gamma = \coprod_{1 \leqslant i \leqslant d} \Gamma\alpha_i = \coprod_{1 \leqslant j \leqslant e} \beta_j\Gamma,$$

(where $\coprod$ denotes disjoint union).
The number d of Γ-left cosets $\Gamma\alpha_i$ in $\Gamma\alpha\Gamma$ is called the degree of $\Gamma\alpha\Gamma$, denoted by $\deg(\Gamma\alpha\Gamma)$ or $d(\alpha)$. It is equal to $[\Gamma : \Gamma \cap \alpha^{-1}\Gamma\alpha]$. We note that e is equal to $[\Gamma : \Gamma \cap \alpha\Gamma\alpha^{-1}]$ or $d(\alpha^{-1})$.

The free $\mathbb{Z}$-module based on all double cosets $\Gamma\alpha\Gamma$, α in $\tilde{\Gamma}$, is denoted by $R = R(\Gamma, \tilde{\Gamma})$. For $x = \Sigma_k c_k(\Gamma\alpha_k\Gamma) \in R$, the degree of x is defined by:

$$\deg(x) = \Sigma_k c_k \deg(\Gamma\alpha_k\Gamma) \quad .$$

A multiplication in R is defined as follows: Let $\alpha, \beta \in \tilde{\Gamma}$ and set:

$$\Gamma\alpha\Gamma = \coprod_i \Gamma\alpha_i \quad , \qquad \Gamma\beta\Gamma = \coprod_j \Gamma\beta_j \quad .$$

The product $(\Gamma\alpha\Gamma).(\Gamma\beta\Gamma)$ is defined to be the sum:

$$\Sigma_k m(\Gamma\alpha\Gamma.\Gamma\beta\Gamma,\Gamma\xi_k\Gamma).(\Gamma\xi_k\Gamma) \qquad .$$

Here $\Gamma\xi_k\Gamma$ ranges over all the distinct double cosets contained in $\Gamma\alpha\Gamma\beta\Gamma$ and the nonnegative integer $m(\Gamma\alpha\Gamma.\Gamma\beta\Gamma,\Gamma\xi_k\Gamma)$ is defined to be the cardinality:

$$\#\{(i,j)\,|\,\Gamma\alpha_i\beta_j \;=\; \Gamma\xi_k\} \qquad .$$

All of these are well defined--independent of the choice of the representatives of the various types of cosets. We extend this product to R through bilinearity over R, turning R into an associative ring (with 1). $R = R(\Gamma,\tilde{\Gamma})$ is called a Hecke ring. Moreover, the degree map: $\deg: R \to \mathbb{Z}$ is a ring homomorphism, i.e., aside from the trivial relation: $\deg(x+y) = \deg(x) + \deg(y)$, we also have:

$$\deg(x.y) \;=\; \deg(x).\deg(y),x,y \in R(\Gamma,\tilde{\Gamma}) \quad .$$

Let Δ be a semigroup with $\Gamma \subset \Delta \subset \Gamma$. The subgroup:

$$R(\Gamma,\Delta) \;=\; \{\Sigma_k c_k(\Gamma\alpha_k\Gamma)\,|\,c_k \in \mathbb{Z},\ \alpha_k \in \Delta\}$$

is actually a subring, also called a Hecke ring.

1.2 Hecke Operators Acting on Group Cohomology

We first note that traditionally Hecke operators act on the right while groups act on the left. Rather than deviating from standard references, we follow this tradition and will call attention to this fact when there is a chance of confusion.

We next recall the main results of Rhie-Whaples [16]. The group cohomology is that of Eilenberg-MacLane. For a group Γ and a (left) Γ-module M, an r-cochain is a map $c:\Gamma \times \ldots \times \Gamma$ $(r+1$ factors$) \to M$ with:

$$c(\gamma\gamma_0,\ldots,\gamma\gamma_r) \;=\; \gamma.c(\gamma_0,\ldots,\gamma_r),\gamma,\gamma_0,\ldots,\gamma_r \in \Gamma \ .$$

The additive group of all r-cochains is denoted by $C^r(\Gamma,M)$. The coboundary map $\delta: C^r(\Gamma,M) \to C^{r+1}(\Gamma,M)$ is the usual one:

$$(\delta c)(\gamma_0,\ldots,\gamma_{r+1}) = \Sigma_j(-1)^j c(\gamma_0,\ldots,\hat{\gamma}_j,\ldots,\gamma_{r+1}) .$$

The corresponding cohomology groups are denoted by $H^r(\Gamma,M)$.

Let $\alpha \in \tilde{\Gamma}$ and form: $\Gamma\alpha\Gamma = \coprod_{1\leqslant i\leqslant d} \Gamma\alpha_i$. For $\gamma \in \Gamma$, we have a permutation of $\{1,\ldots,d\}$ sending i onto $i.\gamma$ so that:

$$\Gamma\alpha_i\gamma = \Gamma\alpha_{i.\gamma} , \quad 1\leqslant i\leqslant d , \quad \gamma\in\Gamma .$$

This is just the permutation (transitive) representation of Γ acting on the right of the Γ-left coset decomposition of $\Gamma\alpha\Gamma$; we read a permutation from left to right, writing permutation to the right of the integer being acted on. This gives us d elements $\rho_i(\gamma) \in \Gamma$ so that: $\alpha_i\gamma = \rho_i(\gamma)\alpha_{i.\gamma}$, $1\leqslant i\leqslant d$ and we have d maps $\rho_i:\Gamma\to\Gamma$, $1\leqslant i\leqslant d$.

Let Δ be a semigroup with $\Gamma\subset\Delta\subset\tilde{\Gamma}$ and let M be a Δ^{-1}-module. For any r-cochain $c\in C^r(\Gamma,M)$, we define c' by:

$$c'(\gamma_0,\ldots,\gamma_r) = \sum_{1\leqslant i\leqslant d} \alpha_i^{-1}c(\rho_i(\gamma_0),\ldots,\rho_i(\gamma_r)) .$$

It can be seen that c' is again an r-cochain. For a fixed choice of α_i, this determines an endomorphism of $C^r(\Gamma,M)$ commuting with δ and hence induces an endomorphism of $H^r(\Gamma,M)$. This endomorphism of $H^r(\Gamma,M)$ is independent of the choice of α_i; see Rhie and Whaples [16]. Extending by linearity to all of $R(\Gamma,\Delta)$ turns $H^r(\Gamma,M)$ into a right $R(\Gamma,\Delta)$-module. The operator defined by $\Gamma\alpha\Gamma$ will be denoted by $(\Gamma\alpha\Gamma)^*$, or simply as $\Gamma\alpha\Gamma$ when there is no chance of confusion. The image of $z=(c)\in H^r(\Gamma,M)$ under $(\Gamma\alpha\Gamma)^*$ will usually be written as:

$$z|\Gamma\alpha\Gamma = (c|\Gamma\alpha\Gamma), \text{ where } c|\Gamma\alpha\Gamma \text{ is defined to be } c'.$$

1.3 Functorial Properties

We continue with the notation of the preceding sections. Let $\varphi:M_1\to M_2$ be a morphism of Δ^{-1}-modules. Restricting to Γ, φ induces a homomorphism: $\varphi_*:H^r(\Gamma,M_1)\to H^r(\Gamma,M_2)$. The following is clear:

<u>Theorem 1.3.1</u>. *The homomorphism* $\varphi_*:H^r(\Gamma,M_1)\to H^r(\Gamma,M_2)$ *commutes with the action of the Hecke ring* $R(\Gamma,\Delta)$.

Let $\Gamma_3 \subset G_3$ be another group pair and let Δ_3 be a semigroup with $\Gamma_3 \subset \Delta_3 \subset \tilde{\Gamma}_3$. Let $\varphi:\Delta \to \Delta_3$ be a morphism of semigroups so that $\Gamma_3 = \varphi(\Gamma)$. We denote this situation by:

$$\varphi:(\Gamma,\Delta) \twoheadrightarrow (\Gamma_3,\Delta_3) \quad .$$

Associated to such a φ, we will construct a homomorphism $\varphi_!$ of Hecke rings:

$$\varphi_!:R(\Gamma,\Delta) \to R(\Gamma_3,\Delta_3) \quad .$$

Lemma 1.3.2. *Let* $\alpha \in \Delta$ *and set* $\xi = \varphi(\alpha) \in \Delta_3$. *Thus,* $\varphi(\Gamma\alpha\Gamma) = \Gamma_3\xi\Gamma_3$. *For any* Γ_3*-left coset* $\Gamma_3\beta$ *contained in* $\Gamma_3\xi\Gamma_3$, *the number of distinct* Γ*-left cosets* $\Gamma\alpha_i$ *in* $\Gamma\alpha\Gamma$ *with* $\varphi(\alpha_i) \in \Gamma_3\beta$ *is a constant* f *independent of the choice of* $\Gamma_3\beta$ *contained in* $\Gamma_3\xi\Gamma_3$.

Proof. The $d(\alpha)$ distinct Γ-left cosets in $\Gamma\alpha\Gamma$ are transitively permuted by Γ under right multiplication. The $d(\xi)$ distinct Γ_3-left cosets in $\Gamma_3\xi\Gamma_3$ are transitively permuted by $\Gamma_3 = \varphi(\Gamma)$, hence by Γ through φ. The desired assertion follows easily by viewing φ as an equivariant map of finite right Γ-sets transitive under Γ. ∎

Corollary 1.3.3. *The integer* f *named in the preceding lemma satisfies:*

$$f = \deg(\Gamma\alpha\Gamma)/\deg(\Gamma_3\xi\Gamma_3) \quad . \tag{1.3.4}$$

The integer f in (1.3.4) will be called the multiplicity of the homomorphism φ at $(\Gamma\alpha\Gamma)$ or at $(\Gamma_3\xi\Gamma_3)$ and written as $f_\varphi(\Gamma\alpha\Gamma)$ or as $f_\varphi(\Gamma_3\xi\Gamma_3)$.

Definition 1.3.5. *For an element* $x = \Sigma_k c_k \Gamma\alpha_k\Gamma \in R(\Gamma,\Delta)$, *set:*

$$\varphi_!(x) = \Sigma_k c_k . f_\varphi(\Gamma\alpha_k\Gamma).(\Gamma_3\xi_k\Gamma_3) \in R(\Gamma_3,\Delta_3) \quad ,$$

where $\xi_k = \varphi(\alpha_k)$ *so that* $\Gamma_3\xi_k\Gamma_3$ *is just* $\varphi(\Gamma\alpha_k\Gamma)$.

Theorem 1.3.6. $\varphi_!:R(\Gamma,\Delta) \to R(\Gamma_3,\Delta_3)$ *is a ring homomorphism. Moreover,* $\deg(\varphi_!(x)) = \deg(x)$ *holds for all* x *in* $R(\Gamma,\Delta)$.

Proof. Additivity as well as the assertion on degree are clear. To see $\varphi_!(x.y) = \varphi_!(x).\varphi_!(y)$, we may assume $x = (\Gamma\alpha\Gamma)$, $y = (\Gamma\beta\Gamma)$. Let:

$$\Gamma\alpha\Gamma = \coprod_{1\leq i\leq d} \Gamma\alpha_i , \qquad \Gamma\beta\Gamma = \coprod_{1\leq j\leq e} \Gamma\beta_j .$$

Then $x.y = \Sigma_k m_k \Gamma\gamma_k\Gamma$, where $\Gamma\gamma_k\Gamma$ ranges over double cosets in $\Gamma\alpha\Gamma\beta\Gamma$ and

$$m_k = m((\Gamma\alpha\Gamma).(\Gamma\beta\Gamma),(\Gamma\gamma_k\Gamma)) = \#\{(i,j) \mid \Gamma\alpha_i\beta_j = \Gamma\gamma_k\} .$$

Now $\varphi_!(x.y) = \Sigma_k m_k \varphi_!((\Gamma\gamma_k\Gamma)) = \Sigma_k m_k f_\varphi(\Gamma\gamma_k\Gamma)\Gamma_3\xi_k\Gamma_3$, where $\xi_k = \varphi(\gamma_k)$. Similarly, we have:

$$\varphi_!(x) = f_\varphi(\Gamma\alpha\Gamma)(\Gamma_3\mu\Gamma_3), \varphi_!(y) = f_\varphi(\Gamma\beta\Gamma)(\Gamma_3\nu\Gamma_3), \mu = \varphi(\alpha), \nu=\varphi(\beta).$$

As a consequence, we have:

$$\varphi_!(x).\varphi_!(y) = f_\varphi(\Gamma\alpha\Gamma) f_\varphi(\Gamma\beta\Gamma)\Sigma_k m((\Gamma_3\mu\Gamma_3).(\Gamma_3\nu\Gamma_3),\Gamma_3\xi_k\Gamma_3).(\Gamma_3\xi_k\Gamma_3)$$

where the summation extends over the distinct double cosets $\Gamma_3\xi_k\Gamma_3$ in $\Gamma_3\mu\Gamma_3\nu\Gamma_3 = \varphi(\Gamma\alpha\Gamma\beta\Gamma)$. The desired assertion is now a consequence of the following equalities:

$$f_\varphi(\Gamma\alpha\Gamma) f_\varphi(\Gamma\beta\Gamma).m((\Gamma_3\mu\Gamma_3).(\Gamma_3\nu\Gamma_3),\Gamma_3\xi_k\Gamma_3)$$

$$= m((\Gamma\alpha\Gamma).(\Gamma\beta\Gamma),\Gamma\gamma_k\Gamma) f_\varphi(\Gamma\gamma_k\Gamma) = \#\{(i,j) \mid \varphi(\alpha_i)\varphi(\beta_j) \in \Gamma_3\xi_k\}. \blacksquare$$

Suppose a Δ_3^{-1}-module M is given, so that it is a Γ_3-module via restriction and a Γ-module via φ. We therefore have the induced homomorphism $\varphi^*: H^r(\Gamma_3,M) \to H^r(\Gamma,M)$, called the inflation map.

<u>Theorem 1.3.7</u>. *The inflation homomorphism* φ^* *described above is compatible with Hecke ring actions as well as with* $\varphi_!$, *i.e., for any cohomology class* $z \in H^r(\Gamma_3,M)$ *and for any element* $x \in R(\Gamma,\Delta)$, *we have:*

$$\varphi^*(z|\varphi_!(x)) = \varphi^*(z)|x .$$

<u>Proof</u>. We may assume $x = (\Gamma\alpha\Gamma) = \coprod_{1\leq i\leq d} \Gamma\alpha_i$. Put $\varphi(\alpha) = \xi$ and let $\Gamma_3\xi\Gamma_3 = \coprod_{1\leq j\leq e} \Gamma_3\xi_j$. Using Lemma 1.3.2 and Corollary 1.3.3, we may assume that ξ_j is equal to $\varphi(\alpha_{(j-1)f+t})$, $1\leq t\leq f$, $1\leq j\leq e$, where $ef = d$, with appropriate choices of the representatives. This

then induces an f to 1 map h on the subscripts, i.e., $h((j-1)f+t)=j$, $1\leqslant t\leqslant f$, $1\leqslant j\leqslant e$. Let $\rho_i:\Gamma\to\Gamma$ be the function with the property:

$$\rho_i(\gamma)\alpha_{i.\gamma} = \alpha_i\gamma, \quad \gamma\in\Gamma, \quad 1\leqslant i\leqslant d \quad .$$

Similarly, let $\tau_j:\Gamma_3\to\Gamma_3$ be the function with the property:

$$\tau_j(\eta)\xi_{j.\eta} = \xi_j\eta, \quad \eta\in\Gamma_3, \ 1\leqslant j\leqslant e \quad .$$

Applying φ to the first of these two, we have:

$$h(i).\varphi(\gamma) = h(i.\gamma), \ \varphi(\rho_i(\gamma)) = \tau_{h(i)}(\varphi(\gamma)), \quad \gamma\in\Gamma \text{ and } 1\leqslant i\leqslant d.$$

In other words, we have compatibility with an appropriate choice of coset representatives.

The desired assertion now follows from direct calculation using the definition of the various maps on the level of cochains and the compatibility properties. We omit the details. ∎

For the rest of this section, we consider an exact sequence of groups:

$$1 \longrightarrow \Gamma_1 \xrightarrow{i} \Gamma \xrightarrow{j} \Gamma_3 \longrightarrow 1 \ .$$

Associated to these groups are the semigroups Δ_1, Δ and Δ_3 where Δ and Δ_3 are as before so that j is the restriction of a morphism from Δ to Δ_3. In addition, we assume that Δ_1 is a subsemigroup of Δ lying inside the commensurator of Γ_1 and contains Γ_1 with i denoting the inclusion map. If M is any Δ_3^{-1}-module, then it will be viewed as a Δ^{-1} as well as Δ_1^{-1}-module through j, so that Γ_1 acts trivially on M.

Lemma 1.3.8. *If the semigroup* Δ_1 *has the properties:*

(a) $\Delta = \Gamma.\Delta_1$,

(b) $\Gamma\alpha\Gamma = \Gamma\alpha\Gamma_1$ *for all* $\alpha\in\Delta_1$, *and*

(c) $\Gamma\alpha\cap\Delta_1 = \Gamma_1\alpha$ *for all* $\alpha\in\Delta_1$,

then the map $i_!$ *sending* $\Gamma_1\alpha\Gamma_1$ *onto* $\Gamma\alpha\Gamma$, $\alpha\in\Delta_1$ *defines an isomorphism of rings:*

$$i_!:R(\Gamma_1,\Delta_1) \cong R(\Gamma,\Delta) \quad .$$

For a proof, see Doi-Miyake [5].

Lemma 1.3.9. *In the situation described above, the action of the Hecke operator* $x \in R(\Gamma_1, \Delta_1)$ *on* $H^r(\Gamma_1, M)$ *and the action of* $i_!(x) \in R(\Gamma, \Delta)$ *on* $H^r(\Gamma, M)$ *are compatible with the restriction homomorphism:*

$$i^*: H^r(\Gamma, M) \to H^r(\Gamma_1, M) \quad .$$

Namely, $i^*(z | i_!(x)) = i^*(z) | x$ *holds for* $z \in H^r(\Gamma, M)$. *Moreover, the image of* i^* *is contained in the* $R(\Gamma_1, \Delta_1)$*-submodule formed by the* Γ_3*-fixed points of* $H^r(\Gamma_1, M)$.

The proof is straightforward and will be omitted; cf. Hochschild-Serre [9].

In general, we can consider the Hochschild-Serre spectral sequence $\{E_r^{a,b}, d_r\}$ with:

$$E_2^{a,b} = H^a(\Gamma_3, H^b(\Gamma_1, M))$$

$$\coprod_{a+b=r} E_\infty^{a,b} \cong Gr(H^r(\Gamma, M))$$

where Gr is the grading functor associated to a suitable filtration of $H^r(\Gamma, M)$. Under the preceding assumptions, and with the identification of the Hecke rings in Lemma 1.3.8, the filtration is preserved by the action of the Hecke rings. As a result, $R(\Gamma_1, \Delta_1)$ $(= R(\Gamma, \Delta))$ acts on the spectral sequence compatibly with the differentials d_r, $r \geqslant 2$. We will primarily be interested in the Hochschild-Serre exact sequence arising from the spectral sequence considerations.

Theorem 1.3.10. *The Hochschild-Serre exact sequence:*

$$0 \to H^1(\Gamma_3, M) \xrightarrow{j^*} H^1(\Gamma, M) \xrightarrow{i^*} H^1(\Gamma_1, M)^{\Gamma_3} \to H^2(\Gamma_3, M)$$

is compatible with the actions of the Hecke rings together with ring homomorphisms:

$$R(\Gamma_3, \Delta_3) \xleftarrow{j_!} R(\Gamma, \Delta) \underset{\cong}{\xleftarrow{i_!}} R(\Gamma_1, \Delta_1) \quad .$$

1.4 DeRham, Hodge, Matsushima-Murakami Theories and Eichler Isomorphism in Fuchsian Group Case

Suppose that Γ is a discrete subgroup of a semisimple Lie group G_1 with a maximal compact subgroup K such that Γ operates on the symmetric space $X = G_1/K$ with compact quotient $U = \Gamma\backslash X$. Let (ρ, M) be a real representation of the Lie group G_1. Then the cohomology group $H^r(\Gamma, M)$ is describable as the space $H^r(X,\Gamma,\rho)$ of M-valued harmonic forms (see Bailey [1], Matsushima-Murakami [14]) by means of a certain hermitian metric on the locally constant vector bundle $\tilde{M}$ over U associated to ρ. Moreover, if X is a hermitian symmetric domain, then we can form the Hodge decomposition:

$$H^r(X,\Gamma,\rho) \otimes \mathbb{C} = \coprod_{p+q=r} H^{(p,q)}(X,\Gamma,\rho) \tag{1.4.1}$$

(Matsushima-Murakami [14], also [13]). Hereafter, we write $H^r(X,\Gamma,\rho)_{\mathbb{C}}$ for the left side of (1.4.1).

In particular, for $X = H^N$ = product of N copies of the upper half plane H and $G_1 = SL(2,\mathbb{R})^N$, the complete description of $H^{(p,q)}(X,\Gamma,\rho)$ is given by Matsushima-Shimura [15].

Let G be a reductive group whose semisimple part is G_1, and let Δ be a semigroup with $\Gamma \subset \Delta \subset \tilde{\Gamma} \subset G$. The Hecke ring $R(\Gamma,\Delta)$ operates on $H^r(X,\Gamma,\rho)$ and $H^{(p,q)}(X,\Gamma,\rho)$ in a natural way, and this action is compatible with the action on $H^r(\Gamma,M)$ through the canonical isomorphism: (see [13])

$$H^r(\Gamma,M) \cong H^r(X,\Gamma,\rho) \tag{1.4.2}$$

We consider the particular case where $X = H$ = upper half plane, $G = GL^+(2,\mathbb{R})$, $G_1 = SL(2,\mathbb{R})$, Γ = a fuchsian group in G_1 with compact fundamental domain, and $G \supset \tilde{\Gamma} \supset \Delta \supset \Gamma$. We take a representation:

$$\rho = \rho_m^* = (\det)^{-m} \otimes \rho_m \tag{1.4.3}$$

of G, defined by:

$$\rho_m^*(g) = \det(g)^{-m} \cdot \rho_m(g), \qquad g \in GL^+(2,\mathbb{R}) \tag{1.4.4}$$

where ρ_m is the symmetric tensor representation of degree m of G.

We note that ρ_m^* and ρ_m are identical on Γ because $\Gamma \subset SL(2,\mathbb{R})$. The following isomorphisms are known:

$$H^1(\Gamma,\rho_m^*)_{\mathbb{C}} \cong H^1(X,\Gamma,\rho_m^*)_{\mathbb{C}} = H^{(1,0)}(X,\Gamma,\rho_m^*) \amalg H^{(0,1)}(X,\Gamma,\rho_m^*) \quad (1.4.5)$$

$$\begin{cases} H^{(1,0)}(X,\Gamma,\rho_m^*) \cong S_{m+2}(\Gamma) \\ H^{(0,1)}(X,\Gamma,\rho_m^*) \cong \overline{S_{m+2}(\Gamma)} \end{cases} \quad (1.4.6)$$

$$\begin{cases} H^0(\Gamma,\rho_m^*)_{\mathbb{C}} = \mathbb{C} = H^2(\Gamma,\rho_m^*)_{\mathbb{C}}, & m = 0 \\ H^0(\Gamma,\rho_m^*)_{\mathbb{C}} = 0 = H^2(\Gamma,\rho_m^*)_{\mathbb{C}}, & m > 0\ . \end{cases} \quad (1.4.7)$$

Here $S_{m+2}(\Gamma)$ is the space of cusp forms of weight $m+2$. The isomorphism (1.4.6) is called the Eichler-Shimura isomorphism; see Eichler [8], Shimura [24], and [13].

The isomorphisms (1.4.5) through (1.4.7) are all compatible with the actions of the Hecke operators. The (right) action of a Hecke operator $x = (\Gamma\alpha\Gamma) = \amalg_{1 \leq i \leq d} \Gamma\alpha_i$ on the space $S_{m+2}(\Gamma)$ is the one defined by Hecke:

$$f(z) \mapsto \sum_{1 \leq i \leq d} f(\alpha_i . z).(\det \alpha_i)^{m+1}/(c_i z + d_i)^{m+2} \quad . \quad (1.4.8)$$

Here the matrix $\begin{pmatrix} a & b \\ c & d \end{pmatrix}$ in $GL^+(2,\mathbb{R})$ sends z onto $(az+b)/(cz+d)$. With obvious interpretation of notation, (1.4.5) becomes:

$$(H^1(\Gamma,\rho_m^*)|x) \cong (S_{m+2}|x) \amalg (\overline{S_{m+2}}|x), \quad x \in R(\Gamma,\Delta) \quad .$$

The notation is adopted to remind us that Hecke operators act on the right.

When Γ is a fuchsian group of the first kind with parabolic elements, the validity of (1.4.5) and (1.4.6) is retained provided that we use parabolic cohomology groups $H^1(\Gamma,\rho_m^*)_{par}$. We then have:

$$H^1(\Gamma,\rho_m^*)_{par}\otimes\mathbb{C}\cong S_{m+2}(\Gamma)\amalg\overline{S_{m+2}(\Gamma)} \tag{1.4.9}$$

$$(H^1(\Gamma,\rho_m^*)_{par}|x)\cong(S_{m+2}|x)\amalg(\overline{S_{m+2}}|x)\ ,\qquad x\in R(\Gamma,\Delta). \tag{1.4.10}$$

See Shimura [24; 223-238] for the definition of parabolic cohomology.

Finally, we mention some relations between $H^1(\Gamma,M)$ and $H^1(\Gamma/\{\pm 1\},M)$ where Γ is a fuchsian subgroup of $SL(2,\mathbb{R})$ with $-1\in\Gamma$ and M is a Γ-module with Γ acting through ρ so that $\rho(-1)=\pm id_M$.

Proposition 1.4.11. *With the preceding notation, we have:*

(a) *if* $\rho(-1)=id_M$ *and* M *has no 2-torsion, then*

$$H^1(\Gamma,M)\cong H^1(\Gamma/\{\pm 1\},M)$$

(b) *if* $\rho(-1)=-id_M$ *and* M *is uniquely 2-divisible, then*

$$H^1(\Gamma,M)\ =\ 0\quad .$$

The proof follows from the Hochschild-Serre exact sequence.

Corollary 1.4.12. *With the preceding notation, assume* M *is a vector space over a field* k *of characteristic not* 2.

(a) *if* $\rho(-1)=id_M$, *then* $H^1(\Gamma,M)\cong H^1(\Gamma/\{\pm 1\},M)$;

(b) *if* $\rho(-1)=-id_M$, *then* $H^1(\Gamma,M)=0$.

We note that the fact $S_{m+2}(\Gamma)=0$ for odd m when $-1\in\Gamma$ agrees with (b) of Corollary 1.4.12. When m is even, $\rho_m(-1)=1$ and the Eichler-Shimura isomorphisms (1.4.6) and (1.4.9) are still valid when Γ is replaced by $\Gamma/\{\pm 1\}$.

1.5 Finite Group Case

We consider the case where Δ is a finite group.

Theorem 1.5.1. *If* Γ *is a normal subgroup of the finite group* Δ, *then the Hecke ring* $R(\Gamma,\Delta)$ *is just the integral group ring* $\mathbb{Z}[\Delta/\Gamma]$ *of the quotient group* Δ/Γ.

Theorem 1.5.2. *Let* Γ *be a central subgroup of the finite group* Δ. *Let* M *be a* Δ*-module with trivial action. Then the action of*

$\Gamma\alpha\Gamma = \Gamma\alpha$, $\alpha \in \Delta$, *on* $H^r(\Gamma, M)$ *is trivial and the action of* x^*, $x \in R(\Gamma,\Delta)$, *is just scalar multiplication by* $\deg(x)$.

In what follows, we specialize to $\Delta = GL(2,\mathbb{F}_q)$ and $\Gamma = SL(2,\mathbb{F}_q)$. Δ operates on $\mathbb{F}_q^2$ through matrix multiplication, therefore it operates on the symmetric m-th power over $\mathbb{F}_q : S^m = S^m(\mathbb{F}_q^2)$. This modular representation of Δ is called the symmetric tensor representation of degree m and written as ρ_m.

Theorem 1.5.3. *Let* $q \geq 5$ *be a prime number. Let* $0 \leq t \leq q-1$.

$$H^1(SL(2,\mathbb{F}_q), S^t(\mathbb{F}_q^2)) = 0 \quad \textit{or} \quad \mathbb{F}_q \quad \textit{according to} \quad t \neq \textit{or} = q-3;$$

$$H^2(SL(2,\mathbb{F}_q), S^t(\mathbb{F}_q^2)) = 0 \quad \textit{or} \quad \mathbb{F}_q \quad \textit{according to} \quad t \neq \textit{or} = 2 \quad .$$

We will sketch the proof. Let B be the Borel subgroup of Γ formed by the upper triangular matrices in Γ. B is then the semi-direct product of the unipotent subgroup $U \cong \mathbb{F}_q$ by the semisimple subgroup $T \cong \mathbb{F}_q^\times$. Since restriction from Δ to B followed by corestriction is multiplication by the index of B in Δ and this index is coprime to the characteristic of $\mathbb{F}_q$, the positive dimensional cohomology of Δ is detected on B. By the Cartan-Eilenberg stability criterion, the image of the restriction map is actually all of the cohomology of B in the present case because two distinct conjugates of B in Δ have only the identity element in common. We can therefore replace Δ by B in our assertion. Since T has order coprime to the characteristic of $\mathbb{F}_q$, the spectral sequence associated to the semidirect product splitting of B, shows that our cohomology groups reduce down to the T-fixed part of the cohomology of U. In the present case, U is a cyclic group so that its cohomology is periodic with period 2. As usual, H^2 and H^1 are respectively:

$$\ker(\sigma-1)/\mathrm{im}\, N_\sigma \quad \text{and} \quad \ker N_\sigma/\mathrm{im}(\sigma-1)$$

where σ is a generator of U and N_σ is $1+\sigma+\cdots+\sigma^{q-1}$. If we let σ be $\begin{pmatrix} 1 & 1 \\ 0 & 1 \end{pmatrix}$, then S^t has an $\mathbb{F}_q$-basis formed by the monomials $X^{t-i}Y^i$, $0 \leq i \leq t$. These are weight vectors for the action of T with weight $t-2i$. If we let $Z = Y/X$ and use the $\mathbb{F}_q$-basis: $X^t.\binom{Z}{i}$, $0 \leq i \leq t$, where $\binom{Z}{i} = Z(Z-1)\ldots(Z-i+1)/i!$, it is immediate that σ has matrix in canonical form with minimal polynomial $(\lambda-1)^{t+1}$ on S^t.

The cohomology of U with coefficients in S^t is therefore $\mathbb{F}_q$ and is a weight space for T with weight t (in case of H^2) or $-t$ (in case of H^1). This description of weight is only valid for the action of T on S^t and does not take into account the action of T on U. The action of T on U converts U into a weight space of dimension 1 over $\mathbb{F}_q$ with weight 2. The correct description of the cohomology groups is then $\mathrm{Hom}_{\mathbb{Z}}(W_2,W_s)$, where the subscript denotes weight. Since $\mathbb{F}_q$ is a prime field, we can replace $\mathbb{Z}$ by $\mathbb{F}_q$ and the cohomology groups are weight spaces of dimension 1 over $\mathbb{F}_q$ with weight $s-2$. Since T is cyclic of order $q-1$, the T-fixed points are 0 when $q-1$ does not divide $s-2$ and $\mathbb{F}_q$ when $q-1$ does divide $s-2$. The desired assertions therefore follow. ∎

The representation ρ_m of Γ is isomorphic to its dual, however this is not the case when viewed as a representation of Δ. The difference is slight. We can describe ρ_m^* by using the same representation space $S^m(\mathbb{F}_q^2)$ through the following formula:

$$\rho_m^*(g) = (\det g)^{-m}\rho_m(g) \quad .$$

The point is that $g \in GL(2,\mathbb{F}_q)$ acts on the dual through the transpose inverse (the contragredient). Since ρ_m is isomorphic to its dual on Γ, we can change basis and assume that ρ_m^* is just ρ_m on Γ. The presence of $(\det g)^{-m}$ is clear from weight considerations. To be more precise, we have an involution of the space $M_2(\mathbb{F}_q)$ of all 2×2 matrices sending $\begin{pmatrix} a & b \\ c & d\end{pmatrix}$ onto $\begin{pmatrix} d & -b \\ -c & a\end{pmatrix} = \begin{pmatrix} a & b \\ c & d\end{pmatrix}^{\rho}$. It is immediate that g^{ρ} is the transpose of the cofactor matrix so that $g.g^{\rho} = (\det g).1$. It is also immediate that g^{ρ} is conjugate to the transpose of g via the matrix $\begin{pmatrix} 0 & -1 \\ 1 & 0\end{pmatrix}$. Thus, $\rho_m^*(g) = \rho_m(g^{\rho})^{-1}$ and ρ_m^* is equivalent to the dual of ρ_m.

Theorem 1.5.4. *Let* $q \geq 5$ *be a prime number. Put* $\Gamma = SL(2,\mathbb{F}_q)$ *and* $\Delta = GL(2,\mathbb{F}_q)$. *Then* $R(\Gamma,\Delta) \cong \mathbb{Z}[\mathbb{F}_q^{\times}]$ *and the action of* $\Gamma\alpha\Gamma$, $\alpha \in \Delta$, *on* $H^1(\Gamma,\rho_{q-3}^*) \cong \mathbb{F}_q$ *is scalar multiplication:*

$$(\Gamma\alpha\Gamma)^* = (\det \alpha)^{-1} \quad .$$

The action of $c = \Sigma_k c_k \Gamma\alpha_k\Gamma$ *is scalar multiplication by* $\Sigma_k c_k (\det \alpha_k)^{-1}$.

This can be seen through the proof of Theorem 1.5.3. In essence, the proof there shows that $H^1(U,\rho_t^*) \cong \mathbb{F}_q$ holds for $0 \leq t \leq q-1$ (this used the assumption that q is a prime twice--once when U is cyclic, second time when $\mathrm{Hom}_{\mathbb{Z}}$ is replaced by $\mathrm{Hom}_{\mathbb{F}_q}$).

We must pinpoint the action of the diagonal subgroup of $GL(2,\mathbb{F}_q)$. Since we are using ρ_t^*, not ρ_t, the action of $\begin{pmatrix}\alpha & 0\\ 0 & \beta\end{pmatrix}$ is multiplication by $\alpha^{-t-1}\beta$. When $t = q-3$, the fact that $\mathbb{F}_q^\times$ is cyclic of order $q-1$ shows that we are multiplying by the determinant $\alpha\beta$. Finally, we note that this description is based on the left action. In order to conform to the right action of Hecke operators, we must use inversion in the group. This explains the inverse determinant action. ∎

1.6 Quaternion Case

In a quaternion algebra B over k (so that B has dimension 4 over the center k), the canonical involution is denoted by ρ. The reduced trace and reduced norm are defined by: $tr(x) = x + x^\rho$, $\nu(x) = x.x^\rho$. The subgroup of the multiplicative group $B^\times$ formed by elements with reduced norm 1 is denoted by $B_1^\times$. When $B = M_2(k)$, the matrix algebra of size 2×2, tr and ν are respectively the trace and determinant of a matrix and $\begin{pmatrix}a & b\\ c & d\end{pmatrix}^\rho = \begin{pmatrix}d & -b\\ -c & a\end{pmatrix}$. If $k = \mathbb{R}$ and $B = \mathbb{H}$ = Hamiltonian quaternions $= \mathbb{R}.1 + \mathbb{R}.i + \mathbb{R}.j + \mathbb{R}.k$, then $\rho(x.1 + y.i + z.j + w.k) = x.1 - y.i - z.j - w.k$, and $\mathbb{H}_1^\times \cong SU(2,\mathbb{C})$, $\mathbb{H}^\times \cong \mathbb{R}^+ \times SU(2,\mathbb{C})$. If $B = M_2(\mathbb{R})$, then $B^\times = GL(2,\mathbb{R})$, $B_1^\times = SL(2,\mathbb{R})$ and $GL^+(2,\mathbb{R})$ is (as before) the subgroup of $GL(2,\mathbb{R})$ formed by matrices with positive determinant.

Let k be a number field of degree n over $\mathbb{Q}$ with ring of all algebraic integers $\mathcal{O}_k$. We have n distinct injections $\varphi(i): k \to \mathbb{C}$. k is said to be totally real if each $\varphi(i)$ has image in $\mathbb{R}$. When this happens, $a \in k$ is called totally positive if $\varphi(i)(a) > 0$ holds for each i and we write $a \gg 0$. Let B be a quaternion algebra over k and let $\mathcal{O}$ be a maximal order of B. Let $d(B) = \mathfrak{p}(1)\ldots\mathfrak{p}(e)$ be the product of all the (maximal) prime ideals of $\mathcal{O}_k$ such that $B \otimes_k k_{\mathfrak{p}(i)}$ is a division algebra over the $\mathfrak{p}(i)$-adic completion $k_{\mathfrak{p}(i)}$ of k. $d(B)$ (occasionally, $d(B)^2$) is sometimes called the discriminant of B.

The set of all nonzero two-sided ideals of $\mathcal{O}$ forms a cancellation semigroup under multiplication with $\mathcal{O}$ as identity. Each such two-sided ideal is a product of powers of maximal two-sided ideals in an essentially unique manner. The maximal two-sided ideals of $\mathcal{O}$ can be described in the following manner:

Take a prime ideal $\mathfrak{p}$ of $\mathcal{O}_k$. There are two possibilities:

<u>Case 1</u>. $\mathfrak{p}\nmid d(B)$. $\mathfrak{p}\mathcal{O}$ is a maximal two-sided ideal of $\mathcal{O}$

<u>Case 2</u>. $\mathfrak{p}|d(B)$, say $\mathfrak{p}=\mathfrak{p}(i)$. There is a unique maximal two-sided ideal $\mathfrak{P}(i)$ of $\mathcal{O}$ with $\mathfrak{p}(i)\mathcal{O}\subset\mathfrak{P}(i)\subset\mathcal{O}$ and $\mathfrak{p}(i)\mathcal{O}=\mathfrak{P}(i)^2$.

These are exactly all the distinct maximal two-sided ideals of $\mathcal{O}$ and any nonzero two-sided ideal A of $\mathcal{O}$ can be written as: $A=\mathfrak{P}(1)^{\varepsilon(1)}\cdot\ldots\cdot\mathfrak{P}(e)^{\varepsilon(e)}\cdot A_1$, $\varepsilon(i)=0$ or 1, A_1 an ideal of $\mathcal{O}_k$. This expression is unique.

With this description, $A^{\rho}=A$ always holds. As a result, $A^2=A.A^{\rho}=A^{\rho}.A=M.\mathcal{O}$ where $M=\mathfrak{p}(1)^{\varepsilon(1)}\cdot\ldots\cdot\mathfrak{p}(e)^{\varepsilon(e)}.A_1^2=\nu(A)$ is an ideal of $\mathcal{O}_k$ called the norm of A. Two two-sided ideals A and B of $\mathcal{O}$ (both nonzero) are said to be coprime, written $(A,B)=1$ if $A+B=\mathcal{O}$. This is equivalent with $\nu(A)$ and $\nu(B)$ being coprime in the Dedekind domain $\mathcal{O}_k$. A non-zero-divisor α of $\mathcal{O}$ and a nonzero two-sided ideal B of $\mathcal{O}$ are said to be coprime, written $(\alpha,B)=1$, if $(\nu(\alpha),\nu(B))=1$.

For an ideal $\mathfrak{q}$ of $\mathcal{O}_k$ ($\mathcal{Q}$ of $\mathcal{O}$), the reduction mod $\mathfrak{q}$ (mod $\mathcal{Q}$) map on $\mathcal{O}_k$ (on $\mathcal{O}$) is denoted by $\varphi_{\mathfrak{q}}$ (by $\varphi_{\mathcal{Q}}$). We abbreviate $\varphi_{\mathfrak{q}\mathcal{O}}$ to $\varphi_{\mathfrak{q}}$, which will not cause any confusion.

If an ideal N of $\mathcal{O}_k$ is coprime with $d(B)$, then $\mathcal{O}/N\mathcal{O}\cong M_2(\mathcal{O}_k/N)$. Fixing an isomorphism $\iota:\mathcal{O}/N\mathcal{O}\to M_2(\mathcal{O}_k/N)$ throughout our discussions, then the map $\iota\circ\varphi_N:\mathcal{O}\to M_2(\mathcal{O}_k/N)$ can be abbreviated to φ_N. If a maximal two-sided ideal $\mathfrak{P}$ with $\mathcal{O}\supset\mathfrak{P}\supset\mathfrak{p}\mathcal{O}=\mathfrak{P}^2$ is given, where $\mathfrak{p}|d(B)$, then $\mathcal{O}/\mathfrak{P}\cong\mathbb{F}_{q^2}$, where $q=\#(\mathcal{O}_k/\mathfrak{p})=N_{k/\mathbb{Q}}(\mathfrak{p})$. We use the same convention concerning abbreviation of $\iota\circ\varphi_{\mathfrak{P}}:\mathcal{O}\to\mathbb{F}_{q^2}$ to $\varphi_{\mathfrak{P}}$.

In a total matrix algebra $M_2(A)$ with entries in the commutative ring A (with 1), consider the subrings:

$$B(A) = \left\{\begin{pmatrix} a & b\\ 0 & d\end{pmatrix}\middle|\, a,\ b,\ d\in A\right\}$$

$$D(A) = \left\{\begin{pmatrix} a & 0\\ 0 & d\end{pmatrix}\middle|\, a,\ d\in A\right\} .$$

The multiplicative groups $B(A)^{\times}$ and $D(A)^{\times}$ then have the following structures:

$$D(A)^{\times}\cong A^{\times}\times A^{\times}$$

$$1\to A_+\to B(A)^{\times}\to D(A)^{\times}\to 1 \qquad \text{is exact.}$$

Here the first map sends $a \in A$ onto $\begin{pmatrix} 1 & a \\ 0 & 1 \end{pmatrix}$ and the second map sends $\begin{pmatrix} a & b \\ 0 & d \end{pmatrix}$ onto $\begin{pmatrix} a & 0 \\ 0 & d \end{pmatrix}$. Actually, we will use the homomorphism ψ sending $\begin{pmatrix} a & b \\ 0 & d \end{pmatrix}$ onto the pair (a, d^{-1}) in $A^\times \times A^\times$.

With the preceding notation, let k be totally real. For x in k, set $|x|_{\varphi(i)} = |\varphi(i)(x)|$ and let $k_{\varphi(i)}$ $(\cong \mathbb{R})$ be the completion of k under the metric $|\ |_{\varphi(i)}$. Take a quaternion algebra B over k and assume:

$$B \otimes_k k_{\varphi(i)} \cong M_2(\mathbb{R}) \quad \text{for} \quad 1 \leqslant i \leqslant a \quad \text{and} \quad \cong \mathbb{H} \quad \text{for} \quad a < i \leqslant n. \tag{1.6.1}$$

It follows that:

$$B \otimes_{\mathbb{Q}} \mathbb{R} \cong M_2(\mathbb{R})^a \times \mathbb{H}^{n-a} \quad . \tag{1.6.2}$$

We fix such an isomorphism ι once for all. The injection of B into the right hand side of (1.6.2) defined by sending x to $\iota(x \otimes 1)$ is denoted by φ. If we let $\varphi(i)$ denote the i-th component of φ, we have an injection: $\varphi(i): B \to M_2(\mathbb{R})$ or $\mathbb{H}$. The restriction of $\varphi(i)$ to the center k of B is just the original injection of k into $\mathbb{R}$ (= center of $M_2(\mathbb{R})$ or of $\mathbb{H}$). These injections are compatible with canonical involutions, therefore with trace and norm maps:

$$\varphi(i) \circ \rho = \rho \circ \varphi(i), \operatorname{tr}(\varphi(i)(x)) = \varphi(i)(\operatorname{tr}(x)), \nu(\varphi(i)(x)) = \varphi(i)(\nu(x)) \ . \tag{1.6.3}$$

Under the injection φ, we have:

$$\varphi: B^\times \to (B \otimes_{\mathbb{Q}} \mathbb{R})^\times \overset{\iota}{=} GL(2,\mathbb{R})^a \times (\mathbb{H}^\times)^{n-a} =_{\text{df.}} G \quad . \tag{1.6.4}$$

The subgroup $SL(2,\mathbb{R})^a \times (\mathbb{H}_1^\times)^{n-a}$ of G is denoted by G_1 and the subgroup $GL^+(2,\mathbb{R})^a \times (\mathbb{H}^\times)^{n-a}$ is denoted by G^+ so that $G \supset G^+ \supset G_1$. An element of G^+ operates on the product H^a of a copies of the upper half plane H through the term-by-term fractional linear action in the first a factors $GL^+(2,\mathbb{R})$ of G^+.

Take an order $\mathcal{O}$ (not necessarily maximal) in B. Set $\mathcal{O}^\times = \{\gamma \in \mathcal{O} \mid \gamma\mathcal{O} = \mathcal{O}\}$, the group of units in $\mathcal{O}$, and set:

$$\Gamma(\mathcal{O},1) = \{\gamma \in \mathcal{O}^\times \mid \nu(\gamma) = 1\} \quad . \tag{1.6.5}$$

We then have:

$$\varphi:\Gamma(\mathcal{O},1)\to G_1 \quad = \quad SL(2,\mathbb{R})^a\times(\mathbb{H}_1^\times)^{n-a} \quad . \tag{1.6.6}$$

It follows that $\Gamma(\mathcal{O},1)$ operates on H^a. The action is properly discontinuous (see Siegel [25]), and the quotient $\Gamma(\mathcal{O},1)\backslash H^a$ is compact if B is division.

Aside from $\Gamma(\mathcal{O},1)$, we also consider various groups commensurable with $\Gamma(\mathcal{O},1)$. For a two-sided ideal A of $\mathcal{O}$, set:

$$\Gamma(\mathcal{O},A) \quad = \quad \{\gamma\in\Gamma(\mathcal{O},1)\,|\,\gamma-1\in A\} \quad . \tag{1.6.7}$$

For an ideal N of $\mathcal{O}_k$, $\Gamma(\mathcal{O},N.\mathcal{O})$ is abbreviated to $\Gamma(\mathcal{O},N)$.

From now on, $\mathcal{O}$ is taken to be a maximal order and N is an ideal of $\mathcal{O}_k$ coprime with $d(B)$. With a fixed isomorphism $\iota:\mathcal{O}/N\mathcal{O}\to M_2(\mathcal{O}_k/N)$ we have the reduction map $\varphi_N:\mathcal{O}\to M_2(\mathcal{O}_k/N)$. The inverse image of $B(\mathcal{O}_k/N)$ under φ_N is denoted by:

$$\mathcal{O}_0(N) \quad = \{\alpha\in\mathcal{O}\,|\,\varphi_N(\alpha)\in B(\mathcal{O}_k/N)\} \quad .$$

This is called the Eichler order of level $d(B)N$. Denote $\Gamma(\mathcal{O}_0(N),1)$ by $\Gamma_0(\mathcal{O},N)$:

$$\Gamma_0(\mathcal{O},N) \quad = \quad \{\gamma\in\mathcal{O}^\times\,|\,\nu(\gamma) \;=\; 1,\quad \varphi_N(\gamma)\in\; B(\mathcal{O}_k/N)\} \quad . \tag{1.6.8}$$

For $\gamma\in\Gamma_0(\mathcal{O},N)$ with $\varphi_N(\gamma)=\begin{pmatrix} a & b\\ 0 & d\end{pmatrix}$, $ad=1$, let $\psi_N(\gamma)=a$, giving a homomorphism:

$$\psi_N:\Gamma_0(\mathcal{O},N)\to(\mathcal{O}_k/N)^\times \quad . \tag{1.6.9}$$

Put $\Gamma_1(\mathcal{O},N)=\ker\psi_N=\{\gamma\in\Gamma(\mathcal{O},1)\,|\,\varphi_N(\gamma)=\begin{pmatrix} 1 & *\\ 0 & 1\end{pmatrix}\}$. We have an exact sequence:

$$1\to\Gamma_1(\mathcal{O},N)\to\Gamma_0(\mathcal{O},N)\to(\mathcal{O}_k/N)^\times\to 1 \quad . \tag{1.6.10}$$

The discussion of the preceding paragraph can be carried out using the subring $D(\mathcal{O}_k/N)$ in place of $B(\mathcal{O}_k/N)$. The analogues of $\mathcal{O}_0(N)$ and $\Gamma_0(\mathcal{O},N)$ will be denoted by $\mathcal{O}_{0,0}(N)$ and $\Gamma_{0,0}(\mathcal{O},N)$, respectively. We then have the exact sequence analogous to (1.6.10):

$$1 \to \Gamma(\mathcal{O},N) \to \Gamma_{0,0}(\mathcal{O},N) \to (\mathcal{O}_k/N)^{\times} \to 1 \quad . \tag{1.6.11}$$

When the choice of $\mathcal{O}$ has been fixed, we will abbreviate the groups $\Gamma(\mathcal{O},1)$, $\Gamma(\mathcal{O},N)$, $\Gamma_0(\mathcal{O},N)$, $\Gamma_1(\mathcal{O},N)$, $\Gamma_{0,0}(\mathcal{O},N)$ to $\Gamma(1)$, $\Gamma(N)$, $\Gamma_0(N)$, $\Gamma_1(N)$, $\Gamma_{0,0}(N)$, respectively.

Next let $\mathfrak{p}$ be a prime ideal of $\mathcal{O}_k$ with $\mathfrak{p}|d(B)$. With $\mathfrak{P}$, q, ι as before, the involution ρ induces an automorphism π of $\mathcal{O}/\mathfrak{P} \cong \mathbb{F}_{q^2}$ corresponding to the Frobenius map sending x to x^q. Namely, we have the commutative diagram:

$$\begin{array}{ccc} \mathcal{O} & \xrightarrow{\varphi_{\mathfrak{P}}} & \mathbb{F}_{q^2} \\ \rho\downarrow & & \downarrow\pi \\ \mathcal{O} & \xrightarrow{\varphi_{\mathfrak{P}}} & \mathbb{F}_{q^2} \end{array} \quad . \tag{1.6.12}$$

It follows that we have the commutative diagram:

$$\begin{array}{ccc} \mathcal{O} & \xrightarrow{\varphi_{\mathfrak{P}}} & \mathbb{F}_{q^2} \\ \nu\downarrow & & \downarrow N \\ \mathcal{O}_k & \xrightarrow{\varphi_{\mathfrak{p}}} & \mathbb{F}_{q} \end{array} \tag{1.6.13}$$

where $N(x) = x.x^q = x^{q+1}$.

From the approximation theorem, it is known that $\varphi_{\mathfrak{P}}(\Gamma(1)) = U(1,\mathbb{F}_{q^2})$, (units of norm 1; see Shimizu [21]), so that we have the exact sequence:

$$1 \to \Gamma(\mathfrak{P}) \to \Gamma(1) \to U(1,\mathbb{F}_{q^2}) \to 1 \quad . \tag{1.6.14}$$

For a two-sided ideal A of $\mathcal{O}$, define semigroups:

$$\begin{cases} \Delta = \Delta(\mathcal{O}) = \{\alpha \in \mathcal{O} | \nu(\alpha) >> 0\} \\ \Delta(A) = \Delta(\mathcal{O},A) = \{\alpha \in \Delta | (\alpha,A) = 1\} \quad . \end{cases} \tag{1.6.15}$$

If $A = N\mathcal{O}$ for an ideal N of $\mathcal{O}_k$, we write $\Delta(N\mathcal{O})$ as $\Delta(N) = \Delta(\mathcal{O},N)$.

Given an ideal N of $\mathcal{O}_k$, express it as $N = AB$, where $(A,d(B))=1$ and B is a suitable product of the prime ideal divisors of $d(B)$ in $\mathcal{O}_k$. Define:

$$\begin{cases} \Delta_0(N) = \Delta_0(\mathcal{O},N) = \mathcal{O}_0(A) \cap \Delta(\mathcal{O},N) \\ \Delta_{0,0}(N) = \Delta_{0,0}(\mathcal{O},N) = \mathcal{O}_{0,0}(A) \cap \Delta(\mathcal{O},N) \quad . \end{cases} \tag{1.6.16}$$

It is then known (compare Lemma 1.3.8):

$$\begin{cases} R(\Gamma_0(N),\Delta_0(Nd(B))) \cong R(\Gamma(1),\Delta(Nd(B))) \\ R(\Gamma(N),\ \Delta_{0,0}(Nd(B))) \cong R(\Gamma(1),\Delta(Nd(B))) \quad . \end{cases} \tag{1.6.17}$$

Take a divisor $\mathfrak{n}$ of N with $(\mathfrak{n},d(B)) = 1$, and consider the reduction: $\varphi_{\mathfrak{n}}:\mathcal{O} \to \mathcal{O}/\mathfrak{n}\mathcal{O}$. It induces:

$$\varphi_{\mathfrak{n}}: \begin{cases} \mathcal{O} \to \mathcal{O}/\mathfrak{n}\mathcal{O} = M_2(\mathcal{O}_k/\mathfrak{n}) \\ \Delta_0(N) \to B(\mathcal{O}_k/\mathfrak{n})^{\times} = \left\{\begin{pmatrix} a & b \\ 0 & d \end{pmatrix} \middle| ad \in (\mathcal{O}_k/\mathfrak{n})^{\times}\right\} \\ \Gamma_0(N) \to \left\{\begin{pmatrix} a & b \\ 0 & d \end{pmatrix} \middle| ad = 1 \text{ in } \mathcal{O}_k/\mathfrak{n}\right\} \\ \Delta_{0,0}(N) \to D(\mathcal{O}_k/\mathfrak{n})^{\times} = \left\{\begin{pmatrix} a & 0 \\ 0 & d \end{pmatrix} \middle| ad \in (\mathcal{O}_k/\mathfrak{n})^{\times}\right\} \\ \Gamma(N) \to \{1\} \quad . \end{cases} \tag{1.6.18}$$

Recall the homomorphism $\psi:B(\mathcal{O}_k/\mathfrak{n})^{\times} \to (\mathcal{O}_k/\mathfrak{n})^{\times} \times (\mathcal{O}_k/\mathfrak{n})^{\times}$ sends $\begin{pmatrix} a & b \\ 0 & d \end{pmatrix}$ onto (a,d^{-1}) and define $\psi_{\mathfrak{n}}$ as $\psi \circ \varphi_{\mathfrak{n}}$. It follows that $\psi_{\mathfrak{n}}(\Delta_0(N)) = ((\mathcal{O}_k/\mathfrak{n})^{\times})^2$ and $\psi_{\mathfrak{n}}(\Gamma_0(N)) = (\mathcal{O}_k/\mathfrak{n})^{\times} =$ diagonal of $((\mathcal{O}_k/\mathfrak{n})^{\times})^2$. We therefore have:

$$\psi_{\mathfrak{n}}:\ (\Gamma_0(N),\Delta_0(Nd(B))) \twoheadrightarrow ((\mathcal{O}_k/\mathfrak{n})^{\times},((\mathcal{O}_k/\mathfrak{n})^{\times})^2) \quad . \tag{1.6.19}$$

For a prime ideal $\mathfrak{n}$ of $\mathcal{O}_k$, $\mathfrak{n} \nmid d(B)$, we have the reduction map $\varphi_{\mathfrak{n}^2}:\mathcal{O} \to \mathcal{O}/\mathfrak{n}^2\mathcal{O} = M_2(\mathcal{O}_k/\mathfrak{n}^2)$. It induces:

$$\varphi_{\mathfrak{n}^2}:\Delta(\mathfrak{n}) \to GL(2,\mathcal{O}_k/\mathfrak{n}^2) \quad . \tag{1.6.20}$$

We then have exact sequences:

$$\begin{aligned} &1 \to M_{\mathfrak{q}^2} \to N_{\mathfrak{q}^2} \to D(\mathcal{O}_k/\mathfrak{q})^{\times} \to 1 \\ &1 \to M_{\mathfrak{q}^2} \to GL(2,\mathcal{O}_k/\mathfrak{q}^2) \to GL(2,\mathcal{O}_k/\mathfrak{q}) \to 1 \end{aligned} \tag{1.6.21}$$

where natural inclusions or identifications exist and,

$$\begin{cases} N_{\mathfrak{q}^2} =_{df.} \varphi_{\mathfrak{q}^2}(\Delta_{0,0}(\mathfrak{q})) \\ M_{\mathfrak{q}^2} =_{df.} \varphi_{\mathfrak{q}^2}(\Gamma(\mathfrak{q})) = \Gamma(\mathfrak{q})/\Gamma(\mathfrak{q}^2) \end{cases} . \tag{1.6.22}$$

The multiplicative group $M_{\mathfrak{q}^2}$ can be naturally identified with the additive group $M_2^0(\mathbb{F}_q)$ of matrices of trace 0 over $\mathbb{F}_q = \mathcal{O}_k/\mathfrak{q}$. We therefore have:

$$\varphi_{\mathfrak{q}^2}: \begin{cases} (\Gamma(\mathfrak{q}),\Delta(\mathfrak{q})) \twoheadrightarrow (M_{\mathfrak{q}^2}, GL(2,\mathcal{O}_k/\mathfrak{q})) \\ (\Gamma(\mathfrak{q}),\Delta_{0,0}(\mathfrak{q})) \twoheadrightarrow (M_{\mathfrak{q}^2}, N_{\mathfrak{q}^2}) \end{cases} . \tag{1.6.23}$$

The conjugation action of $GL(2,\mathcal{O}_k/\mathfrak{q}) = GL(2,\mathbb{F}_q)$ on $M_{\mathfrak{q}^2}$ is just the adjoint action of $GL(2,\mathbb{F}_q)$ on $M_2^0(\mathbb{F}_q)$. In particular, if $\alpha \in \Delta_{0,0}(\mathfrak{q})$ so that $\varphi_{\mathfrak{q}}(\alpha)$ is a scalar matrix, then $\varphi_{\mathfrak{q}}(\alpha)$ acts trivially on $M_{\mathfrak{q}^2}$.

For a prime ideal divisor $\mathfrak{p}$ of $d(B)$, define $\mathfrak{P}$ and q as before. The reduction homomorphism $\varphi_{\mathfrak{P}}: \mathcal{O} \to \mathcal{O}/\mathfrak{P} = \mathbb{F}_{q^2}$ induces:

$$\varphi_{\mathfrak{P}}: (\Gamma(1),\Delta(\mathfrak{P})) \twoheadrightarrow (U(1,\mathbb{F}_{q^2}), \mathbb{F}_{q^2}^{\times}) .$$

With the foregoing notation, the canonical involution ρ of B induces involutions on $\Gamma(1)$, $\Gamma(A)$, $\Gamma_0(N)$, $\Gamma_1(N)$, $\Gamma_{0,0}(N)$, Δ, $\Delta(A)$, $\Delta_0(N)$, $\Delta_{0,0}(N)$. For a double coset $(\Gamma\alpha\Gamma)$, define $(\Gamma\alpha\Gamma)^{\rho}$ to be the ρ-image of $(\Gamma\alpha\Gamma)$. This is just $(\Gamma\alpha^{\rho}\Gamma)$ when Γ is one of the preceding groups. We then extend this action of ρ to any formal integral linear combination of these double cosets through linearity over $\mathbb{Z}$. It is then clear that we have:

Theorem 1.6.24. *If* $\Gamma^{\rho} = \Gamma$ *and* $\Delta^{\rho} = \Delta$, *then* ρ *defines an involution (antiautomorphism of order dividing* 2*) of the Hecke ring* $R(\Gamma,\Delta)$, *where* $\Gamma \subset \Delta$ *denotes any appropriate pair as exhibited.*

A proof of the following result can be found in Shimizu [21].

Theorem 1.6.25. *Let* N *be an ideal in* $\mathcal{O}_k$ *coprime with* $d(B)$. *Then* ρ *induces the identity map on* $R(\Gamma_0(N),\Delta_0(Nd(B)))$ *as well as* $R(\Gamma(N),\Delta_{0,0}(Nd(B)))$. *As a consequence, both of these Hecke rings are commutative. Indeed, both are isomorphic to* $R(\Gamma(1),\Delta(Nd(B)))$ *by* (1.6.17).

From now on, assume the number of real places $\varphi(i)$ where $B\otimes_k \mathbb{R}$ splits $(\cong M_2(\mathbb{R}))$ is 1, and assume Γ, Δ as above so that:

$$\begin{cases} B\otimes_{\mathbb{Q}}\mathbb{R} = M_2(\mathbb{R})\times \mathbb{H}^{n-1} \\ \varphi:\Gamma \xrightarrow{\subset} SL(2,\mathbb{R})\times(\mathbb{H}_1^{\times})^{n-1} \end{cases} \quad . \tag{1.6.26}$$

Γ is then a fuchsian group (of the first kind) operating on H. The quotient space $U=\Gamma\backslash H$ is a Riemann surface. U is compact if and only if B is division; otherwise, $B=M_2(\mathbb{Q})$ and $U=$ (compactification U^*) - (cusps). (k is forced to be $\mathbb{Q}$ because B splits at only one real place.) Given two automorphic forms f, $g\in S_{m+2}(\Gamma)$, their Petersson inner product is defined by: (see [5] or [21])

$$(f,g)_{\text{Petersson}} = (f,g) = \iint_{\Gamma\backslash H} f(z)\overline{g(z)}\,.y^m\,dxdy \ . \tag{1.6.27}$$

When $\Gamma^\rho=\Gamma$, we have: (see Shimizu [21], III, p. 285)

$$((f|x),g) = (f,(g|x^\rho)),\quad x\in R(\Gamma,\Delta) \quad . \tag{1.6.28}$$

As a consequence, if $x^\rho = x$, $(S_{m+2}|x)$ is then a hermitian operator with respect to the Petersson inner product.

2. Fuchsian Groups (Trivial Action)

2.1 Fuchsian Groups with Compact Fundamental Domains

Throughout this section, $G=GL^+(2,\mathbb{R})$, $G_1=SL(2,\mathbb{R})$ and $\Gamma=$ fuchsian group with compact quotient $U=\Gamma\backslash H$. As in Chapter 1, $\Gamma\subset\Delta\subset\tilde{\Gamma}$. Let $S_2=S_2(\Gamma)$ be the space of Γ-automorphic forms of weight 2. We have the Eichler isomorphism:

$$H^1(\Gamma,\mathbb{C}) \cong S_2(\Gamma) \amalg \overline{S_2(\Gamma)} \quad . \tag{2.1.1}$$

(2.1.1) is compatible with the action of the Hecke ring $R(\Gamma,\Delta)$ so that

$$(H^1(\Gamma,\mathbb{C})|x) \cong (S_2|x) \amalg (\overline{S_2}|x), \quad x \in R(\Gamma,\Delta) \quad . \tag{2.1.2}$$

We write $\Gamma^{\#}$ for $\Gamma/(\Gamma \cap \{\pm 1\})$ and let $E_1,\ldots,E_t$ be a complete set of representatives of conjugacy classes of maximal finite (necessarily cyclic) subgroups of Γ with orders greater than 2. Define $E_i^{\#}$ in the same manner and set $m_i = \#(E_i)$, $m_i^{\#} = \#(E_i^{\#})$ so that $m_i^{\#} = m_i$ or $m_i/2$ according to whether -1 is not or is in Γ. Note that if $-1 \in \Gamma$, then $-1 \in E_i$ for each i.

$\Gamma^{\#}$ is then generated by $2g+t$ generators A_i, B_i, $1 \leqslant i \leqslant g$, C_j, $1 \leqslant j \leqslant t$ with the defining relations:

$$A_1B_1A_1^{-1}B_1^{-1}.\cdots.A_gB_gA_g^{-1}B_g^{-1}C_1.\cdots.C_t = 1$$

$$C_j^{e(j)} = 1, \quad e(j) = m_j^{\#}, \quad 1 \leqslant j \leqslant t \tag{2.1.3}$$

$$g = \text{genus of the Riemann surface } U \quad .$$

From this follows:

$$H^1(\Gamma^{\#},\mathbb{Z}) \cong H^1(\Gamma,\mathbb{Z}) \cong \mathbb{Z}^{2g}, \quad H^1(\Gamma,\mathbb{C}) \cong \mathbb{C}^{2g} \quad . \tag{2.1.4}$$

This is due to the fact that H^1 for trivial action is just Hom on the level of groups. We identify $\mathbb{Z}/\ell\mathbb{Z}$ with $\mathbb{F}_\ell$ for the prime number ℓ. This yields:

$$H^1(\Gamma,\mathbb{F}_\ell) \cong H^1(\Gamma^{\#},\mathbb{F}_\ell) \text{ for all primes } \ell \text{ if } -1 \notin \Gamma \text{ and}$$
$$\text{for all odd primes } \ell \text{ if } -1 \in \Gamma \quad . \tag{2.1.5}$$

It is now straightforward that:

$$H^1(\Gamma,\mathbb{F}_\ell) \cong \mathbb{F}_\ell^{2g}, \quad \text{provided that } \ell \nmid m_1.\cdots.m_t \quad . \tag{2.1.6}$$

The natural inclusion: $\iota: \mathbb{Z} \to \mathbb{C}$ induces an injection:

$$0 \to H^1(\Gamma, \mathbb{Z}) \xrightarrow{\iota_*} H^1(\Gamma, \mathbb{C}) \cong S_2 \amalg \overline{S}_2 \quad . \tag{2.1.7}$$

Since the actions are trivial, ι_* is just $\otimes_{\mathbb{Z}} \mathbb{C}$. (2.1.7) is compatible with the Hecke operators described in Chapter 1.

Take a prime number ℓ. Reduction mod ℓ will be denoted by $\eta = \eta(\ell)$ and multiplication by ℓ will be denoted by $\theta = \theta(\ell)$. The short exact sequence:

$$0 \to \mathbb{Z} \xrightarrow{\theta} \mathbb{Z} \xrightarrow{\eta} \mathbb{F}_\ell \to 0 \tag{2.1.8}$$

then induces the long exact sequence: (we have deleted H^0, which gives (2.1.8))

$$\begin{aligned} 0 \to H^1(\Gamma, \mathbb{Z}) &\to H^1(\Gamma, \mathbb{Z}) \to H^1(\Gamma, \mathbb{F}_\ell) \\ &\to H^2(\Gamma, \mathbb{Z}) \to H^2(\Gamma, \mathbb{Z}) \to H^2(\Gamma, \mathbb{F}_\ell) \to \cdots \quad . \end{aligned} \tag{2.1.9}$$

From the explicit descriptions, we have:

$$\begin{array}{ccccccc} 0 \to H^1(\Gamma, \mathbb{Z}) & \xrightarrow{\theta} & H^1(\Gamma, \mathbb{Z}) & \xrightarrow{\eta} & H^1(\Gamma, \mathbb{F}_\ell) & \to 0 & \\ \| & & \| & & \| & & , \ \ell \nmid m_1 \cdot \cdots \cdot m_t \\ 0 \to \ \mathbb{Z}^{2g} & \longrightarrow & \mathbb{Z}^{2g} & \longrightarrow & \mathbb{F}_\ell^{2g} & \to 0 & \end{array} \tag{2.1.10}$$

From now on, we assume that $\ell \nmid m_1 \cdot \cdots \cdot m_t$. From Chapter 1, all the maps above are compatible with the actions of the Hecke operators. With the selection of suitable bases, $(H^1(\Gamma, \mathbb{Z}) \mid x)$ and $(H^1(\Gamma, \mathbb{F}_\ell) \mid x)$ are respectively $2g \times 2g$ matrices over $\mathbb{Z}$ and $\mathbb{F}_\ell$. (2.1.10) translates to:

$$(H^1(\Gamma, \mathbb{F}_\ell) \mid x) \quad = \quad (H^1(\Gamma, \mathbb{Z}) \mid x)^{(\ell)} \tag{2.1.11}$$

where $M^{(\ell)}$ denotes the reduction mod ℓ of the integral matrix M. We have:

<u>Theorem 2.1.12</u>. *Denote the characteristic polynomials of* $(S_2 \mid x)$, $(H^1(\Gamma, \mathbb{Z}) \mid x)$, $(H^1(\Gamma, \mathbb{F}_\ell) \mid x)$ *by:*

$$F(x,\lambda) = \det(\lambda I_g - (S_2|x)) \in \mathbb{C}[\lambda]$$

$$M(x,\lambda) = \det(\lambda I_{2g} - (H^1(\Gamma,\mathbb{Z})|x)) \in \mathbb{Z}[\lambda]$$

$$M_\ell(x,\lambda) = \det(\lambda I_{2g} - (H^1(\Gamma,\mathbb{F}_\ell)|x)) \in \mathbb{F}_\ell[\lambda] \quad .$$

Then,

$$M(x,\lambda) = F(x,\lambda).\bar{F}(x,\lambda) \tag{2.1.13}$$

$$M_\ell(x,\lambda) = M^{(\ell)}(x,\lambda) \tag{2.1.14}$$

where $\bar{F}$ *and* $M^{(\ell)}$ *are obtained from* F *and* M *through complex conjugation and reduction* $\bmod \ell$ *on their coefficients, respectively.*

The proof is just (2.1.2), (2.1.7) and (2.1.11).

Corollary 2.1.15. *If the action* $(S_2|x)$ *of a Hecke operator* $x \in R(\Gamma,\Delta)$ *is representable by a real matrix (with respect to suitable basis), then:*

$$F(x,\lambda) \in \mathbb{R}[\lambda] \quad .$$

As a result, we have:

$$M(x,\lambda) = F(x,\lambda)^2 \quad \text{and} \quad F(x,\lambda) \in \mathbb{Z}[\lambda] \quad . \tag{2.1.16}$$

In order to study the polynomial $M(x,\lambda)$ further, we consider an exact sequence:

$$1 \to \Gamma_1 \xrightarrow{\subset} \Gamma \xrightarrow{\varphi} \Gamma_3 \to 1, \quad \Gamma_3 \text{ a suitable finite group .}$$

We assume that φ can be extended to a homomorphism from Δ to a finite group Δ_3 containing Γ_3. As can be seen from the Hochschild-Serre exact sequence, the inflation map induced by φ is injective:

$$\varphi^*: H^1(\Gamma_3,\mathbb{F}_\ell) \to H^1(\Gamma,\mathbb{F}_\ell) \quad .$$

Suppose that Γ_1 can be chosen so that $H^1(\Gamma_3,\mathbb{F}_\ell) \cong \mathrm{Hom}(\Gamma_3,\mathbb{F}_\ell) \neq 0$. The image of φ^* is then a nontrivial subspace of $H^1(\Gamma,\mathbb{F}_\ell)$ and is stable under the action of $R(\Gamma,\Delta)$. This action of $R(\Gamma,\Delta)$ can be described as in Section 1.5. For example, we have:

Theorem 2.1.17. *Suppose that* $\varphi:\Delta\to\Delta_3$ *is a homomorphism of* Δ *to a finite group* Δ_3 *such that* $\Gamma_3=\varphi(\Gamma)$ *is contained in the center of* Δ_3. *Moreover, assume* $H^1(\Gamma_3,\mathbb{F}_\ell)\neq 0$ *for a prime* $\ell\nmid m_1\cdot\cdots\cdot m_t$. *Then,*

$$|F(\deg(x))|^2 \equiv 0 \mod \ell \quad \textit{for any} \quad x\in R(\Gamma,\Delta) \quad . \tag{2.1.18}$$

If, in addition, $(S_2|x)$ *is real, then* $F(\deg(x))\in\mathbb{Z}$ *and*

$$F(\deg(x)) \equiv 0 \mod \ell \quad . \tag{2.1.19}$$

Here, $F(\lambda)=F(x,\lambda)=\det(\lambda I_g-(S_2|x))$.

Proof. We have the following diagram with exact rows and column (by virtue of the assumption that $\ell\nmid m_1\cdot\cdots\cdot m_t$). All the maps are compatible with the action of Hecke operators.

$$\begin{array}{ccccc} & & 0\to H^1(\Gamma,\mathbb{Z}) \to & H^1(\Gamma,\mathbb{C}) \cong S_2 \amalg \bar{S}_2 \\ & & & \downarrow \eta \\ 0\to H^1(\Gamma_3,\mathbb{F}_\ell) & \xrightarrow{\varphi^*} & & H^1(\Gamma,\mathbb{F}_\ell) \\ & & & \downarrow \\ & & & 0 \end{array} \tag{2.1.20}$$

Applying Theorem 1.3.7 to $(\Gamma,\Delta)\twoheadrightarrow(\Gamma_3,\Delta_3)$, we have:

$$\varphi^*(z|\varphi_!(x)) = \varphi^*(z)|x \quad , \qquad z\in H^1(\Gamma_3,\mathbb{F}_\ell),\ x\in R(\Gamma,\Delta) \ .$$

This means that $\mathrm{im}(\varphi^*)$ is $R(\Gamma,\Delta)$-stable, say of dimension $m>0$ over $\mathbb{F}_\ell$. By Theorem 1.5.2, the action of $\varphi_!(x)$ on $H^1(\Gamma_3,\mathbb{F}_\ell)$ is scalar multiplication by $\deg(\varphi_!(x))=\deg(x)$ (see Theorem 1.3.6 for this equality). It follows that $(H^1(\Gamma,\mathbb{F}_\ell)|x)$ has eigenvalue $\deg(x)$ with multiplicity at least m, $0<m\leq 2g$. Consequently, $M_\ell(x,\deg(x))=0$ or $M(x,\deg(x))\equiv 0 \mod \ell$. Our assertions follow from Theorem 2.1.12 and Corollary 2.1.15. ∎

2.2 Fuchsian Groups with Cusps

In this section we discuss the relevant modifications when the fuchsian group Γ is not cocompact but is of cofinite volume. As mentioned before, (1.4.6) is valid when $S_{m+2}(\Gamma)$ is defined to be the space of cusp forms of weight $m+2$. In order to retain Theorem 2.1.17,

$H^1(\Gamma,*)$ must be replaced by the parabolic cohomology group $H^1_P(\Gamma,*)$ or $H^1(\Gamma,*)_{par}$, see Shimura [24; 223-238].

Let Γ be any group, let $Q\subset\Gamma$ and let M be a Γ-module. For $0\leq i\leq 2$, $H^i_Q(\Gamma,M)$ are then defined by Shimura [24; 224]. When $i=1$, $H^1_Q(\Gamma,M)$ is just the subgroup of $H^1(\Gamma,M)$ formed by the intersection of all the kernels of the restriction homomorphisms induced by restricting from Γ to the cyclic subgroups generated by the elements of Q. When M is Γ-trivial, we have:

$$H^1_Q(\Gamma,M) = \mathrm{Hom}(\Gamma/\langle Q\rangle,M) \tag{2.2.1}$$

where $\langle Q\rangle$ denotes the normal subgroup of Γ generated by Q. For any exact sequence of Γ-trivial modules: $0\to A\to B\to C\to 0$, we then have the long exact sequence:

$$0\to H^1_Q(\Gamma,A)\to H^1_Q(\Gamma,B)\to H^1_Q(\Gamma,C)\to H^2_Q(\Gamma,A)\to \quad . \tag{2.2.2}$$

As in Section 2.1, let $\Gamma\subset SL(2,\mathbb{R})$ be a fuchsian group of the first kind and define $\Gamma^\#$, E_i, $E^\#_i$, m_i, $m^\#_i$, $1\leq i\leq t$ as before. Call a subgroup P of Γ $\pm$-unipotent if its image $P^\#$ in $\Gamma^\#$ consists of parabolic transformations (and 1). Let P_j, $1\leq j\leq u$, be a complete set of representatives of conjugacy classes of maximal $\pm$-unipotent subgroups in Γ. $\Gamma^\#$ is then generated by $2g+t+u$ generators A_i, B_i, $1\leq i\leq g$, C_j, $1\leq j\leq t$, D_s, $1\leq s\leq u$, with defining relations:

$$\begin{gathered} A_1B_1A_1^{-1}B_1^{-1}\cdot\ldots\cdot A_gB_gA_g^{-1}B_g^{-1}C_1\cdot\ldots\cdot C_tD_1\cdot\ldots\cdot D_u = 1 \\ C_j^{e(j)} = 1, \quad e(j) = m^\#_j, \quad 1\leq j\leq t, \quad g = \text{genus of } U \end{gathered} \quad . \tag{2.2.3}$$

The parabolic cohomology is then obtained by taking P to be the set of all $\pm$-unipotent elements in Γ. The image of P in $\Gamma^\#$ will be denoted by $P^\#$. We can also take $Q=\{D_1,\ldots,D_u\}\subset\Gamma^\#$. It follows from (2.2.3) that:

$$\begin{aligned} H^1(\Gamma^\#,\mathbb{C}) &\cong \mathbb{C}^{2g+u-1}, & H^1_Q(\Gamma^\#,\mathbb{C}) &\cong \mathbb{C}^{2g} \\ H^1(\Gamma^\#,\mathbb{Z}) &\cong \mathbb{Z}^{2g+u-1}, & H^1_Q(\Gamma^\#,\mathbb{Z}) &\cong \mathbb{Z}^{2g} \\ H^1(\Gamma^\#,\mathbb{F}_\ell) &\cong \mathbb{F}_\ell^{2g+u-1}, & H^1_Q(\Gamma^\#,\mathbb{F}_\ell) &\cong \mathbb{F}_\ell^{2g}, \ \ell\nmid m^\#_1\cdot\ldots\cdot m^\#_t \end{aligned} \quad . \tag{2.2.4}$$

As in Section 2.1, we also know that:

$$H^1(\Gamma,\mathbb{C}) \cong H^1(\Gamma^\#,\mathbb{C}) , \quad H^1_P(\Gamma,\mathbb{C}) \cong H^1_{P^\#}(\Gamma^\#,\mathbb{C}) = H^1_Q(\Gamma^\#,\mathbb{C})$$

$$H^1(\Gamma,\mathbb{Z}) \cong H^1(\Gamma^\#,\mathbb{Z}) , \quad H^1_P(\Gamma,\mathbb{Z}) \cong H^1_{P^\#}(\Gamma^\#,\mathbb{Z}) = H^1_Q(\Gamma^\#,\mathbb{Z})$$

$$H^1(\Gamma,\mathbb{F}_\ell) \cong H^1(\Gamma^\#,\mathbb{F}_\ell), \quad H^1_P(\Gamma,\mathbb{F}_\ell) \cong H^1_{P^\#}(\Gamma^\#,\mathbb{F}_\ell) = H^1_Q(\Gamma^\#,\mathbb{F}_\ell) ,$$
$$\ell \nmid \#(\Gamma \cap \langle\pm 1\rangle) \quad . \qquad (2.2.5)$$

We therefore have the parabolic versions of (2.1.7) and (2.1.10). According to Shimura [24; 237], the parabolic cohomology is stable for the action of $R(\Gamma,\Delta)$ and $(2.1.7)_{par}$, $(2.1.10)_{par}$ are compatible with this action. If we form the finite quotient group Γ_3 of Γ and define $P_3 = \varphi(P)$, we then have the parabolic version of (2.1.20). Theorem 2.1.17 and its proof go over word by word to the parabolic version. We state this simply:

Theorem $(2.1.17)_{par}$ is valid.

We note that the assumption in Theorem $(2.1.17)_{par}$ includes $H^1_{P_3}(\Gamma_3,\mathbb{F}_\ell) \neq 0$. This is stronger than $H^1(\Gamma_3,\mathbb{F}_\ell) \neq 0$.

2.3 Arithmetic Fuchsian Groups

We now apply the results of the preceding sections to arithmetic fuchsian groups Γ defined in Section 1.6. The notation and assumptions of 1.6 are now in effect. In particular, (1.6.26) is assumed so that the quaternion algebra B over k splits at only one real place and Γ is a fuchsian group operating on H.

We first consider the case where B is division. Thus Γ is cocompact. Consider the pair $(\Gamma,\Delta) = (\Gamma(1),\Delta(\mathfrak{P}))$ where $\mathfrak{P}$ is a maximal 2-sided ideal of a fixed maximal order $\mathcal{O}$ such that $\mathcal{O} \supset \mathfrak{P} \supset \mathfrak{P}^2 = \mathfrak{p}\mathcal{O}$, $\mathfrak{p} \mid d(B)$. As before, let $q = \#(\mathcal{O}_k/\mathfrak{p}) = N_{k/\mathbb{Q}}(\mathfrak{p})$.

<u>Theorem 2.3.1</u>. *Let* ℓ *be a prime number such that* $\ell \mid (q+1)$ *and* $\ell \nmid$ *orders of torsion elements of* $\Gamma(1)$. *Then,*

$$|F(x,\deg(x))|^2 \equiv 0 \bmod \ell, \quad x \in R(\Gamma(1),\Delta(\mathfrak{P})) \quad .$$

Moreover, if $x \in R(\Gamma(1),\Delta(d(B)))$, *then,*

$$F(x,\deg(x)) \equiv 0 \mod \ell \quad .$$

Here $F(x,\lambda) = \det(\lambda I_g - (S_2(\Gamma(1))|x)) \in \mathbb{C}[\lambda]$ *and* $F(x,\lambda)$ *actually has integral coefficients.*

Proof. Under reduction mod $\mathfrak{P}$, we have $\varphi_{\mathfrak{P}}: \mathcal{O} \to \mathbb{F}_{q^2}$ inducing $\varphi_{\mathfrak{P}}: (\Gamma(1), \Delta(\mathfrak{P})) \to\!\!> (U(1, \mathbb{F}_{q^2}), \mathbb{F}_{q^2}^{\times})$. Since $U(1, \mathbb{F}_{q^2}) \cong \mathbb{Z}/(q+1)\mathbb{Z}$, $\mathbb{F}_\ell \cong \mathbb{Z}/\ell\mathbb{Z}$ and $\ell | (q+1)$, we see that:

$$H^1(U(1, \mathbb{F}_{q^2}), \mathbb{F}_\ell) = \mathrm{Hom}(U(1, \mathbb{F}_{q^2}), \mathbb{F}_\ell) \neq 0 \quad .$$

Using $\Gamma_3 = U(1, \mathbb{F}_{q^2})$ in Theorem 2.1.17, we get the first of our congruences. Moreover, if $x \in R(\Gamma(1), \Delta(d(B)))$, then use Theorem 1.6.25 with $N = \mathcal{O}$ to get $x^\rho = x$. As in (1.6.28), $(S_2|x)$ is hermitian with respect to the Petersson inner product. Thus the second of our congruences also follows from Theorem 2.1.17, and the final statement follows from (2.1.16). ∎

We continue with the case B is division and consider the pair $(\Gamma, \Delta) = (\Gamma_0(N), \Delta_0(Nd(B)))$, where N is an ideal of $\mathcal{O}_k$ with $(N, d(B)) = 1$.

Theorem 2.3.2. *Let* ℓ *be a prime number such that* $\ell | \#((\mathcal{O}_k/N)^\times)$ *and* $\ell \nmid$ *orders of torsion elements in* $\Gamma_0(N)$. *Then,*

$$F(x, \deg(x)) \equiv 0 \bmod \ell, \quad x \in R(\Gamma_0(N), \Delta_0(Nd(B))) \quad .$$

Here $F(x,\lambda) = \det(\lambda I_g - (S_2(\Gamma_0(N))|x)) \in \mathbb{Z}[\lambda] \subset \mathbb{C}[\lambda]$.

The proof proceeds in a similar manner. We have the map:

$$\psi_N : (\Gamma_0(N), \Delta_0(Nd(B))) \twoheadrightarrow ((\mathcal{O}_k/N)^\times, ((\mathcal{O}_k/N)^\times)^2)$$

as in (1.6.19) (with $\mathfrak{n} = N$). Composing this with a homomorphism:

$$\kappa : (\mathcal{O}_k/N)^\times \twoheadrightarrow \mathbb{Z}/\ell\mathbb{Z} = \mathbb{F}_\ell$$

we have

$$\kappa \circ \psi_N : (\Gamma_0(N), \Delta_0(Nd(B))) \twoheadrightarrow (\mathbb{F}_\ell, \mathbb{F}_\ell^2) \quad .$$

Applying Theorem 2.1.17 for $\Gamma_3 = \mathbb{F}_\ell$ using $H^1(\mathbb{F}_\ell, \mathbb{F}_\ell) = \mathbb{F}_\ell \neq 0$, the congruences follow as in the second part of Theorem 2.3.1.

We next consider the case where B is not division so that $B = M_2(\mathbb{Q})$ (recall that B is assumed to split at only one real place).

We may assume $\mathcal{O} = M_2(\mathbb{Z})$ and $\Gamma(1) = SL(2,\mathbb{Z})$. For $N = N\mathbb{Z}$, $N \geqslant 1$, we have:

$$\Gamma_0(N) = \Gamma_0(N) = \left\{ \begin{pmatrix} a & b \\ c & d \end{pmatrix} \in SL(2,\mathbb{Z}) \,\middle|\, c \equiv 0 \bmod N \right\}$$

$$\Gamma(N) = \Gamma(N) = \left\{ \begin{pmatrix} a & b \\ c & d \end{pmatrix} \in SL(2,\mathbb{Z}) \,\middle|\, \begin{pmatrix} a & b \\ c & d \end{pmatrix} \equiv \begin{pmatrix} 1 & 0 \\ 0 & 1 \end{pmatrix} \bmod N \right\} .$$

These cases are called "elliptic modular cases" with level N.

<u>Theorem 2.3.3</u>. *Let* ℓ *and* q *be prime numbers with* $\ell \mid (q-1)$, $\ell \geqslant 5$. *Consider the elliptic modular case* $\Gamma = \Gamma_0(q) \subset SL(2,\mathbb{Z})$ *with odd prime* q *as the level. We then have:*

$$F(x, \deg(x)) \equiv 0 \bmod \ell, \quad x \in R(\Gamma_0(q), \Delta_0(q)) .$$

Here $F(x,\lambda) = \det(\lambda I_g - (S_2(\Gamma_0(q)) \mid x)) \in \mathbb{Z}[\lambda] \subset \mathbb{C}[\lambda]$.

The proof proceeds in a similar manner using Theorem $2.1.17_{par}$. The orders of elliptic elements in $\Gamma_0(q)$ are among 3, 4 and 6 because we are inside $SL(2,\mathbb{Z})$. Since q is a prime, the number of cusps of $\Gamma_0(q)$ is 2 and all the $\pm$-unipotent subgroups are conjugate to either P_1 or P_2, where:

$$P_1 = \left\{ \pm\begin{pmatrix} 1 & n \\ 0 & 1 \end{pmatrix} \,\middle|\, n \in \mathbb{Z} \right\}, \quad P_2 = \left\{ \pm\begin{pmatrix} 1 & 0 \\ qn & 1 \end{pmatrix} \,\middle|\, n \in \mathbb{Z} \right\} .$$

Since P_1 and P_2 are both in the kernel of the homomorphism $\kappa \circ \psi_q$ as defined in the proof of Theorem 2.3.2, $H^1(\Gamma_3, \mathbb{F}_\ell)_{par} = H^1(\Gamma_3, \mathbb{F}_\ell)$, where $\Gamma_3 = \mathrm{im}(\kappa \circ \psi_q) = \mathbb{F}_\ell$, we can proceed as before.

<u>Example 2.3.4</u>. Consider the special case of Theorem 2.3.3 with $q = 11$ and $\ell = 5$. Let $x = \Gamma\begin{pmatrix} 1 & 0 \\ 0 & p \end{pmatrix}\Gamma$ with $p \neq 11$ denoting a prime number. We then have:

$$1 - a_p + p \equiv 0 \bmod 5 \tag{2.3.5}$$

where a_p is the coefficient of t^p in the t-expansion of:

$$t \prod_{n \geqslant 1} (1 - t^n)^2 (1 - t^{11n})^2 = \sum_{n \geqslant 1} a_n t^n . \tag{2.3.6}$$

We return to the division case and consider the pair $(\Gamma(\mathfrak{q}), \Delta_{0,0}(\mathfrak{q}d(B)))$ where $\mathfrak{q}$ is a prime ideal of $\mathcal{O}_k$ with $\mathfrak{q} \nmid d(B)$. We have the reduction map $\varphi_{\mathfrak{q}}: \mathcal{O} \to M_2(\mathcal{O}_k/\mathfrak{q})$. Put $N_{k/\mathbb{Q}}(\mathfrak{q}) = q^f$ where q is a prime number and f is the degree of $\mathfrak{q}$ over q.

Theorem 2.3.7. *Suppose that* q *does not divide the orders of any torsion element in* $\Gamma(\mathfrak{q})$. *Let* $x = \Gamma(\mathfrak{q})\alpha\Gamma(\mathfrak{q}) \in R(\Gamma(\mathfrak{q}), \Delta_{0,0}(\mathfrak{q}d(B)))$ *with* $\varphi_{\mathfrak{q}}(\alpha) = a.I_2 \in M_2(\mathcal{O}_k/\mathfrak{q})$, $a \in (\mathcal{O}_k/\mathfrak{q})^{\times}$. *We then have:*

$$F(x, \deg(x)) \equiv 0 \bmod \mathfrak{q}, \qquad F'(x, \deg(x)) \equiv 0 \bmod \mathfrak{q} \quad \text{and}$$

$$F''(x, \deg(x)) \equiv 0 \bmod \mathfrak{q}.$$

Here $F(x, \lambda) = \det(\lambda I_g - (S_2(\Gamma(\mathfrak{q})) | x))$.

The proof uses (1.6.23) and the fact that $\Gamma(\mathfrak{q})/\Gamma(\mathfrak{q}^2)$ has dimension 3 over $\mathcal{O}_k/\mathfrak{q}$. We omit the details.

3. Fuchsian Groups (Nontrivial Action)

3.1 Eichler-Shimura Isomorphism

For convenience, we recall some definitions and notation from Shimura [24; 230-236]. Let W be a finite dimensional vector space over a field k. $S^m(W)$ will denote the m-th symmetric power of W over k. The representation of $GL_k(W)$ on $S^m(W)$ is called the (m-th) symmetric tensor representation and is denoted by $\rho_m = (\rho_m, S^m(W))$.

From now on, assume $\dim_k(W) = 2$. On $S^m(W)$, there is a nondegenerate (assuming $m!$ is invertible in k) bilinear form:

$$\Theta_m : S^m(W) \times S^m(W) \to k \quad ,$$

such that

$$\begin{aligned} \Theta_m(\rho_m(\alpha)x, \rho_m(\alpha)y) &= (\det \alpha)^m \Theta_m(x, y), \quad \alpha \in GL_k(W),\ x, y \in S^m(W) \\ \Theta_m(y, x) &= (-1)^m \Theta_m(x, y), x, y \in S^m(W) \quad . \end{aligned} \tag{3.1.1}$$

Θ_m is unique up to a constant factor. We define the representation $\rho_m^* = (\rho_m^*, S^m(W))$ of $GL_k(W)$ by:

$$\rho_m^*(\alpha) = (\det \alpha)^{-m} \rho_m(\alpha), \quad \alpha \in GL_k(W) \quad . \tag{3.1.2}$$

We note as before, ρ_m^* and ρ_m coincide on $SL_k(W)$.

For a fuchsian group $\Gamma \subset SL(2,\mathbb{R})$, we use ρ_m for $\rho_m|\Gamma$. It will be called the symmetric tensor representation of Γ. The complexification $\rho_m \otimes \mathbb{C}$ will also be written as ρ_m when there is no chance of confusion.

Consider a representation $\Psi = (\Psi, Y)$ of Γ on the finite dimensional $\mathbb{R}$-vector space Y with the following properties:

$\Psi(\Gamma)$ is contained in a compact subgroup of $GL_{\mathbb{R}}(Y)$ (to be identified with $GL(r,\mathbb{R})$ when an $\mathbb{R}$-basis for Y is chosen).

$\Psi(\Gamma)$ is a finite group when Γ has cusps. (3.1.3)

If $-1 \in \Gamma$, then $\Psi(-1) = \varepsilon(\Psi) 1_Y, \varepsilon(\Psi) = \pm 1$.

Such a representation Ψ is called an "admissible" orthogonal representation of Γ. Because $\Psi(\Gamma)$ is contained in a compact subgroup, there is a positive definite symmetric bilinear form P on Y such that:

$$P(\Psi(\alpha)x, \Psi(\alpha)y) = P(x,y), \quad \alpha \varepsilon \Gamma, \ x, \ y \in Y. \tag{3.1.4}$$

Such a P is called an Euclidean metric for Ψ.

For an integer $m \geqslant 0$ and an admissible (orthogonal) representation $\Psi = (\Psi, Y)$, consider holomorphic maps $f: H \to Y \otimes_{\mathbb{R}} \mathbb{C}$, satisfying the following conditions:

$$f(\alpha(z)) = \Psi(\alpha) f(z) . (cz+d)^m, \quad \alpha = \begin{pmatrix} a & b \\ c & d \end{pmatrix} \in \Gamma$$

the components of f (with respect to any basis of $Y \otimes_{\mathbb{R}} \mathbb{C}$) belong to $S_m(\ker \Psi)$ when Γ has cusps. (3.1.5)

The space of all such maps f is denoted by $S_m(\Gamma, \Psi)$. If $-1 \in \Gamma$, then $S_m(\Gamma, \Psi) = 0$ unless $(-1)^m \varepsilon(\Psi) = 1$. Choose an Euclidean metric P for Ψ. We then define the "Petersson inner product" $(\ , \ )$ on $S_m(\Gamma, \Psi)$ by:

$$(f,g) = \iint_{\Gamma \backslash H} P(f(z), \overline{g(z)}) y^{m-2} \, dxdy, \quad f, \ g \in S_m(\Gamma, \Psi) . \tag{3.1.6}$$

Here $P(u, \bar{v})$ is the hermitian form on $Y \otimes_{\mathbb{R}} \mathbb{C}$ induced by P.

Fix an integer $m \geq 0$ and an admissible representation $\Psi = (\Psi, Y)$ with $(-1)^m \varepsilon(\Psi) = 1$. Consider the representation:

$$\Psi \otimes_{\mathbb{R}} \rho_m^* = (\Psi \otimes \rho_m^*, Y \otimes_{\mathbb{R}} S^m(\mathbb{R}^2)) \quad . \tag{3.1.7}$$

Also consider the Γ-invariant bilinear form:

$$P \otimes_{\mathbb{R}} \Theta_m : (Y \otimes_{\mathbb{R}} S^m(\mathbb{R}^2)) \times (Y \otimes_{\mathbb{R}} S^m(\mathbb{R}^2)) \to \mathbb{R} \quad . \tag{3.1.8}$$

The following result is known, see Shimura [24; 234].

Theorem 3.1.9. *There is a canonical isomorphism:*

$$H^1(\Gamma, \Psi \otimes_{\mathbb{R}} \rho_m^*)_{par} \otimes_{\mathbb{R}} \mathbb{C} \cong S_{m+2}(\Gamma, \Psi) \amalg \overline{S_{m+2}(\Gamma, \Psi)} \quad .$$

The pairing (3.1.8) *induces the (cup product) pairing:*

$$\wedge : H^1(\Gamma, \Psi \otimes_{\mathbb{R}} \rho_m^*) \times H^1(\Gamma, \Psi \otimes_{\mathbb{R}} \rho_m^*) \to H^2(\Gamma, \mathbb{R}) \quad . \tag{3.1.10}$$

If $\Gamma \backslash H$ is compact, $H^2(\Gamma, \mathbb{R}) = \mathbb{R}$ and (3.1.10) is nondegenerate. By complexification, this then agrees with the Petersson inner product up to a nonzero constant factor via the isomorphism in Theorem 3.1.9.

Let Δ be a semigroup, $\Gamma \subset \Delta \subset \tilde{\Gamma} \subset GL^+(2, \mathbb{R})$. Assume $\Psi : \Delta^{-1} \to GL_{\mathbb{R}}(Y)$ is a homomorphism extending the admissible representation Ψ of Γ. Then Hecke operators in $R(\Gamma, \Delta)$ act on $S_m(\Gamma, \Psi)$ through: (compare (1.4.8))

$$(f|x)(z) = \sum_{1 \leqslant i \leqslant d} \Psi(\alpha_i^{-1}) f(\alpha_i(z)) . (\det \alpha_i)^{m-1} / (c_i z + d_i)^m \tag{3.1.11}$$

where $x = \Gamma\alpha\Gamma = \amalg_{1 \leqslant i \leqslant d} \Gamma\alpha_i$ and $\alpha_i \in \Delta$ with matrix entries a_i, b_i, c_i, d_i.

Theorem 3.1.12. *Assume* $(-1)^m \varepsilon(\Psi) = 1$ *when* $-1 \in \Gamma$. *The actions of* $R(\Gamma, \Delta)$ *on the two sides of the isomorphism in Theorem* 3.1.9 *are compatible with the isomorphism.*

3.2 Bundles of Lattices (or Tori)

We continue with the notation of Section 3.1 and assume Γ is cocompact.

A representation of the type $\Psi \otimes_{\mathbb{R}} \rho_m^*$ of Γ is called primitive if Ψ is admissible and $(-1)^m \varepsilon(\Psi) = 1$ holds when $-1 \in \Gamma$. In essence, it is a representation of $\Gamma^{\#}$. A (finite) direct sum of primitive representations is called semiprimitive.

We consider a 5-ple $(\pi, V, L, \omega, \chi)$ with the following long list of properties:

(3.2.1) V is a finite dimensional vector space over $\mathbb{R}$.

(3.2.2) L is a lattice in V.

(3.2.3) ω is a bilinear form on V with $\omega(L,L) \subset \mathbb{Z}$.

(3.2.4) (π, V) is a representation of Δ^{-1} with $\pi(\alpha^{-1})L \subset L$, $\alpha \in \Delta$ so that (automatically) $\pi(\gamma)L = L$ for $\gamma \in \Gamma$.

(3.2.5) π and ω are related so that:
$\omega(\pi(\gamma)x, \pi(\gamma)y) = \omega(x,y)$, x, $y \in V$, $\gamma \in \Gamma$;
there is a map $\chi : \Delta \to \mathbb{Z} - \{0\}$ with
$\omega(\pi(\alpha^{-1})x, \pi(\alpha^{-1})y) = \chi(\alpha)\omega(x,y)$, x, $y \in V$, $\alpha \in \Delta$
(automatically, $\chi(\Gamma) = 1$).

(3.2.6) The representation (π, V) of Γ is semiprimitive and admits a decomposition: $(\pi, V) = \coprod_{1 \leq i \leq M} (\Psi_i \otimes_{\mathbb{R}} \rho^*_{m(i)}, V_i \otimes_{\mathbb{R}} S^{m(i)}(\mathbb{R}^2))$
so that ω is an orthogonal sum:

$$\omega = \perp_{1 \leq i \leq M} \omega_i$$

where ω_i has the form: $P_i \otimes \Theta_{m(i)}$ for an Euclidean metric P_i for Ψ_i. We also assume: $m(i)$ are either all even or all odd.

(3.2.7) $V^{\Gamma} = \{x \in V \mid \pi(\gamma)x = x \text{ for all } \gamma \in \Gamma\}$ is 0.

Such a 5-ple will be called a "(polarized) lattice (or torus) bundle for Γ" with homogeneity χ.

From (3.2.6), ω must be nondegenerate and is either symmetric (when all $m(i)$ are even) or alternating (when all $m(i)$ are odd). From the Γ-invariance of ω in (3.2.5), ω induces the (cup product) pairing:

$$\omega^* : H^1(\Gamma, V) \times H^1(\Gamma, V) \to H^2(\Gamma, \mathbb{R}) = \mathbb{R} \quad \text{(recall } \Gamma \text{ is cocompact)}.$$

ω^* is alternating (respectively symmetric) when ω is symmetric (respectively alternating).

In addition, we assume that:

(3.2.8) $\omega^*(c_1|x,c_2) = \omega^*(c_1,c_2|x^\rho)$ holds for $c_1,c_2 \in H^1(\Gamma,V)$ when Γ is defined by a quaternion algebra with $\Gamma^\rho = \Gamma$.

Using (3.2.3), we can define the dual lattice L^* of L by:

$$L^* = \{x \in V | \omega(x,L) \subset \mathbf{Z}\} \quad . \tag{3.2.9}$$

It is immediate that $L^* \supset L$ and L^* enjoys (3.2.2)* and (3.2.4)* where L^* is used in place of L. It follows that the index $|L^*:L|$ is finite and is called the discriminant of (ω,L). We put:

$$T = V/L \quad \text{and} \quad T_* = V/L^* \quad . \tag{3.2.10}$$

These are tori and can be viewed as the Pontrjagin dual of L and L^*, respectively. Δ^{-1} therefore acts naturally on T, T^* with actions denoted by π and π^*, respectively. We have the (Pontrjagin) dual exact sequences:

$$\begin{array}{l} 0 \to L \to V \to T \to 0 \\ 0 \leftarrow T_* \leftarrow V \leftarrow L^* \leftarrow 0 \end{array} \quad . \tag{3.2.11}$$

They induce dual exact sequences:

$$\begin{array}{l} 0 \to H^0(L) \to H^0(V) \to H^0(T) \to H^1(L) \to \cdots \\ 0 \leftarrow H^2(T_*) \leftarrow H^2(V) \leftarrow H^2(L^*) \leftarrow H^1(T_*) \leftarrow \cdots \end{array} \tag{3.2.12}$$

corresponding to duality pairings:

$$\begin{array}{lll} H^i(V) \times H^{2-i}(V) \to H^2(\mathbb{R}/\mathbb{Z}) & = & \mathbb{R}/\mathbb{Z} \\ H^i(T) \times H^{2-i}(L^*) \to H^2(\mathbb{R}/\mathbb{Z}) & = & \mathbb{R}/\mathbb{Z} \\ H^i(L) \times H^{2-i}(T_*) \to H^2(\mathbb{R}/\mathbb{Z}) & = & \mathbb{R}/\mathbb{Z} \end{array} \tag{3.2.13}$$

where $H^i(*)$ is the abbreviation for $H^i(\Gamma,*)$. We note that Γ has a subgroup of finite index with cohomological dimension 2 by virtue of the fact that $\Gamma\backslash H$ is a compact Riemann surface branched over the 2-sphere.

It follows from (3.2.7) that: $L^\Gamma = 0 = L^{*\Gamma}$. Thus $T^\Gamma = H^0(\Gamma,T)$ is a finite group isomorphic to $H^1(\Gamma,L)_{tor}$. Similarly, $T_*^\Gamma = H^0(\Gamma,T_*) \cong H^1(\Gamma,L^*)_{tor}$. From duality, we have $H^2(V) = 0$. As a result, (3.2.12) becomes a dual pair of exact sequences:

$$\begin{gathered} 0 \to T^\Gamma \to H^1(L) \to H^1(V) \to H^1(T) \to H^2(L) \to 0 \\ 0 \leftarrow H^2(L^*) \leftarrow H^1(T_*) \leftarrow H^1(V) \leftarrow H^1(L^*) \leftarrow T_*^\Gamma \leftarrow 0 \end{gathered} \tag{3.2.14}$$

Since T^Γ and T_*^Γ are finite, we have noncanonical isomorphisms:

$$T^\Gamma \cong H^2(L^*) \quad \text{and} \quad T_*^\Gamma \cong H^2(L) \quad \text{as finite abelian groups.} \tag{3.2.15}$$

Let ℓ be a prime number and form the reduction mod ℓ exact sequence:

$$0 \to L \xrightarrow{\theta} L \xrightarrow{\eta} L/\ell L \to 0 \tag{3.2.16}$$

where $\theta = \theta(\ell)$ is multiplication by ℓ and $\eta = \eta(\ell)$ is reduction mod ℓ. We also use superscript (ℓ) to denote the image under reduction mod ℓ. Thus, we have representations: $(\pi^{(\ell)}, L^{(\ell)})$, $(\pi^{*(\ell)}, L^{*(\ell)})$ of Γ over $\mathbb{F}_\ell$.

From (3.2.7) and (3.2.16), we obtain the exact sequence:

$$0 \to (L^{(\ell)})^\Gamma \to H^1(L) \xrightarrow{\theta} H^1(L) \to H^1(L^{(\ell)}) \xrightarrow{\delta} H^2(L) \xrightarrow{\theta} \cdots \tag{3.2.17}$$

and we have:

(3.2.18) $(L^{(\ell)})^\Gamma$ is isomorphic to a finite subgroup of $H^1(L)_{tor} \cong T^\Gamma$.

(3.2.19) $H^1(\Gamma, L^{(\ell)})$ is an $\mathbb{F}_\ell$-vector space whose δ-image lies in the ℓ-th division points of $H^2(\Gamma,L)$, identified with $(T_*)^\Gamma$.

We therefore have:

Lemma 3.2.20. *Let* ℓ *be a prime not dividing the order of* $T^\Gamma \times T_*^\Gamma$. *Then* $(L^{(\ell)})^\Gamma = 0 = \delta(H^1(L^{(\ell)}))$ *and we have exact sequences:*

$$0 \to H^1(L) \xrightarrow{\theta} H^1(L) \xrightarrow{\eta} H^1(L^{(\ell)}) \to 0$$

$$0 \to H^1(L^*) \xrightarrow{\theta} H^1(L^*) \xrightarrow{\eta} H^1(L^{*(\ell)}) \to 0 \quad .$$

Consider an exact sequence:

$$1 \to \Gamma_1 \to \Gamma \xrightarrow{\varphi} \Gamma_3 \to 1 \tag{3.2.21}$$

with the following properties:

(3.2.22) $\Gamma_1 \subset \ker(\pi^{(\ell)})$ and φ extends to a homomorphism from Δ to a finite group $\Delta_3 \supset \Gamma_3$.

We then have the following diagram with exact rows and column:

$$\begin{array}{ccccccc} & & & 0 & & & \\ & & & \downarrow & & & \\ & & & H^1(\Gamma,L) & & & \\ & & & \downarrow & & & \\ 0 \to & T^\Gamma & \to & H^1(\Gamma,L) \to H^1(\Gamma,V)\otimes_{\mathbb{R}}\mathbb{C} \cong S \amalg \bar{S} & & & \\ & & & \downarrow & & & \\ 0 \to H^1(\Gamma_3,L^{(\ell)}) & & \to & H^1(\Gamma,L^{(\ell)}) & \to H^1(\Gamma_1,L^{(\ell)})^{\Gamma_3} & \to & H^2(\Gamma_3,L^{(\ell)}) \\ & & & \downarrow & & & \\ & & & 0 & & & \end{array} \tag{3.2.23}$$

where $S = \amalg_{1\leqslant i\leqslant M} S_{m(i)+2}(\Gamma,\Psi_i)$.

All the maps in (3.2.23) are compatible with the actions of the Hecke operators as we have seen in Chapter 1. We therefore have:

Theorem 3.2.24. *Suppose that* $\Gamma_3 \subset$ *center of* Δ_3 *and for some prime* ℓ *not dividing the order of* $T^\Gamma \times T_*^\Gamma$ *we have* $H^1(\Gamma_3,L^{(\ell)}) \neq 0$. *Then,*

$$\left| \prod_{1\leqslant i\leqslant M} F_{m(i)+2,\Psi_i}(x,\deg(x)) \right|^2 \equiv 0 \mod \ell$$

where $F_{m,\Psi}(x,\lambda)$ *is the characteristic polynomial of* $(S_m(\Gamma,\Psi)|x)$, $x \in R(\Gamma,\Delta)$.

3.3 Quaternion Case Over $\mathbb{Q}$

Let B be an indefinite quaternion division algebra over $\mathbb{Q}$ with discriminant $d(B)$. Let $\mathcal{O}$ be a maximal order of B and define the groups $\Gamma(1) = \Gamma(\mathcal{O},1)$, $\Gamma(N) = \Gamma(\mathcal{O},N\mathcal{O}),\ldots$ as in Section 1.6. Under the identification of $B \otimes_{\mathbb{Q}} \mathbb{R}$ with $M_2(\mathbb{R})$, $\Gamma = \Gamma(1)$, or $\Gamma(N),\ldots,$ are fuchsian subgroups of $SL(2,\mathbb{R})$ with compact quotients $\Gamma\backslash H$.

For positive even m, the $(m+1)$-dimensional $\mathbb{R}$-vector space $V = V_m$, with the action $\pi = \rho_m^*$ of $SL(2,\mathbb{R})$ on it, has a $\Gamma(1)$-invariant lattice $L = L_m$ (see Shimura [24]) with the property that for prime numbers $\ell \nmid d(B)$, $\ell > m$, $L^{(\ell)} = L/\ell L$ is isomorphic to $S^m(\mathbb{F}_\ell^2)$ as $\Gamma(1)/\Gamma(\ell) \cong SL(2,\mathbb{F}_\ell)$ modules:

$$(\pi^{(\ell)}, L^{(\ell)}) \cong (\rho_m, S^m(\mathbb{F}_\ell^2)), \quad (\text{note } \rho_m = \rho_m^* \text{ for } SL(2,*)) . \tag{3.3.1}$$

The semigroup Δ^{-1} acts through $\pi = \rho_m^*$ on V and L. In particular,

$$L \supset \pi(\alpha^{-1})L \qquad \text{and} \qquad [L:\pi(\alpha^{-1})L] = \nu(\alpha)^{m(m+1)/2}, \quad \alpha \in \Delta. \tag{3.3.2}$$

We also have a nondegenerate symmetric bilinear form ω (a non-zero multiple of Θ_m) on V with values in $\mathbb{R}$ with $\omega(L,L) \subset \mathbb{Z}$ (see Shimura [24]). Let (Γ,Δ) be one of $(\Gamma(1),\Delta(d(B)))$, $(\Gamma_0(N),\Delta_0(Nd(B)))$ or $(\Gamma(N),\Delta_{0,0}(Nd(B)))$. We then have the following result:

<u>Proposition 3.3.3</u>. (π,V,L,ω,χ) *is a torus bundle with homogeneity:*

$$\chi(\alpha) = \nu(\alpha)^{m(m+1)} \qquad , \qquad \alpha \in \Delta .$$

The next result is easily checked:

<u>Lemma 3.3.4</u>. *If a prime number* ℓ *satisfies:* $\ell \nmid d(B)$ *and* $\ell > m$, *then* ℓ *does not divide* $|T^\Gamma \times T_*^\Gamma|$.

We therefore have:

<u>Theorem 3.3.5</u>. *If* ℓ *is an odd prime with* $\ell \nmid d(B)$ *and* $\ell > m$, *then for* $\Gamma_3 = \Gamma(1)/\Gamma(\ell) \cong SL(2,\mathbb{F}_\ell)$, *we have the following diagram of maps with exact rows and column with all maps compatible with actions of Hecke operators. Moreover,* $\eta \circ \delta = 0$, $\theta = \ell.\mathrm{id}$, $\eta = \text{reduction mod } \ell$, $S = S_{m+2}(\Gamma(1))$.

$$
\begin{array}{ccccccc}
 & & & & 0 & & \\
 & & & & \downarrow & & \\
0 & \to & T^{\Gamma(1)} & \to & H^1(\Gamma(1),L) & & \\
 & & \cong\downarrow\theta & & \downarrow\theta & & \\
0 & \to & T^{\Gamma(1)} & \xrightarrow{\delta} & H^1(\Gamma(1),L) & \to & H^1(\Gamma(1),V)_{\mathbb{C}} \cong S \amalg \bar{S} \\
 & & & & \downarrow\eta & & \\
0\to & & H^1(\Gamma_3,L^{(\ell)}) & \to & H^1(\Gamma(1),L^{(\ell)}) & \to & H^1(\Gamma(\ell),L^{(\ell)})^{\Gamma_3} \to \\
 & & & & \downarrow & & \\
 & & & & 0 & &
\end{array}
\tag{3.3.6}
$$

Combining the preceding result for $\ell = m+3$ with Theorems 1.5.3, 1.5.4, we have:

<u>Corollary 3.3.7</u>. *If* $\ell \geqslant 5$ *is a prime with* $\ell \nmid d(B)$, *then:*

$$F_{\ell-1}(\Gamma\alpha\Gamma, d(\alpha)\nu(\alpha)^{-1}) \equiv 0 \bmod \ell, \quad (\Gamma\alpha\Gamma) \in R(\Gamma(1),\Delta(\ell d(B))) .$$

Here $\Gamma = \Gamma(1), F_s(x,\lambda)$ *is the characteristic polynomial of the Hecke operator* $x \in R(\Gamma,\Delta)$ *on the space* $S_s(\Gamma)$ *of all automorphic forms of weight* s.

In particular, we can apply this to the double coset T_p defined as follows. Given a prime p with $p \nmid \ell d(B)$, there exists an element $\alpha \in \Delta(\ell d(B))$ such that $\nu(\alpha) = p$ and $T_p = \Gamma\alpha\Gamma$ has degree $p+1$. Thus:

$$d(\alpha)\nu(\alpha)^{-1} = (p+1)p^{-1} \equiv 1 + p^{\ell-2} \bmod \ell .$$

We therefore have:

<u>Corollary 3.3.8</u>. *If* p *is (another) prime with* $p \nmid \ell d(B)$, *then*

$$F_{\ell-1}(T_p, 1+p^{\ell-2}) \equiv 0 \bmod \ell \quad .$$

<u>Remark 3.3.9</u>. Corollary 3.3.8 is not true for $B = M_2(\mathbb{Q})$, $\Gamma = \Gamma(1) = SL(2,\mathbb{Z})$.

3.4 Another Quaternion Case

We continue with the notation and definitions of Section 3.3. Take a prime $\ell \geq 5$ with $\ell \nmid d(B)$. Consider the case $m = \ell - 1$. Since $H^1(\Gamma_3, L_m^{(\ell)}) = H^1(SL(2,\mathbb{F}_\ell), S^{\ell-1}(\mathbb{F}_\ell^2)) = 0$, the argument in Section 3.3 does not work. We will employ another "trick". Put:

$$X = X_\ell = \mathbb{P}^1(\mathbb{F}_\ell) = \mathbb{F}_\ell \cup \{\infty\} \quad . \tag{3.4.1}$$

The group $GL(2,\mathbb{F}_\ell)$ acts on X through fractional linear transformations. Hence the semigroup $\Delta(\ell) = \{\alpha \in \mathcal{O} | \nu(\alpha) > 0,\ (\nu(\alpha),\ell) = 1\}$ acts on X through $GL(2,\mathbb{F}_\ell)$ via the reduction $\varphi_\ell : \Delta(\ell) \to \Delta(\ell)/\Gamma(\ell) = GL(2,\mathbb{F}_\ell)$. As $\Gamma(1)$-spaces, we have the isomorphisms:

$$X \cong \Gamma(1)/\Gamma_0(\ell) \cong SL(2,\mathbb{F}_\ell)/B_\ell \tag{3.4.2}$$

where B_ℓ is the Borel subgroup of $SL(2,\mathbb{F}_\ell)$ formed by all upper triangular matrices.

Let $\mathbb{Z}[X]$ and $\mathbb{F}_\ell[X]$ be free modules over $\mathbb{Z}$ and $\mathbb{F}_\ell$ based on the set X, respectively. Since X is a $\Delta(\ell)$ as well as $GL(2,\mathbb{F}_\ell)$-space, these are $\Delta(\ell)$ as well as $GL(2,\mathbb{F}_\ell)$-modules (in a compatible way). We have the augmentation maps: (on $\mathbb{Z}[X]$ and $\mathbb{F}_\ell[X]$)

$$\kappa\left(\sum_i n_i x_i\right) = \sum_i n_i, \qquad x_i \in X \quad . \tag{3.4.3}$$

The kernels of κ will be denoted by K and $K^{(\ell)}$, respectively ($K^{(\ell)}$ is in fact the reduction mod ℓ of K). It is known that $K \otimes_{\mathbb{Z}} \mathbb{C}$ is an irreducible $SL(2,\mathbb{F}_\ell)$-module (often called the Steinberg representation) and that $K^{(\ell)}$ is an absolutely irreducible modular representation of $\Gamma_3 = SL(2,\mathbb{F}_\ell)$ over $\mathbb{F}_\ell$ (also called the Steinberg representation). If we denote these representations by π and $\pi^{(\ell)}$, then we actually have:

$$(\pi^{(\ell)}, K^{(\ell)}) \cong (\rho_{\ell-1}, S^{\ell-1}(\mathbb{F}_\ell^2)) \quad \text{as} \quad SL(2,\mathbb{F}_\ell)\text{-representations.} \tag{3.4.4}$$

The preceding isomorphism can also be viewed as one of $\Gamma(1)$-representations. Let $U = K \otimes_{\mathbb{Z}} \mathbb{R}$ so that $\Delta(\ell)$ acts on K and U with action denoted by π^*. We can then extend the action π to Δ^{-1} through the formula:

$$\pi(\alpha^{-1}) = \pi^*(\alpha^\rho), \quad \alpha \in \Delta = \Delta(\ell) \quad . \tag{3.4.5}$$

We define a symmetric bilinear form ω_X on $\mathbb{Z}[X] \otimes_{\mathbb{Z}} \mathbb{R} = \mathbb{R}[X]$ by:

$$\omega_X(\Sigma_i m_i x_i, \Sigma_i n_i x_i) = \Sigma_i m_i n_i, \quad x_i \in X \quad . \tag{3.4.6}$$

The restriction of ω_X to U will be denoted by ω. The 5-ple $(\pi, U, K, \omega, 1)$ is then a polarized lattice bundle for $\Gamma = \Gamma(1)$ with trivial homogeneity 1. Note π is of primitive type, isomorphic to the $(\ell-1)$-th symmetric tensor power. In order to apply the diagram (3.2.23), put $T = U/K$. It is then easy to see that $T^{\Gamma(1)} = 0$. Moreover, we have:

Lemma 3.4.7. $S_{\ell+1}(\Gamma(1),\pi) \cong S_2(\Gamma_0(\ell))/S_2(\Gamma(1))$ *as modules for the Hecke ring* $R(\Gamma_0(\ell), \Delta_0(\ell d(B))) \cong R(\Gamma(1), \Delta(\ell d(B)))$.

The proof is essentially Shapiro's Lemma using the fact that $\mathbb{Z}[X]$ is induced.

With these results at hand, (3.4.4) implies:

Theorem 3.4.8. *For a prime* $\ell \geq 5$, $\ell \nmid d(B)$, *we have:*

(a) $\dim S_{\ell+1}(\Gamma(1)) = \dim S_2(\Gamma_0(\ell)) - \dim S_2(\Gamma(1))$.

(b) *There are bases in* $S_{\ell+1}(\Gamma(1))$ *and* $S_2(\Gamma_0(\ell))/S_2(\Gamma(1))$ *so that the actions of the Hecke operator* $x \in R(\Gamma(1), \Delta(\ell d(B)))$ *on* $S_{\ell+1}(\Gamma(1))$ *and on* $S_2(\Gamma_0(\ell))/S_2(\Gamma(1))$ *via Lemma* 3.4.7 *are represented by integral matrices* $M(x)$ *and* $N(x)$ *so that:*

$$M(x) \equiv N(x) \bmod \ell \quad .$$

The proof follows from the following diagram of maps where the rows are exact and $\eta \circ \delta = 0$, $H^i(*)$ is the abbreviation for $H^i(\Gamma(1),*)$:

$$0 \to T^{\Gamma(1)} \xrightarrow{\delta} H^1(L_{\ell-1}) \to H^1(V_{\ell-1} \otimes_{\mathbb{R}} \mathbb{C}) \cong S_{\ell+1}(\Gamma(1)) \amalg \overline{S_{\ell+1}(\Gamma(1))}$$

$$\downarrow \eta$$

$$H^1(K^{(\ell)}) \cong H^1(L^{(\ell)}_{\ell-1}) \qquad \text{(as } \Delta(\ell)\text{-modules)}$$

$$\uparrow \eta$$

$$0 \to H^1(K) \to H^1(U \otimes_{\mathbb{R}} \mathbb{C}) \cong S_{\ell+1}(\Gamma(1),\pi) \amalg \overline{S_{\ell+1}(\Gamma(1),\pi)}$$

$$\cong S_2(\Gamma_0(\ell))/S_2(\Gamma(1)) \amalg \overline{S_2(\Gamma_0(\ell))/S_2(\Gamma(1))} \quad .$$

Remark 3.4.9. The same proof is valid for the case $B = M_2(\mathbb{Q})$, $\Gamma(1) = SL(2,\mathbb{Z})$ provided that we use parabolic H^1's. This yields a theorem of Serre [20]. It states in particular that we have the congruence:

$$t \prod_{n\geq 1} (1-t^n)^2(1-t^{11n})^2 \equiv t \prod_{n\geq 1} (1-t^n)^{24} \bmod 11$$

where both sides are viewed as formal power series in t (i.e., as t-expansions). It should be noted that this congruence also follows directly from the little Fermat theorem.

Remark 3.4.10. For the other primes p with $\dim_{\mathbb{C}}(S_2(\Gamma_0(p))) = 1$, namely $p = 17$ or 19, we obtain the following congruences. For $p = 17$, set

$$A_1 = \begin{pmatrix} 2 & 1 & 0 & 0 \\ 1 & 2 & 1 & 1 \\ 0 & 1 & 12 & -5 \\ 0 & 1 & -5 & 12 \end{pmatrix} \quad A_2 = \begin{pmatrix} 2 & 1 & 1 & 0 \\ 1 & 4 & -1 & 1 \\ 1 & -1 & 6 & 2 \\ 0 & 1 & 2 & 10 \end{pmatrix} \quad A_3 = \begin{pmatrix} 4 & 2 & 0 & 1 \\ 2 & 4 & 1 & 0 \\ 0 & 1 & 6 & -3 \\ 1 & 0 & -3 & 6 \end{pmatrix}$$

and for a column of indeterminates $X = {}^t(X_1, X_2, X_3, X_4)$ and $1 \leqslant i \leqslant 3$, set:

$$Q_i(X) = {}^tX \cdot A_i \cdot X/2 .$$

Then Q_i is a positive definite integral quadratic form in 4 variables with discriminant 17^2 (and all such quadratic forms are $\mathbb{Z}$-equivalent to a unique Q_i). Hence the difference of any two theta-series:

$$\theta_i = \sum_x t^{Q_i(x)}, \qquad x \in \mathbb{Z}^4$$

lies in $S_2(\Gamma_0(17))$. Now for an even positive integer $m \geqslant 4$, define

$$E_m = 1 - 2m.B_m^{-1} \sum_{n\geq 1} \sigma_{m-1}(n)t^n$$

where B_m is the m-th Bernoulli number and $\sigma_{m-1}(n) = \Sigma_{d|n}\, d^{m-1}$. Then E_m is an integral modular form of weight m for $SL(2,\mathbb{Z})$. Thus $D.E_6$ lies in $S_{18}(SL(2,\mathbb{Z}))$, where

$$D = t \prod_{n\geq 1} (1-t^n)^{24}$$

and we have:

$$(\theta_1 - \theta_2)/4 = (\theta_2 - \theta_3)/2 \equiv D.E_6 \bmod 17 \quad .$$

Similarly, for $p = 19$, set

$$A_1 = \begin{pmatrix} 2 & 0 & 0 & 1 \\ 0 & 2 & 1 & 0 \\ 0 & 1 & 10 & 0 \\ 1 & 0 & 0 & 10 \end{pmatrix} \quad A_2 = \begin{pmatrix} 2 & 1 & -1 & 1 \\ 1 & 4 & 0 & -1 \\ -1 & 0 & 6 & 2 \\ 1 & -1 & 2 & 12 \end{pmatrix} \quad A_3 = \begin{pmatrix} 4 & 0 & 2 & 1 \\ 0 & 4 & 1 & -2 \\ 2 & 1 & 6 & 0 \\ 1 & -2 & 0 & 6 \end{pmatrix}$$

We then have:

$$(\theta_1 - \theta_2)/2 = (\theta_2 - \theta_3)/2 \equiv D.E_8 \bmod 19 \quad .$$

4. Notes

1. This work was partially supported by a grant from the National Science Foundation.

5. References

[1] W.L. Baily, Jr., "The decomposition theorem for V-manifolds," *Amer. J. Math.* 78 (1956), 862-888.

[2] E. Cline, B. Parshall, L. Scott, and W. van der Kallen, "Rational and generic cohomology," *Inv. Math.* 39 (1977), 143-163.

[3] P. Deligne, "Formes modulaires et representations ℓ-adiques," Sem. Bourbaki, Lect. Notes in Math. 179 (1971), 139-186.

[4] B. Dodson, "Dirichlet series associated to affine manifolds," Thesis, SUNY Stony Brook, 1976.

[5] K. Doi and T. Miyake, "Automorphic forms and number theory," *Kinokuniya Math. Ser.* 7 (1976), Kinokuniya, Tokyo.

[6] K. Doi and M. Ohta, "On some congruences between cusp forms on $\Gamma_0(N)$," Mod. Func. of One Var. V, Lect. Notes in Math. 601 (1977), 91-105.

[7] M. Eichler, "Quaternare quadratische Formen und die Riemannsche Vermutung für die Kongruenz-Zetafunktion," *Archiv Math.* 5 (1954), 355-366.

[8] M. Eichler, "Eine Verallgemeinerung der Abelschen Integrale," *Math. Zeit.* 67 (1957), 267-298.

[9] G. Hochschild and J.-P. Serre, "Cohomology of group extensions," *Trans. Amer. Math. Soc.* 74 (1953), 110-134.

[10] S. Ishida and M. Kuga, Book Review: *Modern Number Theory* by G. Shimura and Y. Taniyama, Sugaku No Ayumi, 6 (1959), No. 4, (in Japanese).

[11] M. Koike, "Congruences between cusp forms and linear representations of the Galois group," Algebraic Number Theory, Kyoto, 1976, Jap. Soc. for the Promotion of Science, Tokyo, 1977.

[12] M. Koike, "Eigenvalues of Hecke operators mod p," (preprint).

[13] M. Kuga, "Fiber varieties over a symmetric space whose fibers are abelian varieties," II, Lect. Notes, Univ. of Chicago, 1963-4.

[14] Y. Matsushima and S. Murakami, "On vector bundle valued harmonic forms and automorphic forms on symmetric Riemannian manifolds," *Ann. of Math.* 78 (1963), 365-416.

[15] Y. Matsushima and G. Shimura, "On the cohomology groups attached to certain vector valued differential forms on the product of the upper half planes," *Ann. of Math.* 78 (1963), 417-449.

[16] Y.H. Rhie and G. Whaples, "Hecke operators in cohomology of groups," *J. Math. Soc. Japan* 22 (1970), 431-442.

[17] C.H. Sah, "Automorphisms of finite groups," *J. Alg.* 10 (1968), 47-68; 44 (1977), 573-575.

[18] C.H. Sah, "Cohomology of split group extensions," *J. Alg.* 29 (1974), 255-302; 45 (1977), 17-68.

[19] J.-P. Serre, "Valeurs propres des operateurs de Hecke modulo ℓ," *Asterisque* 24-25 (1975), 109-117.

[20] J.-P. Serre, "Formes modulaires et fonctions zeta p-adiques," Mod. Func. of One Var. III, Lect. Notes in Math. 350 (1973), 191-269.

[21] H. Shimizu, "Automorphic functions, I, II, III," *Fund. Math.* (1977-8), Iwanami, Tokyo (in Japanese).

[22] G. Shimura, "Sur les integrales attachées aux formes automorphes," *J. Math. Soc. Japan* 11 (1952), 291-311.

[23] G. Shimura, "A reciprocity law in nonsolvable extensions," *J. Reine Angew. Math.* 221 (1966), 209-220.

[24] G. Shimura, "Introduction to the arithmetic theory of automorphic functions," *Publ. Math. Soc. Japan* 11, 1971, Princeton.

[25] C.L. Siegel, "Discontinuous groups," *Ann. of Math.* 44 (1943), 674-689.

[26] H.P.F. Swinnerton-Dyer, "On ℓ-adic representations and congruences for coefficients of modular forms, I, II," Mod. Func. of One Var. III, V, Lect. Notes in Math. 350 (1973), 1-55; 601 (1977), 63-90.

State University of New York
Stony Brook, N.Y. 11794

(Received November 12, 1980)

ON POISSON BRACKETS OF SEMI-INVARIANTS

Hisasi Morikawa

In the present article we shall treat Gelfand-Dikii's theory of formal calculus of variations [1] from a new point of view, invariant theory of formal power series

$$f_\lambda(\xi|x) = \sum_\ell \frac{(\lambda)_\ell}{\ell!}\xi^{(\ell)}x^\ell ,$$

and we shall give natural explicit expressions of Poisson brackets. In formal calculus of variations Poisson brackets are defined on the quotient module

$$K\left[\cdots,\left(\frac{\partial}{\partial x}\right)^\ell y,\cdots\right]\Big/ K + \frac{\partial}{\partial x}K\left[\cdots,\left(\frac{\partial}{\partial x}\right)^\ell y,\cdots\right] ,$$

in our case, however, they are defined on ring of semi-invariants

$$G = \left\{\varphi\in K[\xi]\,\middle|\, \mathcal{D}\varphi = \sum_\ell \ell\xi^{(\ell-1)}\frac{\partial\varphi}{\partial\xi^{(\ell)}} = 0\right\} .$$

A key stone is so called Gram decomposition

$$K[\xi] = \bigoplus_{\ell=0}^{\infty}\Delta^\ell G ,$$

where

$$\Delta = \sum_{\ell=0}^{\infty}(\lambda-\ell)\xi^{(\ell-1)}\frac{\partial}{\partial\xi^{(\ell)}}$$

and λ satisfies a condition so that Hilbert operator is well-defined.

1. Gram Decomposition

1.1 Let K be a field of characteristic zero containing rational number field $\mathbb{Q}$, and let $\xi = (\xi^{(0)}, \xi^{(1)}, \xi^{(2)}, \ldots)$ be a vector of infinite length whose entries are indeterminants. We choose a nonzero λ in K satisfying:

<u>Condition (C)</u>. The subset in K

$$\{d\lambda - 2p \mid d,\ p = 0,1,2,\ldots;\ (d,p) \neq (0,0)\}$$

has no common element with $\{0,1,2,\ldots\}$.

Degree, weight and index are defined in polynomial algebra $K[\xi]$ by

$$\deg \xi^{(\ell)} = 1, \quad \text{weight } \xi^{(\ell)} = \ell, \quad \text{index } \xi^{(\ell)} = \lambda - 2\ell \ (\ell = 0,1,2,\ldots),$$

then they are related each other by index formula:

$$\text{index } \varphi \ = \ \lambda \deg \varphi - 2 \text{ weight } \varphi$$

for homogeneous isobaric φ.

For each $\lambda \in K[\xi]$ may be regarded as an $\mathfrak{sl}(2,K)$-algebra by means of Cayley-Aronholdt derivations:

$$\begin{cases} \mathcal{D} = \displaystyle\sum_{\ell=0}^{\infty} \ell \xi^{(\ell-1)} \frac{\partial}{\partial \xi^{(\ell)}} , \\ \Delta = \displaystyle\sum_{\ell=0}^{\infty} (\lambda-\ell) \ {}^{(\ell+1)} \frac{\partial}{\partial \xi^{(\ell)}} , \\ H = \displaystyle\sum_{\ell=0}^{\infty} (\lambda-2\ell) \xi^{(\ell)} \frac{\partial}{\partial \xi^{(\ell)}} , \end{cases}$$

as follows

$$\begin{pmatrix} 0 & 0 \\ 1 & 0 \end{pmatrix} \varphi = \mathcal{D}\varphi, \qquad \begin{pmatrix} 0 & 1 \\ 0 & 0 \end{pmatrix} \varphi = \Delta\varphi , \qquad \begin{pmatrix} -1 & 0 \\ 0 & 1 \end{pmatrix} \varphi = H\varphi ,$$

i.e., $[\mathcal{D},\Delta] = H$, $[H,\mathcal{D}] = 2\mathcal{D}$, $[H,\Delta] = -2\Delta$.

Since values of index are nothing else than eigenvalues of H and each monomial is an eigenvector of H, we have the eigenspace decomposition:

$$K[\xi] = \bigoplus_{\mu} K[\xi]^{[\mu]}, \qquad K[\xi]^{[\mu]} = \{\varphi \in K[\xi] \mid H\varphi = \mu\varphi\}$$

with projection operators

$$\pi^{[\mu]} : K[\xi] \to K[\xi]^{[\mu]} \quad .$$

The elements of the kernel of $\mathcal{D}$.

$$G = \{\varphi \in K[\xi] \mid \mathcal{D}\varphi = 0\}$$

are called semi-invariants. Since $\mathcal{D}$ is homogeneous of index 2, we have

$$G = \bigoplus_{\mu} G^{[\mu]} , \qquad G^{[\mu]} = \{\varphi \in G \mid H\varphi = \mu\varphi\}$$

1.2 We mean by x an independent variable and by y a dependent variable with respect to x. We denote

$$f_\lambda(\xi|x) = \sum_{\ell=0}^{\infty} \frac{(\lambda)_\ell}{\ell!} \xi^{(\ell)} x^\ell \tag{1.2}$$

with

$$(\lambda)_\ell = \begin{cases} \lambda(\lambda-1)\cdots(\lambda-\ell+1) & (\ell = 1,2,3,\ldots) \\ 1 & (\ell = 0) \end{cases} , \tag{1.3}$$

and call it formal power series of index λ with coefficient ξ.

There exists a commutative diagram of surjective differential algebra isomorphisms:

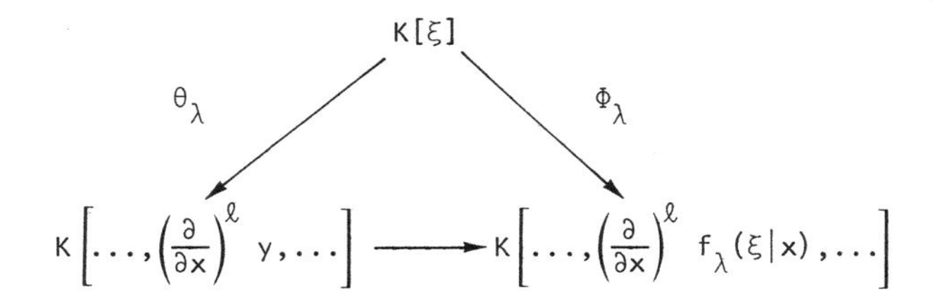

such that

$$\theta_\lambda(\xi^{(\ell)}) = \frac{1}{(\lambda)_\ell}\left(\frac{\partial}{\partial x}\right)^{\ell} y \tag{1.4}$$

$$\Phi_\lambda(\xi^{(\ell)}) = \frac{1}{(\lambda)_\ell}\left(\frac{\partial}{\partial x}\right)^{\ell} f_\lambda(\xi|x) \tag{1.5}$$

$$\frac{\partial}{\partial x}\circ\theta_\lambda = \theta_\lambda\circ\Delta \tag{1.6}$$

$$\frac{\partial}{\partial x}\Phi_\lambda = \Delta\circ\Phi_\lambda = \Phi_\lambda\circ\Delta \tag{1.7}$$

$$\Phi_\lambda(\varphi) = \exp(x\Delta)\varphi \tag{1.8}$$

$$f_\lambda(\xi|x) = \exp(x\Delta)\xi^{(0)} \ . \tag{1.9}$$

1.3 By induction on ℓ we have

$$[\mathcal{D},\Delta^{\ell}] = \sum_{\mu} \ell(\mu-\ell+1)\Delta^{\ell-1}\pi^{[\mu]} \quad , \tag{1.10}$$

$$[\Delta,\mathcal{D}^{\ell}] = \sum_{\mu} \ell(-\mu-\ell+1)\mathcal{D}^{\ell-1}\pi^{[\mu]} \qquad (\ell=1,2,3,\ldots) \ . \tag{1.11}$$

Following Hilbert, we define generalized Hilbert operators by

$$L^{(0)} = \sum_{\mu} \frac{(-1)^{\ell}\Delta^{\ell}\mathcal{D}^{\ell}\pi^{[\mu]}}{\ell!\,(\mu+\ell+1)_\ell} \quad , \tag{1.12}$$

$$L^{(k)} = \sum_{\mu} \frac{L^{(0)}\mathcal{D}^{k}\pi^{[\mu]}}{k!\,(\mu+2k)_k} \qquad (k=1,2,3,\ldots) \ . \tag{1.13}$$

Remark. By virtue of Condition (C) generalized Hilbert operators are all well-defined, because for each $\varphi\in K[\xi]^{[\mu]}$ $\mu+\ell+1=0$ implies $\mathcal{D}^{\ell}\varphi=0$.

From (1.10), (1.11) we have

$$L^{(0)}\Delta = \mathcal{D}L^{(0)} = 0 \quad , \tag{1.14}$$

$$L^{(0)}\varphi = \varphi \qquad (\varphi\in G) \quad . \tag{1.15}$$

Lemma 1.1. *For each* $\varphi \in G^{[\mu]}$ *we have*

$$\mathcal{D}^p \Delta^k \varphi = \begin{cases} \frac{k!}{(k-p)!} (\mu-k+p)_p \Delta^{k-p}\varphi & (k \geqslant p) \\ 0 & (k \lneqq p) \end{cases} \qquad (1.16)$$

Proof. By virtue of (1.10) it follows

$$\mathcal{D}^p \Delta^k \varphi = \mathcal{D}^{p-1}[\mathcal{D},\Delta^k]\varphi = k(\mu-k+1)\mathcal{D}^{p-1}\Delta^{k-1}\varphi = \cdots$$

$$= \frac{k!}{(k-p)!} (\mu-k+1)\cdots(\mu-k+p)\Delta^{k-p}\varphi \quad .$$

Theorem 1.1. *If* λ *satisfies Condition (C), then*

$$K[\xi] = \bigoplus_{\ell=0}^{\infty} \Delta^\ell G \qquad \textit{(Gram decomposition)} , \qquad (1.17)$$

and the projection operators are given by

$$\Delta^k L^{(k)} : K[\xi] \to \Delta^k G \qquad (k = 0,1,2,\ldots) \quad . \qquad (1.18)$$

Proof. From the relation

$$\varphi = L^{(0)}\varphi - \sum_{\ell=1}^{\infty} \frac{(-1)^\ell \Delta^\ell \mathcal{D}^{\ell-1}\mathcal{D}\varphi}{\ell!(\mu+\ell+1)_\ell} \qquad (\varphi \in K[\xi]^{[\mu]})$$

we conclude that $K[\xi]$ is generated by G as an $\mathfrak{sl}(2,K)$-module, because $L^{(0)}\varphi \in G$ and φ is generated by G if and only if $\mathcal{D}\varphi$ is so. Therefore each $\varphi \in K[\xi]^{[\mu]}$ is expressed

$$\varphi = \sum \Delta^k \varphi_k$$

with $\varphi_k \in G^{[\mu+2k]}$. By virtue of (1.16)

$$\frac{\mathcal{D}^k \varphi}{k!} = \sum_h \binom{h}{k} (\mu+h+k)_k \Delta^{h-k}\varphi_h \quad .$$

Applying $L^{(0)}$ on both sides, we have

$$L^{(0)}\mathcal{D}^k \varphi = (\mu+2k)_k \varphi_k \qquad (k = 0,1,2,\ldots) ,$$

and thus

$$\varphi_k = \frac{L^{(0)}\mathcal{D}^k\varphi}{k!(\mu+2k)_k} = L^{(k)}\varphi ,$$

$$\Delta^k\varphi_k = \Delta^k L^{(k)}\varphi .$$

Corollary.

$$K\left[\ldots,\left(\frac{\partial}{\partial x}\right)^{\ell} y,\ldots\right] = \bigoplus_{\ell=0}^{\infty} \left(\frac{\partial}{\partial x}\right)^{\ell} \theta_\lambda(G) , \tag{1.19}$$

$$K\left[\ldots,\left(\frac{\partial}{\partial x}\right)^{\ell} f_\lambda(\xi|x),\ldots\right] = \bigoplus_{\ell=0}^{\infty} \left(\frac{\partial}{\partial x}\right)^{\ell} \Phi_\lambda(G). \tag{1.20}$$

1.4 We shall give an application of Gram decomposition to actions of differential operators on G.

Definition 1.1. *The* k*-th apolar of* $\phi_1 \in G^{[\nu_1]}$ *and* $\phi_2 \in G^{[\nu_2]}$ *is defined by*

$$A_k(\phi_1,\phi_2) = \sum_p (-1)^p \binom{k}{p} \frac{\Delta^p\phi_1}{(\nu_1)_p} \frac{\Delta^{k-p}\phi_2}{(\nu_2)_{k-p}} , \tag{1.21}$$

which is an element of $G^{[\nu_1+\nu_2-2k]}$. *For each* $\phi \in G^{[\nu]}$ *we denote*

$$A_k(\phi) = \sum_\mu \sum_p (-1)^p \binom{k}{p} \frac{\Delta^p\phi}{(\nu)_p} \frac{\Delta^{k-p}\pi^{[\mu]}}{(\mu)_{k-p}} \tag{1.22}$$

so that

$$A_k(\phi,\varphi) = A_k(\phi)\varphi . \tag{1.23}$$

Making apolars is one of popular methods to create new semi-invariant from two known semi-invariants.

Lemma 1.2. *For* $\phi_1 \in G^{[\nu_1]}$ *and* $\phi_2 \in G^{[\nu_2]}$ *we have*

$$L^{(0)}(\phi_1\Delta^k\phi_2) = \frac{(\nu_1)_k(\nu_2)_k(\nu_1+\nu_2+1)_k}{(\nu_1+\nu_2+1)_{2k}} A_k(\phi_1,\phi_2) , \tag{1.24}$$

$$L^{(0)}(\Delta^h\phi_1\Delta^k\phi_2) = \frac{(-1)^h(\nu_1)_{h+k}(\nu_2)_{h+k}(\nu_1+\nu_2+1)_{h+k}}{(\nu_1+\nu_2+1)_{2(h+k)}} A_{h+k}(\phi_1,\phi_2). \tag{1.25}$$

Proof. From the definitions it follows

$$L^{(0)}(\phi_1\Delta^k\phi_2) = \sum_\ell \frac{(-1)^\ell\Delta^\ell\mathcal{D}^\ell(\phi_1\Delta^k\phi_2)}{\ell!(\nu_1+\nu_2-2k+2)\dots(\nu_1+\nu_2-2k+\ell+1)}$$

$$= \sum_\ell (-1)^\ell\binom{k}{\ell}\binom{\ell}{p}\frac{(\nu_2-k+1)\dots(\nu_2-k+\ell)}{(\nu_1+\nu_2-2k+2)\dots(\nu_1+\nu_2-2k+\ell+1)}\Delta^p\phi_1\Delta^{k-p}\phi_2$$

$$= \sum_p\left(\sum_\ell(-1)^{\ell-p}\binom{k-p}{\ell-p}\frac{(\nu_1+\nu_2+1)_{2k-\ell}(\nu_2)_k}{(\nu_1+\nu_2+1)_{2k}(\nu_2)_{k-\ell}}\right)(-1)^p\binom{k}{p}\Delta^p\phi_1\Delta^{k-p}\phi_2.$$

This shows that (1.24) holds, if and only if

$$(*)_k: \sum_\ell (-1)^{\ell-p}\binom{k-p}{\ell-p}\frac{(\nu_1+\nu_2+1)_{2k-\ell}}{(\nu_2)_{k-\ell}} = \frac{(\nu_1)_k(\nu_1+\nu_2+1)_k}{(\nu_1)_p(\nu_2)_{k-p}}$$

It is enough to prove $(*)_{k+1}$ from $(*)_k$:

$$\sum_\ell (-1)^{\ell-p}\binom{k+1-p}{\ell-p}\frac{(\nu_1+\nu_2+1)_{2k-\ell+2}}{(\nu_2)_{k-\ell+1}} = \sum_\ell(-1)^{\ell-p}\left(\binom{k-p}{\ell-p}+\binom{k-p}{\ell-p-1}\right)\cdot\frac{(\nu_1+\nu_2+1)_{2k-\ell+2}}{(\nu_2)_{k-\ell+1}}$$

$$= \sum_\ell (-1)^{\ell-p}\binom{k-p}{\ell-p}\left(\frac{(\nu_1+\nu_2+1)_{2k-\ell+2}}{(\nu_2)_{k-\ell+1}} - \frac{(\nu_1+\nu_2+1)_{2k-\ell+1}}{(\nu_2)_{k-\ell}}\right)$$

$$= \sum_\ell (-1)^{\ell-p}\binom{k-p}{\ell-p}\frac{(\nu_1+\nu_2+1)_{2k-(\ell-1)}(\nu_1-k)}{(\nu_2)_{k-(\ell-1)}}$$

$$= \sum_{\ell} (-1)^{\ell-p}\left(\binom{k-(p-1)}{(\ell-1)-(p-1)} - \binom{k-p}{(\ell-1)-p}\right)\frac{(\nu_1+\nu_2+1)_{2k-(\ell-1)}(\nu_1-k)}{(\nu_2)_{k-(\ell-1)}}$$

$$= \frac{(\nu_1)_{k+1}(\nu_1+\nu_2+1)_k}{(\nu_1)_{p-1}(\nu_2)_{k+1-p}} + \frac{(\nu_1)_{k+1}(\nu_1+\nu_2+1)_k}{(\nu_1)_p(\nu_2)_{k-p}} = \frac{(\nu_1)_{k+1}(\nu_1+\nu_2+1)_{k+1}}{(\nu_1)_p(\nu_2)_{k+1-p}} .$$

Since $L^{(0)}\Delta = 0$ and

$$\Delta^h\phi_1\Delta^k\phi_2 \equiv (-1)^h\phi_1\Delta^{h+k}\phi_2 \bmod \Delta K[\xi] \quad ,$$

we have

$$L^{(0)}(\Delta^h\phi_1\Delta^k\phi_2) \equiv (-1)^h L^{(0)}(\phi_1\Delta^{h+k}\phi_2),$$

and thus (1.24) implies (1.25).

A differential operator $\Sigma_\ell\ \varphi_\ell\Delta^\ell$ is called to be index-homogeneous of index μ if $\varphi_\ell \in K[\xi]^{[\mu+2\ell]}$.

Theorem 1.2. *Let* $\Sigma_\ell\ \varphi_\ell\Delta^\ell$ *be an index-homogeneous of index* μ, *and let*

$$\varphi_\ell = \sum_p \Delta^p\varphi_{\ell,p} \qquad \left(\varphi_{\ell,p} \in G^{[\mu+2\ell+2p]}\right)$$

be the Gram decompositions of coefficients. Then we have

$$L^{(0)}\left(\sum_\ell \varphi_\ell\Delta^\ell\right)L^{(0)}$$

$$= \sum_\nu \sum_{\ell,p} \frac{(-1)^p(\mu+2(\ell+p))_{\ell+p}(\nu)_{\ell+p}(\mu+\nu+2(\ell+p)+1)_{\ell+p}}{(\mu+\nu+2(\ell+p)+1)_{2(\ell+p)}} A_{\ell+p}(\varphi_{\ell,p}) \cdot \pi^{[\nu]}L^{(0)} \quad . \qquad (1.26)$$

This is a direct consequence of Lemma 1.2.

2. Poisson Brackets in G

2.1 Evolution derivations are defined by

$$\Delta_{\varphi} = \sum_{\ell=0}^{\infty} \frac{1}{(\lambda)_{\ell}} \Delta^{\ell}\varphi \frac{\partial}{\partial \xi^{(\ell)}} \qquad (\varphi \in K[\xi]) \quad , \tag{2.1}$$

then

$$[\Delta, \Delta_{\varphi}] = 0 \quad , \tag{2.2}$$

$$[\Delta_{\varphi}, \Delta_{\varphi}] = \Delta_{\Delta_{\varphi}\phi - \Delta_{\phi}\varphi} \quad . \tag{2.3}$$

Evolution derivations form a Lie algebra $\mathrm{Der}_{EV}(K[\xi])$ with composition (2.3).

Examples.

$$\Delta = \Delta_{\Delta\xi^{(0)}} \quad , \qquad \Delta_{\xi^{(0)}} = \sum_{\ell=0}^{\infty} \xi^{(\ell)} \frac{\partial}{\partial \xi^{(\ell)}} \quad .$$

Euler-Lagrange operator is defined by

$$\delta = \sum_{\ell} \frac{(-1)^{\ell}}{(\lambda)_{\ell}} \Delta^{\ell} \frac{\partial}{\partial \xi^{(\ell)}} \quad , \tag{2.4}$$

then, as similarly as formal variational calculus, we have by calculation:

$$\mathrm{Ker}\ \delta = K + \Delta K[\xi] \quad , \tag{2.5}$$

$$\Delta_{\Delta\delta\varphi}\delta\phi - \Delta_{\Delta\delta\phi}\delta\varphi = \delta\Delta_{\Delta\delta\varphi}\phi . \tag{2.6}$$

They are also obtained by the inverse map θ_{λ}^{-1} from the corresponding formulas in formal variational calculus in $K[\ldots,(\partial/\partial x)^{\ell}y,\ldots]$.

Theorem 2.1. *Assume that* λ *satisfies Condition (C). For each pair* (φ, ϕ) *of elements in* G *we define Poisson bracket by*

$$\{\varphi, \phi\} = L^{(0)} \Delta_{\Delta\delta\varphi}\phi \quad , \tag{2.7}$$

then

$$[\Delta_{\Delta\delta\varphi}, \Delta_{\Delta\delta\phi}] = \Delta_{\Delta\delta\{\varphi,\phi\}} \quad . \tag{2.8}$$

Namely the map $\varphi \mapsto \Delta_{\Delta\delta\varphi}$ is a Lie algebra homomorphism of $(G,\{,\})$ into Lie algebra of evolution derivations $DER_{EV}(K[\xi])$. Moreover the kernel of the Lie algebra homomorphism is given by

$$K \oplus K\xi^{(0)} \quad .$$

Proof. Since (2.8) is a direct consequence of (2.2), (2.6), we shall only prove the last assertion. From the definition of evolution derivation we observe that $\Delta_{\Delta\delta\varphi} = 0$ if and only if $\Delta\delta\varphi = 0$. Since Δ is injective on polynomials without constant term and $\Delta K = \{0\}$, hence we conclude that $\Delta\delta\varphi = 0$ if and only if $\delta\varphi \in K$. Let φ be a nonzero semi-invariant homogeneous of degree d and isobaric of weight p such that $\delta\varphi \in K$. Since δ is homogeneous with respect to degree, weight and index:

$$\deg \delta = -1, \quad \text{weight } \delta = 0, \quad \text{index } \delta = -\lambda ,$$

hence by virtue of index-formula it follows

$$(d\lambda - 2p) - \lambda \;=\; \text{index } \delta\varphi \;=\; 0 \quad .$$

By virtue of Condition (C) we conclude $(d,p) = (1,0)$, i.e., $\varphi \in K\xi^{(0)}$.

Lemma 2.1.

$$\{\varphi,\phi\} \;=\; L^{(0)}(\Delta\delta\varphi\delta\phi) \quad . \tag{2.9}$$

Proof. From definitions it follows

$$\{\varphi,\phi\} \;=\; L^{(0)}(\Delta_{\Delta\delta\varphi}\phi) \;=\; L^{(0)}\left(\sum_{\ell} \frac{\Delta^{\ell+1}\delta\varphi}{(\lambda)_{\ell}} \frac{\partial\phi}{\partial\xi^{(\ell)}}\right)$$

$$= \; L^{(0)}\left(\sum_{\ell} (-1)^{\ell} \frac{\Delta\delta\varphi}{(\lambda)_{\ell}} \Delta^{\ell} \frac{\partial\phi}{\partial\xi^{\ell}}\right)$$

$$= \; L^{(0)}(\Delta\delta\varphi\delta\phi) \quad .$$

2.2 We shall give Gram decomposition of $\delta\varphi$.

Lemma 2.2. *For each* $\varphi \in G$ *we have*

$$\mathcal{D}^{\ell} \frac{\partial\varphi}{\partial\xi^{(k)}} \;=\; (-1)^{\ell} \frac{(\ell+k)!}{k!} \frac{\partial\varphi}{\partial\xi^{(\ell+k)}} \qquad (\ell,k = 0,1,2,\ldots) \ . \tag{2.10}$$

Proof. Since $\mathcal{D}$ and $\partial/\partial\xi^{(h)}$ are derivations over K and

$$\left[\frac{\partial}{\partial\xi^{(h)}}, \mathcal{D}\,\xi^{(\ell)}\right] = \begin{cases} h+1 & (\ell = h+1) \\ 0 & (\ell \neq h+1) \end{cases},$$

we have

$$\left[\mathcal{D}, \frac{\partial}{\partial\xi^{(h)}}\right] = -(h+1)\frac{\partial}{\partial\xi^{(h+1)}} .$$

Using this formula successibly, we have (2.10), because

$$\mathcal{D}^{\ell}\frac{\partial\varphi}{\partial\xi^{(k)}} = \mathcal{D}^{\ell-1}\left[\mathcal{D}, \frac{\partial}{\partial\xi^{(k)}}\right]\varphi .$$

Lemma 2.3. *For each* $\varphi \in G^{[\mu]}$ *we have*

$$L^{(0)}\left(\frac{\partial\varphi}{\partial\xi^{(k)}}\right) = \sum_{\ell}\frac{\binom{k+\ell}{\ell}\Delta^{\ell}\frac{\partial\varphi}{\partial\xi^{(\ell+k)}}}{(\mu-\lambda+2k+2)\ldots(\mu-\lambda+2k+\ell+1)} . \tag{2.11}$$

Proof. From (2.10) it follows

$$L^{(0)}\left(\frac{\partial\varphi}{\partial\xi^{(k)}}\right) = \sum_{\ell}\frac{(-1)^{\ell}\Delta^{\ell}\mathcal{D}^{\ell}\frac{\partial\varphi}{\partial\xi^{(k)}}}{\ell!(\mu-\lambda+2k+2)\ldots(\mu-\lambda+2k+\ell+1)}$$

$$= \sum_{\ell}\frac{\binom{k+\ell}{\ell}\Delta^{\ell}\frac{\partial\varphi}{\partial\xi^{(\ell+k)}}}{(\mu-\lambda+2k+2)\ldots(\mu-\lambda+2k+\ell+1)} .$$

Theorem 2.2. *For each* $\varphi \in G^{[\mu]}$ *we have*

$$L^{(k)}(\delta\varphi) = \frac{(-1)^{k}}{(\mu-\lambda+2k)_{k}}\left(\sum_{\ell=0}^{k}\binom{k}{\ell}\frac{(\mu-\lambda+k+\ell)_{\ell}}{(\lambda)_{\ell}}\right)L^{(0)}\left(\frac{\partial\varphi}{\partial\xi^{(k)}}\right), \tag{2.12}$$

$$\delta\varphi = \sum_{k}\frac{(-1)^{k}}{(\mu-\lambda+2k)_{k}}\left(\sum_{\ell=0}^{k}\binom{k}{\ell}\frac{(\mu-\lambda+k+\ell)_{\ell}}{(\lambda)_{\ell}}\right)\Delta^{k}L^{(0)}\left(\frac{\partial\varphi}{\partial\xi^{(k)}}\right). \tag{2.13}$$

Proof. From $L^{(0)}\Delta = 0$ and

$$[\mathcal{D}^k,\Delta] = \sum_{\mu} k(\mu+k-1)\mathcal{D}^{k-1}\pi^{[\mu]}$$

it follows

$$L^{(0)}\left(\mathcal{D}^k\Delta^{\ell}\frac{\partial\varphi}{\partial\xi^{(\ell)}}\right) = L^{(0)}\left([\mathcal{D}^k,\Delta]\Delta^{\ell-1}\frac{\partial\varphi}{\partial\xi^{(\ell)}}\right)$$

$$= k(\mu-\lambda+k+1)L^{(0)}\left(\mathcal{D}^{k-1}\Delta^{\ell-1}\frac{\partial\varphi}{\partial\xi^{(\ell)}}\right).$$

Hence by induction on k we have

$$L^{(0)}\left(\mathcal{D}^k\Delta^{\ell}\frac{\partial\varphi}{\partial\xi^{(\ell)}}\right) = 0 \qquad (k \lneq \ell) ,$$

and for $k \geq \ell$

$$L^{(0)}\left(\mathcal{D}^k\Delta^{\ell}\frac{\partial\varphi}{\partial\xi^{(\ell)}}\right) = \frac{k!}{(k-\ell)!}(\mu-\lambda+k+1)\ldots(\mu-\lambda+k+\ell)L^{(0)}\left(\mathcal{D}^{k-\ell}\frac{\partial\varphi}{\partial\xi^{(\ell)}}\right)$$

$$= (-1)^{k-\ell}k!\binom{k}{\ell}(\mu-\lambda+k+\ell)_{\ell}L^{(0)}\left(\frac{\partial\varphi}{\partial\xi^{(k)}}\right).$$

Finally we have

$$L^{(k)}\delta\varphi = \frac{1}{k!(\mu-\lambda+2k)_k}\sum_{\ell}\frac{(-1)^{\ell}L^{(0)}\left(\mathcal{D}^k\Delta^{\ell}\frac{\partial\varphi}{\partial\xi^{(\ell)}}\right)}{(\lambda)_{\ell}}$$

$$= \frac{(-1)^k}{(\mu-\lambda+2k)_k}\left(\sum_{\ell=0}^{k}\binom{k}{\ell}\frac{(\mu-\lambda+k+\ell)_{\ell}}{(\lambda)_{\ell}}L^{(0)}\left(\frac{\partial\varphi}{\partial\xi^{(k)}}\right)\right) .$$

Combining Lemma 1.2 and Theorem 2.2, we have an explicit expression of Poisson bracket, because $\{\varphi,\phi\} = L^{(0)}(\Delta\delta\varphi\delta\phi)$.

Theorem 2.3. *For* $\varphi \in G^{[\mu]}$ *and* $\phi \in G^{[\nu]}$ *we have*

$$\{\varphi,\phi\} = \sum_{h,k} \frac{(-1)^{k+1}(\mu-\lambda+2k)_{h+k+1}(\nu-\lambda+2k)_{h+k+1}(\mu+\nu-2\lambda+2(h+k)+1)_{h+k+1}}{(\mu-\lambda+2h)_h(\nu-\lambda+2k)_k(\mu+\nu-2\lambda+2(h+k)+1)_{2(h+k+1)}}$$

$$\cdot \sum_p^h \binom{h}{p} \frac{(\mu-\lambda+h+p)_p}{(\lambda)_p} \sum_{q=0}^k \binom{k}{q} \frac{(\nu-\lambda+k+q)_p}{(\lambda)_q}$$

$$\cdot \quad A_{h+k+1}\left(L^{(0)}\left(\frac{\partial\varphi}{\partial\xi^{(h)}}\right),\quad L^{(0)}\left(\frac{\partial\varphi}{\partial\xi^{(k)}}\right)\right) \quad . \tag{2.14}$$

3. Generic Formal Solutions of Evolution Equations

3.1 We mean by t, x independent variables, and by u a dependent variable with respect to t and x. An evolution equation of differential polynomial type is a partial differential equation

$$\frac{\partial u}{\partial t} = \varphi\left(\ldots, \frac{1}{(\lambda)_\ell}\left(\frac{\partial}{\partial x}\right)^\ell u, \ldots\right) \tag{3.1}$$

with an element $\varphi(\ldots,\xi^{(\ell)},\ldots) \in K[\xi]$. If $\varphi \in K[\xi]^{[\mu]}$, we call (3.1) to be index-homogeneous of index μ.

We denote

$$u_\varphi(\xi|t,x) = \exp(t\Delta_\varphi + x\Delta)\xi^{(0)} = \sum_{h,\ell=0}^{\infty} \frac{\Delta_\varphi^h \Delta^\ell \xi^{(0)}}{h!\ell!} t^h h^\ell \quad , \tag{3.2}$$

and call it a generic formal solution of evolution equation (3.1). There exists a differential algebra isomorphism

$$\Psi_\lambda : K[\xi] \to K\left[\ldots, \left(\frac{\partial}{\partial x}\right)^\ell u_\varphi(\xi|t,x), \ldots\right]$$

such that

$$\Psi_\lambda(\phi) = \exp(t\Delta_\varphi + x\Delta)\phi \quad , \tag{3.3}$$

$$\frac{\partial}{\partial t} \circ \Psi_\lambda = \Delta_\varphi \circ \Psi_\lambda = \Psi_\lambda \circ \Delta_\varphi \quad , \tag{3.4}$$

$$\frac{\partial}{\partial x} \circ \Psi_\lambda = \Delta \circ \Psi_\lambda = \Psi_\lambda \circ \Delta \quad . \tag{3.5}$$

Theorem 3.1. *A formal power series* $g(t,x)$ *is a formal solution of an evolution equation*

$$\frac{\partial u}{\partial t} = \varphi\left(\ldots, \frac{1}{(\lambda)_\ell}\left(\frac{\partial}{\partial x}\right)^\ell u, \ldots\right) \quad ,$$

if and only if

$$g(t,x) = u_\varphi(\xi|t,x)\Big|_{\xi=\alpha}$$

with $\alpha = (\alpha^{(0)}, \alpha^{(1)}, \alpha^{(2)}, \ldots)$.

Proof. From $\Delta_\varphi \xi^{(0)} = \varphi$ and (3.4) it follows

$$\frac{\partial}{\partial t} u_\varphi(\xi|t,x) = \Delta_\varphi \Psi_\lambda(\xi^{(0)}) = \Psi_\lambda(\varphi)$$

$$= \varphi\left(\ldots, \frac{1}{(\lambda)_\ell}\left(\frac{\partial}{\partial x}\right)^\ell u_\varphi(\xi|t,x), \ldots\right) \quad .$$

This means that for each pair (h,ℓ) the coefficients $a_{\ell,h}(\xi)$, $b_{\ell h}(\xi)$ of $t^h x^\ell$ in

$$\frac{\partial}{\partial t} u_\varphi(\xi|t,x)$$

and

$$\varphi\left(\ldots, \frac{1}{(\lambda)_\ell}\left(\frac{\partial}{\partial x}\right)^\ell u_\varphi(\xi|t,x), \ldots\right)$$

are the same polynomial, and thus every specialization $u_\varphi(\xi|t,x)\big|_{\xi=\alpha}$ is also a formal solution. Conversely, if $g(t,x)$ is a formal solution, then there exists $\alpha = (\alpha^{(0)}, \alpha^{(1)}, \alpha^{(2)}, \ldots)$ such that

$$g(0,x) = f_\lambda(\alpha|x) = \sum_{\ell=0}^{\infty} \frac{(\lambda)_\ell}{\ell!} \alpha^{(\ell)} x^\ell$$

and

$$g(t,x) = u_{\varphi}(\xi|t,x)\Big|_{\xi=\alpha} .$$

4. References

[1] I.M. Gelfand, and L.A. Dikii, "The structure of Lie algebras in formal calculus of variations," *Funkts. Analiz Prilozhen.*, 10, No. 1. 2836 (1976).

[2] H. Morikawa, "Some analytic and geometric application of the invariant theoretic method," *Nagoya Math. J.* 80, 1-47 (1980).

Nagoya University
Chigusaku, Nagoya 464
Japan

(Received October 30, 1980)

SOME STABILITIES OF GROUP AUTOMORPHISMS

Akihiko Morimoto

0. Introduction

Let $\varphi: M \to M$ be a homeomorphism of a metric space (M,d) with distance function d. A (double) sequence $\{x_i\}_{i \in Z}$ of points $x_i \in M$ is called a δ-pseudo-orbit of φ iff $d(\varphi(x_i), x_{i+1}) \leq \delta$ for every $i \in Z$, where $\delta > 0$ is a constant (*cf.* [2]). Given $\varepsilon > 0$, a δ-pseudo-orbit $\{x_i\}$ is called to be ε-traced by a point $y \in M$ iff $d(\varphi^i(y), x_i) \leq \varepsilon$ for every $i \in Z$. We shall call φ stochastically stable, iff for any $\varepsilon > 0$ there exists $\delta > 0$ such that every δ-pseudo-orbit of φ can be ε-traced by some point $y \in M$.

R. Bowen [2] proves that every Anosov diffeomorphism φ of a compact manifold is stochastically stable by using the stable manifold theory. K. Kato [5] proves this result without using the stable manifold theory: He generalizes the result of P. Walters [11] concerning the topological stability of Anosov diffeomorphisms.

The main purpose of this paper is to characterize group automorphisms of R^n and of a torus T^n to be stochastically stable and to clarify the relations to other stabilities (Thms. 3 and 4). Moreover we shall see that every isometry of a Riemannian manifold M ($\dim M \geq 1$) is not stochastically stable (Cor. 1) and, using this, we shall prove that if there is a stochastically stable group automorphism of a compact connected Lie group G, then G is necessarily a torus (Thm. 5).

The main parts of this paper were announced in the Symposium on Dynamical Systems held in 1977 at the Research Institute for Mathematical Sciences, Kyoto University.

1. Definitions and Preparatory Lemmas

Let $\varphi: M \to M$ be a homeomorphism of a metric space (M,d). We denote by $H(M)$ the group of all homeomorphisms of M.

Definition 1. We call φ *topologically stable* iff for any $\varepsilon > 0$ there exists $\delta > 0$ with the property that for any $\psi \in H(M)$ with $d(\varphi(x),\psi(x)) < \delta$ for every $x \in M$ there is a continuous map $h: M \to M$ such that (i) $h \circ \psi = \varphi \circ h$, (ii) $d(h(x),x) < \varepsilon$ for every $x \in M$.

Definition 2. A sequence of points $\{x_i\}_{i \in (a,b)}$ ($-\infty \leqslant a < b \leqslant +\infty$, $i \in Z$, $a, b \in Z \cup \{\pm\infty\}$) is called a *$\delta$-pseudo-orbit* of φ iff $d(\varphi(x_i),x_{i+1}) \leqslant \delta$ for $i \in (a+1,b-2)$. If $a > -\infty$ and $b < \infty$, this sequence will be called a *finite δ-pseudo-orbit* of φ and if $a = -\infty$ and $b = +\infty$, the sequence will be (sometimes) called an (infinite) δ-pseudo-orbit of φ. $\{x_i\}$ is called to be *ε-traced by* $x \in M$ (with respect to φ) iff $d(\varphi^i(x),x_i) \leqslant \varepsilon$ holds for $i \in (a,b)$.

φ is called *stochastically stable* iff for any $\varepsilon > 0$ there exists $\delta > 0$ such that any (infinite) δ-pseudo-orbit of φ can be ε-traced by some point $x \in M$.

Definition 3. We denote by $\mathrm{Orb}^{\delta}(\varphi)$ the set of all (finite or infinite) δ-pseudo-orbit of φ and $\mathrm{Tr}^{\varepsilon}(\{x_i\},\varphi) = \mathrm{Tr}^{\varepsilon}(\{x_i\})$ the set of all $y \in M$ such that $\{x_i\}$ is ε-traced by y.

Assumption. In the sequel we assume that every subset of M is relatively compact unless otherwise stated.

Lemma 1. *Let* $h \in H(M)$ *be a homeomorphism of* M *such that* h *and* h^{-1} *are both uniformly continuous. Take* $\varphi \in H(M)$ *and set* $\psi = h \circ \varphi \circ h^{-1}$. *Then* φ *is stochastically stable if and only if* ψ *is.*

Lemma 2. *Let* $\varphi \in H(M)$ *be stochastically stable. Then, for any integer* $k > 0$, φ^k *is also stochastically stable.*

Lemma 3. *Let* $\varphi \in H(M)$ *be uniformly continuous, and fix an integer* $N > 0$. *Then for any* $\varepsilon > 0$, *there is* $\delta > 0$ *such that if* $\{x_i\}_{i=0}^{N} \in \mathrm{Orb}^{\delta}(\varphi)$ *then* x_0 *ε-traces* $\{x_i\}_{i=0}^{N}$.

Lemma 4. *Let* $\varphi \in H(M)$ *be uniformly continuous. If* φ *is stochastically stable, then* φ^{-1} *is.*

Proofs of Lemma 1~4 are omitted, since they are more or less standard.

Lemma 5. *Let* $\varphi \in H(M)$ *be uniformly continuous. Suppose* φ^k *is stochastically stable for some integer* $k > 0$. *Then* φ *is stochastically stable.*

Proof. By using Lemma 3, we see that for any $\varepsilon > 0$, there is $\varepsilon_1 > 0$ with $\varepsilon > \varepsilon_1$ such that

(i) $\{x_i\}_{i=0}^k \in Orb^{\varepsilon_1}(\varphi)$ implies $x_0 \in Tr^{\varepsilon/2}(\{x_i\}_o^k, \varphi)$, and

(ii) $d(x,y) \leq \varepsilon_1$ implies $d(\varphi^i(x), \varphi^i(y)) < \varepsilon/2$ for $i = 1,2,\ldots,k$.

For this $\varepsilon_1 > 0$, there exists $\delta_1 > 0$ such that $\{x_i\} \in Orb^{\delta_1}(\varphi^k)$ implies $Tr^{\varepsilon_1}(\{x_i\}, \varphi^k) \neq \emptyset$. For this $\delta_1 > 0$ there is $\delta > 0$ with $\varepsilon_1 > \delta$ such that $\{z_i\}_{i=0}^k \in Orb^\delta(\varphi)$ implies $z_0 \in Tr^{\delta_1}(\{z_i\}, \varphi)$. Now, we show $\{y_i\} \in Orb^\delta(\varphi)$ implies $Tr^\varepsilon(\{y_i\}, \varphi) \neq \emptyset$. Set $x_i = y_{ki}$ for $i \in Z$. Since $\{y_{ki+j}\}_{j=0}^k \in Orb^\delta(\varphi)$ implies $y_{ki} \in Tr^{\delta_1}(\{y_{ki+j}\}_{j=0}^k, \varphi)$, we get $d(\varphi^k(x_i), x_{i+1}) = d(\varphi^k(y_{ki}), y_{ki+k}) \leq \delta_1$, which means $\{x_i\} \in Orb^{\delta_1}(\varphi^k)$. Hence there is $y \in Tr^{\varepsilon_1}(\{x_i\}, \varphi^k)$, that is, we have

$$d(\varphi^{ki}(y), y_{ki}) \leq \varepsilon_1 \tag{1.1}$$

for $i \in Z$. Next, $\{y_{ki+j}\}_{j=0}^k \in Orb^\delta(\varphi) \subset Orb^{\varepsilon_1}(\varphi)$ implies $y_{ki} \in Tr^{\varepsilon/2}(\{y_{ki+j}\}_{j=0}^k, \varphi)$, namely:

$$d(\varphi^j(y_{ki}), y_{ki+j}) \leq \varepsilon/2 \tag{1.2}$$

for $j = 0,1,\ldots,k$. On the other hand (1.1) and (ii) imply

$$d(\varphi^{ki+j}(y), \varphi^j(y_{ki})) \leq \varepsilon/2 \tag{1.3}$$

for $j = 0,1,\ldots,k$. Hence (1.2) and (1.3) imply $d(\varphi^{ki+j}(y), y_{ki+j}) \leq \varepsilon$ for $j = 0,1,\ldots,k$ and $i \in Z$, from which we get $d(\varphi^i(y), y_i) \leq \varepsilon$ for $i \in Z$ and hence $y \in Tr^\varepsilon(\{y_i\}, \varphi)$. This completes the proof of Lemma 5.■

Lemma 6. *Let* $\varphi \in H(M)$, *and* $\psi \in H(M')$. *The direct product* $M \times M'$ *is a metric space by the distance function* $d((x,y),(x',y')) = Max\{d(x,x'), d(y,y')\}$ *for* $x,x' \in M$ *and* $y,y' \in M'$. *Then* $\varphi \times \psi$ *is stochastically stable if and only if,* φ *and* ψ *are stochastically stable.*

Proof omitted.

Lemma 7. *Let* $\varphi \in H(M)$, *where* M *is a differentiable manifold of dimension* $\geqslant 1$. *Assume* φ *is topologically stable. Then for any integer* $k > 0$, $M - \mathrm{Fix}(\varphi^k)$ *is dense in* M, *where* $\mathrm{Fix}(\varphi^k) = \{x \in M | \varphi^k(x) = x\}$.

Proof. Induction on k. First, we prove the lemma for $k = 1$.

To prove that $M - \mathrm{Fix}(\varphi)$ is dense in M, we assume that there is an open set $U \neq \emptyset$ such that $U \subset \mathrm{Fix}(\varphi)$. We can suppose that U is a coordinate neighborhood of a point $x_0 \in U$ with coordinate system $(x_1, \ldots, x_n)$. Take $\varepsilon_1 > 0$ such that $Q_{\varepsilon_1} \subset U$, where $Q_{\varepsilon_1} = Q_{\varepsilon_1}(x_0)$ means the cubic neighborhood with center x_0 and of breadth $2\varepsilon_1$. Take $\varepsilon > 0$ such that $4\varepsilon < \varepsilon_1$. For this $\varepsilon > 0$, we can find $\delta > 0$ with the property in Definition 1. Now, take a differentiable function α on M such that $\alpha(x) = 1$ for $x \in Q_{3\varepsilon}$, $\alpha(x) = 0$ for $x \notin Q_{4\varepsilon}$. Define a differentiable vector field Y on M by $Y(x) = \delta_1 \cdot \alpha(x) \cdot (\partial/\partial x_1)_x$ for $x \in Q_{\varepsilon_1}$ and $Y(x) = 0$ for $x \notin Q_{\varepsilon_1}$, where $\delta_1 > 0$ is a constant. Let $\{\eta_t\}$ be the one-parameter group of diffeomorphisms η_t of M generated by Y and put $\eta = \eta_1$. It is clear that if $\delta_1 < \delta$ is sufficiently small, then we have $d(\eta(x), x) < \delta$ for $x \in M$. Set $\psi = \eta \circ \varphi$, then we have $d(\varphi(x), \psi(x)) < \delta$ for $x \in M$ and hence there is a continuous map $h: M \to M$ such that $h \circ \psi = \varphi \circ h$ and $d(h(x), x) < \varepsilon$ for $x \in M$. Since $\alpha = 1$ on $Q_{3\varepsilon}$, we see that there is a sufficiently large integer $k > 0$ such that $\psi^k(x_0) = \eta^k(x_0) \notin Q_{3\varepsilon}$ and hence $h(\psi^k(x_0)) \notin Q_{2\varepsilon}$. On the other hand, since $d(h(x), x) < \varepsilon$ we have $h(x_0) \in Q_\varepsilon \subset U \subset \mathrm{Fix}(\varphi)$, and so we have $h(\psi^k(x_0)) = \varphi^k(h(x_0)) = h(x_0) \in Q_\varepsilon$, which is a contradiction. Thus we have proved the Lemma for $k = 1$.

Assume that $k \geqslant 2$ and that the Lemma is true for any $k' \leqslant k - 1$. Suppose that $M - \mathrm{Fix}(\varphi^k)$ is not dense. Then there will be a non-empty open set $U \subset \mathrm{Fix}(\varphi^k)$. Since $M - \mathrm{Fix}(\varphi^i)$ is dense in M for $i \leqslant k - 1$ there exists $x_0 \in U$ such that $\varphi^i(x_0) \neq x_0$ for any $i \leqslant k - 1$. Hence we can assume that U is a coordinate neighborhood of x_0 with coordinate system $(x_1, \ldots, x_n)$, $n = \dim M$ and that $\{\varphi^i(U)\}_{i=0}^{k-1}$ is disjoint. Take ε_1 such that $U \supset Q_{\varepsilon_1}(x_0)$, and take $\varepsilon > 0$ with $4\varepsilon < \varepsilon_1$. Since φ is topologically stable there exists a $\delta > 0$ with the property in Definition 1. For this $\delta > 0$ we can find a diffeomorphism $\eta: M \to M$ such that $\eta(U) = U$, $d(\eta(x), x) \leqslant \delta$ $(x \in M)$, $\eta(x) = x$ $(x \notin U)$ and that $\eta|_{Q_{4\varepsilon}}$ is a parallel translation along the x_1-axis as in the proof of the Lemma for $k = 1$. Define $g \in H(M)$ by $g(x) = \varphi(x)$ for $x \notin \varphi^{k-1}(U)$

and $g(x)=\eta\circ\varphi(x)$ for $x\in\varphi^{k-1}(U)$. Since $U=\varphi^k(U)$, g is in fact a homeomorphism of M and $d(g,\varphi)\leqslant\delta$ holds. Therefore, there is a continuous map $h:M\to M$ such that $h\circ g=\varphi\circ h$ and $d(h(x),x)<\varepsilon$ $(x\in M)$ holds. We see easily that $g^k(x)=\eta(x)$ for $x\in U$. Hence we can find a sufficiently large integer $m>0$ such that $g^{km}(x_0)=\eta^m(x_0)\notin Q_{3\varepsilon}$. On the other hand we get $h\circ g^{km}(x_0)=\varphi^{km}(h(x_0))=h(x_0)\in Q_\varepsilon$ since $h(x_0)\in U$ and $d(h(x_0),x_0)<\varepsilon$. Hence we have $g^{km}(x_0)\in Q_{2\varepsilon}$, which is a contradiction. This completes the proof of Lemma 7. ∎

Lemma 8. *Let* $\varphi:M\to M$ *be a homeomorphism of a manifold* M *with* $\dim M\geqslant 1$ *and suppose* $M-\mathrm{Fix}(\varphi)$ *is dense in* M. *Take and fix a constant* $\delta_1>0$ *and an integer* $k>0$. *Then for any* $\{x_i\}\in \mathrm{Orb}^{\delta_1}(\varphi)$ *and* $\varepsilon_1>0$, *there is* $\{x_i'\}\in \mathrm{Orb}^{3\delta_1}(\varphi)$ *such that* (i) $d(x_i,x_i')<\varepsilon_1$ *for* $i=0,1,\ldots,k$ *and* (ii) $X_i=\{\varphi(x_i'),x_{i+1}'\}$ $(i=0,1,\ldots,k-1)$ *are disjoint.*

Proof omitted.

Lemma 9. *Let* M *be a differentiable manifold of* $\dim M\geqslant 2$. *Let* $X_i=\{p_i,q_i\}$ $(i=1,\ldots,k)$ *be a subset of* M *consisting of at most two points* p_i *and* q_i *with* $d(p_i,q_i)<\delta$. *Suppose* $X_i\cap X_j=\emptyset$ *for* $i\neq j$. *Then there is a diffeomorphism* $\eta:M\to M$ *such that* $d(\eta(x),x)<2\pi\delta$ *for* $x\in M$ *and that* $\eta(p_i)=q_i$ *for* $i=1,2,\ldots,k$.

Proof. Lemma 13 in [7].

Lemma 10. *If* d *and* d' *are equivalent metrics on* M *then* $\varphi\in H(M)$ *is stochastically stable with respect to* d *if and only if* φ *is so with respect to* d'.

Lemma 11. *Let* φ *be a stochastically stable diffeomorphism of a compact Riemannian manifold* M. *Then* φ *is stochastically stable with respect to any Riemannian metric on* M.

Proof. Easily verified by Lemma 10.

Lemma 12. *Let* $\varphi:M\to M$ *be a homeomorphism of a metric space* M *whose bounded set is always relatively compact. Assume that for any* $\varepsilon>0$ *there exists* $\delta>0$ *such that for any integer* $k>0$ *and any* $\{x_i\}_{i=0}^k\in \mathrm{Orb}^\delta(\varphi)$ *we have* $\mathrm{Tr}^\varepsilon(\{x_i\}_0^k,\varphi)\neq\emptyset$. *Then* φ *is stochastically stable.*

Proof omitted.

2. Topological and Stochastic Stabilities

The essential part of the following theorem was independently proved by P. Walters [12].

Theorem 1. *Let* M *be a differentiable (metric) manifold of* $\dim M \geqslant 2$ *and assume that there exists* $\varepsilon_0 > 0$ *such that* ε_0*-neighborhood* $U_{\varepsilon_0}(x)$ *of any point* $x \in M$ *is relatively compact. Let* $\varphi : M \to M$ *be a topologically stable homeomorphism of* M. *Then* φ *is stochastically stable. In particular, if* M *is compact or* $M = R^n$ *is the Euclidean space then the topological stability implies the stochastic stability.*

Proof. Since φ is topologically stable, for any $\varepsilon > 0$ there is $\delta > 0$ with the property in Definition 1. We can assume $\delta < \mathrm{Min}(\varepsilon, \varepsilon_0)$.

First, we shall prove, for any $\{x_i\} \in \mathrm{Orb}^{\delta/6\pi}(\varphi)$ and any integer $k > 0$, that $\mathrm{Tr}^{2\varepsilon}(\{x_i\}_0^k, \varphi) \neq \emptyset$.

By Lemma 8 we can find $\{x_i'\} \in \mathrm{Orb}^{\delta/2\pi}(\varphi)$ such that $d(x_i, x_i') < \delta$ $(i = 0, \ldots, k)$ and that the sets $\{\varphi(x_i'), x_{i+1}'\}$ are disjoint for $i = 0, 1, \ldots, k-1$. By Lemma 9, there is a $\eta \in H(M)$ such that $d(\eta(x), x) < \delta$ for $x \in M$ and $\eta(\varphi(x_i')) = x_{i+1}'$ for $i = 0, 1, \ldots, k-1$. Put $\psi = \eta \circ \varphi$, then $d(\varphi(x), \psi(x)) < \delta$. Hence by the property for $\delta > 0$, we can find a continuous map $h : M \to M$ such that $h \circ \psi = \varphi \circ h$ and $d(h(x), x) < \varepsilon$ for $x \in M$. Put $y = h(x_0')$. Now, we have for $i = 0, 1, \ldots, k$

$$d(\varphi^i(y), x_i) = d(\varphi^i(h(x_0')), x_i) = d(h(\psi^i(x_0')), x_i)$$

$$\leqslant d(h(x_i'), x_i') + d(x_i', x_i) < \varepsilon + \delta < 2\varepsilon \quad ,$$

which shows $y \in \mathrm{Tr}^{2\varepsilon}(\{x_i'\}_0^k, \varphi)$. Thus we have proved that φ satisfies the condition in Lemma 12, which concludes that φ is stochastically stable.

Definition 4. $\varphi \in H(M)$ is called *expansive*, iff there exists $\varepsilon_0 > 0$ (called an *expansiveness constant* of φ) with the property that for any $x, y \in M$ with $x \neq y$, there is $n \in Z$ such that

$$d(\varphi^n(x), \varphi^n(y)) \geqslant \varepsilon_0 \quad .$$

Theorem 2. *Let* M *be a metric space such that every bounded set is relatively compact. Let* $\varphi : M \to M$ *be stochastically stable homeomorphism of* M. *If* φ *is expansive, then* φ *is topologically stable.*

<u>Proof</u>. For any $\varepsilon > 0$, there is $\delta > 0$ such that $\{x_i\} \in \mathrm{Orb}^{\delta}(\varphi)$ implies $\mathrm{Tr}^{\varepsilon}(\{x_i\},\varphi) \neq \emptyset$. We can assume $\varepsilon < \varepsilon_0/4$, where ε_0 is an expansiveness constant of φ. Take a $\psi \in H(M)$ with $d(\psi(x),\varphi(x)) < \delta$ for $x \in M$. We shall prove that there exists a continuous map $h: M \to M$ having the property in Definition 1.

Take a point $x \in M$. It is easy to see that $\{\psi^i(x)\}_{i\in Z} \in \mathrm{Orb}^{\delta}(\varphi)$. Hence there is $y \in \mathrm{Tr}^{\varepsilon}(\{\psi^i(x)\},\varphi)$, i.e.,

$$d(\varphi^i(y),\psi^i(x)) \leqslant \varepsilon$$

for $i \in Z$. If $y' \in \mathrm{Tr}^{\varepsilon}(\{\psi^i(x)\},\varphi)$, then we have

$$d(\varphi^i(y),\varphi^i(y')) \leqslant d(\varphi^i(y),\psi^i(x)) + d(\psi^i(x),\varphi^i(y')) \leqslant 2\varepsilon < \varepsilon_0 \quad ,$$

which implies $y = y'$. Thus by putting $h(x) = y$, we get a well-defined map $h: M \to M$ with the property

$$d(\varphi^i(h(x)),\psi^i(x)) \leqslant \varepsilon \qquad \text{for} \quad i \in Z . \tag{1.4}$$

Putting $i = 0$ in (1.4) we get

$$d(h(x),x) \leqslant \varepsilon \qquad \text{for} \quad x \in M . \tag{1.5}$$

Next, we have, again by (1.4) for x and $\psi(x)$,

$$\begin{aligned} d(\varphi^i(\varphi(h(x))),\varphi^i(h(\psi(x))) &\leqslant d(\varphi^i(\varphi(h(x))),\psi^i(\psi(x))) \\ &+ d(\varphi^i(h(\psi(x))),\psi^i(\psi(x))) \leqslant 2\varepsilon < \varepsilon_0 \end{aligned}$$

for every $i \in Z$, which implies

$$\varphi(h(x)) \quad = \quad h(\psi(x)) \quad . \tag{1.6}$$

Finally we shall prove the continuity of h. Assume that h is not continuous at $x_0 \in M$. Then, there is a sequence $x_\nu \to x_0$ $(\nu \to \infty)$ such that $y_\nu = h(x_\nu)$ does not tend to $y_0 = h(x_0)$ as $\nu \to \infty$. Since $\{x_\nu\}$ is bounded and $d(h(x_\nu),x_\nu) \leqslant \varepsilon$ for $\nu > 0$, the set $\{h(x_\nu)\}$ is also bounded. Hence we can assume, by taking a subsequence if necessary, that $y_\nu \to y'_0 \neq y_0$ $(\nu\to\infty)$. Since φ is expansive, there is $k \in Z$ such that $d(\varphi^k(y'_0),\varphi^k(y_0)) \geq \varepsilon_0$.

Fixing k, we can find $\nu_0 > 0$ such that

$$d(\varphi^k(y_\nu),\varphi^k(y'_0)) < \varepsilon_0/4 \tag{1.7}$$

for $\nu \geq \nu_0$, since φ^k is continuous and $y_\nu \to y'_0$ $(\nu\to\infty)$. We can assume

$$d(\psi^k(x_\nu),\psi^k(x_0)) < \varepsilon \tag{1.8}$$

for $\nu \geq \nu_0$, since ψ^k is continuous and $x_\nu \to x_0$ $(\nu\to\infty)$. Now, we have, using (1.5) and (1.8),

$$\begin{aligned} d(\varphi^k(y_\nu),\varphi^k(y_0)) &= d(\varphi^k(h(x_\nu)),\varphi^k(h(x_0))) \\ &= d(h(\psi^k(x_\nu)),h(\psi^k(x_0))) \leq d(h(\psi^k(x_\nu)),\psi^k(x_\nu)) \\ &\quad + d(\psi^k(x_\nu),\psi^k(x_0)) + d(\psi^k(x_0),h(\psi^k(x_0))) < 3\varepsilon \end{aligned}$$

and hence by (1.7) we obtain

$$\begin{aligned} \varepsilon_0 \leq d(\varphi^k(y'_0),\varphi^k(y_0)) &\leq d(\varphi^k(y'_0),\varphi^k(y_\nu)) \\ &\quad + d(\varphi^k(y_\nu),\varphi^k(y_0)) \leq \varepsilon_0/4 + 3\varepsilon < \varepsilon_0 \quad , \end{aligned}$$

which is a contradiction. This completes the proof of Theorem 2. ∎

3. Stochastic Stability of Linear Automorphisms

In this section we shall characterize affine transformations of R^n to be stochastically stable.

Proposition 1. *Let* $\varphi: R^n \to R^n$ *be a linear automorphism of* R^n. *Then* φ *is stochastically stable if and only if* φ *is hyperbolic, i.e., if* λ *is an eigenvalue of* φ *then* $|\lambda| \neq 1$.

Proof. Assume φ is stochastically stable. Consider the complexification $\varphi^C: C^n \to C^n$. Identifying C^n with $R^n \times R^n$, we can identify φ^C with $\varphi\times\varphi$. By virtue of Lemmas 6 and 11, φ is stochastically stable if and only if φ^C is. Since a linear map is uniformly continuous, it follows from Lemma 1 that φ^C is stochastically stable if and only if every factor of the Jordan canonical form of φ^C is stochastically stable. Now it suffices to show that if

$$\psi = \begin{pmatrix} \lambda & 1 & & \\ & \lambda & 1 & \\ & & \ddots & \ddots \\ & & & \ddots & 1 \\ & & & & \lambda \end{pmatrix} : C^n \to C^n \qquad (\text{resp. } \psi_0 = \lambda \cdot 1_{C^1})$$

is stochastically stable, then $|\lambda| \neq 1$. Suppose $|\lambda| = 1$. Set $z_j = j \cdot \lambda^j \cdot \delta$ for $j \in Z$. Since

$$d(\psi_0(z_j), z_{j+1}) = |\psi_0(z_j) - z_{j+1}| = |j \cdot \lambda^{j+1}\delta - (j+1)\lambda^{j+1}\delta| = \delta ,$$

we have $\{z_j\} \in \mathrm{Orb}^\delta(\psi_0)$. However, since

$$d(\psi_0^n(\xi), z_n) = |\lambda^n \xi - \lambda^n n\delta| = |\xi - n\delta| ,$$

there is no ξ such that $d(\psi_0^n(\xi), z_n)$ is bounded for any small $\delta > 0$. In particular, for any $\delta > 0$ we have $\mathrm{Tr}^1(\{z_j\}, \psi_0) = \emptyset$. Hence ψ_0 is not stochastically stable.

Similarly, considering the vector $v_j = (0, \ldots, 0, z_j)$ for $j \in Z$, we see that $\{v_j\} \in \mathrm{Orb}^\delta(\psi)$ and that $\mathrm{Tr}^1(\{v_j\}, \psi) = \emptyset$, which means ψ is not stochastically stable. Thus we have proved that ψ is hyperbolic.

Conversely, assume that φ is hyperbolic. Then it is well-known that there are subspaces E^s and E^u of R^n and constants $C > 0$, $0 < \lambda < 1$ such that

(i) $R^n = E^s \oplus E^u$

(ii) $\varphi(E^\sigma) = E^\sigma$, $\sigma = s, u$,

(iii) $\|\varphi^n v\| \leqslant C\lambda^n \|v\|$, $\quad v \in E^s$

$\|\varphi^{-n} w\| \leqslant C\lambda^n \|w\|$, $\quad w \in E^u$

for $n \geqslant 0$. Set $\psi = \varphi|_{E^s}$ and $\eta = \varphi|_{E^u}$, then identifying R^n with $E^s \times E^u$ we can identify φ with $\psi \times \eta$. By virtue of Lemma 6, it suffices to show that ψ and η are stochastically stable.

First consider $\eta : E^u \to E^u$ and take $\varepsilon > 0$. Put $\delta = (1 - \lambda)\varepsilon / C$. We assert that $\{x_i\} \in \mathrm{Orb}^\delta(\eta)$ implies $\mathrm{Tr}^\varepsilon(\{x_i\}, \eta) \neq \emptyset$. For $k \in Z$ we set $\alpha_k = x_{k+1} - \eta(x_k) \in E^u$. Then we have $\|\alpha_k\| \leqslant \delta$ for $k \in Z$. By induction we see that for $k > 0$

$$x_k = \eta^k(x_0) + \eta^{k-1}(\alpha_0) + \eta^{k-2}(\alpha_1) + \cdots + \alpha_{k-1}$$

holds. Put $\xi = \eta^{-1}$. Then we have $\|\xi^k\| \leqslant c\lambda^k$ for $k \geqslant 0$ (cf. (iii)). We have also:

$$\begin{aligned} x_k &= \eta^k(x_0 + \xi(\alpha_0) + \xi^2(\alpha_1) + \cdots + \xi^k(\alpha_{k-1})) \\ &= \eta^k(x_0 + v_k) \end{aligned},$$

where we put $v_k = \xi(\alpha_0) + \xi^2(\alpha_1) + \cdots + \xi^k(\alpha_{k-1})$ for $k > 0$.

We shall show that $\{v_k\}_{k=1}^{\infty}$ is a Cauchy sequence. In fact, for any $p > k > 0$, we have

$$\begin{aligned} \left\| v_p - v_k \right\| &= \left\| \sum_{i=k+1}^{p} \xi^i(\alpha_{i-1}) \right\| \leqslant \sum_{i=k+1}^{p} \left\| \varphi^{-i}(\alpha_{i-1}) \right\| \\ &\leqslant c \cdot \sum_{i=k+1}^{p} \lambda^i \|\alpha_{i-1}\| \leqslant c \cdot \delta \cdot \lambda^{k+1}/(1-\lambda) \to 0 \qquad (k \to \infty) . \end{aligned}$$

Hence there is $\beta \in E^u$ such that $\lim_{k\to\infty} v_k = \beta$. Put $y = x_0 + \beta$. Then we have

$$\begin{aligned} \eta^k(y) - x_k &= \eta^k(x_0 + \beta - x_o - v_k) \\ &= \eta^k\left(\sum_{i=k+1}^{\infty} \xi^i(\alpha_{i-1}) \right) = \sum_{i=k+1}^{\infty} \xi^{i-k}(\alpha_{i-1}) \end{aligned},$$

and hence we have

$$d(\eta^k(y), x_k) = \left\| \eta^k(y) - x_k \right\| \leqslant \sum \left\| \xi^{i-k} \right\| \delta \leqslant c\delta/(1-\lambda) = \varepsilon.$$

By Lemma 12, we see that η is stochastically stable. Similarly, we conclude that ψ^{-1} is stochastically stable. By Lemma 4, ψ is also stochastically stable. Thus we have proved that φ is stochastically stable.

Theorem 3. *Let* $\varphi: R^n \to R^n$ *be a linear automorphism. Then the following conditions are mutually equivalent.*

(1) φ *is hyperbolic*

(2) φ *is expansive*

(3) φ *is structurally stable*

(4) φ *is stochastically stable*

(5) φ *is topologically stable.*

Proof. Equivalence (1) $\leftrightarrow$ (3) was proved by Hartman (see Theorem 2.3 [10] for details).

(1) $\leftrightarrow$ (2) is standard.

(1) $\leftrightarrow$ (4) is by Proposition 1.

(5) $\to$ (4) for $n \geqslant 2$ is by Theorem 1. For $n = 1$, $\varphi : R^1 \to R^1$ is given by $\varphi(x) = \lambda \cdot x$ for some $\lambda \neq 0$. If φ is topologically stable, then $\lambda \neq \pm 1$. For if $\lambda = \pm 1$, then $\varphi^2 = 1_R$ and $Fix(\varphi^2) = R^1$, which contradicts Lemma 7. Thus $|\lambda| \neq 1$, which means φ is hyperbolic and so stochastically stable. Finally (4) $\to$ (5), since (4) $\to$ (2) and so we can apply Theorem 2. This completes the proof of Theorem 3. ■

Proposition 2. *Let* $f : R^n \to R^n$ *be a linear automorphism and* $\xi \in R^n$ *a fixed vector. Define the affine transformation* $\varphi : R^n \to R^n$ *by*

$$\varphi(x) = f(x) + \xi$$

for $x \in R^n$. *Then* φ *is stochastically stable if and only if* f *is.*

Proof. Let $\{x_i\} \in Orb^\delta(\varphi)$. Put

$$x_i' = x_i - (f^{i-1}(\xi) + f^{i-2}(\xi) + \cdots + \xi)$$

for $i \in Z$. We see that $\{x_i'\} \in Orb^\delta(f)$. It is easy to verify that $\{x_i\} \to \{x_i'\}$ is a one-one correspondence between $Orb^\delta(\varphi)$ and $Orb^\delta(f)$ and that $Tr^\varepsilon(\{x_i\},\varphi) = Tr^\varepsilon(\{x_i'\},f)$ for every $\varepsilon > 0$. Thus φ is stochastically stable if and only if f is.

4. Stochastic Stability of Toral Automorphisms

In this section we shall characterize group automorphisms of the torus $T^n = R^n/Z^n$ to be stochastically stable.

Proposition 3. *Let* M *and* $\tilde{M}$ *be metric spaces and* $\pi : \tilde{M} \to M$ *be a locally isometric covering map of* $\tilde{M}$ *onto* M. *Assume that* M *is compact and that every* ε*-neighborhood* $U_\varepsilon(x)$ *of* $x \in M$ *is connected for small* $\varepsilon > 0$. *Let* $f \in H(\tilde{M})$ *and* $\varphi \in H(M)$ *such that*

$\pi \circ f = \varphi \circ \pi$. *Then,* f *is stochastically stable if and only if* φ *is.*

Proof. Since M is compact, and π is locally isometric covering map, it is easy to find a $\delta_0 > 0$ with the property that for any $x \in M$, $\pi|_{U_{\delta_0}(x)} : U_{\delta_0}(x) \to U_{\delta_0}(\pi(x))$ is an isometry, where $U_\varepsilon(x)$ denotes the ε-neighborhood of x and that $U_{\delta_0}(x)$ is connected.

First, we shall prove that f is uniformly continuous on $\tilde{M}$.

For any $\varepsilon > 0$, we must find $\delta > 0$ such that $d(p,q) < \delta, p,q \in M$ imply $d(f(p),f(q)) < \varepsilon$. We can assume $\varepsilon < \delta_0$. Since M is compact, we can find $\delta > 0$ such that $\delta < \delta_0$ and that $x,y \in M$, $d(x,y) < \delta$ imply $d(\varphi(x),\varphi(y)) < \varepsilon$. Now take $p,q \in \tilde{M}$ with $d(p,q) < \delta$. Since $\delta < \delta_0$, we have $d(\pi(p),\pi(q)) = d(p,q) < \delta$, and hence we get $d(\pi f p, \pi f q) = d(\varphi\pi p, \varphi\pi q) < \varepsilon < \delta_0$. We denote by W the connected component of $\pi^{-1}(U_{\delta_0}(\varphi\pi p))$ containing $f(p)$. Then $\pi : W \to U_{\delta_0}(\varphi\pi p)$ is a homeomorphism. The set $f(U_\delta(p))$ is connected and is contained in W since $\pi f(U_\delta(p)) \subset \varphi\pi U_\delta(p) \subset \varphi U_\delta(\pi p) \subset U_\varepsilon(\varphi\pi p) \subset U_{\delta_0}(\varphi\pi p)$. Since $f(q) \in f(U_\delta(p)) \subset W$ by our assumption, we see that $W = U_{\delta_0}(f(p))$ and that $\pi : W \to U_{\delta_0}(\varphi\pi p)$ is an isometry. Therefore, we have $d(fq,fp) = d(\pi f q, \pi f p) < \varepsilon$. Thus we have proved the uniform continuity of f.

Now, suppose f is stochastically stable. For any $\varepsilon > 0$, there is $\delta > 0$ such that $\{p_i\} \in \mathrm{Orb}^\delta(f)$ implies $\mathrm{Tr}^\varepsilon(\{p_i\},f) \neq \emptyset$.

Take $\{y_i\} \in \mathrm{Orb}^\delta(\varphi)$. There is $p_0 \in \tilde{M}$ such that $\pi(p_0) = y_0$. We shall define p_i $(i > 0)$ by induction. Consider $f(p_i)$, then we have

$$\pi f(p_i) = \varphi\pi p_i = \varphi(y_i) .$$

Since $y_{i+1} \in U_\delta(\varphi(y_i))$, there exists uniquely a point $p_{i+1} \in U_\delta(f(p_i))$ such that $\pi(p_{i+1}) = y_{i+1}$. Thus we have defined $\{p_i\}_{i \geq 0}$. Clearly $\{p_i\}_{i\geq 0} \in \mathrm{Orb}^\delta(f)$. We can find $y \in \mathrm{Tr}^\varepsilon(\{p_i\}_{i\geq 0}, f)$, i.e.,

$$d(f^i(y),p_i) \leqslant \varepsilon < \delta_0$$

holds for $i \geqslant 0$. Put $z = \pi(y)$. Since $d(f^i(y),p_i) < \delta_0$, we have

$$d(\varphi^i(z),y_i) = d(\varphi^i \pi y, y_i) = d(\pi f^i(y), \pi(p_i)) = d(f^i(y),p_i) \leqslant \varepsilon$$

for $i \geqslant 0$, which means $z \in \mathrm{Tr}^\varepsilon(\{y_i\},\varphi)$. By Lemma 12, φ is stochastically stable.

Conversely, suppose φ is stochastically stable. For any $\varepsilon > 0$, there is $\delta > 0$ such that $\{y_i\} \in \mathrm{Orb}^{\delta}(\varphi)$ implies $\mathrm{Tr}^{\varepsilon}(\{y_i\},\varphi) \neq \emptyset$. We can assume $\varepsilon < \delta_0/2$ and $\delta < \delta_0/2$. Moreover, since f is uniformly continuous, we can assume that $p,q \in \tilde{M}$, $d(p,q) < \varepsilon$ imply $d(fp,fq) < \delta_0/2$. Take $\{x_i\} \in \mathrm{Orb}^{\delta}(f)$ and put $y_i = \pi(x_i)$ for $i \in Z$.

We assert that $\{y_i\} \in \mathrm{Orb}^{\delta}(\varphi)$. In fact, we have $d(\varphi(y_i),y_{i+1}) = d(\pi f(x_i),\pi(x_{i+1}))$. Since $d(f(x_i),x_{i+1}) \leqslant \delta < \delta_0$, and π is locally isometric we have $d(\varphi(y_i),y_{i+1}) = d(f(x_i),x_{i+1}) \leqslant \delta$, which means $\{y_i\} \in \mathrm{Orb}^{\delta}(\varphi)$. There is $z \in \mathrm{Tr}^{\varepsilon}(\{y_i\},\varphi)$. We have, in particular, $d(z,y_0) \leqslant \varepsilon < \delta_0$, and hence we can find $y \in \tilde{M}$ such that $\pi(y) = z$ and $y \in U_{\delta_0}(x_0)$. We shall show $y \in \mathrm{Tr}^{\varepsilon}(\{x_i\},f)$. It suffices to show

$$d(f^i(y),x_i) \leqslant \varepsilon \tag{4.1}$$

for $i \geqslant 0$. First, we have

$$d(\pi f^i(y),\pi x_i) = d(\varphi^i \pi(y),y_i) = d(\varphi^i(z),y_i) \leqslant \varepsilon .$$

We prove (4.1) by induction on $i \geqslant 0$. To do this, assume $d(f^{i-1}(y), x_{i-1}) \leqslant \varepsilon$. Then we have $d(ff^{i-1}(y),f(x_{i-1})) < \delta_0/2$, and $d(f(x_{i-1}), x_i) \leqslant \delta_0/2$ and hence $d(f^i(y),x_i) \leqslant \delta_0$. Since $\pi(f^i(y)) = \varphi^i(z)$ and $\pi(x_i) = y_i$ and since π is locally isometric, we get $d(f^i(y),x_i) \leqslant \varepsilon$. Therefore, we have $y \in \mathrm{Tr}^{\varepsilon}(\{x_i\}_{i \geqslant 0},f)$. By Lemma 12, we see f is stochastically stable.

Lemma 13. *Let* f *be a linear automorphism and* $\varphi : T^n \to T^n$ *be a group automorphism of* T^n *such that* $\pi \circ f = \varphi \circ \pi$, *where* $\pi : R^n \to T^n = R^n/Z^n$ *is the projection. Then* f *is expansive if* φ *is.*

Proof. There is $\varepsilon_0 > 0$ such that if $x,y \in T^n$, $x \neq y$, then there is $n \in Z$ such that $d(\varphi^n(x),\varphi^n(y)) \geqslant \varepsilon_0$. Since $\pi : R^n \to T^n$ satisfies the condition for $\pi : \tilde{M} \to M$ in Proposition 3, we can find $\delta_0 > 0$ satisfying the property in Proposition 3. We can assume $\varepsilon_0 \leqslant \delta_0$, $\varepsilon_0 \leqslant 1$. Take $p,q \in R^n$ with $p \neq q$. In the case $\pi p \neq \pi q$, there is $n \in Z$ such that $d(\varphi^n \pi p,\varphi^n \pi q) \geqslant \varepsilon_0$. We have now

$$d(f^n p,f^n q) \geqslant d(\pi f^n p,\pi f^n q) = d(\varphi^n \pi p,\varphi^n \pi q) \geqslant \varepsilon_0 \quad .$$

In the case $\pi p = \pi q$, we have $d(p,q) \geqslant 1 \geqslant \varepsilon_0$. Thus, f is expansive.

Theorem 4. *Let* $\varphi: T^n \to T^n$ *be a group automorphism of the torus* T^n. *Then the following conditions are mutually equivalent:*

(1) φ *is an Anosov diffeomorphism,*

(2) φ *is expansive,*

(3) φ *is structurally stable,*

(4) φ *is stochastically stable,*

(5) φ *is topologically stable,*

(6) φ *satisfies Axiom* A *and the strong transversality condition.*

Proof. (1) → (5) is proved by Walters [14].

(5) → (4) is proved by Theorem 1 for case $n \geqslant 2$. In case $n = 1$, if $\varphi: T^1 \to T^1$ is a group automorphism $\varphi^2 = 1_{T^1}$ and so φ is not topologically stable by Lemma 7.

To prove (4) → (1), we denote by $f: R^n \to R^n$ the linear automorphism covering φ, i.e., $\pi \circ f = \varphi \circ \pi$. Since φ is stochastically stable, f is also so by Proposition 3. Hence by Theorem 3, f is hyperbolic. Then φ is clearly an Anosov diffeomorphism.

(1) → (3) is proved by Anosov [1].

(3) → (1), since $T_0\varphi$ (the differential of φ at the neutral element of T^n) is hyperbolic by a result of Franks [3], and hence is an Anosov diffeomorphism.

(1) → (2) is proved also by Anosov [1].

(2) → (1), since f is expansive by Lemma 12, and hence f is hyperbolic by Theorem 3 and so φ is Anosov.

(1) → (6) is verified by the very definition and a result of Anosov [1].

(6) → (3) is proved by Robbin [11].

This completes the proof of Theorem 4. ∎

5. Isometries of Riemannian Manifolds and Group Automorphisms of Compact Connected Lie Groups

In this section we shall prove that any isometry of a connected Riemannian manifold M of $\dim M \geqslant 1$ is not stochastically stable and using this we shall give the result mainly due to H. Urakawa, which says that if there is a stochastically stable group automorphism of a compact connected Lie group G, then G is a torus.

Proposition 4. *Let* (M,d) *be a metric space such that the identity map* 1_M *of* M *is not stochastically stable. Let* $\varphi: M \to M$

be a homeomorphism such that the set $\{\varphi^i \mid i \in Z\}$ *is uniformly equicontinuous, i.e., for any* $\varepsilon > 0$ *there is* $\delta > 0$ *such that* $x, y \in M$, $d(x,y) < \delta$ *imply* $d(\varphi^i(x), \varphi^i(y)) < \varepsilon$ *for every* $i \in Z$. *Then,* φ *is not stochastically stable.*

Proof. Since 1_M is not stochastically stable, there is $\varepsilon > 0$ such that for any $\delta > 0$ there is $\{x_i\} \in \mathrm{Orb}^\delta(1_M)$, which cannot be ε-traced by 1_M. By the assumption for φ, for the above ε, there is $\varepsilon_1 > 0$ such that $d(x,y) < \varepsilon_1$ implies $d(\varphi^i(x), \varphi^i(y)) < \varepsilon$ for any $i \in Z$. We shall prove that for any $\delta > 0$, there is $\{z_i\} \in \mathrm{Orb}^\delta(\varphi)$ such that $\{z_i\}$ cannot be ε_1-traced by φ, which proves the Proposition. Now, for $\delta > 0$, there is $\eta > 0$ such that $d(x,y) < \eta$ implies $d(\varphi^i(x), \varphi^i(y)) < \delta$ for any $i \in Z$. For this η, there is $\{y_i\} \in \mathrm{Orb}^\eta(1_M)$ such that $\{y_i\}$ cannot be ε-traced by 1_M. Put $z_i = \varphi^i(y_i)$ for $i \in Z$. Then, $\{z_i\} \in \mathrm{Orb}^\delta(\varphi)$, since we have $d(\varphi(z_i), z_{i+1}) = d(\varphi^{i+1}(y_i), \varphi^{i+1}(y_{i+1})) < \delta$. We assert that $\{z_i\}$ cannot be ε_1-traced by φ. For, if there were $w \in M$ such that $\{z_i\}$ is ε_1-traced by w with respect to φ, we would have $d(\varphi^i(w), z_i) < \varepsilon_1$ $(i \in Z)$, hence $d(w, y_i) = d(\varphi^{-i}(\varphi^i(w)), \varphi^{-i}(z_i)) < \varepsilon$, which means that $\{y_i\}$ is ε-traced by w with respect to 1_M, contradicting to the property of $\{y_i\}$. This completes the proof of Proposition 4. ∎

Corollary 1. *Let* $\varphi: M \to M$ *be an isometry of a Riemannian manifold* M *of dimension* ≥ 1. *Then,* φ *is not stochastically stable.*

Proof. We can easily see that 1_M is not stochastically stable. Hence the Corollary follows from Proposition 4.

Corollary 2. *For any compact connected manifold* M *of* $\dim M \geq 1$ *there is a nonempty open set in* $\mathrm{Diff}^1(M)$ *with* C^1*-topology with the property that for any* $\varphi \in U$*, any integer* $k > 0$ *and any Riemannian metric* g *on* M*,* φ^k *is not an isometry of* g.

Proof. Let U be the set of all Morse-Smale diffeomorphisms of M. We know that U is a nonempty open set in $\mathrm{Diff}^1(M)$ and that any $\varphi \in U$ is topologically stable (cf. [6],[12]). If $\dim M \geq 2$, then by Theorem 1 and Lemma 2 we see that φ^k is stochastically stable and hence, by Corollary 1, φ^k is not an isometry of g. If $\dim M = 1$, then, consider $\varphi \times \varphi$, which is a Morse-Smale diffeomorphism of $M \times M$ and so we can prove by Lemma 6 that φ^k is not an isometry of g. This completes the proof of Corollary 2. ∎

Theorem 5. *Let* G *be a compact connected Lie group. Suppose that there is a group automorphism* $\varphi: G \to G$, *which is stochastically stable with respect to some (and so any) Riemannian metric on* G. *Then,* G *is isomorphic to a torus.*

Proof. Let A (resp. S) be the maximal abelian (resp. semi-simple) normal subgroup of G, and set $Z = A \cap S$. Then we know (cf. [4]) that $G = A \cdot S$ and Z is a finite group. It is well-known that $\varphi(A) = A$ and $\varphi(S) = S$. Put $\xi = \varphi|_A$ and $\eta = \varphi|_S$. Since $\pi: A \times S \to G$ defined by $\pi(a,x) = a \cdot x$ for $a \in A$, $x \in S$ is a (finite) covering map and since $\pi \circ (\xi \times \eta) = \varphi \circ \pi$, we see, by Proposition 3, that $\xi \times \eta$ and hence by Lemma 6, η is stochastically stable. Since η is an automorphism of S, η leaves invariant the Killing form β of the Lie algebra of the Lie group S, which is negative definite and so η is an isometry of the invariant Riemannian metric on S induced by $-\beta$. By Corollary 1 $\dim S = 0$, and hence $G = A$ is a torus.

6. References

[1] D.V. Anosov, "Geodesic flows on closed Riemannian manifolds of negative curvature," *Proc. Steklov Inst. Math.* 90 (1967).

[2] R. Bowen, "ω-limit sets for Axiom A diffeomorphisms," *J. Diff. Eq.* 18 (1975), 333-339.

[3] J. Franks, "Necessary conditions for stability of diffeomorphisms," *Trans. Amer. Math. Soc.* 158 (1971), 301-308.

[4] G. Hochschild, *The Structure of Lie Groups*, Holden-Day, 1965.

[5] K. Kato, "Stochastic stability of Anosov diffeomorphisms," *Nagoya Math. J.* 69 (1978), 121-129.

[6] Z. Nitecki, "On semi-stability for diffeomorphisms," *Inv. Math.* 14 (1971), 83-122.

[7] Z. Nitecki-M. Shub, "Filtrations, decompositions and explosions," *Amer. J. Math.* 97 (1976), 1029-1047.

[8] J. Robbin, "Topological conjugacies and structural stabilities for discrete dynamical systems," *Bull. Amer. Math. Soc.* 78 (1972), 923-952.

[9] J. Robbin, "A structural stability theorem," *Ann. of Math.* 94 (1971), 447-493.

[10] S. Smale, "Differentiable dynamical systems," *Bull. Amer. Math. Soc.* 73 (1967), 747-817.

[11] P. Walters, "Anosov diffeomorphisms are topologically stable," *Topology* 9 (1970), 71-78.

[12] P. Walters, "On the pseudo-orbit tracing property and its relationship to stability," Springer Lect. Notes. 668 (1979), 231-244.

Nagoya University
Chigusaku, Nagoya 464
Japan

(Received December 24, 1980)

A NOTE ON COHOMOLOGY GROUPS OF HOLOMORPHIC LINE BUNDLES OVER A COMPLEX TORUS

Shingo Murakami[1]

Let $E = V/L$ be an n-dimensional complex torus, where V is an n-dimensional complex vector space and L a lattice of V. Let F be a holomorphic line bundle over E. In this note, we shall show that the q-th cohomology group $H^q(E,\underline{F})$ $(q \geqq 0)$ of E with coefficients in the sheaf $\underline{F}$ of germs of holomorphic sections of F can be completely determined by applying harmonic theory. The results have been obtained by Mumford [3] and Kempf [1] by an algebraico-geometric way, and later Umemura [4] and Matsushima [2] showed that harmonic theory can be used to get the results provided that the line bundle F has nondegenerate Chern class. Our purpose is to note that the method found by Umemura and Matsushima may be modified so as to prove a structure theorem in the general case.

1. Statement of Theorem

First let us recall some known results on complex line bundles over a complex torus. Let $E = V/L$ be as in the introduction. A factor $J: L \times V \to \mathbb{C}$ is a map such that $J(g,v)$ is holomorphic in $v \in V$ for each $g \in L$ and that $J(g+h,v) = J(g,v+h)J(h,v)$ for $g,h \in L$ and $v \in V$. A factor J defines an action of L on $V \times \mathbb{C}$ by the rule

$$(v,\xi)g = (v+g, J(g,v)\xi)$$

where $(v,\xi) \in L \times \mathbb{C}$ and $g \in L$. The action of L on $V \times \mathbb{C}$ is free and the quotient of $V \times \mathbb{C}$ by L has a natural structure of holomorphic line bundle over the complex torus $E = V/L$.

The following is known as theorem of Appell-Humbert (cf. [3, Chap. 1]).

<u>Proposition 1</u>. *Any holomorphic line bundle* F *over the complex torus* $E = V/L$ *is defined by one and only one factor* $J: L \times V \to \mathbb{C}$ *of the following form.*

$$J(g,v) = \psi(g)\exp\pi\left\{H(v,g) + \frac{1}{2}H(g,g)\right\}, \tag{1.1}$$

where $(g,v) \in L \times V$, H *is a hermitian form on* V *such that* $A = \mathrm{Im}\, H$, *the imaginary part of* H, *takes integral values on* $L \times L$, *and* ψ *is a semi-character of* L *with respect to* A, *namely, a map* $\psi: L \to \mathbb{C}_1 = \{z \in \mathbb{C}; |z| = 1\}$ *such that*

$$\psi(g+h) = \psi(g)\psi(h)\exp\pi i A(g,h)$$

for $g, h \in L$.

The hermitian form H represents the Chern class of F.

The notation being as in Proposition 1, assume that a line bundle F on E be given. Let V_0 be the kernel of the hermitian form H, namely $V_0 = \{v \in V; H(v,V) = 0\}$. Put $L_0 = L \cap V_0$, $\hat{V} = V/V_0$ and let $\hat{L}$ be the image of L under the projection $V \to \hat{V}$. We know that L_0 (resp. $\hat{L}$) is a lattice in V_0 (resp. $\hat{V}$) (cf. [5, Chap. 6]). Put $\hat{E} = \hat{V}/\hat{L}$ and $E_0 = V_0/L_0$. Since V_0 coincides with the kernel of the skew-symmetric form A, the semi-character ψ defines a character ψ_0 of L_0.

Suppose that the character ψ_0 is trivial. In this case, it is easy to verify that the value of $J(g,v)$ depends only on the cosets g mod. L_0 and v mod. V_0. Therefore, J defines a factor $\hat{J}: \hat{L} \times \hat{V} \to \mathbb{C}$ of the form (1.1) for which the hermitian form is the nondegenerate hermitian form $\hat{H}$ on $\hat{V}$ induced from H on V. We shall denote by $\hat{F}$ the holomorphic line bundle over the complex torus $\hat{E}$ defined by $\hat{J}$.

Returning to the general case, since F is defined by the factor J, the vector space $H^0(E,\underline{F})$ of all holomorphic sections of F is canonically isomorphic to the space of all holomorphic theta functions on V for the factor J, i.e., holomorphic functions f on V such that

$$f(v+g) = J(g,v)f(v)$$

for $v \in V$ and $g \in L$. According to the classical theory of theta functions (cf. [3, Chap. 1], [5, Chap. 6]), we have

Proposition 2. *The notation being as above,*

(i) *If* $H^0(E,\underline{F}) \neq (0)$, *then the hermitian form* H *is positive semi-definite.*

(ii) *Suppose* H *be positive semi-definite.*

(a) *If the character* ψ_0 *of* L_0 *is nontrivial, then* $H^0(E,\underline{F}) = (0)$.

(b) *If the character* ψ_0 *is trivial, then every holomorphic theta function on* V *for the factor* J *is induced from a holomorphic theta function on* $\hat{V}$ *for the factor* $\hat{J}$ *by the projection* $V \to \hat{V}$, *and so we obtain the isomorphism* $H^0(E,\underline{F}) \cong H^0(\hat{E},\hat{\underline{F}})$. *Moreover,* $\dim H^0(E,\underline{F})$ *is equal to the reduced Phaffian of* A *relative to* L.

Now we state

Theorem. *The notation being as above, let* r *(resp.* s*) be the number of negative (resp. positive) eigenvalues of the hermitian form* H.

(i) *If the character* ψ_0 *of* L_0 *defined by the semi-character* ψ *is nontrivial, then* $H^q(E,\underline{F}) = (0)$ *for all* $q \geqq 0$.

(ii) *If the character* ψ_0 *is trivial, then*

$$H^q(E,\underline{F}) = (0) \qquad \text{for} \quad q < r \quad \text{or} \quad q > n-s \quad ,$$

$$H^q(E,\underline{F}) = H^r(E,\underline{F}) \otimes H^{q-r}(E_0,\underline{C}) \qquad \text{for} \quad r \leqq q \leqq n-s \quad ,$$

where $\underline{C}$ *denotes the trivial line bundle over* E_0. *Moreover,* $H^r(E,\underline{F}) \cong H^r(\hat{E},\hat{\underline{F}})$ *and* $\dim H^r(E,\underline{F})$ *is equal to the reduced Phaffian of* A *relative to* L.

This theorem may be considered as an analytic version of a theorem of Kempf [1].

2. Harmonic Forms

We retain the notation introduced in the previous section. For the later use we give a survey of the harmonic theory for holomorphic line bundles over a complex torus. We refer to [2, Appendix 2] for details.

Let $A^q(E,F)$ be the vector space of all F-valued differential forms of type $(0,q)$ on E. Then the direct sum $A(E,F) = \Sigma_q A^q(E,F)$ forms a complex with coboundary operator d''. The cohomology group $H^q(E,\underline{F})$ is isomorphic, *via* the Dolbeault's isomorphism, to the q-th cohomology group $H^{0,q}(E,F)$ of this complex.

We apply harmonic theory to study the cohomology groups $H^{0,q}(E,F)$ as in [4] and [2]. In general, a hermitian metric on E, together with a system of hermitian metrics in the fibers of F, defines canonically a hermitian inner product $(\, , \,)$ in the vector space $A(E,F)$, the operator δ adjoint to d'' and the Laplacian $\square = d''\delta + \delta d''$. A form $\varphi \in A(E,F)$ is said to be harmonic if $\square\varphi = 0$. Let $\mathcal{H}^q(E,F)$ be the vector space of all harmonic forms belonging to $A^q(E,F)$. The harmonic theory asserts that the cohomology group $H^{0,q}(E,F)$ is isomorphic to $\mathcal{H}^q(E,F)$. So we have

$$H^q(E,\underline{F}) \cong \mathcal{H}^q(E,F) \quad . \tag{2.1}$$

Let now G be a positive definite Hermitian form on the complex vector space V. Then G defines in an obvious way a hermitian metric on the complex torus E. On the other hand, we shall make use of the system of hermitian metrics in the fibers of F defined as follows. The positive function $\exp - \pi H(v,v)$ on V defines in an obvious way hermitian metrics in the fibers of the trivial line bundle $V \times \mathbb{C}$ over V, and this being invariant under the action of L defined by the factor J, induces hermitian metrics in the fibers of F.

It is known that a system of hermitian metrics in the fibers of F defines uniquely a connection of type $(1,0)$ preserving the metrics in the fibers of F. In the following we consider the connection defined by the above system of hermitian metrics in the fibers of F. For each vector field X on E, the connection defines the covariant derivation D_X which is an operator acting on the space $A^0(E,F)$ of C^∞ sections of F.

Since F is defined by the factor J, we may identify an F-valued $(0,q)$-form on E with a complex-valued $(0,q)$-form φ on V such that

$$T_g^*\varphi = J(g,\cdot)\varphi \tag{2.2}$$

for $g \in L$, where T_g denotes the translation of V by the element g. We shall identify the space $A^q(E,F)$ with the space consisting of all $(0,q)$-forms φ on V satisfying the condition (2.2). In particular, $A^0(E,F)$ will be identified with the space of all C^∞ theta functions on V for the factor J.

Let us now introduce a basis $\{v_1,\ldots,v_n\}$ in V and let $\{z_1,\ldots,z_n\}$ be the coordinates with respect to this basis. In the following, the indices i and j shall run over 1 to n. We may express the hermitian forms G and H as follows:

$$G(v,v) = \sum_i \sum_j g_{ij} z_i \bar{z}_j$$

$$H(v,v) = \sum_i \sum_j h_{ij} z_i \bar{z}_j$$

where $v=(z_1,\ldots,z_n)$, (g_{ij}) is a positive definite hermitian matrix and (h_{ij}) a hermitian matrix. On the other hand, a form $\varphi \in A^q(E,F)$ $(q \geqq 1)$ being regarded as a $(0,q)$-form on V, is written uniquely in the form

$$\varphi = \frac{1}{q!} \sum \varphi_I \, d\bar{z}_I$$

where the sum runs over all multi-indices $I=(i_1,\ldots,i_q)$, $d\bar{z}_I = d\bar{z}_{i_1} \wedge \ldots \wedge d\bar{z}_{i_q}$, and φ_I are functions, called components of φ, belonging to $A^0(E,F)$ and which are alternating in the indices I. Let D'_i (resp. D''_j) denote the covariant derivation D_X with $X = \partial/\partial z_i$ (resp. $\partial/\partial \bar{z}_j$), X being regarded as complex vector field on E. Then we can prove the following formulas.

$$(D'_i f)(v) = \frac{\partial f}{\partial z_i}(v) - \pi \left(\sum_j h_{ij} \bar{z}_j \right) f(v) \quad , \qquad (2.3)$$

$$(D''_j f)(v) = \frac{\partial f}{\partial \bar{z}_j}(v) \quad , \qquad (2.4)$$

$$(D'_i f,g) + (f,D''_i g) = 0 \quad , \qquad (2.5)$$

where $v=(z_1,\ldots,z_n)$ and $f,g \in A^0(E,F)$. Let (g^{ij}) be the inverse matrix of (g_{ij}). For a multi-index $I=(i_1,\ldots,i_q)$, we denote by I_u $(1 \leqq u \leqq q)$ the multi-index $(i_1,\ldots,i_{u-1},i_{u+1},\ldots,i_q)$ and by jI the multi-index $(j,i_1,\ldots,i_q)$. Under this notation, a form $\varphi \in A^q(E,F)$ $(q \geqq 1)$ being given, the components of the forms $d''\varphi \in A^{q+1}(E,F)$, $\delta\varphi \in A^{q-1}(E,F)$ and $\square\varphi \in A^q(E,f)$ are given as follows:

$$(d''\varphi)_I = \sum_{u=1}^{q+1} (-1)^{u-1} D''_{i_u} \varphi_{I_u} \qquad (2.6)$$

$$(\delta\varphi)_I = -\sum_i \sum_j g^{ij} D'_i \varphi_{jI} \tag{2.7}$$

$$(\Box\varphi)_I = -\sum_i \sum_j g^{ij} D'_i D''_j \varphi_I$$

$$+ \pi \sum_{u=1}^{q} (-1)^{u-1} \sum_i \sum_j g^{ij} h_{ji_u} \varphi_{iI_u} \quad . \tag{2.8}$$

3. Proof of Theorem

We retain the notation settled in the previous sections. We may choose the basis $\{v_1,\ldots,v_n\}$ of V so that the hermitian form H is represented as

$$H(v,v) = |z_1|^2 + \cdots + |z_s|^2 - |z_{s+1}|^2 - \cdots - |z_{s+r}|^2 \tag{3.1}$$

for $v = (z_1,\ldots,z_n)$, where $s \geqq 0$ (resp. $r \geqq 0$) is the number of positive (resp. negative) eigenvalues of H. We shall assume that the positive definite form G is such that

$$G(v,v) = \frac{1}{a}\left(|z_1|^2 + \cdots + |z_s|^2\right) + |z_{s+1}|^2 + \cdots + |z_n|^2$$

for $v = (z_1,\ldots,z_n)$, where a is a positive number $> r$ (cf. [2],[4]). Then the matrices (g_{ij}), (g^{ij}) and (h_{ij}) are all diagonal matrices. Put $\alpha(i) = g^{ii} h_{ii}$. We see immediately

$$\alpha(i) = \begin{cases} a & \text{for} \quad 1 \leqq i \leqq s \\ -1 & \text{for} \quad s+1 \leqq i \leqq s+r \\ 0 & \text{for} \quad s+r \leqq i \leqq n \end{cases} \quad . \tag{3.2}$$

For each multi-index $I = (i_1,\ldots,i_q)$, we put

$$\alpha(I) = \sum_{u=1}^{q} \alpha(i_u) \quad . \tag{3.3}$$

The formula (2.8) is reduced to

$$(\Box\varphi)_I = -\sum_j g^{jj} D'_j D''_j \varphi_I + \pi\alpha(I)\varphi_I \quad . \tag{3.4}$$

By (2.5) the hermitian inner product between $(\Box\varphi)_I$ and φ_I in $A^0(E,F)$ is then given by

$$((\Box\varphi)_I, \varphi_I) = \sum_j g^{jj}(D''_j\varphi_I, D''_j\varphi_I) + \pi\alpha(I)(\varphi_I,\varphi_I) \quad .$$

Therefore we have

$$((\Box\varphi)_I, \varphi_I) \geqq \pi\alpha(I)(\varphi_I,\varphi_I) \quad . \tag{3.5}$$

We proceed now to the proof of our Theorem. For each multi-index $I=(i_1,\ldots,i_q)$ such that $i_1<\cdots<i_q$, we denote by $A^q_I(E,F)$ the subspace of $A^q(E,F)$ consisting of elements of the form $fd\bar{z}_I$ $(f\in A^0(E,F))$. Then we have the direct sum decomposition

$$A^q(E,F) = \sum A^q_I(E,F) \quad ,$$

where the sum runs over all multi-indices $I=(i_1,\ldots,i_q)$ such that $i_1<\cdots<i_q$. For simplicity, we write H^q for $H^q(E,F)$ and put $H^q_I = H^q \cap A^q_I(E,F)$. We see by (3.4) that the Laplacian $\Box$ maps each $A^q_I(E,F)$ into itself. We get therefore the decomposition

$$H^q = \sum H^q_I \tag{3.6}$$

where the sum runs over all increasing multi-indices I. On the other hand, we see by (3.5) that $H^q_I = (0)$ if $\alpha(I)>0$. Since $a>r$, (3.2) implies that this is the case if $I\cap\{1,\ldots,s\}\neq\emptyset$. Thus we get

$$H^q_I = (0) \qquad \text{if} \quad I\cap\{1,\ldots,s\}\neq\emptyset \ . \tag{3.7}$$

When I contains more than $n-s$ indices, then we have necessarily $I\cap\{i,\ldots,s\}\neq\emptyset$. Therefore, by (3.6) and (3.7), $H^q=(0)$ for $q>n-s$. Then, by (2.1) we see

$$H^q(E,\underline{F}) = (0) \qquad \text{for} \quad q > n-s \quad . \tag{3.8}$$

Now, by the Serre's duality, $H^q(E,\underline{F}) \cong H^{n-q}(E,\underline{K\otimes F^*})$, where K is the canonical line bundle of E and F^* denotes the dual bundle of F. Since K is the trivial bundle and since the Chern class of F^* is represetned by $-H$, it follows that we have

$$H^q(E,\underline{F}) = (0) \qquad \text{for} \quad q < r \quad . \tag{3.9}$$

Remark. If H is nondegenerate, i.e., if $s+r=n$, (3.8) and (3.9) mean that $H^q(E,\underline{F}) = (0)$ for $q \neq r$. This has been proved in [2], [3] and [4].

Now we shall study $H^q(E,\underline{F})$ for $r \leqq q \leqq n-s$. In view of (2.1), (3.6) and (3.7), we consider $\mathcal{H}^q_I$ when $I \subset I_0$, where $I_0 = \{s+1,\ldots,n\}$. For such an I, put

$$K = I \cap \{s+1,\ldots,s+r\} \quad .$$

Then we see $\alpha(I) = \alpha(K)$ by (3.2) and (3.3). It follows from (3.4) that for a form $fd\bar{z}_I \in A^q(E,F)$,

$$fd\bar{z}_I \in \mathcal{H}^q_I \qquad \text{if and only if} \qquad fd\bar{z}_K \in \mathcal{H}^p_K \quad , \tag{3.10}$$

where p is the cardinality of K. But, $\mathcal{H}^p_K = (0)$ for $p < r$ by (2.1) and (3.9). Therefore, we see $\mathcal{H}^q_I = (0)$ unless $I \supset K_0$, where $K_0 = \{s+1,\ldots,s+r\}$ ($= \emptyset$, if $r=0$). Combining this with (3.7), we get by (3.6)

$$\mathcal{H}^q = \sum \mathcal{H}^q_I \quad , \tag{3.11}$$

where the sum runs over the increasing multi-indices I such that $I_0 \supset I \supset K_0$. Let V_0^C be the complexification of the vector space V_0 considered over reals. The complex structure of V_0 defines the decomposition $V_0^C = V_0^+ + V_0^-$, where V_0^+ (resp. V_0^-) is the complex subspace of vectors of type $(1,0)$ (resp. $(0,1)$). Let $(V_0^-)^*$ be the dual vector space of V_0^-. Then $\{d\bar{z}_{s+r+1},\ldots,d\bar{z}_n\}$ may be considered as a basis of $(V_0^-)^*$. By (3.10), $\mathcal{H}^q_I$ $(I_0 \supset I \supset K_0)$ consists of elements of the form $fd\bar{z}_{K_0} \wedge d\bar{z}_M$ where $fd\bar{z}_{K_0} \in \mathcal{H}^r_{K_0}$ and $M = I \cap \{s+r+1,\ldots,n\}$.

By (3.11) we get therefore

$$H^q \cong H^r_{K_0} \otimes \overset{q-r}{\wedge} (V_0^-)^* \tag{3.12}$$

for $r \leqq q \leqq n-s$.

It is well-known that we have

$$H^{q-r}(E_0, \underline{C}) \cong \overset{q-r}{\wedge} (V_0^-)^* ,$$

where C denotes the trivial line bundle over the complex torus E_0 ([3, Chap. 1]). On the other hand, it follows from (3.11) that $H^r = H^r_{K_0}$. Therefore, by (2.1), and (3.12), we get

$$H^q(E, \underline{F}) = H^r(E, \underline{F}) \otimes H^{q-r}(E_0, \underline{C}) \tag{3.13}$$

for $r \leqq q \leqq n-s$.

To prove the remaining part of the Theorem, we may assume that $r > 0$; otherwise, the Theorem follows from (3.8), (3.13) and Proposition 2. We shall study the space $H^r_{K_0}$ and for this purpose we first follow Matsushima [2] who establishes an isomorphism between the space $H^r(E, \underline{F})$ and the space of holomorphic theta functions of certain kind. As Matsushima is concerned exclusively with the case where H is non-degenerate, we should extend his results to our general case, which we shall do in the following.

Let $\{v_1, \dots, v_n\}$ be the basis of V chosen before by which H is represented as (3.1). Let V_1 and V_2 be the complex subspace of V spanned by $\{v_1, \dots, v_s\}$ and $\{v_{s+1}, \dots, v_{s+r}\}$, respectively. We have the direct sum decomposition

$$V = V_1 + V_2 + V_0 .$$

Let W be the underlying 2n-dimensional real vector space of V and I the complex structure of W defining the complex vector space V. The above decomposition gives rise to the direct sum decomposition

$$W = V_1 + V_2 + V_0 \tag{3.14}$$

where V_1, V_2 and V_0 are all considered over reals. We define a linear transformation I_b of W by requiring

$$I_b = \begin{cases} I & \text{on} \quad V_1 + V_0 , \\ -I & \text{on} \quad V_2 . \end{cases}$$

Then I_b defines a complex structure on W. The complex vector space defined by W and I_b will be denoted by V_b. Put

$$H_b(u,v) = A(I_b u,v) + iA(u,v)$$

for $u, v \in V_b$. Then H_b is a positive semi-definite hermitian form on V_b whose kernel coincides with V_0.

Consider the complex torus $E_b = V_b/L$. We may define a factor $J_b\colon L \times V_b \to C$ by

$$J_b(g,v) = \psi(g)\exp \pi\left\{H_b(v,g) + \frac{1}{2} H_b(g,g)\right\}$$

where $(g,v) \in L \times V_b$. This factor defines a holomorphic line bundle F_b over the complex torus E_b.

Let p_2 be the projection $W \to V_2$ with respect to the decomposition (3.14), and define a function ϕ_b on W by

$$\phi_b(v) = \exp\{-\pi H_b(p_2(v),p_2(v))\}$$

for $v \in W$. By a direct calculation, we can verify that the map $f \to \phi_b^{-1} f$ defines a bijection between the vector spaces $A^0(E,F)$ and $A^0(E_b,F_b)$ of C^∞ theta functions for the factors J and J_b, respectively. Notice that we have a bijection $fd\bar{z}_{K_0} \to f$ between $A^r_{K_0}(E,F)$ and $A^0(E,F)$. Let $\varphi = fd\bar{z}_{K_0} \in A^r_{K_0}(E,F)$. Then φ is harmonic if and only if $d''\varphi = \delta\varphi = 0$. But the components of the forms $d''\varphi$ and $\delta\varphi$ are given as follows:

$$(d''\varphi)_I = \begin{cases} \dfrac{\partial f}{\partial \bar{z}_i} & \text{if} \quad I = iK_0 \\ 0 & \text{if} \quad I \not\supset K_0 , \end{cases}$$

$$(\delta\varphi)_I = \begin{cases} -g^{ii}\left(\dfrac{\partial f}{\partial z_i} + \pi \bar{z}_i f\right) & \text{if} \quad K_0 = iI \\ 0 & \text{if} \quad I \not\subset K_0 \end{cases} .$$

These formulas follow from (2.3), (2.4), (2.6) and (2.7). Therefore φ is harmonic if and only if

$$\begin{cases} \dfrac{\partial f}{\partial \bar{z}_i} = 0 & \text{for} \quad i \notin K_0 , \\ \dfrac{\partial f}{\partial z_i} + \pi \bar{z}_i f = 0 & \text{for} \quad i \in K_0 . \end{cases} \tag{3.15}$$

Applying this criterion, we can verify that $\varphi = f d\bar{z}_{K_0}$ is harmonic if and only if the function

$$\theta = \phi_b^{-1} f$$

is a holomorphic theta function on V_b for the factor J_b (cf. [2]). Therefore we get the isomorphism

$$H^r_{K_0} \cong H^0(E_b, \underline{F}_b) . \tag{3.16}$$

We apply now Proposition 2 to study $H^0(E_b, \underline{F}_b)$. The character of L_0 defined by J_b being the character ψ_0 defined by J, if ψ_0 is nontrivial, $H^0(E_b, F_b) = (0)$. By (3.8), (3.9), (3.12) and (3.16), it follows that if ψ_0 is nontrivial, $H^q(E, \underline{F}) = (0)$ for all $q \geqq 0$, which proves the assertion (i) of Theorem.

Suppose now the character ψ_0 be trivial. Then, by Proposition 2, every holomorphic theta function θ on V_b is invariant under the translations of V_b by elements of V_0. Since the function ϕ_b is also invariant under these translations, the same holds for the function $f = \phi_b \theta$. The function f is then induced from a function $\hat{f}$ on $\hat{V} = V/V_0 : f = \hat{f} \circ \lambda$, λ denoting the projection $V \to \hat{V}$. Let $m = s + r$. Then, $\dim \hat{V} = m$ and $\{\hat{v}_1, \ldots, \hat{v}_m\}$ is a basis of $\hat{V}$ where $\hat{v}_i = \lambda(v_i)$. Let $\{w_1, \ldots, w_m\}$ be the coordinates of $\hat{V}$ with respect to this basis. By the above observation, every form $\varphi = f d\bar{z}_{K_0} \in H^r_{K_0}$ is induced from the form $\hat{\varphi} = \hat{f} d\bar{w}_{K_0}$. On the other hand, it follows from (3.15) that the function $\hat{f}$ satisfies

$$\begin{cases} \dfrac{\partial \hat{f}}{\partial w_i} = 0 & \text{for} \quad 1 \leqq i \leqq s \\ \dfrac{\partial \hat{f}}{\partial w_i} + \pi \overline{w_i}\, f = 0 & \text{for} \quad s+1 \leqq i \leqq s+r \end{cases} \tag{3.17}$$

Conversely, if a function $\hat{f}$ on $\hat{V}$ is a C^∞ theta function for the factor $\hat{J}$ and satisfies (3.17), then the form $\hat{\varphi} = \hat{f} d\bar{w}_{K_0}$ defines a form $\varphi = \lambda^* \hat{\varphi}$ on V which belongs to $H^r_{K_0}$.

Consider the hermitian form $\hat{H}$ on $\hat{V}$ induced from H on V. We have

$$\hat{H}(\hat{v},\hat{v}) = |w_1|^2 + \cdots + |w_s|^2 - |w_{s+1}|^2 - \cdots - |w_m|^2$$

for $\hat{v} = (w_1, \ldots, w_m)$. Let $\hat{G}$ be the positive definite hermitian form on $\hat{V}$ defined by

$$\hat{G}(\hat{v},\hat{v}) = \frac{1}{a}(|w_1|^2 + \cdots + |w_s|^2) + |w_{s+1}|^2 + \cdots + |w_m|^2$$

for $\hat{v} = (w_1, \ldots, w_m)$, where $a > r$. Taking these hermitian forms $\hat{H}$ and $\hat{G}$ for H and G, we can develop harmonic theory for $\hat{F}$-valued forms on $\hat{E}$, and we get analogous results as before in this new situation. In particular, let $\hat{H}^r_{K_0}$ denote the set of all harmonic forms of the form $\hat{f} d\bar{w}_{K_0}$ on $\hat{V}$ where $\hat{f}$ is a C^∞ theta function on $\hat{V}$ for the factor $\hat{J}$. Then, we have

$$H^r(\hat{E}, \hat{\underline{F}}) \cong \hat{H}^r_{K_0}$$

and $\hat{H}^r_{K_0}$ consists of all forms $\hat{f} d\bar{w}_{K_0}$ where $\hat{f}$ is a C^∞ theta function on $\hat{V}$ for the factor $\hat{J}$ which satisfies the condition (3.17). By what we have seen above, it follows that the projection $\lambda: V \to \hat{V}$ induces an isomorphism $H^r_{K_0} \cong \hat{H}^r_{K_0}$ and therefore we get

$$H^r(E, \underline{F}) \cong H^r(\hat{E}, \hat{\underline{F}}) \quad .$$

Finally, we see $\dim H^r(E,\underline{F}) = \dim H^r_{K_0} = \dim H^0(E_b, \underline{F}_b)$ by (3.16). But, since the imaginary part of H_b coincides with the imaginary part A of H, $\dim H^0(E_b, \underline{F}_b)$ is equal to the reduced Phaffian of A relative to L (Proposition 2). Our Theorem is thus completely proved. ∎

4. References

[1] G. Kempf, Appendix to D. Mumford's article: "Varieties defined by quadratic equations," *Questions in Algebraic Varieties*, C.I.M.E., 1969.

[2] Y. Matsushima, "On the intermediate cohomology group of a holomorphic line bundle over a complex torus," *Osaka J. Math.* 16 (1979), 617-632.

[3] D. Mumford, *Abelian Varieties*, Tata Institute Studies in Math., Oxford Univ. Press, 1970.

[4] H. Umemura, "Some results in the theory of vector bundles," *Nagoya Math. J.* 52 (1973), 97-128.

[5] A. Weil, *Variétés Kählériennes*, Hermann, Paris, 1958.

5. Notes

1. Supported by Grant-in-Aid for Scientific Research.

Osaka University
Toyonaka, Osaka 560
Japan

(Received July 1, 1980)

PERIODIC POINTS ON NILMANIFOLDS

Minoru Nakaoka

1. Introduction

Shub and Sullivan [13] proves that every C^1-map $f:M \to M$ of a compact smooth manifold has infinitely many periodic points if the Lefschetz numbers $L(f^k)$, $k = 1,2,\ldots,$ are unbounded. This is not generally true if f is a continuous map, and even if f is a homeomorphism (see [11]).

Identify the torus T^n with the coset space R^n/Z^n. Then a diffeomorphism $g:T^n \to T^n$ covered by a linear map $\tilde{g}:R^n \to R^n$ is called a *toral automorphism*. If $\tilde{g}$ is hyperbolic (i.e., has no eigenvalues of absolute value 1), g is called a *hyperbolic toral automorphism*. Every C^1-map $f:T^n \to T^n$ homotopic to a hyperbolic toral automorphism g has infinitely many periodic points, because $|L(f^k)| = |L(g^k)| \to \infty$ as $k \to \infty$. Shub and Sullivan ask whether every continuous map $f:T^n \to T^n$ homotopic to a hyperbolic toral automorphism g must have infinitely many periodic points (see p. 140 of Hirsch [6]).

Halpern [5] shows that the theorem of Shub and Sullivan is true for every continuous map if $M = T^n$, and hence that the above problem has a positive answer.

This paper is concerned with generalization of these results of Halpern.

Let N be a simply connected nilpotent Lie group, and let Γ be a discrete subgroup which is finitely generated, torsion-free and nilpotent. Then the coset space N/Γ is compact, and it is called a *nilmanifold* (see [7]).

Theorem 1. *Let* $f:N/\Gamma \to N/\Gamma$ *be a homotopy equivalence of a nilmanifold. If the Lefschetz numbers* $L(f^k)$, $k = 1,2,\ldots,$ *are unbounded, then* f *has infinitely many periodic points.*

A diffeomorphism $g:M \to M$ of a compact smooth manifold is called an *Anosov diffeomorphism* in Smale [14] if there exists a continuous splitting $TM = E^s \oplus E^u$ of the tangent bundle of M such that the derivative dg preserves the splitting and, relative to some Riemannian metric $\| \ \|$ on TM, there exists $c' > 0$, $c'' > 0$ and $\lambda (0 < \lambda < 1)$ satisfying

$$\|dg^k(v')\| \leqq c'\lambda^k \|v'\| \quad ,$$

$$\|dg^k(v'')\| \geqq c''\lambda^{-k} \|v''\| \quad ,$$

where $v' \in E^s$, $v'' \in E^u$ and $k = 1, 2, \ldots$.

A hyperbolic toral automorphism is the simplest example of Anosov diffeomorphisms.

Theorem 2. *If* $g: N/\Gamma \to N/\Gamma$ *is an Anosov diffeomorphism of a nilmanifold, then* $|L(g^k)| \to \infty$ *as* $k \to \infty$.

The notion of nilmanifold is generalized as follows. Let N be a simply connected nilpotent Lie group, and let A be a finite group of automorphisms of N. Consider N as acting on itself by left translation, and take the product $N \cdot A$ in the group of automorphisms of N. Let Γ be a torsion-free discrete subgroup of $N \cdot A$ such that the coset space $N \cdot A/\Gamma$ is compact. Define an action of $N \cdot A$ to N by $(x,a) \cdot y = xa(y)$ $(x, y \in N,\ a \in A)$. Then Γ acts freely on N, and the orbit space N/Γ is compact. N/Γ is called an *infranilmanifold* (see [4], [12]).

For an infranilmanifold, Theorems 1 and 2 hold in a rather weak form (see Theorems 1' and 2' in Section 6).

Theorem 3. *Let* $g: N/\Gamma \to N/\Gamma$ *be an Anosov diffeomorphism of an infranilmanifold. Then, for every continuous map* $f: N/\Gamma \to N/\Gamma$ *homotopic to* g, *we have*

$$\overline{\lim_{k \to \infty}} \ \frac{1}{k} \ \log N_k(f) > 0 \quad ,$$

where $N_k(f)$ *denotes the number of fixed points of* f^k. *In particular,* f *has infinitely many periodic points.*

2. Nielsen Numbers

Let $f: M \to M$ be a continuous map of a compact manifold, and let $\mathrm{Fix}(f)$ denote the set of fixed points of f. Define $x, x' \in \mathrm{Fix}(f)$ to be equivalent if there is a path ℓ in M from x to x' such that

$f \circ \ell$ is homotopic to ℓ leaving end points fixed. It is known that each equivalence class $F_i \subset M$ is compact and the set $\{F_i\}$ is finite (see [2]). The *Nielsen number* $\nu(f)$ is defined to be the number of F_i such that the fixed point index $I(f, F_i)$ of f around F_i is not zero. Obviously f has at least $\nu(f)$ distinct fixed points. We know that if $f, f' : M \to M$ are homotopic then $\nu(f) = \nu(f')$ (see [2]).

The following generalizes Proposition 1 of [5].

Proposition 1. *Let* G *be a Lie group, and* Γ *be a discrete subgroup such that* G/Γ *is compact. Let* $h : G \to G$ *be a Lie group endomorphism such that* $h(\Gamma) \subset \Gamma$, $\mathrm{Fix}(h) = \{e\}$ *and* 1 *is not an eigenvalue of* $(dh)_e : T_e(G) \longrightarrow T_e(G)$, *where* e *is the unit of* G. *Then, for the map* $\bar{h} : G/\Gamma \to G/\Gamma$ *induced by* h, *we have*

$$\nu(\bar{h}) = \#\mathrm{Fix}(\bar{h}) = |L(\bar{h})| .$$

Proof. Let $\pi : G \to G/\Gamma$ denote the covering map, and let $\bar{x} = \pi(x) = \Gamma x$ $(x \in G)$ be a fixed point of $\bar{h}$. Then, for the left translation $L_x : G \to G$ by x, we have $\bar{h} \circ \pi \circ L_x = \pi \circ L_x \circ h$ and hence a commutative diagram

$$\begin{array}{ccc} T_e(G) & \xrightarrow{(dh)_e} & T_e(G) \\ \Big\downarrow{\scriptstyle d(\pi \circ L_x)_e} & & \Big\downarrow{\scriptstyle d(\pi \circ L_x)_e} \\ T_{\bar{x}}(G/\Gamma) & \xrightarrow{(d\bar{h})_{\bar{x}}} & T_{\bar{x}}(G/\Gamma) \end{array}$$

Since $d(\pi \circ L_x)_e$ is an isomorphism, it follows that

$$\det(1 - (d\bar{h})_{\bar{x}}) = \prod_{i=1}^{n} (1 - \lambda_i) ,$$

where λ_i's are the eigenvalues of $(dh)_e$ (counted with multiplicity). Hence $\det(1 - (d\bar{h})_{\bar{x}}) \neq 0$ by our assumption. Therefore each fixed point $\bar{x}$ of $\bar{h}$ is isolated, and the fixed point index $I(\bar{h}, \bar{x})$ of $\bar{h}$ around $\bar{x}$ is 1 or -1 according as $\prod(1 - \lambda_i)$ is positive or negative. Thus it follows from the Lefschetz fixed point formula that $\#\mathrm{Fix}(\bar{h}) = |L(\bar{h})|$, and to complete the proof it suffices to show that if $\bar{x}, \bar{x}' \in \mathrm{Fix}(\bar{h})$

and there exists a path $\bar{\ell}$ such that $\bar{\ell} \simeq \bar{h}\circ\bar{\ell}: (I,0,1) \to (G/\Gamma,\bar{x},\bar{x}')$ then $\bar{x} = \bar{x}'$.

Take a lift $\ell:(I,0,1) \to (G,x,x')$ of $\bar{\ell}$. Since $\pi h(x) = \pi(x)$, there exists $c \in \Gamma$ such that

$$x \quad = \quad ch(x) \quad . \tag{2.1}$$

It holds that $\pi \circ L_c = \pi$ and hence $\pi \circ L_c \circ h \circ \ell = \bar{h}\circ\bar{\ell}$. We have also $L_c h\ell(0) = x$. Therefore $L_c \circ h \circ \ell$ is a lift of $\bar{h}\circ\bar{\ell}$ starting from x. Since $\bar{\ell} \simeq \bar{h}\circ\bar{\ell}:(I,0,1) \to (G/\Gamma,\bar{x},\bar{x}')$, we have $\ell(1) = L_c h\ell(1)$ and hence

$$x' \quad = \quad ch(x') \quad . \tag{2.2}$$

Since $x'^{-1}x = h(x'^{-1}x)$ by (2.1) and (2.2), we get $x'^{-1}x = e$ by our assumption. Thus $\bar{x} = \bar{x}'$ holds. ∎

3. Nilmanifold Endomorphisms

Let N/Γ be a nilmanifold, and let $h:N \to N$ be a Lie group endomorphism such that $h(\Gamma) \subset \Gamma$. Then the map $\bar{h}:N/\Gamma \to N/\Gamma$ induced by h is called a *nilmanifold endomorphism.*

Proposition 2. *Let* $\bar{h}:N/\Gamma \to N/\Gamma$ *be a nilmanifold endomorphism such that* 1 *is not an eigenvalue of* $(dh)_e:T_e(N) \to T_e(N)$, *and let* $f:N/\Gamma \to N/\Gamma$ *be a continuous map homotopic to* $\bar{h}$. *Then we have*

$$\nu(f) \quad = \quad |L(f)| \quad .$$

Proof. Since N is a simply connected nilpotent Lie group, the exponential map $\exp:T_e(N) \to N$ is a homeomorphism (see [7]). Since h is a Lie group homomorphism, the diagram

$$\begin{array}{ccc} T_e(N) & \xrightarrow{(dh)_e} & T_e(N) \\ \exp \downarrow & & \downarrow \exp \\ N & \xrightarrow{h} & N \end{array}$$

commutes. Therefore, if there exists $x \in N$ such that $h(x) = x \neq e$, then there exists $X \in T_e(N)$ such that $(dh)_e(X) = X \neq 0$, and hence $(dh)_e$ has 1 as an eigenvalue. By the assumption this shows

$\mathrm{Fix}(h) = \{e\}$. Thus we have by Proposition 1 $\nu(\bar{h}) = |L(\bar{h})|$. Since $\nu(f) = \nu(\bar{h})$ and $L(f) = L(\bar{h})$, this completes the proof. ■

Let N/Γ be a nilmanifold, and let

$$\{e\} = N_0 \subset N_1 \subset \cdots \subset N_{c-1} \subset N_c = N$$

be the upper central series of N. Put $\Gamma_i = N_i \cap \Gamma$ for $i = 0, 1, \ldots, c$. Then it is known that

$$\{e\} = \Gamma_0 \subset \Gamma_1 \subset \cdots \subset \Gamma_{c-1} \subset \Gamma_c = \Gamma$$

is the upper central series of Γ, and Γ_i/Γ_{i-1} is a free abelian group for $i = 1, 2, \ldots, c$ (see [6], [9]).

A nilmanifold endomorphism $\bar{h}: N/\Gamma \to N/\Gamma$ is called a *nilmanifold automorphism* if it is induced by an automorphism $h: N \to N$ such that $h(\Gamma) = \Gamma$. In this case h induces automorphisms $\phi_i: \Gamma_i/\Gamma_{i-1} \to \Gamma_i/\Gamma_{i-1}$ $(i = 1, 2, \ldots, c)$.

The following proposition is proved in Manning [9], [10].

Proposition 3. *Let* $\bar{h}: N/\Gamma \to N/\Gamma$ *be a nilmanifold automorphism induced by* $h: N \to N$. *Then the totality of the eigenvalues of* $\phi_1, \phi_2, \ldots, \phi_c$ *coincides with the eigenvalues of* $(dh)_e: T_e(N) \to T_e(N)$, *and if we denote them by* $\lambda_1, \lambda_2, \ldots, \lambda_n$ *it holds*

$$L(\bar{h}) = \prod_{i=1}^{n} (1 - \lambda_i) \quad .$$

4. Proof of Theorem 1

Theorem 1 is a direct consequence of the following.

Proposition 4. *If* $f: N/\Gamma \to N/\Gamma$ *is a homotopy equivalence of a nilmanifold, then we have*

$$\# \mathrm{Fix}(f) \geqq |L(f)| \quad .$$

Proof. Identifying $\pi_1(N/\Gamma)$ with Γ, let $f_*: \Gamma \to \Gamma$ denote the automorphism induced by f on $\pi_1(N/\Gamma)$. (f_* is only determined up to an inner automorphism of Γ, but that is sufficient for our purpose.) By a result in [7], the automorphism $f_*: \Gamma \to \Gamma$ can be uniquely extended to a Lie group automorphism $h: N \to N$. Consider the nilmanifold automorphism $\bar{h}: N/\Gamma \to N/\Gamma$. It follows that $\bar{h}_* = f_*: \Gamma \to \Gamma$. Since N/Γ is an

Eilenberg-MacLane space $K(\Gamma,1)$, we see that $\bar{h}, f: N/\Gamma \to N/\Gamma$ are homotopic. Therefore, by Proposition 3, it holds

$$L(f) = L(\bar{h}) = \prod_{i=1}^{n} (1-\lambda_i) \quad , \tag{4.1}$$

where λ_i's are the eigenvalues of $(dh)_e: T_e(N) \to T_e(N)$.

If $L(f) = 0$ then the conclusion is trivial. Therefore we assume $L(f) \neq 0$. Then (4.1) shows that $(dh)_e$ does not have 1 as eigenvalue. Hence by Proposition 2 we have $\nu(f) = |L(f)|$ which proves the desired result. ■

Remark. Let $f: T^n \to T^n$ be a continuous map. Regarding $T^n = R^n/Z^n$ and $Z^n = H_1(T^n)$, consider the linear map $h: R^n \to R^n$ which extends $f_*: H_1(T^n) \to H_1(T^n)$. Then $\bar{h}: T^n \to T^n$ is homotopic to f. Since it holds that $L(f) = \Pi_{i=1}^n (1-\lambda_i)$ for the eigenvalues λ_i's of $f_*: H_1(T^n) \to H_1(T^n)$, the similar argument as in the proof of Proposition 4 shows that $\#\mathrm{Fix}(f) \geqq |L(f)|$ for every continuous map $f: T^n \to T^n$. This proves the theorem of Halpern stated in the Introduction.

5. Proof of Theorem 2

A nilmanifold automorphism $\bar{h}: N/\Gamma \to N/\Gamma$ is said to be *hyperbolic* if the linear map $(dh)_e: T_e(N) \to T_e(N)$ is hyperbolic.

Manning [10] proves that every Anosov diffeomorphism of a nilmanifold is topologically conjugate to a hyperbolic nilmanifold automorphism. In view of this fact, to prove Theorem 2 it is sufficient to show

$$\lim_{k\to\infty} |L(\bar{h}^k)| = \infty \tag{5.1}$$

for a hyperbolic nilmanifold automorphism $\bar{h}: N/\Gamma \to N/\Gamma$.

Let $\lambda_1, \lambda_2, \ldots, \lambda_n$ be the eigenvalues of $(dh)_e: T_e(N) \to T_e(N)$. Since $\bar{h}$ is hyperbolic, we have

$$|\lambda_i| \neq 1 \qquad (i = 1, 2, \ldots, n) \quad . \tag{5.2}$$

It follows from Proposition 3 that

$$L(\bar{h}^k) = \prod_{i=1}^{n} (1-\lambda_i^k) \quad , \tag{5.3}$$

$$\prod_{i=1}^{n} \lambda_i = \prod_{i=1}^{c} \det \phi_i = \pm 1 \quad . \tag{5.4}$$

By (5.2) and (5.4) there exists λ_i such that $|\lambda_i| > 1$. Since

$$\lim_{k\to\infty} |1 - \lambda_i^k| = \infty \quad \text{or} \quad 1$$

according as $|\lambda_i| > 1$ or < 1, we obtain (5.1) from (5.3). ■

6. Modification of Theorems 1 and 2

Theorem 1'. *Let* $f:N/\Gamma \to N/\Gamma$ *be a homotopy equivalence of an infranilmanifold. If* $|L(f^k)| \to \infty$ *as* $k \to \infty$, *then* f *has infinitely many periodic points.*

Proof. It is known that $N/N \cap \Gamma$ is a nilmanifold and the natural map $p:N/N \cap \Gamma \to N/\Gamma$ is a finite regular covering with fiber $\Gamma/N \cap \Gamma$ (see [1]). We may identify the inclusion $i:N \cap \Gamma \to \Gamma$ with $p_*:\pi_1(N/N \cap \Gamma) \to \pi_1(N/\Gamma)$. It is known that $N \cap \Gamma$ is the maximal normal nilpotent subgroup of Γ (see [1]). Therefore the automorphism $f_*:\Gamma \to \Gamma$ induced by f on $\pi_1(N/\Gamma)$ takes $N \cap \Gamma$ onto itself. Consequently we have a commutative diagram

$$\begin{array}{ccc} \pi_1(N/N \cap \Gamma) & \xrightarrow{f_*|N \cap \Gamma} & \pi_1(N/N \cap \Gamma) \\ \downarrow p_* & & \downarrow p_* \\ \pi_1(N/\Gamma) & \xrightarrow{f_*} & \pi_1(N/\Gamma) \end{array}$$

which shows that $f:N/\Gamma \to N/\Gamma$ has a lift $\tilde{f}:N/N \cap \Gamma \to N/N \cap \Gamma$ such that $\tilde{f}_* = f_*|N \cap \Gamma$. It follows that $\tilde{f}$ is a homotopy equivalence.

Put $m = \#(\Gamma/N \cap \Gamma)$. Then we see that $z \in \mathrm{Fix}(f^{mk})$ implies $p^{-1}(z) \subset \mathrm{Fix}(\tilde{f}^{mk})$, where k is a positive integer. Therefore, we have

$$L(\tilde{f}^{mk}) = mL(f^{mk})$$

(see (8.19) of [3]), which implies

$$\lim_{k\to\infty} |L(\tilde{f}^{mk})| = \infty ,$$

because $|L(f^{mk})| \to \infty$ as $k \to \infty$ by the assumption. Thus it holds by Theorem 1 that $\tilde{f}^m$ has infinitely many periodic points and so does $\tilde{f}$. Since $p: N/N\cap\Gamma \to N/\Gamma$ is a finite covering we conclude that f have infinitely many periodic points. ∎

Theorem 2'. *If* $g: N/\Gamma \to N/\Gamma$ *is an Anosov diffeomorphism of an infranilmanifold, then* $L(g^k)$, $k = 1,2,\ldots,$ *are unbounded.*

Proof. Take a lift $\tilde{g}: N/N\cap\Gamma \to N/N\cap\Gamma$. It follows that $\tilde{g}$ is an Anosov diffeomorphism. Therefore, by Theorem 2 we have

$$\lim_{k\to\infty} |L(\tilde{g}^k)| = \infty .$$

Since $L(\tilde{g}^{km}) = mL(g^{km})$ with $m = \#(\Gamma/N\cap\Gamma)$, we get

$$\lim_{k\to\infty} |L(g^{km})| = \infty$$

which proves Theorem 2'. ∎

7. Proof of Theorem 3

By Proposition 4 we have

Theorem 1". *If* $f: N/\Gamma \to N/\Gamma$ *is a homotopy equivalence of a nilmanifold, then*

$$\overline{\lim_{k\to\infty}} \frac{1}{k} \log N_k(f) \geqq \overline{\lim_{k\to\infty}} \frac{1}{k} \log |L(f^k)| .$$

From the facts stated in the proof of Theorem 2, we see

Theorem 2". *If* $g: N/\Gamma \to N/\Gamma$ *is an Anosov diffeomorphism of a nilmanifold, then*

$$\overline{\lim_{k\to\infty}} \frac{1}{k} \log |L(g^k)| > 0 .$$

We can now prove Theorem 3 as follows.

Theorem 3 for nilmanifold is a direct consequence of Theorems 1'' and 2''. In the case of infranilmanifold, take lifts $\tilde{f}, \tilde{g}: N/N \cap \Gamma \to N/N \cap \Gamma$ of f, g. Then $\tilde{f}$ is homotopic to $\tilde{g}$ which is Anosov, and hence we have

$$\overline{\lim_{k\to\infty}} \frac{1}{k} \log N_k(\tilde{f}) > 0 \quad . \tag{7.1}$$

Since

$$\frac{1}{m} N_k(\tilde{f}) \leqq N_k(f)$$

with $m = \#(\Gamma/N \cap \Gamma)$, it follows that

$$\overline{\lim_{k\to\infty}} \frac{1}{k} \log N_k(\tilde{f}) \leqq \overline{\lim_{k\to\infty}} \frac{1}{k} \log N_k(f) \quad .$$

This and (7.1) imply the desired result. ■

Remark. From the fact stated in Remark of Section 4, we see that

$$\overline{\lim_{k\to\infty}} \frac{1}{k} \log N_k(f) \geqq \overline{\lim_{k\to\infty}} \frac{1}{k} \log |L(f^k)|$$

holds for every continuous map $f: T^m \to T^m$.

Note added in proof.

The following propositions hold for a nilmanifold N/Γ:

(I) *Any endomorphism of* Γ *extends to a Lie group endomorphism of* N.

(II) *For a nilmanifold endomorphism* $\bar{h}: N/\Gamma \to N/\Gamma$ *induced by* $h: N \to N$ *we have* $L(\bar{h}) = \Pi_{i=1}^{n} (1-\lambda_i)$, *where* λ_i*'s are the eigenvalues of* $(dh)_e$.

(I) can be proved by using the Malcev polynomials (see 5.1 of P. Buser and H. Karcher: Gromov's almost flat manifolds, to appear in Asterisque). (II) can be proved by expressing N/Γ as a sequence of torus extensions associated with the lower central series of N (see [9]). In virtue of (I) and (II) we see that Proposition 4 is valid for any continuous map $f: N/\Gamma \to N/\Gamma$, and so are Theorems 1, 1', and 1''.

8. References

[1] L. Auslander, "Bieberbach's theorems on space groups and discrete uniform subgroups of Lie groups," *Ann. of Math.* 71 (1960), 579-590.

[2] R. Brown, *The Lefschetz Fixed Point Theorem*, Scott, Foresman and Company, 1971.

[3] A. Dold, "The fixed point transfer of fibre-preserving maps," *Math. Z.* 148 (1976), 215-244.

[4] J. Franks, *"Anosov diffeomorphisms"*, Global Analysis, Proc. Symp. Pure Math., 14, Amer. Math. Soc. (1970), 61-93.

[5] B. Halpern, "Periodic points on tori," *Pacific J. of Math.* 83, (1979), 117-133.

[6] M. Hirsch, *Differential Topology*, Springer-Verlag, 1976.

[7] A. Malcev, "On a class of homogeneous spaces," *Amer. Math. Soc. Translation* (1) 9 (1962), 276-307.

[8] A. Malcev, "Nilpotent torsion-free groups," *Izv. Akad. Nauk SSSR Ser. Math.* 13 (1949), 201-212 (Russian).

[9] A. Manning, "Anosov diffeomorphisms on nilmanifolds," *Proc. Amer. Soc.* 38 (1973), 423-426.

[10] A. Manning, "There are no new Anosov diffeomorphisms on tori," *Amer. J. Math.* 96 (1974), 422-429.

[11] C. Pugh *et al.*, "On the entropy conjecture," Lecture Notes in Math. 468, Springer-Verlag, (1975), 257-261.

[12] M. Shub, "Endomorphisms of compact differentiable manifolds," *Amer. J. Math.* 91 (1969), 175-199.

[13] M. Shub and D. Sullivan, "A remark on the Lefschetz fixed point formula for differentiable maps," *Topology* 13 (1974), 189-191.

[14] S. Smale, "Differentiable dynamical systems," *Bull. Amer. Math. Soc.* 73 (1967), 747-817.

Osaka University
Toyonaka, Osaka 560
Japan

(Received July 31, 1980)

ISOGENIES AND CONGRUENCE SUBGROUPS

M. S. Raghunathan

1. Introduction

Throughout this paper k will denote a number field and V its set of valuations. Let S be any finite set of valuations including ∞, the set of archimedean valuations of k. For each $v \in V$, k_v will denote the completion of k with respect to v and O_v the ring of integers in k_v. We denote by A (resp. $A(S)$) the ring of integers (resp. S-integers) in k (so that $A = A(\infty)$). Let G, H be linear reductive algebraic groups defined over k and $f: G \to H$ be a k-isogeny. We investigate in this paper the images of S-congruence subgroups in G under f. Recall that a subgroup $\Gamma \subset G(k)$ is said to be S-*arithmetic* if for one (and hence any) realization $i: G \hookrightarrow GL(n)$ of G as a k algebraic subgroup of $GL(n)$, $\Gamma \cap i^{-1}\{GL(n, A(S))\}$ has finite index in Γ as well as $i^{-1}\{GL(n,A(S))\}$. The group Γ is an S-*congruence* subgroup if there is a nonzero ideal $\mathfrak{a} \subset A(S)$ such that $\Gamma \supset i^{-1}(GL(n, \mathfrak{a}))$ where $GL(n,\mathfrak{a}) = \{x \in GL(n,A(S)) \mid x \equiv 1 (\mathrm{mod}\ \mathfrak{a})\}$. It is easily seen that the inverse image of an S-arithmetic (resp. S-congruence) subgroup of H is an S-arithmetic (resp. S-congruence) subgroup of G. On the other hand, it is a theorem of Borel-Harish Chandra that the image of an S-arithmetic subgroup under f is an S-arithmetic subgroup. Further images of S-congruence subgroups of G are in general not S-congruence subgroups of H. For instance the following *necessary* condition we establish in the *special* case when $f: G \to H$ is the isogeny of a simply connected semisimple G onto its adjoint group enables one to give a general method of construction of such S-congruence subgroups in G: if $\Gamma \subset G$ is an S-congruence subgroup with $f(\Gamma)$ an S-congruence subgroup, then $\Gamma.C(k)$ (C = Center of G) contains a finite intersection of maximal arithmetic subgroups of G. The main thrust of the present paper is in the converse direction: We will give a measure of the extent to which a finite intersection of maximal arithmetic subgroups in G fails to have for its image an S-congruence subgroup.

To formulate our main result we need some further notation. Let Γ be an S-arithmetic subgroup of a reductive group B. Then we denote by $\hat{\Gamma}$ the intersection of *all* the S-congruence subgroups containing Γ; then $\hat{\Gamma}$ is the minimal S-congruence subgroup containing Γ. In the sequel we will say that an S-arithmetic subgroup is a *special S-arithmetic* subgroup if Γ is normal in $\hat{\Gamma}$. If Γ is a special S-arithmetic subgroup its *deviation group* is the group $\hat{\Gamma}/\Gamma$; it is denoted by $d(\Gamma):d(\Gamma)$ measures in some sense by how much Γ fails to be an S-congruence subgroup. With this notation our main result is

<u>Theorem</u>. *Let* $\Gamma \subset G$ *be a finite intersection of maximal* S-*arithmetic subgroups of* G. *Let* $\Gamma_1^* = f(\Gamma)$. *Then* Γ_1^* *contains* Γ^*, *a special arithmetic subgroup of* H *such that* $\tilde{\Gamma}^* = \Gamma^*.C_H(k)$, *where* C_H *is the center of* H, *is again special arithmetic with deviation group* $d(\tilde{\Gamma}^*)$ *of index* $\leqslant 2^r$ *(*r *= number of* k-*simple factors of* H*) and exponent* $\leqslant 2$; *(consequently)* $|d(\Gamma^*)| \leqslant 2^r.|C_H(k)|$. *Further we can find* $\Gamma_o \subset G^*$ *a finite intersection of maximal arithmetic subgroups with* $\Gamma_o \subset \Gamma$ *and an* S-*congruence subgroup* $\Gamma_o' \subset \Gamma_o$ *such that* $f(\Gamma_o) = f(\Gamma_o')$ *and* $\Gamma_o' \cap \ker f = 1$. *Also,* $d(\Gamma^*) = \{1\}$ *if* G *has no* k-*simple factors of outer type* ${}^2A_n$, $4|(n+1)$ or ${}^2D_n$, n *odd.*

This will be referred to as the main theorem in the sequel. The main theorem in the special case when $G \to H$ is the isogeny of a simply connected semi-simple group on its adjoint group will be handled first with the aid of some Galois cohomological results. The general case will then be deduced for this. The reason why the special case is easy to handle is that the kernel of the isogeny viz. the center C_G of G has a reasonably easy structure leading to pleasant consequences for its Galois cohomology.

Section 2 deals with some Galois cohomology results--in essence a kind of Hasse principle for special finite abelian groups over k. Section 3 takes up the case of the adjoint isogeny. Section 4 completes the general case and also establishes the partial "converse" stated above.

2. Two Lemmas on Centers of Simply Connected Semisimple Groups

Throughout this section G will denote a connected simply connected semisimple algebraic group defined over k. Such a group is a direct product of its k-simple factors. Each k-simple factor is of the form $R_{L/K}B$ where B is an absolutely almost simple algebraic group defined over L, a finite extension of k and $R_{L/k}$ is the Weil restriction of scalar functor. We make use of these comments to reduce the proof of the following two lemmas on the center C of G to the case of absolutely almost simple G.

2.1 Lemma. *The natural map* $h:H^1(k,C) \to \Pi_{v\notin S} H^1(k_v,C)$ *has for kernel a group of exponent* $\leqslant 2$ *and index* $\leqslant 2^r$ *where* r *is the number of* k*-simple factors of* G.

The reduction to the case when G is absolutely almost simple follows from the natural isomorphism

$$H^1(k,R_{L/k}B) \simeq H^1(L,B)$$

for any L-algebraic group B.

2.2 Lemma. *Let* $S' \subset S$ *be any set of valuations. Then there exist finitely many valuations* $v_1,\ldots,v_r$ *such that* $C(k)$ *in its diagonal imbedding is a direct summand of* $\Pi_{i=1}^r C(k_{v_i})$.

This lemma evidently can also be reduced to the case when G is absolutely almost simple; and in that case Lemma 2.2 is implied by the following statement. *Let* p *be a prime dividing the order* C. *Then there exists infinitely many valuations* v *of* k *such that the inclusion* $C(k) \hookrightarrow C(k_v)$ *induces an isomorphism in the* p*-torsion subgroups.* Our proof involves first an examination of this center C as a k-group. As is well-known there is a unique quasi-split group G_o over k such that G is an "inner" form of G. Now a Galois twist by an inner cocycle obviously cannot change the k-structure of the center C_o of G_o so that we have

2.3 Lemma. C *is* k*-isomorphic to* C_o, *the center of the unique quasi-split* k*-group* G_o *of which* G *is an inner form.*

The structure of quasi-split groups is of course well-known and their centers afford reasonably simple descriptions. The table below furnishes information on the structure of $C \simeq C_o$ needed in the sequel. In the column type we use the notation of Tits in his classification. $K = K(G)$ denotes the minimal extension of k over which G_o splits. $T(G) = T(G_o)$ will denote the split torus G_m if $(K(G):k) = 1$; if $(K(G):k) \neq 1$, $T(G_o)$ denotes the k torus of norm 1 elements in K/k. We denote by π the Galois group of K over k. For any torus S

	Type of G	$(K:k)$	Information on C C_o
1	A_n	1	$C \simeq T(G)_{n+1}$ over k $(T(G) \underset{k}{\approx} G_m)$
2	C_n, B_n, E_7	1	$C \simeq T(G)_2$ over k $(T(G) \underset{k}{\approx} G_m)$
3	D_n n odd	1	$C \simeq T(G)_4$ over k $(T(G) \underset{k}{\approx} G_m)$
4	D_n n even	1	$C \simeq T(G)_2 \oplus T(G)_2$ over k $(T(G) \underset{k}{\approx} G_m)$
5	E_6	1	$C \simeq T(G)_3$ over k $(T(G) \underset{k}{\approx} G_m)$
6	E_8, F_4, G_2	1	C is trivial
7	${}^2A_n$	2	$C \simeq T(G)_{n+1}$ over k $(T(G) \underset{K}{\approx} G_m)$
8	${}^2D_n$ n odd	2	$C \simeq T(G)_4$ over k $(T(G) \underset{K}{\approx} G_m)$
9	${}^2D_n$ n even	2	$C \simeq R_{K/k}(G_m)2$
10	${}^2E_6$	2	$C \simeq T(G)_3$ over k $(T(G) \underset{K}{\approx} G_m)$
11	${}^3D_4$	3	$C \simeq (G_m)_2 \oplus (G_m)_2$ over K; as a π-module $C(K)$ is irreducible
12	${}^6D_4$	6	$C \simeq (G_m)_2 \oplus (G_m)_2$ over K; if $L \subset K$ is a cubic extension of k, $C \simeq R_{K/L}(G_m)_2$ over L

and integer $r > 0$, $S(r)$ will denote the group of elements of order r in S. The table shows that C is k-isomorphic to one of the following list of groups:

1. $(G_m)_r$ for some integer r
2. $(G_m)_2 \oplus (G_m)_2$

3. $R_{K/k}(G_m)_2$ with $(K:k) = 2$
4. T_r where T is the torus of norm 1 elements in a quadratic extension K of k
5. A group isomorphic to $(G_m)_2 \oplus (G_m)_2$ or $R_{K/L}(G_m)_2$ over $L \subset K$, $(K:L) = 2$ (L is *cubic* extension of k).

We will take these cases up one by one.

Case 1. $C \simeq (G_m)_r$. The group C is imbedded in the exact sequence

$$1 \to C \to G_m \xrightarrow{u} G_m \to 1$$

where $u(x) = x^r$. Since $H^1(k, G_m) = 0$, the Galois cohomology exact sequence yields an isomorphism $H^1(k,C) \simeq k^*/(k^*)^r$. Our Lemma 2.1 follows then from the following well-known result from class field theory: if $x \in k^*$ is not a r-th power in k, there are *infinitely* many valuations v of k such that x is not an r-th power in k_v: in fact we have that h *is injective*.

Case 2. $C \simeq (G_m)_2 \oplus (G_m)_2$, $R_{K/L}(G_m)_2$ over a *cubic* extension L of k. Since C is a 2-torsion group, the map $H^1(k,C) \to H^1(L,C)$ is injective. When $C \simeq (G_m)_2 \oplus (G_m)_2$ over L, by the arguments of Case 1 (with L replacing k) we have the injectivity of $H^1(L,C) \to \Pi_{v \notin S_L} \cdot H^1(L_v, C)$ where S_L denotes the set of valuations of L lying over S. If $C \simeq R_{K/L}(G_m)_2$, again from standard facts about Galois cohomology we have $H^1(L,C) \simeq H^1(K,(G_m)_2)$ and we can appeal to Case 1 and conclude again that in the cases h *is injective*.

Case 3. $C \simeq T_r$, T the (1-dimensional) torus of norm 1 element in the quadratic extension K of k with 4 *not dividing* r. Evidently in this case either r is odd or $T_r \simeq T_{r/2} \oplus T_2$; and $T_2 \simeq (G_m)_2$. It follows that h is injective if we show that $H^1(k,T_r) \to \Pi_{v \notin S} H^1(k_v, T_r)$ is injective for *odd* r. Since $T_r \simeq (G_m)_r$ over K and $H^1(k,T_r) \to H^1(K,T_r)$ is injective for odd r, we conclude that h *is injective* in this case also exactly as we did in Case 2.

Case 4. $C \simeq T_r$, T the 1-dimensional torus of norm 1 elements in quadratic extension K of k with 4 dividing r. We decompose now r into a product $r = p.q$ with p odd and q a power of 2; then $T_r = T_p \oplus T_q$. We have already seen that $H^1(k,T_p) \to \Pi_{v \notin S} H^1(k_v,T_p)$ is injective. Thus we have to deal with the case when $r = q$ is a power of 2. Here again over a quadratic extension $T_q \simeq (G_m)_q$ so that if $h(x) = 0$ for $x \in H^1(k,T_r)$, x has trivial image in $H^1(K,T_r)$. It follows that x belongs to the subgroup $H^1(\pi,T_r(K)) \subset H^1(k,T_r)$. It suffices thus to show that $H^1(\pi,T_r(K))$ is either trivial or isomorphic to $Z/2$. Since the torus T is isomorphic to G_m over K, $T_r(K)$ is the group of r-th roots of 1 in K. The Galois group is a group of order 2. $T_r(K)$ is a cyclic group with a generator α, say. There are only four possible automorphisms of this cyclic subgroup of K^* which are of order 2. The effects of the generator of π on α are given by the four following possibilities

$$\sigma(\alpha) = \begin{matrix} \alpha \\ \alpha^{-1} \\ -\alpha \ (\neq \alpha^{-1}) \\ -\alpha^{-1} \ (\neq \alpha) \end{matrix}$$

Taking note of the fact that $\sqrt{-1}$ is an *even* power in the last case, a simple computation shows that $H^1(\pi,T_r(K))$ is $Z/2$ in the first two cases while it is trivial in the last. We see thus even here h is sometimes injective. However, it must be noted that all four possibilities described above for the Galois action do indeed occur for suitable choices of k, K (and G).

We now take up the proof of Lemma 2.2. Here again we argue case by case.

Case 1. $C \simeq (G_m)_r$. Let p be any prime and p^a the order of the p-torsion subgroup of $(G_m)_r$. Let p^b the order of the p-torsion subgroup of $C(k)$. Let L be the extension of k obtained by adjoining a $p^{(b+1)}$-th root of 1 to k. Then L/k is cyclic of order t, say ($t = p$ if k contains p-th root of 1; otherwise t divides $(p-1)$). Now by the Cebatarev theorem there exist infinitely many $v \in V$ such that L is linearly disjoint from k_v. It follows that $L_v = L \otimes_k k_v$ is a cyclic extension of k_v of degree t and $C(k_v)$ has p-torsion subgroup of order p^b. This proves the result in Case 1.

Case 2. $C \simeq (G_m)_2 \oplus (G_m)_2$. This follows from Case 1 immediately.

Case 3. $C \simeq R_{K/k}(G_m)_2$. There are infinitely many v such that $K_v = K \otimes_k k_v$ is an unramified quadratic extension of k_v. Over any such k_v we have again $C \simeq R_{K_v/k_v}(G_m)_2$ leading to $C(k) \simeq C(k_v) \simeq Z/2$.

<u>Case 4</u>. $(K:k) = 3$ or 6. In these cases a cyclic subgroup π' (of π) of order 3 acts nontrivially irreducibly on $C(K)$. If we choose v to be one of the infinitely many unramified valucations in which π' becomes the Galois group, clearly $C(K_v) = C(K)$ will remain an irreducible module over the Galois group. Thus $C(K_v)$ has no Galois group invariants, i.e., $C(k_v)$ is trivial; this proves the desired result.

This leaves us to discuss the last

<u>Case 5</u>. $(K:k) = 2$, $C \simeq T_r$ with T the group of norm 1 elements in K. Let α be a generator of the p-torsion in $T_r(K) \simeq (G_m)_r(K)$: α may be regarded as a root of 1 in K^*. Let β be a p-th root of α. Suppose first that $\alpha \notin C(k)$. Then under the action of the Galois group $\sigma(\alpha) \neq \alpha$. Let v be one of the infinitely many valuations such that $K_v = K \otimes_k k_v$ is a quadratic extension of k_v. If $\beta \in C(K_v)$ does not belong to $C(K)$, then α is necessarily a power of β. Thus $\sigma \in \mathrm{Gal}(K_v/k_v) = \mathrm{Gal}(K/k)$ operates *nontrivially* on any such β. It follows that p-torsion in $C(k_v) \subset C(K)$ and hence $c(k_v)$ and $C(k)$ have the same p torsion. Now consider the case when $\alpha \in C(k)$. In this case let v be the restriction of a valuation w of K such that $C(K) = (G_m)_r(K)$ and $(G_m)_r(K_w) = C(K_w)$ have the same p-torsion subgroup. (Case 1 shows that infinitely many such w exist.) Clearly $C(k)$ and $C(k_v)$ then have the same p-torsion subgroup. This settles Case 5 and Lemma 2.2 is completely proved.

3. The Case of the Adjoint Isogeny

In this section we prove the main theorem for the natural isogeny of a connected *simply connected* semisimple k-group G onto its adjoint group G^*. We denote the adjoint isogeny by f. Note that in this special case $C_{G^*} = 1$ so that the main theorem takes a simpler form viz.: if Γ is a finite intersection of maximal S-arithmetic groups $f(\Gamma)$ is special arithmetic and $d(f(\Gamma))$ has exponent ≤ 2 and order $\leq 2^r$ where r is the number of k simple factors of G.

Consider first a maximal S-arithmetic subgroup $\Gamma \subset G(k)$. Let $\Gamma' \subset G^*$ be a maximal S-arithmetic subgroup of G^* containing $\Gamma_1^* = f(\Gamma)$. The group Γ' is finitely generated. The sequence $1 \to C_G \to G \to G^* \to 1$ gives rise to Galois cohomology exact sequence $(*) G(k) \to G^*(k) \xrightarrow{\delta} H^1(k,C)$. Evidently, $\delta(\Gamma')$ is a finite group. We can find a finite set S', $S' \cap S = \emptyset$ such that the kernel of $h: \delta(\Gamma') \to \Pi_{v \notin S} H^1(k_v, C)$ is the same as the kernel of $\delta(\Gamma') \to \Pi_{v \in S'} H^1(k_v, C)$. The kernel of h is of exponent $\leqslant 2$ and order $\leqslant 2^r$ (Lemma 2.1). For $v \in V$, let P_v denote the image of $G(k_v)$ in $G^*(k_v)$; P_v is an open subgroup of the latter. Let $\tilde{\Gamma}^* = \{x \in \Gamma' | x \in P_v$ for all $v \in S'\}$. Then $\tilde{\Gamma}^*$ is an S-congruence subgroup of G^*. Let $\Gamma^* = \{x \in \tilde{\Gamma}^* | \delta(x) = 0\}$. Then by the exactness of $*$, $\Gamma^* \subset f(G(k))$. $\Gamma^* f(f^{-1}(\Gamma'))$. But $f^{-1}(\Gamma') \supset \Gamma$ and hence equals Γ by maximality of Γ. Thus $\Gamma^* \subset f(\Gamma)$. Evidently Γ^* is normal in $\tilde{\Gamma}^*$ and $\tilde{\Gamma}^*/\Gamma^*$ has exponent $\leqslant 2$ and index $\leqslant 2^r$. Now $\hat{\Gamma}^* \subset \tilde{\Gamma}^*$ so that $\hat{\Gamma}^*/\Gamma^*$ also has exponent $\leqslant 2$ and index $\leqslant 2^r$.

Suppose now that $\Gamma = \Phi_1 \cap \ldots \cap \Phi_q$ with Φ_i maximal arithmetic in G. Choose $\tilde{\Phi}_i^*$ to be congruence subgroups such that $\delta(\tilde{\Phi}_i^*) \subset$ kernel h and $\Phi_i^* = \{x \in \Phi_i^* | \delta(x) = 0\} \subset f(\Phi_i)$. Since each Φ_i contains the kernel of f, we see that $f(\Gamma) \supset \Gamma^* = \cap_{i=1}^q \Phi_i^*$. Now let $\tilde{\Gamma}^* = \cap_{i=1}^q \Phi_i^*$; if $x \in \tilde{\Gamma}^*$ with $\delta(x) = 0$, then $x \in \Gamma^*$. Hence Γ^* is normal in $\tilde{\Gamma}^*$ and $\tilde{\Gamma}^*/\Gamma^*$ has exponent $\leqslant 2$ and order $\leqslant 2^r$. It follows that $\hat{\Gamma}^*/\Gamma^*$ also has exponent $\leqslant 2$ and order $\leqslant 2^r$, i.e., exponent $d(\Gamma^*\ ) \leqslant 2$, $|(d(\Gamma^*))| \leqslant 2^r$.

To obtain the last assertion we make some observations most of which are known from the work of Platanov on strong approximation (see Platanov, 1969, 1971, Gopal Prasad). A maximal arithmetic subgroup of G is obtained as follows. Fix a realization of G as a k-subgroup of $GL(n)$ and set for $v \in V$ non-archimedean $G(O_v) = G(k_v) \cap GL(n, O_v)$. We assume, as we may, that the realization of G as a subgroup of $GL(n)$ is such that $M_v = G(O_v)$ is a *maximal compact* (open) subgroup of $G(k_v)$ for all (non-archimedian) v. Then any maximal compact subgroup of S-adele group $G(\mathbb{A}(S))$ of G is of the form $\Pi_{v \notin S} M'_v$ where M'_v is a maximal compact subgroup of $G(k_v)$ and $M'_v = M_v$ for almost all v. The intersection of a maximal compact subgroup of $G(\mathbb{A}(S))$ with $G(k)$ is a maximal S-arithmetic subgroup of $G(k)$. Conversely if Γ is a maximal S-arithmetic subgroup of $G(k)$ the closure of Γ in $G(A(S))$ is a maximal compact subgroup of $G(A(S))$ whose intersection with $G(k)$ is precisely Γ. It follows that $U \mapsto U \cap G(k)$ sets up a bijection between finite intersections of maximal compact subgroups in $G(\mathbb{A}(S))$ and finite intersections of maximal S-arithmetic subgroups in $G(k)$. From the description of the maximal compact subgroups in $G(\mathbb{A}(S))$, it is clear that finite intersections of such groups take the form of a product $\Pi_{v \notin S} B_v$ where each B_v is a compact open subgroup of $G(k_v)$ which is a finite intersection of maximal compact subgroups of $G(k_v)$ and $B_v = M_v$ for almost all v.

Now the intersection of *all* the maximal compact subgroups of $G(k_v)$ is clearly the center $C(k_v)$ of $G(k_v)$. It follows that if $B'_v \subset G(k_v)$ is *any compact open* subgroup of $G(k_v)$, $B'_v.C(k_v)$ contains a finite intersection of maximal compact subgroup. If B'_v is torsion free-- such compact open subgroups exist--then $B'_v.C(k_v)$ is a direct product; if we choose any finite intersection B_v of maximal compact subgroups contained in $B'_v.C(k_v)$, then such a B_v decomposes into a direct product $(B_v \cap B'_v).C(k_v)$ as well. In the sequel we fix a torsion-free open compact subgroup B'_v for each $v \in V$, v non-archimedean.

Suppose now that S_1 is a finite subset of V with $S_1 \cap S = \emptyset$ and $C(k)$ is a direct summand of $\Pi_{v \in S_1} C(k_v)$ (Lemma 2.2). Let Γ_1 be any finite intersection of maximal arithmetic subgroups of $G(k)$. Then Γ_1 contains a subgroup Γ which is again a finite intersection of maximal arithmetic subgroups and whose closure $\bar{\Gamma}$ in $G(\mathbb{A}(S))$ is a product $\Pi_{v \notin S} B_v$ where B_v is a maximal compact subgroup of $G(k_v)$ for almost all v and for $v \in S_1$, $B_v = D'_v.C(k_v)$, a direct product with D'_v torsion-free. Now let $B = \Pi_{v \in S_1} B_v$. Then $B = \Pi_{v \in S_1} D'_v \cdot \Pi_{v \in S_1} C(k_v)$, a direct product. Now $C(k)$ is a direct summand of $\Pi_{v \in S_1} C(k_v)$. It follows that B is a direct product of the subgroup $C(k)$ and a compact open subgroup D of $\Pi_{v \in S_1} G(k_v)$. Let $\Gamma' = \{x \in \Gamma | x \in D$--when x is regarded as an element of $\Pi_{v \in S_1} G(k_v)\}$. Suppose now that $x \in \Gamma$, then we can write $x = x'.\theta$ where $x' \in D$ and $\theta \in C(k)$. Since $C(k) \subset \Gamma$, $x' \in \Gamma$ and hence $x' \in \Gamma'$ and $f(x) = f(x')$. Since x was arbitrary, $f(\Gamma) = f(\Gamma')$. Since $\Gamma' \subset D$, $\Gamma' \cap C(k) = 1$ proving the required result.

4. General Isogenies

We consider now a general isogeny between reductive groups G and $H: f: G \to H$. G is an almost direct product $A.M$ of a torus A and a semisimple group M. Similarly H is an almost direct product $B.N$ with B a torus and N semisimple. Also $f(A) = B$ and $f(M) = N$. Now we claim that an S-arithmetic subgroup $\Phi \subset H$ is an S-congruence subgroup if and only if $\Phi \cap N$ is an S-arithmetic subgroup. Once the claim is granted, it is clear that the main theorem for reductive G is a consequence of the main theorem for semisimple groups.

To establish the claim, we observe first that because of a theorem of Chevalley every S-arithmetic subgroup of a torus is an S-congruence subgroup. Now consider the natural map $\rho: H \to H/N$; H/N is a torus.

Let Γ be an S-arithmetic subgroup in H with $\Gamma \cap N$ an S-congruence subgroup. We assume $H \subset GL(n)$ as a k-subgroup and $\Gamma \subset H \cap GL(n,A(S))$ contains all $x \in N \cap GL(n,A(S))$, $x \equiv 1 \pmod{\mathfrak{a}}$ for some nonzero ideal $\mathfrak{a}$ in $A(S)$. Next by Chevalley's theorem we know that there is a non-zero ideal $\mathfrak{b} \subset \mathfrak{a}$ of $A(S)$ such that $B \cap \Gamma \supset \{x \in B \cap GL(n,A(S)) \mid x \equiv 1 \pmod{b}\} = \Psi$, say. Let $\mathfrak{c} \subset \mathfrak{b}$ be an ideal such that the group $\{x \in H \cap GL(n,A(S)) \mid x \equiv 1 \pmod{c}\} = \Gamma'$ maps into the image of Ψ under ρ. We claim that $\Gamma' \subset \Gamma$. In fact if $x \in \Gamma'$ we can find $y \in \Psi$ such that $\rho(x) = \rho(y)$. Thus $xy^{-1} \in N$. Clearly, $xy^{-1} \in GL(n,\mathfrak{b}) = \{x \in GL(n, A(S)) \mid x \equiv 1 \pmod{\mathfrak{b}}\}$ as well. Thus $xy^{-1} \in N \cap \Gamma$. Hence $x \in N \cap \Gamma$. $B \cap \Gamma \subset \Gamma$. It follows that $\Gamma' \subset \Gamma$.

The claim having been established we can now restrict our attention to isogenies between semisimple group G, H. Let $f: G \to H$ be the isogeny and $p: \tilde{G} \to G$ the universal covering of G. Let $u = f \circ p$. Since the inverse image under p of a maximal arithmetic group is a maximal arithmetic subgroup of $\tilde{G}$ and for $\Gamma \subset \tilde{G}$ with $\Gamma \cap C_{\tilde{G}}(k) = 1$ we have $p(\Gamma) \cap C_G(k) = \{1\}$, one sees immediately that the main theorem for the isogeny $f \circ p$ implies the main theorem for f itself. In other words there is no loss of generality in assuming that $\tilde{G}$ is simply connected.

From now on we assume that G is simply connected. Let $q: H \to G^*$ be the unique isogeny of H on the adjoint group of G so that $q \circ f$ is the natural isogeny of G on G^*. Let Γ, Γ_o, Γ_o' be as in the main theorem applied to the isogeny $q \circ f: G \to G^*$ (for which the main theorem is proved already). Let $\Phi^* \subset q \circ f(\Gamma_o) = q \circ f(\Gamma_o')$ be a special arithmetic subgroup in G^* with $\hat{\Phi}^*/\Phi^*$ of index $\leqslant 2^r$ and exponent $\leqslant 2$. Let $\tilde{\Gamma}^* = q^{-1}(\Phi^*)$ and $\tilde{\Gamma}_1^* = q^{-1}(\hat{\Phi}^*)$. Then $\tilde{\Gamma}_1^*$ is an S-congruence subgroup in which $\tilde{\Gamma}^*$ is a normal subgroup with $|\tilde{\Gamma}_1^*/\Gamma^*| \leqslant 2^r$. Now consider the group $f(\Gamma)$. If $x \in \tilde{\Gamma}^*$ we can find $y \in \Gamma$ such $q(f(y)^{-1}x) = 1$, i.e., $f(y)^{-1}x \in C_H(k)$. Thus we see that $f(\Gamma)C_H(k)$ contains $\tilde{\Gamma}^*$. Hence if we set $\Gamma^* = f(\Gamma) \cap \tilde{\Gamma}^*$, we have $\Gamma^* C_H(k) = \tilde{\Gamma}^*$ and $d(\tilde{\Gamma}^*) \leqslant 2^r$ so that $d(\Gamma^*) \leqslant 2^r . |C_H(k)|$. This completes the proof of the main theorem.

We now establish the necessary condition for $f(\Gamma)$ to be a congruence group in the case of adjoint isogenies. Let $\Gamma \subset G(k)$, G simply connected be the inverse image of an S-congruence subgroup Γ^* of G^*. We assume as we may that Γ^* is of the following form. Fix a realization $G^* \subset GL(n)$ of G^* as a k-subgroup of $GL(n)$. Let $M_v^* = G^* \cap GL(n, O_v)$ and $M_v \subset G(k_v)$ the inverse image of M_v^* in $G(k_v)$;

then there are open compact subgroups $B_v^* \subset M_v^*$ such that $B_v^* = M_v^*$ for almost all v and

$$\Gamma^* = \{x \in G^*(k) \mid x \in B_v^* \quad \text{for all} \quad v \notin S\} \ .$$

Let $B_v = f^{-1}(B_v^*)$ in $G(k_v)$. Evidently then $\Gamma = \{x \in G(k) \mid x \in B_v$ for all $v \notin S\}$. Now B_v is a maximal compact subgroup for almost all v (coherent with respect to k-structure). In addition every B_v contains $C(k_v)$ the center of $G(k_v)$. Now for fixed v, each $B(v)$ contains a finite intersection D_v of maximal compact subgroups in $G(k_v): D_v \subset B_v$; also we take $D_v = B_v$ for almost all v. Then $\Pi_{v \notin S} D_v$ is a finite intersection of maximal compact subgroups in $G(\mathbb{A}(S))$ and its intersection Γ' with $G(k)$ is (because of strong approximation) an intersection of maximal S-arithmetic subgroups. Since $\Gamma' \subset \Gamma$ the necessity of our condition is proved.

To exhibit an S-arithmetic subgroup Γ such that $\Gamma \cdot C(k)$, $(C = C_G)$ is not an intersection of maximal arithmetic subgroups is not difficult. Start with a group Γ which is a finite intersection of maximal arithmetic subgroups. Assume Γ so chosen that the following property holds. There exists two finite sets S_1, S_2 with $S_1 \cap S_2 \neq \emptyset$ such that S_1, S_2 and the closures $\bar{\Gamma}_i$ of Γ in $\Pi_{v \in S_i} G(k_v) = G_i$, $i = 1,2$ have the following property: $C(k)$ is a direct factor of $\bar{\Gamma}_i : \bar{\Gamma}_i = D_i \times C(k)$ with D_i open and compact; $\bar{\Gamma}_i = D_i' \, \Pi_{v \in S_i} C(k_v)$ with $D_i' \subset D_i$, D_i' torsion free. Such a choice of S_1 and S_2 is possible (Lemma 2.2 and well-known facts about maximal compact subgroups in p-adic semisimple groups lead to this). Let $C(k) \neq \{1\}$; set $\Gamma' = \{x \in \Gamma \mid x \in D_i', \ i = 1,2\}$. We claim that $f(\Gamma')$ cannot be an S-congruence subgroup. The closure of $\Gamma'.C(k)$ in $G_1 \times G_2$ is contained in $\bar{\Gamma}_1 \times \bar{\Gamma}_2 = D_1 \times D_2 \times C(k) \times C(k)$. The closure of $\bar{\Gamma}' \cdot C(k)$ goes over under this isomorphism into $D_1 \times D_2 \times$ diagonal in $\{C(k) \times C(k)\}$. And this last group does not contain $C(k) \times C(k)$; if $C(k) \neq 1$, it follows that the closure of $\Gamma' \cdot C(k)$ is not a finite intersection of maximal compact subgroups in $G_1 \times G_2$; hence $\Gamma'.C(k)$ is not a finite intersection of maximal arithmetic subgroups. When $C(k) = \{1\}$ we can argue with in fact a single $v \notin S$ with $C(k_v) \neq \{1\}$. If M_v is a compact open subgroup with $M_v \cap C(k_v) = \{1\}$ and Γ' is an S-congruence group whose closure in $G(k_v)$ is contained in M_v, then $\Gamma = \Gamma' C(k)$ cannot be the finite intersection of maximal arithmetic subgroups of G.

For instance when $k = \mathbb{Q}$, $G = SL(3)$, the adjoint image of $SL(3,7) = \{x \in SL(3,\mathbb{Z}) \mid x \equiv 1 \pmod 7\}$ is not a congruence group in the adjoint group of $SL(3)$. In $SL(2)$, on the other hand for every prime p, $SL(2,p^n)$ maps onto a congruence subgroup in the adjoint group but if p, q are distinct odd primes $SL(2,pq)$ does not map onto a congruence subgroup.

Remark. It is possible now to formulate a suitable result for any *surjective* k-morphism of algebraic groups. The formulation is straightforward; on the other hand it is somewhat cumbersome. So we will not go into it here. The essential difficulties are in dealing with isogenies of reductive groups.

5. References

A. Borel and Harish Chandra, "Arithmetic subgroups of algebraic groups," *Annals of Mathematics* 75 (1962), 485-535.

F. Bruhat and J. Tits, "Groupes algebriques simple sur un groupe local," in Proc. Conf. Local Fields (Driebergen, 1966), Springer, Berlin 1967, 23-36.

C. Chevalley, "Deux theoremes d'arithmetique," *J. of Math. Soc. Japan* 3 (1951), 36-44.

G. Prasad, "Strong approximation for semisimple groups over function fields," *Ann. Math.* 105 (1977), 553-572.

V.P. Platonov, "The problem of strong approximation and the Kneser-Tits conjecture for algebraic groups," *Math USSR-Izventiya* 3 (1969), 1139-1147 (English translation).

V.P. Platonov, "On the maximality problem of arithmetic groups," *Soviet. Math. Doklady* 12 (1971), 1431-1435 (English translation).

M.S. Raghunathan, "On the congruence subgroup problem," *Publ. Math. IHES*, No. 46 (1976), 107-161.

J. Tits, "Classification of algebraic semisimple groups," Proc. Symp. Pure Math. 9 (1966), AMS.

Tata Institute of Fundamental Research
Bombay 400 005
India

(Received January 5, 1981)

ON COMPACT EINSTEIN KÄHLER MANIFOLDS WITH ABUNDANT HOLOMORPHIC TRANSFORMATIONS

Yusuke Sakane

0. Introduction

Let (M,J,g) be a compact connected Kähler manifold and let $Ric(g)$ denote the Ricci tensor. A compact Kähler manifold (M,J,g) is said to be *Einstein* if $Ric(g) = kg$ for some $k \in \mathbb{R}$. If we denote by γ the Ricci form of (M,J,g) $(\gamma(X,Y) = Ric(g)(X,JY))$ and by ω the Kähler form, (M,J,g) is Einstein if and only if $\gamma = k\omega$ $(k \in \mathbb{R})$. Let $H^2(M,\mathbb{R})$ denote the 2nd cohomology group with the coefficients in $\mathbb{R}$. It is known that the first Chern class $c_1(M)$ of a compact Kähler manifold (M,J,g) is given by

$$c_1(M) = \frac{1}{2\pi}[\gamma] \in H^2(M,\mathbb{R}) .$$

The first Chern class $c_1(M)$ of a compact complex manifold (M,J) is said to be *positive* (resp. *negative*) if it contains a closed real 2-form of type $(1,1)$ such that the associated hermitian form is positive definite (resp. negative definite). Thus if (M,J,g) is Einstein Kähler, we have

a) $c_1(M) > 0$ if $k > 0$

b) $c_1(M) < 0$ if $k < 0$

c) $c_1(M) = 0$ if $k = 0$.

Now we can ask whether the converse is true or not.

<u>Theorem</u> (Aubin [1], Yau [27]). *If* (M,J,g) *is a compact Kähler manifold with* $c_1(M) < 0$, *then there is a unique Einstein Kähler metric* $\tilde{g}$ *on* (M,J) *up to constant multiple.*

<u>Theorem</u> (Yau [27]). *If* (M,J,g) *is a compact Kähler manifold with* $c_1(M) = 0$, *then there exists an Einstein Kähler metric* $\tilde{g}$ *on* (M,J).

But in the case of $c_1(M) > 0$, it is known that there are compact Kähler manifolds which do not admit any Einstein Kähler metric (see [26],[10]). In this note we shall consider compact Einstein Kähler manifolds with $c_1(M) > 0$ using a theorem of Matsushima (see Theorem 3.5). In Section 1 we consider connected closed subgroups of $SU(n)$. Let $Aut(M,J,g)$ denote the Lie group of all holomorphic isometries of (M,J,g). In Section 2 we determine n-dimensional compact Kähler manifolds (M,J,g) such that $\dim Aut(M,J,g) \geqq n^2+1$ $(n \geqq 2)$. Let $Aut(M,J)$ denote the complex Lie group of all holomorphic transformations of (M,J). In Section 3 we determine n-dimensional compact Einstein Kähler manifolds with $c_1(M) > 0$ such that $\dim_{\mathbb{C}} Aut(M,J) \geqq n^2 - 2n + 8$ $(n \geqq 6)$. Our results in this Section depend on the classification of projective algebraic manifolds endowed with regular actions of a certain algebraic group due to Mabuchi [18]. In Section 4, as an application we show that a projective algebraic submanifold M defined by a hyperplane section of the Segre imbedding

$$P^1(\mathbb{C}) \times P^n(\mathbb{C}) \longrightarrow P^{2n+1}(\mathbb{C}) \qquad (n \geqq 2)$$

does not admit any Einstein Kähler metric and has a positive first Chern class $c_1(M) > 0$.

1. Connected Closed Subgroups of $SU(n)$

We consider connected closed proper subgroups of $SU(n)$. Following lemma is essentially due to Uchida ([25] Lemma 1.5).

Lemma 1.1. *Let* H *be a connected closed proper subgroup of* $SU(n)$ $(n \geqq 4)$.

(a) *For* $n \geqq 5$, $n \neq 6$, *if* $\dim H \geqq n^2 - 4n + 7$, H *is conjugate to* $SU(n-1)$, $N(SU(n-1),SU(n))$ *the normalizer of* $SU(n-1)$ *in* $SU(n)$ *or* $N(SU(n-2),SU(n))$ *the normalizer of* $SU(n-2)$ *in* $SU(n)$.

(b) *For* $n = 6$, *if* $\dim H \geqq n^2 - 4n + 7 = 19$, H *is conjugate to* $SU(n-1)$, $N(SU(n-1),SU(n))$, $N(SU(n-2),SU(n))$ *or* $Sp(3)$ $(\dim Sp(3) = 21)$.

(c) *For* $n = 4$, *if* $\dim H \geqq 9$, H *is conjugate to* $N(SU(n-1),SU(n))$ *or* $Sp(2)$ $(\dim Sp(2) = 10)$.

Proof. We can prove our lemma by the same way as Lemma 1.5 in [25], but we repeat the proof for convenience sake. The inclusion

$\rho: H \to SU(n)$ gives an n-dimensional complex representation of H.

(1) First we consider the case when the representation ρ is *irreducible* and hence H is semisimple. If H is not simple, then there are integers $p \geqq q \geqq 2$ with $pq = n$, such that H is conjugate to a subgroup of the tensor product $SU(p) \otimes SU(q)$ in $SU(n)$, by considering the induced representation of the universal covering group of H. Therefore

$$\dim H \leqq p^2 + q^2 - 2 \leqq \left(\frac{n}{2}\right)^2 + 2 < n^2 - 4n + 7 \qquad \text{for} \quad n \geqq 5$$

and $\dim H \leqq 6$ for $n = 4$. Hence if $\dim H \geqq n^2 - 4n + 7$ for $n \geqq 5$ or $\dim H \geqq 9$ for $n = 4$, H is simple. If H is simple but not one of the type A_k, D_{2k+1} and E_6, then H is conjugate to a subgroup of $SO(n)$ or $SP(n/2)$ (see [7], p. 336 Theorem 0.20 or [6], p. 132). Since $\dim SO(n) = n(n-1)/2$, $\dim SP(n/2) = n(n+1)/2$, we have $\dim H \leqq n(n+1)/2$. Note that $n^2 - 4n + 7 \geqq n(n+1)/2$ for $n \geqq 8$, $n^2 - 4n + 7 = n(n+1)/2$ for $n = 7$, $n^2 - 4n + 7 = 19$ for $n = 6$ and $n^2 - 4n + 7 = 12$ for $n = 5$. For $n = 7$ we have $\dim H < n(n+1)/2$ and for $n = 6$ we have $\dim SO(6) = 15$ and $\dim Sp(3) = 21$. Note also the maximal dimension of closed proper subgroups of $Sp(3)$ is 13. For $n = 5$ we have $\dim SO(5) = 10$ and for $n = 4$ we have $\dim SO(4) = 6$ and $\dim Sp(2) = 10$. Note that the maximal dimension of closed proper subgroup of $Sp(2)$ is 6. Therefore if H is simple but not one of the type A_k, D_{2k+1} and E_6, and $\dim H \geqq n^2 - 4n + 7$ $(n \geqq 5)$, then $n = 6$ and $H = Sp(3)$. Moreover under the same assumption if $n = 4$ and $\dim H \geqq 9$ then $H = Sp(2)$. If H is of type D_{2k+1} $(k \geqq 2)$, then the lowest dimensional nontrivial irreducible complex representation is $(4k+2)$-dimensional. Therefore, $4k + 2 \leqq n$ and hence $n \geqq 10$ and $\dim H = \dim SO(4k+2) = (2k+1)(4k+1) \leqq n(n-1)/2$. If H is of type E_6, then $n \geqq 27$. Therefore $\dim H = 78 \leqq 3n \leqq n(n+1)/2 < n^2 - 4n + 7$. If H is of type A_{k-1} $(k < n)$, then $k(k-1)/2 \leqq n$ by the Weyl's formula for the dimension of irreducible representation. Thus $\dim H = \dim SU(k) = k^2 - 1 \leqq 3n - 2 < n^2 - 4n + 7$ $(n \geqq 6)$. For $n = 5$ we see easily that $k \leqq 3$ and hence $\dim H \leqq 8 < n^2 - 4n + 7$. For $n = 4$ we have $\dim H = \dim SU(k) \leqq 8$. Note also that if $\dim H \geqq (n-1)^2$ $(n \geqq 4)$, then $n = 4$ and $H = Sp(2)$.

(2) Now we consider the case when the representation ρ is *reducible*. Let $N(SU(n-p), SU(n))$ $(1 \leqq p \leqq n/2)$ denote the normalizer

of $SU(n-p)$ in $SU(n)$. Then H is conjugate to a subgroup of $N(SU(n-p),SU(n))$. But $\dim N(SU(n-p),SU(n)) = (n-p)^2 + p^2 - 1 \leqq (n-2)^2 + 2^2 - 1 = n^2 - 4n + 7$ $(n \geqq 4)$, for $2 \leqq p \leqq n/2$ and the equality holds if and only if $p = 2$. If $\dim H > n^2 - 4n + 7$, then H is conjugate to a subgroup H' of $N(SU(n-1),SU(n))$. If $H' \neq N(SU(n-1),SU(n))$, then

$$\dim H' \leqq \dim H'' + 1$$

where $H'' = H' \cap SU(n-1)$, by the isomorphism

$$N(SU(n-1),SU(n))/SU(n-1) = S^1 ,$$

if $H'' = SU(n-1)$, then $H' = H'' = SU(n-1)$. If $H'' \neq SU(n-1)$, then

$$\dim H'' \leqq (n-2)^2 = \dim N(SU(n-2),SU(n-1)) \quad \text{for} \quad n \geqq 6$$

$$\dim H'' \leqq 10 \quad \text{for} \quad n = 5 ,$$

by making use of the first part of the proof of this lemma for $SU(n-1)$ instead of $SU(n)$, and hence

$$\dim H' \leqq (n-2)^2 + 1 = n^2 - 4n + 3 < n^2 - 4n + 7 \qquad (n \geqq 6)$$

$$\dim H' \leqq 10 + 1 = 11 < 12 = n^2 - 4n + 7 \qquad (n = 5) .$$

Therefore we see that H is conjugative to $SU(n-1)$ or $N(SU(n-1), SU(n))$ $(n \geqq 5)$. For $n = 4$, $\dim SU(3) = 8$ and hence we have $H = N(SU(n-1),SU(n))$ if $\dim H \geqq 9$. ∎

Now we have the following corollary (cf. [19], p. 539).

Corollary 1.2. *Let* H *be a connected closed proper subgroup of* $SU(n)$ $(n \geqq 2)$. *If* $\dim H \geqq (n-1)^2$, *then* H *is conjugate to* $N(SU(n-1), SU(n))$ *for* $n \neq 4$ *and* $N(SU(n-1),SU(n))$ *or* $Sp(2)$ *for* $n = 4$.

Proof. For $n \geqq 4$, our claim is easy from Lemma 1.1. For $n = 2$, it is easy to see that $H = S^1$ (1-dimensional torus). For $n = 3$, we have $\operatorname{rank} H \leqq 2$ and $\dim H \leqq 4$, and hence $H = N(SU(2),SU(3))$. ∎

2. Compact Kähler Manifolds (M,J,g) with $\dim \mathrm{Aut}(M,J,g) \geqq n^2+1$

Let (M,J,g) denote an n-dimensional hermitian manifold with the complex structure J and the hermitian metric g and let $\mathrm{Aut}(M,J,g)$ denote the group of all holomorphic isometries of (M,J,g). It is well-known that

$$\dim \mathrm{Aut}(M,J,g) \leqq n^2+2n \qquad \text{(cf. [24])} .$$

Moreover the following Theorem is known (cf. [24]).

Theorem 2.1. *If* $\dim \mathrm{Aut}(M,J,g) = n^2+2n$, *then* (M,J,g) *is holomorphically isometric to one of the following*:

(a) $P^n(\mathbb{C})$ *complex projective space with a Fubini-Study metric.*

(b) $\mathbb{C}^n$ *complex Euclidean space.*

(c) D^n *open ball with a Kähler metric of negative constant holomorphic sectional curvature.*

Let (M,g) denote a Riemannian manifold with a metric g and G a connected closed subgroup of the group $\mathrm{Aut}(M,g)$ of all isometries of (M,g).

Lemma 2.2. *If* $G(x)$ *is a* G-*orbit of highest dimension, then* G *acts essentially effectively on* $G(x)$, *i.e., those elements of* G *which act trivially on* $G(x)$ *are discrete.*

Proof. See [15] Lemma 2.1, p. 198.

From now on let K denote the connected component of the identity of $\mathrm{Aut}(M,J,g)$ of an hermitian manifold (M,J,g).

Proposition 2.3 (cf. [11]). *Let* (M,J,g) *be an* n-*dimensional hermitian manifold. If* $\dim K \geqq n^2+1$, *then* K *acts transitively on* M.

Proof. Consider an orbit $K(x)$ of the highest dimension. Let H denote the isotropy subgroup of K at $x \in M$. Then H is regarded as a closed subgroup of $U(n)$ under the linear isotropy representation. Moreover the Lie algebra $\mathfrak{h}$ of H is a Lie subalgebra of $\mathfrak{o}(k)$ where $k = \dim K(x)$ by Lemma 2.2. We claim that if K is not transitive then $\dim K \leqq n^2$. Let W denote the tangent space $T_x(K(x))$ of $K(x)$ at $x \in M$.

Case (a): $\qquad d = \dim_{\mathbb{C}}(W + J_x W) < n$.

Since the complex subspace $W + J_x W$ of $T_x(M)$ is an H-invariant subspace of $T_x(M)$, we have $\dim H \leqq d^2 + (n-d)^2$. Thus

$$\begin{aligned} \dim K &= \dim H + \dim K(x) \leqq d^2 + (n-d)^2 + 2d \\ &= n^2 + 2d(d+1-n) \leqq n^2 \end{aligned}$$

Case (b): $\qquad \ell = \dim_{\mathbb{C}}(W \cap J_x W) > 0 \qquad (\ell < n)$.

We may assume that $W + J_x W = T_x(M)$. Since the complex subspace $W \cap J_x W$ is an H-invariant subspace of $T_x(M)$ and the Lie algebra $\mathfrak{h}$ of H is contained in $\mathfrak{n}(k)$ $(k \leqq 2n-1)$, we have $\dim H < \ell^2 + (n-\ell)^2$. Thus $\dim K = \dim H + \dim K(x) < \ell^2 + (n-\ell)^2 + 2n - 1 \leqq n^2 + 1$.

Case (c): $\qquad T_x(M) = W \oplus J_x W \quad$ (direct sum) .

Since $\dim_{\mathbb{R}} W = n$, we have $\dim H \leqq n(n-1)/2$. Thus $\dim K = \dim H + \dim K(x) \leqq n(n-1)/2 + n = n(n+1)/2 \leqq n^2$. ∎

Proposition 2.4. *Let* (M,J,g) *be an* n*-dimensional hermitian manifold* $(n \geqq 2)$. *If* K *contains a closed subgroup of dimension* ℓ *for* $\ell > n^2 + 2$ *if* $n \neq 4$, $\ell > 19$ *if* $n = 4$, *then* $\dim K = n^2 + 2n$ *or* $n^2 + 2n - 1$.

Proof. Let H be the isotropy subgroup of K at $x \in M$. Since K acts transitively on M,

$$\dim H = \dim K - 2n > n^2 - 2n + 2 \qquad \text{if } n \neq 4$$

$$\dim H = \dim K - 2n > 19 - 8 = 11 \qquad \text{if } n = 4 ,$$

Thus the identity component H_0 of H is $SU(n)$ or $U(n)$ by Corollary 1.2, and hence $\dim K = n^2 + 2n$ or $n^2 + 2n - 1$. ∎

Now we have the following structure theorem on homogeneous Kähler manifolds on which a reductive Lie group acts transitively.

Theorem 2.5 (Matsushima [20]). *Let* (M,J,g) *be a Kähler manifold and* G *a connected reductive Lie subgroup of* $\mathrm{Aut}(M,J,g)$ *acting transitively on* M. *Thus* M *is identified with* G/H, *where* H *is the isotropy subgroup of* G *at a point of* M. *Then we have*

(1) *G is the direct product $C \times G_0$ of the connected center C of G and the largest connected normal semisimple subgroup G_0 of G, and H is compact connected and coincides with the centralizer in G_0 of a toral subgroup of G_0. Thus G_0 has no center and M is diffeomorphic to the direct product $C \times M_0$ of submanifolds C and $M_0 = G_0/H$. Furthermore M_0 is simply connected and G_0 acts effectively on M_0.*

(2) *The complex structure J of M induces complex structures J_1 on C and J_0 on M_0, and the Kähler metric g on M induces a homogeneous flat Kähler metric h on C and a homogeneous Kähler metric g_0 on M_0 such that (M,J,g) is the direct product:*

$$(M,J,g) = (C,J_1,h) \times (M_0,J_0,g_0)$$

of Kähler manifolds (C,J_1,h) and (M_0,J_0,g_0).

Theorem 2.6. *Let (M,J,g) be a compact Kähler manifold of complex dimension n $(n \geqq 2)$. If $\dim K \geqq n^2+1$, then we have the following:*

(1) *$\dim K = n^2+2n$ and M is holomorphically isometric to $P^n(\mathbb{C})$.*

(2) *$\dim K = n^2+2$ and M is holomorphically isometric to $P^{n-1}(\mathbb{C}) \times P^1(\mathbb{C})$.*

(3) *$\dim K = n^2+1$ and either M is holomorphically isometric to $P^{n-1}(\mathbb{C}) \times T^1$ where T^1 denotes 1-dimensional complex torus, or $n = 3$ and M is $P^2(\mathbb{C}) \times T^1$ or a complex quadric $Q_3(\mathbb{C})$.*

Proof. Since K acts transitively on M by Proposition 2.3 and K is a compact Lie group, we see that $K = K_0 \times C$ where K_0 is semisimple and C is a complex torus and that $M = M_0 \times T^\ell$ where M_0 is a simply connected compact homogeneous Kähler manifold and T^ℓ is a complex torus of complex dimension ℓ. Since $\dim K = \dim K_0 + 2\ell \leqq (n-\ell)^2 + 2(n-\ell) + 2\ell = n^2 - 2n\ell + \ell^2 + 2n$, we have $n^2+1 \leqq n^2 - 2n\ell + \ell^2 + 2n$, that is, $0 \geqq (\ell-1)(2n-\ell-1)$. Since $2n-\ell-1 > 0$, we have $\ell = 0$ or 1.

Case (1): $\ell = 1$.

Since $\dim K \leqq n^2+1$, we see that $\dim K = n^2+1$ and M_0 is $(n-1)$-dimensional compact Kähler manifold with $\dim \mathrm{Aut}(M_0,J_0,g_0) = n^2-1$. Hence M_0 is holomorphically isometric to $P^{n-1}(\mathbb{C})$ by Theorem 2.1.

Case (2): $\ell = 0$.

Note that $M = K/H$ is a simply connected compact homogeneous Kähler manifold and the Lie algebra $\mathfrak{h}$ of H has nontrivial center. Moreover, note that rank K = rank H. First we consider the case when $\dim K > n^2 + 2$. By Proposition 2.4, we have $\dim K = n^2 + 2n$ or $n^2 + 2n - 1$ if $n \neq 4$, and $\dim K = n^2 + 2n$, $n^2 + 2n - 1$ or 19 if $n = 4$. In the case of $\dim K = n^2 + 2n - 1$, $\mathfrak{h}$ is isomorphic to $\mathfrak{su}(n)$ and hence this case cannot occur. In the case of $\dim K = 19$ and $n = 4$, $\mathfrak{h}$ is isomorphic to $B_2 \times T_1$ where B_2 denotes the compact simple Lie algebra of type B and of rank 2 and T_1 denotes 1-dimensional Lie algebra. Since rank K = rank H, we have rank $K = 3$ and $\dim K = 19$. But there is no compact Lie group K such that rank $K = 3$ and $\dim K = 19$. Thus this case cannot occur. Therefore if $\dim K > n^2 + 2$, then $\dim K = n^2+2n$ and hence M is holomorphically isometric to $P^n(\mathbb{C})$ by Theorem 2.1.

Now we consider the case when $\dim K = n^2 + 2$. We have $\dim H = n^2 - 2n + 2$ and hence $\mathfrak{h}$ is isomorphic to $A_{n-2} \times T_1 \times T_1$ where A_{n-2} denotes the compact simple Lie algebra of type A_{n-2}. Since rank K = rank $H = n$, we see that K is not simple Lie group by using the classification of compact simple Lie algebra. If K is not simple, we have $K = K_1 \times K_2$ and $M = K_1/H_1 \times K_2/H_2$ where K_1/H_1, K_2/H_2 are simply connected compact homogeneous Kähler manifolds and K_i acts effectively on K_i/H_i $(i = 1,2)$ (cf. [20]). Put $\ell = \dim_{\mathbb{C}} K_1/H_1$. Then we have $\dim K = \dim K_1 + \dim K_2 \leqq \ell^2 + 2\ell + (n-\ell)^2 + 2(n-\ell) = \ell^2 + (\ell-n)^2 + 2n \leqq n^2 + 2$. Therefore $\ell = 1$ or $n - 1$ and M is holomorphically isometric to $P^{n-1}(\mathbb{C}) \times P^1(\mathbb{C})$.

In the case when $\dim K = n^2 + 1$, we have $\dim H = n^2 - 2n + 1$ and hence $\mathfrak{h}$ is isomorphic to $A_{n-2} \times T_1$. Since rank K = rank H and $\mathfrak{k}$ is simple, we see that $n = 3$ and $\mathfrak{k} = B_2$ by using the classification of compact simple Lie algebra. Therefore M is a complex quadric $Q_3(\mathbb{C})$. ■

3. Compact Einstein Kähler Manifolds M with $\dim_{\mathbb{C}} \mathrm{Aut}(M,J) \geqq n^2-2n+8$ $(n \geqq 6)$

Let (M,J,g) be an hermitian manifold with $\dim K \geqq n^2 - 2n + 9$ $(n \geqq 5)$.

Proposition 3.1. *If* $\dim K \geqq n^2 - 2n + 9$ $(n \geqq 5)$, *then*

(1) *if* $n \neq 6$ *then the Lie algebra* $\mathfrak{k}$ *of* K *contains a Lie subalgebra which is isomorphic to* A_{n-2} *and* $\dim K \geqq n^2 - 2$ *for* $n \geqq 7$.

(2) *if* $n = 6$, *either* $\dim K \geqq n^2 - 2$ *and the Lie algebra* $\mathfrak{k}$ *of* K *contains a Lie subalgebra which is isomorphic to* A_{n-2}, *or* $\dim K = 33, 34$ *and* $\mathfrak{k}$ *contains a Lie subalgebra which is isomorphic to* C_3 *or* $C_3 \times T_1$.

Proof. Consider an orbit $K(x)$ of the highest dimension. Let H denote the isotropy subgroup of K at $x \in M$. Then H is regarded as a closed subgroup of $U(n)$. By our assumption, $\dim H = \dim K - \dim K(x) \geqq n^2 - 4n + 9$. By Lemma 1.1, we see that the Lie algebra $\mathfrak{h}$ contains a Lie subalgebra which is isomorphic to A_{n-2} and $\dim \mathfrak{h} \geqq n^2 - 2n$ for $n \neq 6$, and for $n = 6$, $\mathfrak{h}$ contains a Lie subalgebra which is isomorphic to A_{n-2} and $\dim \mathfrak{h} \geqq n^2 - 2n$ or $\mathfrak{h}$ is isomorphic to C_3 or $C_3 \times T_1$. Put $\ell = \dim K(x)$. Then $\mathfrak{h}$ is also regarded as a subalgebra of $\mathfrak{u}(\ell)$ by Lemma 2.2. Since $\mathfrak{su}(n-1) \subset \mathfrak{h} \subset \mathfrak{u}(\ell)$, we have $\ell \geqq 2(n-1)$ and hence $\dim K = \dim H + \dim K(x) \geqq n^2 - 2$ for $n \geqq 7$. For $n = 6$ and $\mathfrak{h}$ is isomorphic to C_3 or $C_3 \times T_1$, if K does not act transitively on M, then H has a nontrivial invariant subspace of $T_x(M)$ (cf. proof of Proposition 2.3). Since $Sp(3) \subset SU(6)$ is irreducible, we see that K acts transitively on M. Therefore $\dim K = 33$ or 34. ∎

Lemma 3.2. *Let* (M,J,g) *be a compact Kähler manifold. Then* $n = 6$ *and the case* $\dim K = 33$ *or* 34 *and* $\mathfrak{k}$ *contains a Lie subalgebra which is isomorphic to* C_3 *or* $C_3 \times T_1$ *does not occur.*

Proof. Since K acts transitively on M, the case of $\mathfrak{h} = C_3$ cannot occur by Theorem 2.5. Now consider the case of $\mathfrak{h} = C_3 \times T_1$. Since $\operatorname{rank} K = \operatorname{rank} H = 4$, K is a compact simple Lie group of rank 4 and $\dim K = 34$. But there is no compact simple Lie group such that $\dim K = 34$ and $\operatorname{rank} K = 4$. ∎

From now on let (M,J,g) be a compact Kähler manifold with $c_1(M) > 0$.

Proposition 3.3. *Let* (M,J,g) *be an* n-*dimensional compact Kähler manifold with* $c_1(M) > 0$. *Then* $\operatorname{rank} K \leqq n$.

Proof. See [2], [16].

Proposition 3.4. *Let* (M,J,g) *be an* n-*dimensional compact Kähler manifold with* $c_1(M) > 0$. *If* $\dim K \geqq n^2 - 2n + 9$ $(n \geqq 5)$, *then either*

$\dim K \geqq n^2+1$ *or the Lie algebra* $\mathfrak{k}$ *of* K *contains a subalgebra which is isomorphic to* A_{n-1}.

Proof. By Proposition 3.1 and Lemma 3.2, we may assume that $\dim K \geqq n^2-2$ for $n \geqq 6$ and $\operatorname{rank} K \geqq n-2$. Note that for $n=5$, $\dim K = n^2-2n+9 = n^2-1$. Thus $\operatorname{rank} K = n-2$, $n-1$ or n by Proposition 3.3. Note that the Lie algebra $\mathfrak{k}$ of a compact Lie group K can be expressed in the following form

$$\mathfrak{k} = \mathfrak{k}_1 \times \cdots \times \mathfrak{k}_s \times T_\ell \quad \text{(direct sum)}$$

where $\mathfrak{k}_j$ denotes a compact simple Lie algebra of $\dim \mathfrak{k}_j \geqq 3$ and T_ℓ denotes an abelian Lie algebra of ℓ-dimension and hence one of $\mathfrak{k}_j$ $(j=1,\dots,s)$ contains a Lie subalgebra which is isomorphic to A_{n-2}.

Case 1. $\operatorname{rank} \mathfrak{k} = n-2$.

Since $\operatorname{rank} \mathfrak{k} = \operatorname{rank} A_{n-2}$, we see that $\mathfrak{k}$ is simple. By a theorem of Borel-Siebenthal [3], we have $\mathfrak{k} = A_{n-2}$, E_7 or E_8. Since $\mathfrak{k} \geqq n^2-2$, we see that $\mathfrak{k} = E_7$ or E_8. Since $\dim E_7 = 133$, and $\dim E_8 = 248$, $\dim \mathfrak{k} \geqq n^2+1$.

Case 2. $\operatorname{rank} \mathfrak{k} = n-1$.

We may assume that $\mathfrak{k}$ is simple. Consider the dimensions of compact simple Lie algebras of rank $n-1$ $(\geqq 4)$.

Table 1

type	dimension	n^2+1
A_{n-1}	n^2-1	
B_{n-1}	$(n-1)(2n-1)$	
C_{n-1}	$(n-1)(2n-1)$	
D_{n-1}	$(n-1)(2n-3)$	
F_4	52	26
E_6	78	50
E_7	133	65
E_8	248	82

Since $(n-1)(2n-3) \geqq n^2+1$ $(n \geqq 5)$, we see that $\dim K \geqq n^2+1$ or $\mathfrak{k}$ is isomorphic to A_{n-1}.

Case 3. $\operatorname{rank} \mathfrak{k} = n$.

We may assume that $\mathfrak{k}$ is simple. Then it is easy to see that $\dim \mathfrak{k} \geqq n^2+1$ by the table above. ∎

Theorem 3.5 (Matsushima [21]). *Let* (M,J,g) *be a compact Einstein Kähler manifold with* $c_1(M) > 0$ *and* $\mathfrak{g}(M)$ *the complex Lie algebra of all holomorphic vector fields on* M *and* $\mathfrak{k}(M,g)$ *the Lie algebra of all Killing vector fields on* (M,g). *Then*

$$\mathfrak{g}(M) = \mathfrak{k}(M,g) \oplus J\mathfrak{k}(M,g) \quad (\textit{direct sum}) .$$

From now on let G denote the identity component of $\operatorname{Aut}(M,J)$ the group of all holomorphic transformations of a compact complex manifold (M,J).

Lemma 3.6. *Let* (M,J) *be a compact complex manifold with* $c_1(M) > 0$. *Then there is a holomorphic imbedding of* M *into a complex projective space* $P^N(\mathbb{C})$ *such that every element of* G *is induced by a unique projective linear transformation of* $P^N(\mathbb{C})$.

Proof. See [13].

Theorem 3.7 (Mabuchi [18]). *Let* M *be an* n*-dimensional projective algebraic manifold endowed with an essentially effective regular action of the algebraic group* $SL(n,\mathbb{C})$ $(n \geqq 5)$. *Then* M *is holomorphically isomorphic to one of the following:*

(1) *The complex projective space* $P^n(\mathbb{C})$.

(2) $P^{n-1}(\mathbb{C}) \times N$, *where* N *is a compact complex manifold of dimension* 1.

(3) *The projective bundle* $P(1 \oplus \xi^d)$ $(d \in \mathbb{N})$ *associated with the vector bundle* $1 \oplus \xi^d$ *over* $P^{n-1}(\mathbb{C})$, *where* ξ *denote the hyperplane line bundle.*

Theorem 3.8. *Let* (M,J,g) *be an* n*-dimensional compact Einstein Kähler manifold (with* $c_1(M) > 0$*). If* $\dim_{\mathbb{C}} G \geqq n^2 - 2n + 8$ $(n \geqq 6)$ *then* M *is holomorphically isometric to* $P^n(\mathbb{C})$, $P^1(\mathbb{C}) \times P^{n-1}(\mathbb{C})$ *or* $P^2(\mathbb{C}) \times P^{n-2}(\mathbb{C})$. *If* $\dim_{\mathbb{C}} G \geqq 24$ $(n = 5)$ *then* M *is holomorphically isometric to* $P^5(\mathbb{C})$ *or* $P^1(\mathbb{C}) \times P^4(\mathbb{C})$.

Proof. By Theorem 3.5, $\dim K \geq n^2 - 2n + 8$. First we consider the case when $\dim K \geqq n^2 - 2n + 9$ $(n \geqq 5)$. By Proposition 3.4, either (1) $\dim K \geqq n^2 + 1$ or (2) the Lie algebra $\mathfrak{k}$ of K contains a subalgebra which is isomorphic to A_{n-1}. In the case when $\dim K \geqq n^2 + 1$, we see that M is holomorphically isometric to $P^n(\mathbb{C})$ or $P^1(\mathbb{C}) \times P^{n-1}(\mathbb{C})$ by Theorem 2.6. In the case (2) we see that $SL(n, \mathbb{C})$ acts essentially effectively on M as an algebraic linear group by Theorem 3.5 and Lemma 3.6. By Theorem 3.7, we have

(1) M is biholomorphic to $P^n(\mathbb{C})$,

(2) M is biholomorphic to $P^1(\mathbb{C}) \times P^{n-1}(\mathbb{C})$ since $c_1(M) > 0$ or

(3) M is biholomorphic to $P(1 \oplus \xi^d)$ $(d \in \mathbb{N})$.

On the other hand we know that $P(1 \oplus \xi^d)$ does not admit any Einstein Kähler metric (cf. [10]). But in our case we can prove this in the following way. We have

$$\dim_{\mathbb{C}} \mathrm{Aut}(P(1 \oplus \xi^d), J) = n^2 - 1 + \binom{n-1+d}{d} + 1 \geqq n^2 + n \quad \text{(cf. [10])}.$$

Hence if $P(1 \oplus \xi^d)$ admits an Einstein Kähler metric, we have $\dim K \geqq n^2 + n$ by Theorem 3.5 and hence it is holomorphically isometric to $P^n(\mathbb{C})$ by Theorem 2.6. This is a contradiction. Therefore M is biholomorphic to $P^n(\mathbb{C})$ or $P^1(\mathbb{C}) \times P^{n-1}(\mathbb{C})$. Now by a Theorem of Matsushima [22] the Einstein Kähler metric on a simply connected compact complex homogeneous manifold which admits a Kähler metric is essentially unique, and thus M is holomorphically isometric to $P^n(\mathbb{C})$ or $P^1(\mathbb{C}) \times P^{n-1}(\mathbb{C})$.

Now consider the case when $\dim K = n^2 - 2n + 8$ $(n \geqq 6)$. Consider an orbit $K(x)$ at $x \in M$ and let H be the isotropy subgroup of K at $x \in M$. Then $\dim H = \dim K - \dim K(x) \geqq n^2 - 4n + 8$. If $\dim H \geqq n^2 - 4n + 9$, then $\dim K \geqq n^2 - 2$ (cf. proof of Proposition 3.1 and Lemma 3.2). Thus $\dim H = n^2 - 4n + 8$ and K acts transitively on M. By Lemma 1.1, the Lie algebra $\mathfrak{h}$ is isomorphic to $A_1 \times A_{n-3} \times T_1 \times T_1$. Since M is simply connected ([14]), rank K = rank $H = n$ by Theorem 2.5. Moreover using the classification of compact simple Lie group, we see that K is not simple. Then $M = K_1/H_1 \times K_2/H_2$ where K_1/H_1, K_2/H_2 are simply connected compact homogeneous Kähler manifolds (cf. [20]). Since the Lie algebras $\mathfrak{h}_i$ of H_i $(i = 1, 2)$ has nontrivial center by Theorem 2.5, we have the following possibilities:

(1) $\mathfrak{h}_1 = T_1$, $\mathfrak{h}_2 = A_1 \times A_{n-3} \times T_1$

(2) $\mathfrak{h}_1 = A_1 \times T_1$, $\mathfrak{h}_2 = A_{n-3} \times T_1$

Case (1). In this case rank $\mathfrak{k}_1$ = rank $\mathfrak{h}_1 = 1$ and hence $\dim \mathfrak{k}_1 = 3$ and $\dim \mathfrak{k}_2 = n^2 - 2n + 5$ and $\mathfrak{k}_2$ is a compact simple Lie algebra of rank $n-1$. Using the classification of compact simple Lie algebra, we see that there is no compact simple Lie algebra $\mathfrak{k}$ such that rank $\mathfrak{k} = n-1$ and $\dim \mathfrak{k} = n^2 - 2n + 5$.

Case (2). Since rank $\mathfrak{k}_1 = 2$, $\dim_{\mathbb{C}} K_1/H_1 \geqq 2$ and hence $\dim K = \dim K_1 + \dim K_2 \leqq 2^2 + 2\cdot 2 + (n-2)^2 + 2(n-2) = n^2 - 2n + 8$. Thus $K_1/H_1 = P^2(\mathbb{C})$ and $K_2/H_2 = P^{n-2}(\mathbb{C})$ by Theorem 2.1. Therefore M is holomorphically isometric to $P^2(\mathbb{C}) \times P^{n-2}(\mathbb{C})$. ∎

4. An Application

Let $f: P^m(\mathbb{C}) \times P^n(\mathbb{C}) \to P^{mn+m+n}(\mathbb{C})$ be the Segre imbedding and M a compact complex submanifold of $P^m(\mathbb{C}) \times P^n(\mathbb{C})$ defined by a hyperplane section. Let H_m denote the holomorphic line bundle over $P^m(\mathbb{C})$ defined by a hyperplane of $P^m(\mathbb{C})$ and let $p_1: P^m(\mathbb{C}) \times P^n(\mathbb{C}) \to P^m(\mathbb{C})$ and $p_2: P^m(\mathbb{C}) \times P^n(\mathbb{C}) \to P^n(\mathbb{C})$ denote the projections. We denote by F a holomorphic line bundle $p_1{*}H_m \otimes p_2{*}H_n$ over $P^m(\mathbb{C}) \times P^n(\mathbb{C})$. Then F is very ample and defines the Segre imbedding $f: P^m(\mathbb{C}) \times P^n(\mathbb{C}) \to P^{mn+m+n}(\mathbb{C})$. Moreover the canonical line bundle $K(P^m(\mathbb{C}) \times P^n(\mathbb{C}))$ of $P^m(\mathbb{C}) \times P^n(\mathbb{C})$ is given by $(p_1{*}H_m)^{-(m+1)} \otimes (p_2{*}H_n)^{-(n+1)}$. We denote by $\{M\}$ the holomorphic line bundle over $P^m(\mathbb{C}) \times P^n(\mathbb{C})$ defined by a compact complex submanifold M of $P^m(\mathbb{C}) \times P^n(\mathbb{C})$. Then we have $\{M\} = F$. Let $j: M \to P^m(\mathbb{C}) \times P^n(\mathbb{C})$ denote the inclusion. Since the canonical line bundle $K(M)$ of M is given by $j{*}(K(P^m(\mathbb{C}) \otimes P^n(\mathbb{C})) \times \{M\}^{-1})$,

$$K(M) = j{*}((p_1{*}H_m)^{-m} \otimes (p_2{*}H_n)^{-n}) .$$

Thus the first Chern class $c_1(M)$ of M is given by $mj{*}c_1(p_1{*}H_m) + nj{*}c_1(p_2{*}H_n)$ and hence $c_1(M)$ is positive.

Lemma 4.1. *Let $f: P^m(\mathbb{C}) \times P^n(\mathbb{C}) \to P^{mn+m+n}(\mathbb{C})$ be the Segre imbedding and let M^{m+n-1} be a compact complex submanifold defined by a hyperplane section. Then*

$$\dim_{\mathbb{C}} \mathrm{Aut}(M,J) \geqq m^2 + n^2 - mn + m + n .$$

We shall give a proof of Lemma 4.1 in Section 5.

Theorem 4.2. *Let* $f:P^1(\mathbb{C})\times P^n(\mathbb{C})\to P^{2n+1}(\mathbb{C})$ *be the Segre imbedding and let* M^n *be a compact complex submanifold defined by a hyperplane section. Then* M $(n\geqq 2)$ *has a positive first Chern class* $c_1(M)>0$ *but does not admit any Einstein Kähler metric.*

Proof. By Lemma 4.1, we have

$$\dim_{\mathbb{C}}\mathrm{Aut}(M,J)\geqq n^2+2 \quad .$$

If M admits an Einstein Kähler metric, we have $\dim_{\mathbb{R}}\mathrm{Aut}(M,J,g)\geqq n^2+2$ by Theorem 3.5. Thus M is holomorphically isometric to $P^n(\mathbb{C})$ or $P^1(\mathbb{C})\times P^{n-1}(\mathbb{C})$ by Theorem 2.6. Since the second Betti number $b_2(M)$ of M is 2 by the Lefschetz theorem of hyperplane section (cf. [23]). M is holomorphically isometric to $P^1(\mathbb{C})\times P^{n-1}(\mathbb{C})$. We claim that M is not biholomorphic to $P^1(\mathbb{C})\times P^{n-1}(\mathbb{C})$. Let F be the holomorphic line bundle on $P^1(\mathbb{C})\times P^n(\mathbb{C})$ as above. Put $g=c_1(p_1{}^*H_1)$ and $h=c_1(p_1{}^*H_n)$. Then g, h are the generators of $H^2(P^1(\mathbb{C})\times P^n(\mathbb{C}),\mathbb{Z})$. Now consider the Chern numbers $c_1(j^*F)^n[M]$. Since $c_1(F)=g+h\in H^2(P^1(\mathbb{C})\times P^n(\mathbb{C}),\mathbb{Z})$, we have

$$\begin{aligned} c_1(j^*F)^n[M] &= (g+h)(g+h)^n[P^1(\mathbb{C})\times P^n(\mathbb{C})] \\ &= (n+1)gh^n[P^1(\mathbb{C})\times P^n(\mathbb{C})] \\ &= n+1 \quad . \end{aligned}$$

Now take a holomorphic line bundle on $P^1(\mathbb{C})\times P^{n-1}(\mathbb{C})$ and consider the Chern number. For a holomorphic line bundle L on $P^1(\mathbb{C})\times P^{n-1}(\mathbb{C})$, the Chern class $c_1(L)$ of L is given by

$$c_1(L) = a\alpha+b\beta \;(a,b\in\mathbb{Z})$$

where α,β is the generators of $H^2(P^1(\mathbb{C})\times P^{n-1}(\mathbb{C}),\mathbb{Z})$. Now we have

$$\begin{aligned} c_1(L)^n[P^1(\mathbb{C})\times P^{n-1}(\mathbb{C})] &= (a\alpha+b\beta)^n[P^1(\mathbb{C})\times P^{n-1}(\mathbb{C})] \\ &= nab^{n-1}gh^{n-1}[P^1(\mathbb{C})\times P^{n-1}(\mathbb{C})] \\ &= nab^{n-1} \quad . \end{aligned}$$

Suppose that M is biholomorphic to $P^1(\mathbb{C})\times P^{n-1}(\mathbb{C})$, then j^*F is a holomorphic line bundle on $P^1(\mathbb{C})\times P^{n-1}(\mathbb{C})$ and hence $0\neq c_1(j^*F)^n[M]$ is a multiple of n. Since $c_1(j^*F)^n[M]=n+1$, this is a contradiction.

Hence M is not biholomorphic to $P^1(\mathbb{C}) \times P^{n-1}(\mathbb{C})$. Therefore M cannot admit any Einstein Kähler metric. ■

5. A Proof of Lemma 4.1

We can prove Lemma 4.1 by using Borel-Weil-Bott Theorem by the same way as in [12]. We recall the known facts on Kähler C-spaces to fix our notations. A compact simply connected homogeneous complex manifold is called a C-space. A C-space is said to be Kähler if it has a Kähler metric. Let X be a Kähler C-space and let G denote the identity component of $\mathrm{Aut}(X,J)$. Then G is a connected complex semi-simple Lie group without the center. Fix a point $x \in X$ and put $U = \{g \in G | g(x) = x\}$. Then U is a closed connected complex Lie subgroup of G and $X = G/U$. Let $\mathfrak{g}$ denote the Lie algebra of G and denote by $(\ ,\ )$ the Killing form of $\mathfrak{g}$. Now $\mathfrak{u}$ the Lie algebra of U is a parabolic Lie subalgebra of $\mathfrak{g}$ and described as follows. Take a Cartan subalgebra $\mathfrak{h}$ of $\mathfrak{g}$ contained in $\mathfrak{u}$. Denote by $\mathfrak{h}_{\mathbb{R}}$ the real part of $\mathfrak{h}$, and the root system Σ of $\mathfrak{g}$ relative to $\mathfrak{h}$ is identified with a subset of $\mathfrak{h}_{\mathbb{R}}$ by means of the duality defined by the Killing form. Then there exist a lexicographic order $>$ on $\mathfrak{h}_{\mathbb{R}}$ and a subset Π_0 of the fundamental root system Π with the following property. If we set $\Sigma_0 = \Sigma \cap \mathbb{Z}\Pi_0$ and $\Sigma^+_{\mathfrak{m}} = \{\alpha \in \Sigma - \Sigma_0 | \alpha > 0\}$, then $\mathfrak{u}$ is given by

$$\mathfrak{u} = \mathfrak{h} + \sum_{\alpha \in \Sigma_0 \cup \Sigma^+_{\mathfrak{m}}} \mathfrak{g}_\alpha ,$$

where $\mathfrak{g}_\alpha$ stand for the root space of $\alpha \in \Sigma$. Let $\{\Lambda_\alpha ; \alpha \in \Pi\} \subset \mathfrak{h}_{\mathbb{R}}$ be the fundamental weights corresponding to Π. We put

$$\mathfrak{c} = \{H \in \mathfrak{h}_{\mathbb{R}} ; (H, \Pi_0) = (0)\}$$

and

$$Z_{\mathfrak{c}} = \left\{ \Lambda \in \mathfrak{c} ; \ \frac{2(\Lambda,\alpha)}{(\alpha,\alpha)} \in \mathbb{Z} \quad \text{for each} \quad \alpha \in \Sigma \right\}$$

which is a lattice of $\mathfrak{c}$ generated by $\{\Lambda_\alpha ; \alpha \in \Pi - \Pi_0\}$. Let $\tilde{G}$ be the universal covering group of G and $\tilde{U}$ the connected complex Lie subgroup of G corresponding to $\mathfrak{u}$. Then we have also an identification:

$X = \tilde{G}/\tilde{U}$. For each $\Lambda \in Z_{\mathfrak{r}}$ there exists a unique holomorphic character χ_Λ of $\tilde{U}$ such that $\chi_\Lambda(\exp H) = \exp(\Lambda, H)$ for each $H \in \mathfrak{h}$. Then the correspondence $\Lambda \mapsto \chi_\Lambda$ gives an isomorphism of $Z_{\mathfrak{r}}$ to the group of holomorphic characters of $\tilde{U}$. Let F_Λ denote the holomorphic line bundle on X associated to the principal bundle $\tilde{U} \to \tilde{G} \to X$ by χ_Λ. The correspondence $\Lambda \mapsto F_\Lambda$ induces a homomorphism of $Z_{\mathfrak{r}}$ to the group $H^1(X, \theta^*)$ of isomorphism classes of holomorphic line bundles on X. The correspondence $F \mapsto c_1(F)$ also defines a homomorphism of $H^1(X,\theta^*)$ to $H^2(X, \mathbb{Z})$.

Fact 1. *Both of these homomorphisms*

$$Z_{\mathfrak{r}} \xrightarrow{F} H^1(X,\theta^*) \xrightarrow{c_1} H^2(X,\mathbb{Z})$$

are isomorphisms (see [9]).

Now we define Lie algebras $\mathfrak{g}_1$, $\mathfrak{m}^+$ of $\mathfrak{g}$ as follows:

$$\mathfrak{g}_1 = \mathfrak{h} + \sum_{\alpha \in \Sigma_0} \mathfrak{g}_\alpha$$

$$\mathfrak{m}^+ = \sum_{\alpha \in \Sigma^+} \mathfrak{g}_\alpha \quad .$$

Put $Z = \{\Lambda \in \mathfrak{h}_{\mathbb{R}};\ 2(\Lambda,\alpha)/(\alpha,\alpha) \in \mathbb{Z}$ for each $\alpha \in \Sigma\}$. We denote by D (resp. D_1) the set of dominant integral forms of $\mathfrak{g}$ (resp. $\mathfrak{g}_1$), by W the Weyl group of $\mathfrak{g}$ and by Σ^+ (resp. Σ^-) the set of all positive (resp. negative) roots of Σ. Note that $D_1 = \{\Lambda \in Z; (\Lambda,\alpha) \geqq 0$ for each $\alpha \in \Sigma^+ \cap \Sigma_0\}$ and hence $Z_{\mathfrak{r}} \subset D_1$. We define a subset W^1 of W by

$$W^1 = \{\sigma \in W;\ \sigma(D) \subset D_1\}$$

and the index $n(\sigma)$ of $\sigma \in W$ by

$$n(\sigma) = \#(\sigma(\Sigma^+) \cap \Sigma^-) \quad .$$

Put also $\delta = 1/2\ \Sigma_{\alpha \in \Sigma^+} \alpha$. For $\xi \in D_1$ consider the irreducible representation $(\rho^1_{-\xi}, W_{-\xi})$ of $\mathfrak{g}_1$ with the lowest weight $-\xi$. We may extend it to a representation of $\mathfrak{u}$ so that its restriction to $\mathfrak{m}^+$ is trivial and denote it by $(\rho_{-\xi}, W_{-\xi})$. Since each irreducible representation of $\mathfrak{u}$ is trivial on $\mathfrak{m}^+$, we may call $(\rho_{-\xi}, W_{-\xi})$ the irreducible representation of $\mathfrak{u}$ with the lowest weight $-\xi$. Moreover there exist a representation of $\tilde{U}$ which induces the representation $(\rho_{-\xi}, W_{-\xi})$ of $\mathfrak{u}$ and we also denote it by $(\rho_{-\xi}, W_{-\xi})$. Let $E_{-\xi}$ denote the holomorphic vector bundle on X associated to the principal bundle $\tilde{U} \to \tilde{G} \to X$ by the representation $(\rho_{-\xi}, W_{-\xi})$ of $\tilde{U}$.

Theorem of Bott ([4] cf. [17]). *Let* $H^j(X, E_{-\xi})$ *denote the* j-*th cohomology group over* X *with coefficients in the sheaf of germs of local holomorphic sections of* $E_{-\xi}$. *For* $\xi \in D_1$, *if* $\xi + \delta$ *is not regular,*

$$H^j(X, E_{-\xi}) = (0) \qquad \textit{for all} \quad j = 0, 1, \ldots,$$

and if $\xi + \delta$ *is regular,* $\xi + \delta$ *is expressed uniquely as* $\xi + \delta = \sigma(\lambda + \delta)$, *where* $\lambda \in D$ *and* $\sigma \in W^1$, *and*

$$H^j(X, E_{-\xi}) = (0) \qquad \textit{for all} \quad j \neq n(\sigma) \quad ,$$

$$\dim H^{n(\sigma)}(X, E_{-\xi}) = \dim V_{-\lambda} \quad ,$$

where $(\rho_{-\lambda}, V_{-\lambda})$ *denotes the irreducible representation of* $\tilde{G}$ *with the lowest weight* $-\lambda$.

For each integer $p \geqq 0$, put

$$W^1(p) = \{\sigma \in W^1, \ n(\sigma) = p\} \quad .$$

Fact 2. *Let* X *be an hermitian symmetric space of compact type. For* $\Lambda \in Z_{\mathfrak{r}}$ *let* $H^q(X, \Omega^p(F_\Lambda))$ *denote the cohomology group of type* (p,q) *over* X *with coefficients in the sheaf of germs of local holomorphic sections of* F_Λ. *Then*

$$\dim H^q(X, \Omega^p(F_\Lambda)) = \sum_{\sigma \in W^1(p)} \dim H^q(X, E_{-(\sigma\delta - \delta - \Lambda)})$$

for $q = 0, 1, \ldots$. (See [12] Theorem 2, [17]).

Now we consider the case of $X = P^m(\mathbb{C}) \times P^n(\mathbb{C})$. We see that $\tilde{G} = SL(m+1, \mathbb{C}) \times SL(n+1, \mathbb{C})$ and the fundamental root system Π is given by

$$\Pi = \{\alpha_1, \ldots, \alpha_m, \beta_1, \ldots, \beta_n\} \quad ,$$

the subset Π_0 is given by

$$\Pi_0 = \{\alpha_2,\ldots,\alpha_m,\beta_2,\ldots,\beta_n\}$$

and

$$\Sigma^+_{\mathfrak{m}} = \{\alpha_1,\alpha_1+\alpha_2,\ldots,\alpha_1+\cdots+\alpha_m,\beta_1,\beta_1+\beta_2,\ldots,\beta_1+\cdots+\beta_n\} .$$

Moreover $Z = \{a\Lambda_{\alpha_1} + b\Lambda_{\beta_1};\ a,b \in \mathbb{Z}\}$ where $\{\Lambda_{\alpha_j},\Lambda_{\beta_k};\ j=1,\ldots,m, k=1,\ldots,n\}$ is the fundamental weights corresponding to Π. Now by Fact 1 we see that the holomorphic line bundle F over $P^m(\mathbb{C}) \times P^n(\mathbb{C})$ defined in Section 4 is given by $F = F_{-(\Lambda_{\alpha_1}+\Lambda_{\beta_1})}$, and the canonical line bundle $K(P^m(\mathbb{C}) \times P^n(\mathbb{C}))$ of $P^m(\mathbb{C}) \times P^n(\mathbb{C})$ is given by

$$K(P^m(\mathbb{C}) \times P^n(\mathbb{C})) = F_{(m+1)\Lambda_{\alpha_1} + (n+1)\Lambda_{\beta_1}} .$$

We claim that

$$H^q\left(P^m(\mathbb{C}) \times P^n(\mathbb{C}), \Omega^1\left(F_{m\Lambda_{\alpha_1} + n\Lambda_{\beta_1}}\right)\right) = (0)$$

For each $q \geqq 0$. It is enough to see that

$$\dim H^q\left(P^m(\mathbb{C}) \times P^n(\mathbb{C}), E_{-(\sigma\delta-\delta-(m\Lambda_{\alpha_1}+n\Lambda_{\beta_1}))}\right) = 0$$

for each $\sigma \in W^1(1)$, by Fact 2.

Let τ_α denote the reflection corresponding to $\alpha \in \Pi$. Since

$$n(\sigma) = \min\left\{k;\ \sigma=\tau_{\gamma_1}\cdots\tau_{\gamma_k},\ \gamma_j \in \Pi\right\}$$

we see that $W^1(1) = \{\tau_{\alpha_1},\tau_{\beta_1}\}$.

Noting that

$$\delta = \sum_{i=1}^{m} \Lambda_{\alpha_i} + \sum_{j=1}^{n} \Lambda_{\beta_j} ,$$

$$\sigma\delta - \left(m\Lambda_{\alpha_1} + n\Lambda_{\beta_1}\right)$$

$$= \begin{cases} \sum_{i=1}^{m} \Lambda_{\alpha_i} - \alpha_1 + \sum_{j=1}^{n} \Lambda_{\beta_j} - \left(m\Lambda_{\alpha_1} + n\Lambda_{\beta_1}\right) & \text{for} \quad \sigma = \tau_{\alpha_1} \\ \sum_{i=1}^{m} \Lambda_{\alpha_i} + \sum_{j=1}^{n} \Lambda_{\beta_j} - \beta_1 - \left(m\Lambda_{\alpha_1} + n\Lambda_{\beta_1}\right) & \text{for} \quad \sigma = \tau_{\beta_1} \end{cases} .$$

Since

$$\left(\sum_{i=1}^{m} \Lambda_{\alpha_1} - \alpha_1 + \sum_{j=1}^{n} \Lambda_{\beta_j} - \left(m\Lambda_{\alpha_1} + n\Lambda_{\beta_1}\right), \quad \beta_1 + \cdots + \beta_n\right) = 0$$

and

$$\left(\sum_{i=1}^{m} \Lambda_{\alpha_i} + \sum_{j=1}^{n} \Lambda_{\beta_j} - \beta_1 - \left(m\Lambda_{\alpha_1} + n\Lambda_{\beta_1}\right), \quad \alpha_1 + \cdots + \alpha_m\right) = 0 ,$$

$(\sigma\delta - \delta - (m\Lambda_{\alpha_1} + n\Lambda_{\beta_1})) + \delta$ is not regular and hence we get our claim by a Theorem of Bott. Now we can prove our Lemma 4.1 as follows.

For simplicity put $X = P^m(\mathbb{C}) \times P^n(\mathbb{C})$. Consider the following exact sequences of sheaves:

$$0 \to T(X) \otimes \{M\}^{-1} \to T(x) \to T(X)|_M \to 0 \quad ,$$

$$0 \to T(M) \to T(X)|_M \to \{M\}|_M \to 0 \quad .$$

Then we have the following exact sequences of cohomology groups:

$$0 \to H^0(X, T(x) \otimes \{M\}^{-1}) \to H^0(X, T(X))$$
$$\to H^0(M, T(X)|_M) \to H^1(X, T(X) \otimes \{M\}^{-1})$$
$$\to H^1(X, T(X)) \to H^1(M, T(X)|_M)$$
$$\to H^2(X, T(X) \otimes \{M\}^{-1}) \to H^2(X, T(X)) \to \cdots .$$

$$0 \to H^0(M, T(M)) \to H^0(M, T(X)|_M) \to H^0(M, \{M\}|_M)$$
$$\to H^1(M, T(M)) \to H^1(M, T(X)|_M) \to \cdots \quad .$$

By a theorem of Bott [4], $H^j(X, T(X)) = (0)$ for $j \geqq 1$. Hence we see that

$$0 \to H^0(X,T(X)\otimes\{M\}^{-1}) \to H^0(X,T(X))$$

$$\to H^0(M,T(X)|_M) \to H^1(X,T(X)\otimes\{M\}^{-1}) \to 0 \quad ,$$

$$0 \to H^1(M,T(X)|_M) \to H^2(X,T(X)\otimes\{M\}^{-1}) \to 0 \quad .$$

If $H^j(X,T(X)\otimes\{M\}^{-1}) = (0)$ for each $j \geqq 0$, we have

$$0 \to H^0(X,T(X)) \to H^0(M,T(X)|_M) \to 0$$

$$H^1(M,T(X)|_M) = (0)$$

and hence

$$0 \to H^0(M,T(M)) \to H^0(M,T(X)|_M) \to H^0(M,\{M\}|_M)$$

$$\to H^1(M,T(M)) \to 0 \quad .$$

Thus

$$\begin{aligned} \dim_{\mathbb{C}} \mathrm{Aut}(M,J) &= \dim H^0(M,T(M)) \\ &= \dim H^0(X,T(X)) - \dim H^0(M,\{M\}|_M) + \dim H^1(M,T(M)) \\ &\geqq \dim H^0(X,T(X)) - \dim H^0(M,\{M\}|_M) \quad . \end{aligned}$$

From the exact sequence

$$0 \to \theta \to \{M\} \to \{M\}|_M \to 0 \ ,$$

we have

$$0 \to H^0(X,\theta) \to H^0(X,\{M\}) \to H^0(M,\{M\}|_M)$$

$$\to H^1(X,\theta) \to \cdots$$

and hence $\dim H^0(M,\{M\}|_M) = \dim H^0(X,\{M\}) - 1$ since $H^1(X,\theta) = (0)$. Therefore

$$\dim_{\mathbb{C}} \mathrm{Aut}(M,J) \geqq \dim H^0(X,T(X)) - \dim H^0(X,\{M\}) - 1$$

if $H^j(X,T(X)\otimes\{M\}^{-1}) = (0)$ for each $j \geqq 0$.

By the Serre duality, we have

$$H^j(X,T(X)\otimes\{M\}^{-1}) \cong H^{m+n-j}(X,\Omega^1(K\otimes\{M\})) \quad .$$

Since

$$K\otimes\{M\} = F_{(m+1)\Lambda_{\alpha_1}+(n+1)\Lambda_{\beta_1}} \otimes F_{-(\Lambda_{\alpha_1}+\Lambda_{\beta_1})} = F_{m\Lambda_{\alpha_1}+n\Lambda_{\beta_1}} \quad ,$$

$H^j(X,T(X)\otimes\{M\}^{-1}) = (0)$ for each $j \geqq 0$. Since

$$\dim H^0(X,T(X)) = \dim(SL(m+1,\mathbb{C})\times SL(n+1,\mathbb{C})) = m^2+2m+n^2+2n$$

and

$$\dim H^0(X,\{M\}) = \dim H^0(X,F) = (m+1)(n+1) \quad ,$$

we get $\dim_{\mathbb{C}} Aut(M,J) \geqq m^2+n^2-mn+m+n$.

6. References

[1] T. Aubin, "Equations du type Monge-Ampère sur les variétiés Kähleriennes compacts," *C.R. Acad. Sci. Paris* 283 (1976), 119-121.

[2] A. Borel, *Linear Algebraic Groups*, Benjamin, New York, 1969.

[3] A. Borel and J. De Siebenthal, "Les sous-groupes férmes des rang maximum des groupes de Lie clos," *Comment. Math. Helv.* 23 (1949), 200-221.

[4] R. Bott, "Homogeneous vector bundles," *Ann. of Math.* 66 (1975), 203-248.

[5] N. Bourbaki, *Groupes et Algebres de Lie*, Chap. 4,5 et 6, Hermann, Paris, 1968.

[6] N. Bourbaki, *Groupes et Algebres de Lie*, Chap. 7 et 8, Hermann, Paris, 1975.

[7] E.B. Dynkin, "The maximal subgroups of the classical groups," *Amer. Math. Soc. Transl.* (2), 6 (1957), 245-378.

[8] F. Hirzebruch, *Topological Methods in Algebraic Geometry*, Springer Verlag, Berlin, 1966.

[9] M. Ise, "Some properties of complex analytic vector bundles over compact complex homogeneous spaces," *Osaka Math. J.* 12 (1960), 217-252.

[10] K. Ishikawa and Y. Sakane, "On complex projective bundles over a Kähler C-space," *Osaka Math. J.* 16 (1979), 121-132.

[11] W. Kaup, "Reelle Transformationsgruppen und invariante Metriken auf komplexen Räumen," *Invent. Math.* 3 (1967), 43-70.

[12] Y. Kimura, "On the hypersurfaces of hermitian symmetric spaces of compact type," *Osaka J. Math.* 16 (1979), 79-119.

[13] S. Kobayashi, *Transformation Groups in Differential Geometry*, Springer-Verlag, Berlin, 1972.

[14] S. Kobayashi, "On compact Kähler manifolds with positive definite Ricci tensor," *Ann. of Math.* 74 (1961), 570-574.

[15] S. Kobayashi and T. Nagano, "Riemannian manifolds with abundant isometries," *Differential Geometry*, in honor of K. Yano, Kinokuniya, Tokyo (1972), 195-219.

[16] S. Kobayashi and T. Ochiai, "Three-dimensional compact Kähler manifolds with positive holomorphic bisectional curvature," *J. Math. Soc. Japan* 24 (1972), 465-480.

[17] B. Kostant, "Lie algebra cohomology and the generalized Borel-Weil theorem," *Ann. of Math.* 74 (1961), 329-387.

[18] T. Mabuchi, "On the classification of essentially effective SL(n,C)-actions on algebraic n-folds," *Osaka J. Math.* 16 (1979), 745-758.

[19] L.N. Mann, "Gaps in the dimensions of transformation group," *Illinois J. Math.* 10 (1966), 532-546.

[20] Y. Matsushima, "Sur les espaces homogènes Kählériennes d'un groupe de Lie reductif," *Nagoya Math. J.* 11 (1957), 53-60.

[21] Y. Matsushima, "Sur la structure du groupe d'homéomorphismes analytiques d'une certaine variété Kählérienne," *Nagoya Math. J.* 11 (1957), 145-150.

[22] Y. Matsushima, "Remarks on Kähler-Einstein manifolds," *Nagoya Math. J.* 46 (1972), 161-173.

[23] J. Milnor, "Lectures on Morse theory," Annals Math. Studies No. 51, Princeton Univ. Press, 1963.

[24] S. Tanno, "The automorphism groups of almost hermitian manifolds," *Trans. Amer. Soc.* 137 (1969), 269-275.

[25] F. Uchida, "Smooth actions of special unitary groups on cohomology complex projective spaces," *Osaka J. Math.* 12 (1975), 375-400.

[26] S.T. Yau, "On the curvature of compact Hermitian manifolds," *Invent. Math.* 25 (1974), 213-239.

[27] S.T. Yau, "On the Ricci cúrvature of a compact Kähler manifold and the complex Monge-Ampère equation I," *Comm. Pure Appl. Math.* 31 (1978), 339-411.

Osaka University
Toyonaka, Osaka 560
Japan

(Received December 17, 1980)

SPECIAL VALUES OF ZETA FUNCTIONS ASSOCIATED WITH SELF-DUAL HOMOGENEOUS CONES

I. Satake

0. Introduction

To explain the main idea of this paper, and also to fix some notations, we start with reviewing the classical case of Riemann zeta function. As usual we set

$$\zeta(s) = \sum_{n=1}^{\infty} n^{-s} \qquad (\mathrm{Re}\ s > 1) \quad ,$$

$$\Gamma(s) = \int_0^{\infty} x^{s-1} e^{-x}\, dx \qquad (\mathrm{Re}\ s > 0) \quad .$$

Then, for $\mathrm{Re}\ s > 1$, one obtains

$$\begin{aligned} \Gamma(s)\zeta(s) &= \sum_{n=1}^{\infty} n^{-s} \int_0^{\infty} x^{s-1} e^{-x}\, dx \\ &= \sum_{n=1}^{\infty} \int_0^{\infty} x'^{\,s-1} e^{-nx'}\, dx' \qquad (x = nx') \\ &= \int_0^{\infty} \frac{x^{s-1}}{e^x - 1}\, dx \quad . \end{aligned}$$

We put

$$b(x,y) = \frac{e^{xy}}{e^x - 1} = \sum_{\nu=0}^{\infty} \frac{B_\nu(y)}{\nu!} x^{\nu-1} \qquad (|x| < 2\pi) \quad , \tag{0.1}$$

with the Bernoulli polynomials:

$$B_\nu(y) = \sum_{\mu=0}^{\nu} \binom{\nu}{\mu} b_\mu y^{\nu-\mu} \qquad (\nu = 0,1,2,\dots) \quad ,$$

where the b_μ are the Bernoulli numbers:

$$b_0 = 1, \qquad b_1 = -\frac{1}{2}, \qquad b_\nu = \begin{cases} (-1)^{\frac{\nu}{2}-1} B_{\nu/2} & (\nu \text{ even}, \geqslant 2), \\ 0 & (\nu \text{ odd}, \geqslant 3). \end{cases}$$

Then, the above integral can be transformed into a contour integral of the form

$$\Gamma(s)\zeta(s) = (e^{2\pi i s} - 1)^{-1} \int_{I(\varepsilon,\infty)} x^{s-1} b(x,0)\, dx \quad , \tag{0.2}$$

where $I(\varepsilon,\infty)$ denotes the contour consisting of the half-line $[\varepsilon,\infty)$ taken twice in opposite directions and of a (small) circle of radius ε about the origin taken in the counterclockwise direction. The contour integral is absolutely convergent for all $s \in \mathbb{C}$, so that the function $\Gamma(s)\zeta(s)$ can be analytically continued to a meromorphic function on $\mathbb{C}$. Moreover, in virtue of the functional equation of the gamma function:

$$\Gamma(s)\Gamma(1-s) = \frac{\pi}{\sin \pi s} = 2\pi i \frac{e^{\pi i s}}{e^{2\pi i s}-1} \quad , \tag{0.3}$$

one obtains

$$\zeta(s) = e^{-\pi i s}\Gamma(1-s) \cdot \frac{1}{2\pi i} \int_{I(\varepsilon,\infty)} x^{s-1} b(x,0)\, dx \quad . \tag{0.4}$$

This shows that $\zeta(s)$ is holomorphic for $\mathrm{Re}\, s < 1$. In particular, for $s = 1-m$, $m \in \mathbb{Z}_+$ (positive integers), the contour integral reduces to the residue of $x^{-m}b(x,0)$ at $x = 0$, i.e., $b_m/m!$. Thus one obtains

$$\zeta(1-m) = (-1)^{m-1}(m-1)!\, \frac{b_m}{m!} = (-1)^{m-1} \frac{b_m}{m} \quad , \tag{0.5}$$

which shows that $\zeta(1-m)$ $(m \in \mathbb{Z}_+)$ is rational. In particular,

$$\zeta(0) = -\frac{1}{2}, \qquad \zeta(-1) = -\frac{1}{12} \quad ,$$

$$\zeta(-2\mu) = 0 , \qquad \zeta(1-2\mu) = (-1)^{\mu} \frac{B_\mu}{2\mu} \quad (\mu \geqslant 1) \quad .$$

This result has been generalized by Hecke, Klingen and Siegel (see [13]) to the case of Dedekind zeta functions of totally real number fields. More recently, Shintani [11] gave a proof based on a direct generalization of the classical method explained above. Now, the "partial" Dedekind zeta function corresponding to an ideal class may be viewed as a zeta function attached to an angular domain. Zeta functions attached to self-dual homogeneous cones have been studied by Siegel [12] in a special case of quadratic cones, and by Sato-Shintani [8] in a more general context of "prehomogeneous spaces." (Cf. also Shintani [9], [10].) On the other hand, the gamma functions attached to self-dual homogeneous cones were introduced by Koecher [5], and studied by Gindikin [3] and others (cf. e.g., Resnikoff [6]).

In this paper, we try to extend Shintani's method (i.e., the classical method) to examine the rationality of the special values of zeta functions attached to a self-dual homogeneous cone Ω. We will compute the special values of the "partial" zeta functions corresponding to (closed) simplicial cones contained in the interior of Ω and show that, when d (see (1.10)) is even, these values can be expressed as $\mathbb{Q}$-linear combinations of certain integrals $I((\nu_{ij}))$ (see (2.11)) taken over the maximal compact subgroup of $G(\Omega)^{\circ}$ (Theorem 1); when d is odd, the situation becomes a little more complicated, so that we omit the discussion here. For the case $r=2$, i.e., the case of quadratic cones (including the case of odd d), one can obtain more precise results, which we hope to discuss elsewhere; in this case, there is also an independent work of A. Kurihara [14]. It should be noted, however, that our results so far are still incomplete in two essential points: first, our method does not seem applicable to the case where the simplicial cone has edges contained in the boundary of Ω, and second, we do not know anything about the nature of $I((\nu_{ij}))$ when some of ν_{1j}'s are negative. The author would like to thank the late T. Shintani for his useful comments on the first draft of this paper.

<u>Notation</u>. We denote by $\mathbb{R}_+$ the set of positive real numbers and put $\mathbb{Z}_+ = \mathbb{Z} \cap \mathbb{R}_+$, $\mathbb{Q}_+ = \mathbb{Q} \cap \mathbb{R}_+$. (As usual, $\mathbb{Z}$, $\mathbb{Q}$ are the ring of integers and the field of rational numbers, respectively.) For any ring R, $R^{\times}$ denotes the group of invertible elements in R. For $\xi \in \mathbb{C}$, we set $\mathbf{e}(\xi) = \exp(2\pi i \xi)$. For $\alpha, \beta \in \mathbb{C}^{\times}$, the relation $\alpha \underset{\mathbb{Q}}{\sim} \beta$ means that $\alpha/\beta \in \mathbf{Q}^{\times}$. For a Lie group G, G° and $\mathrm{Lie}\ G$ stand for the identity connected component of G and the Lie algebra of G, respectively. For a vector space V, the dual space of V is denoted by V^*.

1. The Gamma Function of a Self-Dual Homogeneous Cone

1.1 Let U be a real vector space of dimension n, endowed with a positive definite inner product $\langle\ \rangle$. By a "cone" in U we always mean a nondegenerate open convex cone in U with vertex at the origin, i.e., a nonempty open set Ω in U such that

$$x,y \in \Omega, \quad \lambda,\mu \in \mathbb{R}_+ \Rightarrow \lambda x + \mu y \in \Omega$$

and such that Ω does not contain any straight lines. A cone Ω in U is called *homogeneous* if the group of linear automorphisms

$$G(\Omega) = \{g \in GL(U) \mid g(\Omega) = \Omega\}$$

is transitive on Ω; and Ω is called *self-dual* if the "dual" of Ω

$$\Omega^* = \{x \in U \mid \langle x,y\rangle > 0 \quad \text{for all} \quad y \in \bar{\Omega} - \{0\}\}$$

coincides with Ω.

Let Ω be a self-dual homogeneous cone in U and $G = G(U)^\circ$. Then it is well-known (e.g., Satake [7]) that the Zariski closure of G (in $GL(U)$) is a reductive algebraic group, containing $G(\Omega)$ as a subgroup of finite index, and $g \mapsto {}^t g^{-1}$ is a Cartan involution of G; moreover, the corresponding maximal compact subgroup $K = G \cap O(U)$ coincides with the isotropy subgroup of G at a "base point" $e \in \Omega$ (which is not unique, but will be fixed once and for all). Let

$$\mathfrak{g} = \mathfrak{k} + \mathfrak{p}$$

be the Cartan decomposition of $\mathfrak{g} = \mathrm{Lie}\ G$ with $\mathfrak{k} = \mathrm{Lie}\ K$. Then for $T \in \mathfrak{g}$ one has

$$T \in \mathfrak{k} \Longleftrightarrow {}^t T = -T \Longleftrightarrow Te = 0 . \tag{1.1}$$

It follows that, for each $u \in U$, there exists a uniquely determined element $T_u \in \mathfrak{p}$ such that $T_u e = u$. It is well-known that the vector space U endowed with a product

$$u \circ u' = T_u u' \qquad (u,u' \in U)$$

becomes a formally real Jordan algebra (cf. Braun-Koecher [2], or Satake [7]).

We define the (regular) *trace* of $u \in U$ by

$$\tau(u) = tr(T_u) \quad . \tag{1.2}$$

For the given (Ω, e), one may assume (by Schur's lemma) that the inner product $< >$ is so normalized that one has

$$<u,u'> = \tau(u \circ u') \qquad (u,u' \in U) \quad . \tag{1.3}$$

Next, let $u \in \Omega$. Then, since G is transitive on Ω, there exists $g_1 \in G$ such that $u = g_1 e$. We define the (regular) *norm* $N(u)$ by

$$N(u) = \det(g_1) \quad ,$$

which is clearly independent of the choice of g_1 satisfying the above condition. Alternatively, one may choose a unique element $u_1 \in U$ such that $u = \exp u_1$ (which is defined to be $(\exp T_{u_1})e$), to define $N(u)$ by

$$N(u) = \det\left(\exp T_{u_1}\right) = \exp(\tau(u_1)) \quad . \tag{1.4}$$

[In terms of the "quadratic multiplication" $P(u) = 2T_u^2 - T_{u^2}$, one can also write $N(u) = \det(P(u))^{1/2}$]. By the definition, it is clear that

$$N(e) = 1, \quad N(gu) = \det(g)\, N(u) \qquad (g \in G(\Omega), u \in \Omega), \tag{1.5}$$

which characterizes the norm uniquely. Denoting the Euclidean measure on U by du, we see that $d_\Omega(u) = N(u)^{-1} du$ is an invariant measure on Ω.

Example. Let $U = Sym_r(\mathbb{R})$ (the space of real symmetric matrices of degree r) and $\Omega = P_r(\mathbb{R})$ (the cone of positive definite elements in U). Then one has

$$T_u(u') = \frac{1}{2}(uu' + u'u)$$

and so

$$\tau(u) = \frac{r+1}{2} tr(u) \quad , \quad N(u) = \det(u)^{\frac{r+1}{2}} \quad .$$

1.2 We define the *gamma function* of the cone Ω by the integral

$$\Gamma_\Omega(s) = \int_\Omega N(u)^{s-1}\exp(-\tau(u))\,du \quad , \tag{1.6}$$

which converges absolutely for $\mathrm{Re}\, s$ sufficiently large (actually for $\mathrm{Re}\, s > 1 - r/n$ as we will see later on).

Lemma 1. *Suppose that the inner product* $\langle\ \rangle$ *is normalized by* (1.3). *Then one has for any* $v \in \Omega$

$$\int_\Omega N(u)^{s-1}\exp(-\langle u,v\rangle)\,du = \Gamma_\Omega(s)N(v)^{-s} \quad . \tag{1.7}$$

Proof. Let $v = g_1 e$ with $g_1 \in G$ and put $u' = {}^t g_1 u$. Then one has

$$\langle u,v\rangle = \langle u,g_1 e\rangle = \langle u',e\rangle = \tau(u') \quad .$$

Hence by (1.5) the left-hand side of (1.7) is equal to

$$\int_\Omega N(u)^s\exp(-\langle u,v\rangle)\,d_\Omega(u) = \int_\Omega (\det(g_1)^{-1}N(u'))^s\exp(-\tau(u'))\,d_\Omega(u')$$

$$= N(v)^{-s}\Gamma_\Omega(s) \quad . \qquad \blacksquare$$

It is known that the function $\Gamma_\Omega(s)$ can be expressed as a product of ordinary gamma functions (cf. e.g., Resnikoff [6]). For convenience of the reader, we sketch a proof. First, it is clear that, if

$$\Omega = \Omega_1 \times \cdots \times \Omega_m$$

is the decomposition of Ω into the direct product of irreducible (self-dual homogeneous) cones, then one has

$$\Gamma_\Omega(s) = \Gamma_{\Omega_1}(s) \ldots \Gamma_{\Omega_m}(s) \quad .$$

Hence, for our purpose, we may assume that Ω is irreducible (i.e., the semi-simple part G^s of G is simple).

We need the root structure of $\mathfrak{g}$, which can be determined as follows. Let

$$e = \sum_{i=1}^{r} e_i \quad , \qquad e_i e_j = \delta_{ij} e_i \tag{1.8}$$

be a decomposition of the unit element e (in the Jordan algebra U) into the sum of mutually orthogonal primitive idempotents. ("Primitive" means that each e_i is $\neq 0$ and cannot be decomposed into the sum of mutually orthogonal non-zero idempotents any more.) Then we obtain the following decomposition of U into the direct sum of subspaces ("Peirce decomposition"):

$$U = \bigoplus_{1\leq i\leq j\leq r} U_{ij} \quad , \tag{1.9}$$

where

$$U_{ii} = \left\{ u \in U \mid e_i u = u \right\} \quad ,$$

$$U_{ij} = \left\{ u \in U \mid e_i u = e_j u = \frac{1}{2} u \right\} \qquad (i \neq j) \quad ,$$

and for $u \in U_{ij}$ one has $e_k u = 0$ for all $k \neq i,j$. Moreover

$$\dim U_{ii} = 1, \quad \dim U_{ij} = d \qquad (i \neq j) \quad , \tag{1.10}$$

where d is a positive integer depending on the irreducible cone Ω. (E.g., one has $d = 1$ for $\Omega = \mathcal{P}_r(\mathbb{R})$.) From (1.9), (1.10) one has the relation

$$n = r + \frac{1}{2} r(r-1)d, \quad \text{or} \quad d = \frac{2(n-r)}{r(r-1)} \quad . \tag{1.11}$$

It follows that

$$\tau(e_i) = \mathrm{tr}(T_{e_i}) = 1 + \frac{1}{2}(r-1)d = \frac{n}{r} \quad . \tag{1.12}$$

Put

$$\mathfrak{a} = \left\{ T_{e_i} \ (1 \leq i \leq r) \right\}_{\mathbb{R}} \quad . \tag{1.13}$$

Then $\mathfrak{a}$ is an abelian subalgebra of $\mathfrak{g}$ of dimension r contained in $\mathfrak{p}$. We denote by (λ_i) the basis of $\mathfrak{a}^*$ (the dual space of $\mathfrak{a}$) dual to (T_{e_i}); then one has

$$T = \sum_{i=1}^{r} \lambda_i(T)\, T_{e_i} \qquad (T \in \mathfrak{a}) \quad .$$

We put

$$\alpha_{ij} = \frac{1}{2}(\lambda_i - \lambda_j) \qquad \text{for} \quad i \neq j \ .$$

Then we obtain

<u>Proposition 1</u>. $\mathfrak{a}$ *is a maximal abelian subalgebra in* $\mathfrak{p}$, *and the root system of* $\mathfrak{g}$ *relative to* $\mathfrak{a}$ *is given by*

$$\Phi = \{\alpha_{ij} \quad (i \neq j)\} \ .$$

The corresponding root spaces $\mathfrak{g}(\alpha_{ij})$ *are given by*

$$\mathfrak{g}(\alpha_{ij}) = \left\{ T_u + \left[T_{e_i - e_j}, T_u \right] \middle| \ u \in U_{ij} \right\} \ . \tag{1.14}$$

This can be verified by a straightforward computation; see e.g., Ash *et al.* [1] Ch. II, §3. Proposition 1 implies that the $\mathbb{R}$-rank of $\mathfrak{g}$ is equal to r and the root system Φ is of type (A_{r-1}).

1.3 Next we determine the Haar measure of G. Put

$$\mathfrak{n} = \sum_{i<j} \mathfrak{g}(\alpha_{ij})$$

and let A, N denote the analytic subgroups of G corresponding to $\mathfrak{a}$, $\mathfrak{n}$, respectively. Then one has an Iwasawa decomposition $G = NA \cdot K$ $(\approx N \times A \times K)$, which gives rise to the following formula for (the volume element of) a (bi-invariant) Haar measure on G:

$$dg = c_1 \exp(-2\rho(\log a)) \, d\nu \, da \, dk \tag{1.15}$$

for $g = \nu ak$ with $\nu \in N$, $a \in A$, $k \in K$, where $d\nu$, da, dk denote Haar measures on N, A, K, respectively, c_1 is a positive constant depending on the normalization of the Haar measures, and ρ is a linear form on $\mathfrak{a}$ defined by

$$\rho(T) = \frac{1}{2} \operatorname{tr}(\operatorname{ad} T|\mathfrak{n}) \qquad (T \in \mathfrak{a}) \ ;$$

then by Proposition 1 one has

$$\rho = \frac{d}{2} \sum_{i<j} \alpha_{ij} = \frac{d}{2} \sum_{i=1}^{r} (r - 2i + 1)\lambda_i \ . \tag{1.16}$$

[A direct proof of (1.15) will be given later.] The Haar measure of K is always normalized by $\int_K dk = 1$. We make an identification $A = (\mathbb{R}_+)^r$ by the correspondence $a \leftrightarrow (t_i)$ defined by the relation $a = \exp(\Sigma\, \lambda_i T_{e_i})$, $\lambda_i = \log t_i$; then one has $da = \Pi\, d\lambda_i = \Pi(dt_i/t_i)$. Moreover one has by (1.12), (1.16)

$$\begin{cases} \det(a) = \exp\left(\frac{n}{r}\sum \lambda_i\right) = \left(\prod_{i=1}^{r} t_i\right)^{\frac{n}{r}} , \\ a\cdot e = \sum(\exp \lambda_i) e_i = \sum_{i=1}^{r} t_i e_i , \\ \exp(2\rho(\log a)) = \prod_{i=1}^{r} t_i^{\frac{d}{2}(r-2i+1)} . \end{cases} \tag{1.17}$$

Since $\Omega = G/K$, we can normalize the Haar measure of G by the relation $dg = d_\Omega(u)\cdot dk$ where $u = ge$. Then by (1.15), (1.16), (1.17) one has

$$\begin{aligned} \Gamma_\Omega(s) &= \int_G N(ge)^s \exp(-\tau(ge))\, dg \\ &= c_1 \int_A \det(a)^s \exp(-2\rho(\log a))\, da \int_N \exp(-\tau(\nu a e))\, d\nu \\ &= c_1 \int_0^\infty \cdots \int_0^\infty \prod_{i=1}^{r} \left(t_i^{\frac{n}{r}s - \frac{d}{2}(r-2i+1)-1}\, dt_i\right) \\ &\quad \times \int_N \exp\left(-\tau\left(\nu\left(\sum_{i=1}^{r} t_i e_i\right)\right)\right) d\nu . \end{aligned} \tag{1.18}$$

To compute the integral over N, we introduce more notations. For $x = \Sigma_{i<j}\, x_{ij}$ with $x_{ij} \in U_{ij}$, we put

$$\begin{aligned} \nu(x) &= \exp\left(\frac{1}{2}\left(T_x + \sum_{i<j}\left[T_{e_i - e_j}, T_{x_{ij}}\right]\right)\right) \quad (\in N) \\ \varepsilon_{ij}(x) &= \sum_{m=1}^{\infty} \frac{1}{m!} \sum_{i<k_1<\cdots<k_{m-1}<j} x_{ik_1} x_{k_1k_2} \cdots x_{k_{m-1}j} \quad (\in U_{ij}) \quad \text{for } i \neq j . \end{aligned} \tag{1.19}$$

Then, by a straightforward computation in the Jordan algebra U, one obtains

Lemma 2. *In the above notations, one has*

$$\nu(x)\left(\sum_{i=1}^{r} t_i e_i\right) = \sum_{i=1}^{r}\left(t_i + \frac{1}{4}\sum_{k>i} t_k \varepsilon_{ik}(x)^2\right)e_i$$

$$+ \frac{1}{2}\sum_{i<j}\left(t_j \varepsilon_{ij}(x) + \sum_{k>j} t_k \varepsilon_{ik}(x)\varepsilon_{jk}(x)\right). \quad (1.20)$$

(This may be regarded as a generalization of the so-called "Jacobi transformation.") It follows that

$$\tau\left(\nu(x)\left(\sum_i t_i e_i\right)\right) = \frac{n}{r}\sum_i t_i + \frac{1}{8}\sum_{i<k} \tau(\varepsilon_{ik}(x)^2)\, t_k . \quad (1.21)$$

We denote the Euclidean measure on U_{ij} $(i<j)$ (relative to the inner product $<\ >$) by dx_{ij} and define the Haar measure on N by $d\nu = \Pi_{i<j}\, dx_{ij}$ for $\nu = \nu(x)$. Since the map $x \mapsto x' = \Sigma_{i<j}\, \varepsilon_{ij}(x)$ is a bijection of $\Sigma_{i<j}\, U_{ij}$ onto itself with jacobian equal to one, one has

$$d\nu = \prod_{i<j} dx_{ij} = \prod_{i<j} dx'_{ij} \qquad (\nu = \nu(x),\ x'_{ij} = \varepsilon_{ij}(x)) .$$

Hence by (1.21) one has

$$\int_N \exp\left(-\tau\left(\nu\left(\sum_i t_i e_i\right)\right)\right) d\nu = \exp\left(-\frac{n}{r}\sum_i t_i\right) \prod_{i<j} \int_{U_{ij}} \exp\left(-\frac{t_j}{8}\tau(x'^2_{ij})\right) dx'_{ij}$$

$$= \exp\left(-\frac{n}{r}\sum_i t_i\right) \prod_{i<j}\left(\frac{8\pi}{t_j}\right)^{\frac{d}{2}}$$

$$= (8\pi)^{\frac{n-r}{2}} \prod_{j=1}^{r}\left(t_j^{-\frac{d}{2}(j-1)} \exp\left(-\frac{n}{r} t_j\right)\right) .$$

Inserting this in (1.18), one obtains

$$\Gamma_\Omega(s) = c_1(8\pi)^{\frac{n-r}{2}} \prod_{j=1}^{r} \left(\int_0^\infty t_j^{\frac{n}{r}s - \frac{d}{2}(r-j)-1} \exp\left(-\frac{n}{r}t_j\right) dt_j \right)$$

$$= c_1(8\pi)^{\frac{n-r}{2}} \prod_{j=1}^{r} \left(\left(\frac{n}{r}\right)^{-\frac{n}{r}s + \frac{d}{2}(r-j)} \Gamma\left(\frac{n}{r}s - \frac{d}{2}(r-j)\right)\right)$$

$$= c_1(8\pi)^{\frac{n-r}{2}} \left(\frac{n}{r}\right)^{-ns + \frac{n-r}{2}} \prod_{j=1}^{r} \Gamma\left(\frac{n}{r}s - \frac{d}{2}(j-1)\right) .$$

The constant c_1 can be determined by the following observation. We denote by dx_0 the Euclidean measure on $U_0 = \Sigma_{i=1}^{r} U_{ii} = \{e_1,\ldots,e_r\}_{\mathbb{R}}$ (relative to $< >$). Then, since $<e_i,e_j> = n/r\ \delta_{ij}$, the bijection $A \to U_0$ defined by $a = \exp T_{x_0}$ gives the relation $dx_0 = (n/r)^{r/2}\, da$. Hence by (1.20) the absolute value of the jacobian of the map

$$(\mathbb{R}_+)^r \times \left(\sum_{i<j} U_{ij}\right) \ni (t,x) \mapsto u = \nu(x)\left(\sum_i t_i e_i\right) \in U$$

is given by

$$2^{r-n}\left(\frac{n}{r}\right)^{\frac{r}{2}} \prod_{j=1}^{r} t_j^{(j-1)d} .$$

It follows that

$$d_\Omega(u) = 2^{r-n}\left(\frac{n}{r}\right)^{\frac{r}{2}} \prod_{j=1}^{r} \left(t_j^{(j-1)d - \frac{n}{r}} dt_j\right) dx ,$$

which, in view of (1.11) and (1.16), implies (1.15) and the relation

$$c_1 = 2^{r-n}\left(\frac{n}{r}\right)^{\frac{r}{2}} . \tag{1.22}$$

Thus we obtain the formula

$$\Gamma_\Omega(s) = (2\pi)^{\frac{n-r}{2}} \left(\frac{n}{r}\right)^{n\left(\frac{1}{2}-s\right)} \prod_{j=1}^{r} \Gamma\left(\frac{n}{r}s - \frac{d}{2}(j-1)\right) . \tag{1.23}$$

Our computation also shows that the integral for $\Gamma_\Omega(s)$ converges absolutely for $\mathrm{Re}\, s > 1 - r/n$.

From the relation (0.3) one obtains

$$\Gamma_\Omega(s)\Gamma_\Omega(1-s) = (2\pi)^{n-r} \prod_{j=1}^{r} \left(\Gamma\left(\frac{n}{r}s - \frac{d}{2}(j-1)\right)\Gamma\left(\frac{n}{r}(1-s) - \frac{d}{2}(r-j)\right)\right)$$

$$= (2\pi)^{n-r}(2\pi i)^r \prod_{j=1}^{r} \frac{\mathbf{e}\left(\frac{1}{2}\left(\frac{n}{r}s - \frac{d}{2}(j-1)\right)\right)}{\mathbf{e}\left(\frac{n}{r}s - \frac{d}{2}(j-1)\right) - 1} .$$

Since one has by (1.11)

$$n - r = d\,\frac{r(r-1)}{2} \equiv \begin{cases} 0 \pmod 2 & \text{for } d \text{ even} \\ \left[\frac{r}{2}\right] \pmod 2 & \text{for } d \text{ odd}, \end{cases}$$

one has

$$\prod_{j=1}^{r} \mathbf{e}\left(-\frac{d}{4}(j-1)\right) = (-i)^{d\frac{r(r-1)}{2}} = \begin{cases} i^{n-r} & \text{for } d \text{ even} \\ (-1)^{\left[\frac{r}{2}\right]} i^{n-r} & \text{for } d \text{ odd.} \end{cases}$$

Hence one obtains the following functional equation:

$$\Gamma_\Omega(s)\Gamma_\Omega(1-s) = (2\pi i)^n\, \mathbf{e}\left(\frac{1}{2}ns\right) \begin{cases} \left(\mathbf{e}\left(\frac{n}{r}s\right) - 1\right)^{-r} & (d \text{ even}) \\ \left(\mathbf{e}\left(\frac{n}{r}s\right) - 1\right)^{-\left[\frac{r+1}{2}\right]}\left(\mathbf{e}\left(\frac{n}{r}s\right) + 1\right)^{-\left[\frac{r}{2}\right]} & (d \text{ odd}). \end{cases} \tag{1.24}$$

2. Zeta Functions of a Self-Dual Homogeneous Cone

2.1 We fix a $\mathbb{Q}$-structure on U and assume that (the Zariski closure of) G is defined over $\mathbb{Q}$ and $e \in U_{\mathbb{Q}}$; then (the Zariski closure of) K is also defined over $\mathbb{Q}$. We also fix a lattice L in U compatible with this $\mathbb{Q}$-structure, i.e., such that $U_{\mathbb{Q}} = L \otimes_{\mathbb{Z}} \mathbb{Q}$, and an arithmetic subgroup Γ fixing L, i.e., a subgroup of

$G_L = \{g \in G \mid gL = L\}$ of finite index; for simplicity we assume that Γ has no fixed point in Ω. Then we define the *zeta function* associated with Ω, Γ, and L as follows:

$$\zeta_\Omega(s;\Gamma,L) = \sum_{u \in \Gamma\backslash\Omega\cap L} N(u)^{-s} , \tag{2.1}$$

the summation being taken over a complete set of representatives of $\Omega \cap L$ modulo Γ. It can be shown easily that this series is absolutely convergent for $\operatorname{Re} s > 1$.

By the reduction theory, Γ has a fundamental domain in Ω which is a rational polyhedral cone. More precisely, there exists a finite set of simplicial cones

$$\begin{aligned} C^{(i)} &= \left\{v_1^{(i)}, \dots, v_{\ell_i}^{(i)}\right\}\mathbb{R}_+ \\ &= \left\{ \sum_{j=1}^{\ell_i} \lambda_j v_j^{(i)} \,\middle|\, \lambda_j \in \mathbb{R}_+ \right\} \qquad (1 \leqslant i \leqslant m) , \end{aligned}$$

where $v_1^{(i)}, \dots, v_{\ell_i}^{(i)}$ are linearly independent elements in $\bar{\Omega} \cap L$, such that

$$\Omega = \coprod_{\substack{\gamma \in \Gamma \\ 1 \leqslant i \leqslant m}} \gamma C^{(i)} .$$

It follows that

$$\zeta(s;\Gamma,L) = \sum_{i=1}^{m} \sum_{u \in C^{(i)} \cap L} N(u)^{-s} .$$

For any set of linearly independent vectors $v_1, \dots, v_\ell \in L$, we put

$$R((v_j),L) = \left\{ \sum_{j=1}^{\ell} \lambda_j v_j \,\middle|\, 0 < \lambda_j \leqslant 1 \right\} \cap L ,$$

which is finite. Then $u \in C^{(i)} \cap L$ can be written uniquely in the form

$$u = v_0 + \sum_{j=1}^{\ell_i} m_j v_j^{(i)} , \qquad v_0 \in R((v_j^{(i)}),L), \quad m_j \in \mathbb{Z}, \quad m_j \geqslant 0 .$$

For a set of linearly independent vectors $v_1,\dots,v_\ell \in \bar{\Omega} \cap V_{\mathbb{Q}}$ and $v_0 = \Sigma_j \alpha_j v_j \in \Omega$ $(\alpha_j \in \mathbb{Q}_+)$, we define a "partial zeta function" by

$$\zeta_\Omega(s;(v_j),v_o) = \sum_{m_1,\dots,m_\ell=0}^{\infty} N\left(v_o + \sum_{j=1}^{\ell} m_j v_j\right)^{-s} , \tag{2.2}$$

which will also be written as $\zeta_\Omega(s;(v_j),(\alpha_j))$. Then the zeta function (2.1) can be written as a finite sum of partial zeta functions as follows:

$$\zeta_\Omega(s;\Gamma,L) = \sum_{i=1}^{m} \sum_{v_o \in R((v_j^{(i)}),L)} \zeta_\Omega(s;(v_j^{(i)}),v_o) . \tag{2.3}$$

Hence, in principle, the study of special values of $\zeta_\Omega(s;\Gamma,L)$ is reduced to that of the partial zeta functions of the form (2.2).

2.2 For the sake of simplicity, we again assume that Ω is irreducible. Let (v_j) and v_o be as above. Then by (1.7) one obtains

$$\begin{aligned}
\Gamma_\Omega(s)\zeta_\Omega(s;(v_j),v_o) &= \sum_{m_1,\dots,m_\ell=0}^{\infty} \Gamma_\Omega(s) N\left(v_o + \sum_{j=1}^{\ell} m_j v_j\right)^{-s} \\
&= \sum_{m_1,\dots,m_\ell=0}^{\infty} \int_\Omega N(u)^{s-1} \exp\left(- \sum_j (\alpha_j + m_j)\langle v_j,u\rangle\right) du \\
&= \int_\Omega N(u)^s \prod_{j=1}^{\ell} b(\langle v_j,u\rangle, 1-\alpha_j) \; d_\Omega(u) \\
&= \int_G \det(g)^s \prod_{j=1}^{\ell} b(\langle v_j,ge\rangle, 1-\alpha_j) \; dg .
\end{aligned}$$

In the notation of Section 1, but this time using the decomposition $G = KAK$, one has

$$dg = c\,\Delta(a)\,dk\cdot da\cdot dk' \tag{2.4}$$

for $g = kak'$, $k, k' \in K$, $a \in A$. Here c is a positive constant and

$$\Delta(a) = \prod_{\alpha\in\phi_+} (\exp(\alpha(\log a)) - \exp(-\alpha(\log a)))^d$$

$$= \left(\prod_{i=1}^{r} t_i\right)^{-\frac{d}{2}(r-1)} |\Delta(t_1,\ldots,t_r)|^d \quad ,$$

where $\Delta(t_1,\ldots,t_r) = \Pi_{i<j}(t_i-t_j)$ (cf. Helgason [4], Ch. X, §1). Hence in view of (1.11) and (1.17) one has

$$\Gamma_\Omega(s)\zeta_\Omega(s;(v_j),(\alpha_j)) = c \int_0^\infty \cdots \int_0^\infty \left(\prod_{i=1}^{r} t_i\right)^{\frac{n}{r}(s-1)} |\Delta(t)|^d \, F(t) \prod_i dt_i \ , \tag{2.5}$$

where

$$F(t_1,\ldots,t_r) = \int_K \prod_{j=1}^{\ell} b\left(\langle v_j, k \sum_i t_i e_i\rangle,\ 1-\alpha_j\right) dk \ .$$

It is clear that $F(t_1,\ldots,t_r)$ is holomorphic for $\mathrm{Re}\ t_i > 0$ $(1 \leqslant i \leqslant r)$.

Since K contains an element which induces any given permutation of $e_1,\ldots,e_r$, the function F is symmetric. Hence, denoting by E_1 an open simplicial cone in $\mathbb{R}^r$ defined by $t_1 > \cdots > t_r > 0$, one has

$$\Gamma_\Omega(s)\zeta_\Omega(s;(v_j),(\alpha_j)) = c\ r! \int_{E_1} \left(\prod_i t_i\right)^{\frac{n}{r}(s-1)} \Delta(t)^d \, F(t) \prod_i dt_i . \tag{2.5'}$$

Still following Shintani [11], we make a change of variables $(t_i) \to (t_1,\tau_2,\ldots,\tau_r)$ with $\tau_i = t_i/t_{i-1}$ $(2 \leqslant i \leqslant r)$. Then E_1 can be expressed as

$$E_1 = \left\{ (t_i) \,\middle|\, t_i = t_1 \prod_{j=2}^{i} \tau_j,\ 0 < t_1 < \infty,\ 0 < \tau_i < 1 \right\} \quad .$$

Putting $\tau_1 = t_1$, one has

$$\frac{\partial(t_1,\ldots,t_r)}{\partial(\tau_1,\ldots,\tau_r)} = \prod_{i=1}^{r} \tau_i^{r-i} \quad ,$$

$$\prod_{i=1}^{r} t_i = \prod_{i=1}^{r} \tau_i^{r-i+1} ,$$

$$\Delta(t) = \prod_{i=1}^{r} \tau_i^{\frac{1}{2}(r-i+1)(r-i)} \prod_{2\leqslant i\leqslant j\leqslant r} (1-\tau_i \cdots \tau_j) .$$

Hence one has

$$\Gamma_\Omega(s)\zeta_\Omega(s;(v_j),(\alpha_j)) = c\, r! \int_0^\infty t_1^{ns-1}\, dt_1$$
$$\times \int_0^1 \cdots \int_0^1 \left(\prod_{i=1}^{r} \tau_i^{(r-i+1)\left\{\frac{n}{r}s - \frac{d}{2}(i-1)\right\}-1} \right) \cdot \tilde{F}(t_1,\tau) \prod_{i=2}^{r} d\tau_i , \quad (2.6)$$

where

$$\tilde{F}(t_1,\tau) = \left(\prod_{2\leqslant i<j\leqslant r} (1-\tau_i \ldots \tau_j) \right)^d F(t_1, t_1\tau_2, \ldots, t_1\tau_2\cdots\tau_r) . \quad (2.7)$$

2.3 We now assume that all v_j's are in Ω (not on the boundary of Ω). (In the situation explained in 2.1, this assumption holds for all $C^{(i)}$ if and only if G^s is $\mathbb{Q}$-anisotropic, i.e., the $\mathbb{Q}$-rank of G is equal to 1.) Then for any $v \in \bar{\Omega} - \{0\}$, one has $\langle v_j, v\rangle > 0$; in particular,

$$\langle v_j, ke_j\rangle > 0 \qquad \text{for all } k\in K,\ 1\leqslant i\leqslant r,\ 1\leqslant j\leqslant \ell .$$

Put

$$\xi_j = \langle v_j,\ k\sum_{i=1}^{r} t_i e_i\rangle$$
$$= t_1 \langle v_j, k\left(e_1 + \sum_{i=2}^{r} \tau_2\cdots\tau_r e_i\right)\rangle$$
$$= t_1 \langle v_j, ke_1\rangle \left(1 + \sum_{i=2}^{r} \tau_2\cdots\tau_i \frac{\langle v_j, ke_i\rangle}{\langle v_j, ke_1\rangle}\right) . \quad (2.8)$$

For the fixed e_i, v_j $(1 \leq i \leq r,\ 1 \leq j \leq \ell)$, choose $\rho, \rho_1 > 0$ in such a way that

$$\begin{cases} \displaystyle\sum_{i=2}^{r} \rho^{i-1} \frac{\langle v_j, ke_i\rangle}{\langle v_j, ke_1\rangle} < 1 & \text{for all } k \in K,\ 1 \leq j \leq \ell\ , \\ \rho_1 < \dfrac{\pi}{\langle v_j, ke_1\rangle} & \text{for all } 1 \leq j \leq \ell\ . \end{cases} \tag{2.9}$$

Then for

$$0 < |t_1| < \rho_1\ , \quad |\tau_i| < \rho \qquad (2 \leq i \leq r) \qquad , \tag{2.10}$$

one has $0 < |\xi_j| < 2\pi$ and so $b(\xi_j, 1-\alpha_j)$ is holomorphic. Hence the function $\tilde{F}(t_1,\tau)$ has a Laurent expansion in $t_1, \tau_2, \ldots, \tau_r$ in the domain defined by (2.10).

We denote the coefficients of $t_1^{m_1} \Pi_{i=2}^{r} \tau_i^{m_i}$ in the Laurent expansion of $F(t_1, \ldots, t_1\tau_2 \ldots \tau_r)$ and $\tilde{F}(t_1,\tau)$ by $c_{(m_i)}$ and $\tilde{c}_{(m_i)}$, respectively. For $(\nu_{ij}) \in M_{r,\ell}(\mathbb{Z})$ $(r \times \ell$ integral matrices), we put

$$I((\nu_{ij})) = \int_K \prod_{\substack{1\leq i\leq r\\ 1\leq j\leq \ell}} \langle v_j, ke_i\rangle^{\nu_{ij}}\, dk\ . \tag{2.11}$$

[It can be shown that $I((\nu_{ij}))$ is rational if all ν_{ij} are nonnegative.] For a given $(\mu_i) \in \mathbb{Z}^r$, we denote by $M_{r,\ell}((\mu_i))$ the set of all $(\nu_{ij}) \in M_{r,\ell}(\mathbb{Z})$ satisfying the following conditions:

$$\nu_{ij} \geq 0 \qquad \text{for} \quad i \geq 2 \qquad , \tag{2.12a}$$

$$\sum_{i=1}^{r} \nu_{ij} = -1 \qquad \text{if} \quad \nu_{1j} < 0 \qquad , \tag{2.12b}$$

$$\sum_{j=1}^{\ell} \nu_{ij} = \mu_i \qquad \text{for all} \quad 1 \leq i \leq r \qquad . \tag{2.12c}$$

It is easy to see that the set $M_{r,\ell}((\mu_i))$ is always finite (possibly empty). In these notations, we obtain

PROPOSITION 2. *The coefficient* $c_{(m_i)}$ *in the Laurent expansion of* $F(t_1,\dots,t_1\tau_2\dots\tau_r)$ *is a* $\mathbb{Q}$*-linear combination of* $I((\nu_{ij}))$ *for* $(\nu_{ij}) \in M_{r,\ell}((m_i - m_{i+1}))$, *where we put* $m_{r+1} = 0$.

In fact, by (0.1) and (2.8) one has

$$b(\xi_j, 1-\alpha_j) = \xi_j^{-1} + \sum_{\nu=0}^{\infty} \frac{B_{\nu+1}(1-\alpha_j)}{(\nu+1)!}\, \xi_j^{\nu}$$

$$= b_-^{(j)}(t_1,\tau) + b_+^{(j)}(t_1,\tau) \qquad ,$$

where

$$\begin{cases} b_-^{(j)}(t_1,\tau) = t_1^{-1} \langle v_j, ke_1\rangle^{-1} \displaystyle\sum_{\mu=0}^{\infty} \left(- \sum_{i=2}^{r} \tau_2\cdots\tau_i \frac{\langle v_j, ke_i\rangle}{\langle v_j, ke_1\rangle} \right)^{\mu} , \\ b_+^{(j)}(t_1,\tau) = \displaystyle\sum_{\nu=0}^{\infty} \frac{B_{\nu+1}(1-\alpha_j)}{(\nu+1)!}\, t_1^{\nu} \langle v_j, ke_1\rangle^{\nu} \left(1 + \sum_{i=2}^{r} \tau_2\cdots\tau_1 \frac{\langle v_j, ke_i\rangle}{\langle v_j, ke_1\rangle} \right)^{\nu} . \end{cases} \tag{2.13}$$

Therefore

$$\prod_{j=1}^{\ell} b(\xi_j, 1-\alpha_j) = \sum_{J\subset\{1,\dots,\ell\}} \left(\prod_{j\in J} b_-^{(j)}(t_1,\tau) \prod_{j\notin J} b_+^{(j)}(t_1,\tau) \right) , \tag{2.14}$$

where the summation is taken over all subsets J of $\{1,\dots,\ell\}$. The coefficient of $t_1^{m_1} \prod \tau_i^{m_i}$ in the term corresponding to J in (2.14) is given by

$$\sum_{(\nu_{ij})} (-1)^{\Sigma\, \nu_1^{(j)}} \prod_{j\in J} \frac{\nu_1^{(j)}!}{\langle v_j, ke_1\rangle^{\nu_1^{(j)}+1}} \prod_{j\notin J} \frac{B_{\nu^{(j)}+1}(1-\alpha_j)}{\nu^{(j)}+1}$$

$$\times \prod_{\substack{2\leqslant i\leqslant r \\ j\in J}} \frac{\langle v_j, ke_i\rangle^{\nu_{ij}}}{\nu_{ij}!} \prod_{\substack{1\leqslant i\leqslant r \\ j\notin J}} \frac{\langle v_j, ke_i\rangle^{\nu_{ij}}}{\nu_{ij}!} , \tag{2.15}$$

where

$$\nu_1^{(j)} = \sum_{i=2}^{r} \nu_{ij} \quad (j \in J) \; , \quad \nu^{(j)} = \sum_{i=1}^{r} \nu_{ij} \quad (j \notin J)$$

and the summation on (ν_{ij}) is taken over all integers ν_{ij} satisfying the conditions:

$$\begin{cases} \nu_{ij} \geqslant 0 & \text{for } i \geqslant 2,\ j \in J \quad \text{or} \quad i \geqslant 1,\ j \notin J \; , \\ -|J| + \sum_{j \notin J} \nu^{(j)} = m_1 & (|J| \text{ denotes the cardinality of } J) \; , \\ \sum_{\substack{k \geqslant i \\ 1 \leqslant j \leqslant \ell}} \nu_{kj} = m_i & \text{for } 2 \leqslant i \leqslant r \; . \end{cases}$$

Putting $\nu_{1j} = -1 - \nu_1^{(j)}$ for $j \in J$ and $\mu_i = m_i - m_{i+1}$, $m_{r+1} = 0$, these conditions can be rewritten as

$$\begin{cases} \nu_{ij} \geqslant 0 & \text{unless} \quad i = 1,\ j \in J \; , \\ \sum_{i=1}^{r} \nu_{ij} = -1 & \text{for} \quad j \in J \; , \\ \sum_{j=1}^{\ell} \nu_{ij} = \mu_i & \text{for all} \quad 1 \leqslant i \leqslant r \; . \end{cases} \tag{2.12}_J$$

We denote by $M_{r,\ell}(J,(\mu_i))$ the set of all $(\nu_{ij}) \in M_{r,\ell}(\mathbb{Z})$ satisfying $(2.12)_J$. Then, clearly we have

$$M_{r,\ell}((\mu_j)) = \bigcup_{J \subset \{1,\ldots,\ell\}} M_{r,\ell}(J,(\mu_i))$$

and

$$c(m_i) = \sum_{J \subset \{1,\ldots,\ell\}} \; \sum_{(\nu_{ij}) \in M_{\ell,r}(J,\mu_i)} \frac{\prod_{j \in J} (-1)^{\nu_1^{(j)}} \nu_1^{(j)}! \prod_{j \notin J} \dfrac{B_{\nu^{(j)}+1}(1-\alpha_j)}{\nu^{(j)}+1}}{\prod_{\substack{i \geqslant 2 \\ \text{or } j \notin J}} \nu_{ij}!} \cdot I((\nu_{ij})) \; , \tag{2.16}$$

which proves our assertion.

In the expansion of $\Pi_{2\leqslant i<j\leqslant r}(1-\tau_i\cdots\tau_j)^d$, there appear only those terms $\Pi_{i=2}^r \tau_i^{p_i}$ for which the exponents p_i satisfy the conditions:

$$\begin{cases} 0\leqslant p_i \leqslant (i-1)(r-i+1)d \\ (i-r)d \leqslant p_i - p_{i+1} \leqslant (i-1)d \qquad (p_1 = p_{r+1} = 0) \quad . \end{cases} \tag{2.17}$$

Hence Proposition 2 implies

Corollary. *The coefficient* $\tilde{c}_{(m_i)}$ *in the Laurent expansion of* $\tilde{F}(t_1,\tau)$ *is a* $\mathbf{Q}$*-linear combination of* $I((\nu_{ij}))$ *for* $(\nu_{ij}) \in M_{r,\ell}(\mathbb{Z})$ *satisfying the conditions* (2.12a, b) *and*

$$\mu_i - (i-1)d \leqslant \sum_{j=1}^{\ell} \nu_{ij} \leqslant \mu_i + (r-i)d \qquad \text{for all} \quad 1\leqslant i\leqslant r \tag{2.12$\tilde{c}$}$$

instead of (2.12c).

Note that (ν_{ij}) appearing in $\tilde{c}_{(m_i)}$ should also satisfy the condition

$$\sum_{\substack{1\leqslant i\leqslant r \\ 1\leqslant j\leqslant \ell}} \nu_{ij} = m_1 \quad . \tag{2.18}$$

2.4 Let $I(\varepsilon,1)$ denote the contour consisting of the line segment $[\varepsilon,1]$ taken twice in opposite directions and of a (small) circle of radius ε about the origin taken in the counterclockwise direction. When the τ_i $(2\leqslant i\leqslant r)$ are on $I(\varepsilon,1)$, one has by the Schwartz inequality and (1.12)

$$\left|\left\langle v_j, k\left(e_1 + \sum_{i=2}^r \tau_2\cdots\tau_i e_i\right)\right\rangle\right| \leqslant |v_j| \cdot \sum_{i=1}^r |e_i| = \sqrt{nr}\,|v_j|$$

and

$$\begin{aligned} \mathrm{Re}\left\langle v_j, k\left(e_1 + \sum_{i=2}^r \tau_2\cdots\tau_i e_i\right)\right\rangle &= \langle v_j, ke_1\rangle + \sum_{i=2}^r \mathrm{Re}(\tau_2\cdots\tau_i)\langle v_j, ke_i\rangle \\ &\geqslant \langle v_j, ke_1\rangle - \varepsilon|v_j| \sum_{i=2}^r |e_i| \\ &= \langle v_j, ke_1\rangle - \varepsilon(r-1)\sqrt{\frac{n}{r}}\,|v_j| \quad . \end{aligned}$$

We choose ε so that one has

$$\varepsilon\sqrt{rn}|v_j| < \mathrm{Min}\{2\pi, \langle v_j, ke_1\rangle \quad (k\in K)\} \qquad \text{for all} \quad 1\leqslant j\leqslant \ell . \tag{2.19}$$

Then the above inequalities show that the $\langle v_j, k(e_1+\Sigma_{i=2}^r \tau_2\cdots\tau_1 e_i)\rangle$ $(1\leqslant j\leqslant \ell)$ belong to the domain

$$\left\{z\in\mathbb{C}\;\middle|\; |z|<\frac{2\pi}{\varepsilon}\quad, \mathrm{Re}\ z>\varepsilon\sqrt{\frac{n}{r}}\,|v_j|\right\} .$$

It follows that, if t_1 is on the contour $I(\varepsilon,\infty)$ and the τ_i $(2\leqslant i\leqslant r)$ are on $I(\varepsilon,1)$, one has

$$0<|\xi_j|<2\pi \qquad \text{or} \qquad \mathrm{Re}\ \xi_j>0 \quad ,$$

so that the function $b(\xi_j, 1-\alpha_j)$ is holomorphic.

From this observation, it is clear that the integral on the right-hand side of (2.6) is equal to the contour integral

$$(\mathbf{e}(ns)-1)^{-1}\int_{t_1\in I(\varepsilon,\infty)} \cdot \prod_{i=2}^{r}\left(\left(\mathbf{e}\left(\frac{r-i+1}{r}\,ns\right)-1\right)^{-1}\int_{\tau_i\in I(\varepsilon,1)}\right)$$

which is independent of the choice of ε satisfying (2.19). It can be shown that the contour integral converges for all $s\in\mathbb{C}$. Therefore, the above integral, viewed as a function of s, can be continued to a meromorphic function on the whole plane, with possible poles at the points $r\nu/(r-i+1)n$ $(\nu\in\mathbf{Z},\ 1\leqslant i\leqslant r)$.

3. The Special Values of the Zeta Functions

3.1 As a preliminary, we check the rationality of the constant c in (2.4). For that purpose, we compute $\Gamma_\Omega(s)$ by using the decomposition $G=KAK$.

$$\begin{aligned}\Gamma_\Omega(s) &= \int_\Omega N(u)^s \exp(-\tau(u))\, d_\Omega(u)\\ &= \int_G N(ge)^s \exp(-\tau(ge))\, dg\\ &= c\int_A \det(a)^s \exp(-\tau(ae))\Delta(a)\, da\\ &= c\int_0^\infty\cdots\int_0^\infty \left(\prod_i t_i\right)^{\frac{n}{r}(s-1)} |\Delta(t)|^d \exp\left(-\frac{n}{r}\sum_i t_i\right)\prod_i dt_i\,. \end{aligned} \tag{3.1}$$

We make another change of variables:

$$t = \sum_{i=1}^r t_i, \qquad t'_i = t_i/t$$

with

$$\frac{\partial(t_1,\ldots,t_r)}{\partial(t,t'_1,\ldots,t'_{r-1})} = (-1)^{r-1}\, t^{r-1} \quad .$$

Then one obtains

$$\Gamma_\Omega(s) = c\cdot\gamma(s)\beta(s) \quad , \tag{3.2}$$

where

$$\left\{\begin{aligned}\gamma(s) &= \int_0^\infty t^{ns-1}\exp\left(-\frac{n}{r}\,t\right) dt = \left(\frac{r}{n}\right)^{ns}\Gamma(ns) \quad ,\\ \beta(s) &= \int\!\!\cdots\!\!\int_{\substack{t'_i>0\\ \Sigma t'_i<1}} \left\{t'_1\cdots t'_{r-1}\left(1-\sum_i t'_i\right)\right\}^{\frac{n}{r}(s-1)}\\ &\quad\times\left|\Delta\left(t'_1,\ldots,t'_{r-1},1-\sum_i t'_i\right)\right|^d \prod_i dt'_i\,. \end{aligned}\right. \tag{3.3}$$

For $s=1$, one has

$$\Gamma_\Omega(1) = c\cdot\gamma(1)\beta(1) = c\left(\frac{r}{n}\right)^n (n-1)!\beta(1) ,$$

$$\beta(1) = \underset{\substack{t_i'>0 \\ \Sigma t_i'<1}}{\int\cdots\int} \left|\Delta\left(t_1',\ldots,t_{r-1}',1-\sum_i t_i'\right)\right|^d \prod_i dt_i' \in \mathbb{Q}_+ .$$

By (1.23) one has

$$\Gamma_\Omega(1) = (2\pi)^{\frac{n-r}{2}} \left(\frac{r}{n}\right)^{\frac{n}{2}} \prod_{j=1}^{r} \Gamma\left(1+\frac{d}{2}(j-1)\right)$$

$$\underset{\mathbb{Q}}{\sim} \begin{cases} \pi^{\frac{n-r}{2}} & (d \text{ even}) \\ \pi^{\frac{1}{2}\left(n-\left[\frac{r+1}{2}\right]\right)} & (d \text{ odd}) , \end{cases} \tag{3.4}$$

(where $a \underset{\mathbb{Q}}{\sim} b$ means that $a/b \in \mathbb{Q}^\times$). Thus one has

$$c = \frac{(2\pi)^{\frac{n-r}{2}} \left(\frac{n}{r}\right)^{\frac{n}{2}} \prod_{j=1}^{r} \Gamma\left(1+\frac{d}{2}(j-1)\right)}{(n-1)!\ \beta(1)} \underset{\mathbb{Q}}{\sim} \Gamma_\Omega(1) . \tag{3.5}$$

Since $\Gamma_\Omega(1) \underset{\mathbb{Q}}{\sim} \Gamma_\Omega(1+(r/n)\nu)$ for $\nu \in \mathbb{Z}_+$, one obtains

$$c\Gamma_\Omega\left(1+\frac{r}{n}\nu\right) \underset{\mathbb{Q}}{\sim} \Gamma_\Omega(1)^2 \underset{\mathbb{Q}}{\sim} \begin{cases} \pi^{n-r} & (d \text{ even}) \\ \pi^{n-\left[\frac{r+1}{2}\right]} & (d \text{ odd}) . \end{cases} \tag{3.6}$$

3.2 We are interested in the value of ζ_Ω at $s = -(r/n)\nu$ ($\nu = 0, 1, \ldots$). Here we consider only the case where d is even. In this case, one has $n \equiv r \pmod 2$, $r|n$, and

$$\Gamma_\Omega(s)\Gamma_\Omega(1-s) = (2\pi i)^n \mathbf{e}\left(\frac{n}{2}s\right)\left(\mathbf{e}\left(\frac{n}{r}s\right)-1\right)^{-r} .$$

Hence one has

$$\zeta_\Omega(s;(v_j),(\alpha_j)) = \frac{c\Gamma_\Omega(1-s)}{(2\pi i)^{n-r}\mathbf{e}\left(\frac{n}{2}s\right)} \cdot R(s) \quad , \tag{3.7}$$

where

$$R(s) = \left(\frac{\mathbf{e}\left(\frac{n}{r}s\right)-1}{2\pi i}\right)^r \cdot r! \int_{E_1} \left(\prod_i t_i\right)^{\frac{n}{r}(s-1)} \Delta(t)^d F(t) \prod_i dt_i$$

$$= \prod_{j=1}^{r} \frac{\mathbf{e}\left(\frac{n}{r}s\right)-1}{\mathbf{e}\left(\frac{r-j+1}{r}ns\right)-1} \cdot \frac{r!}{(2\pi i)^r} \int_{I(\varepsilon,\infty)} t_1^{ns-1}\, dt_1$$

$$\times \int_{I(\varepsilon,1)^{r-1}} \prod_{i=2}^{r} \tau_i^{(r-i+1)\left\{\frac{n}{r}s-\frac{d}{2}(i-1)\right\}-1} \tilde{F}(t_1,\tau) \prod_i d\tau_i \ .$$

The first factor in the right-hand side of (3.7) is holomorphic for $\mathrm{Re}\, s < r/n$ and by (3.6) the value at $s=-(r/n)\nu$ is rational:

$$\frac{c\Gamma_\Omega\left(1+\frac{r}{n}\nu\right)}{(2\pi i)^{n-r}\mathbf{e}\left(-\frac{r}{2}\nu\right)} = (-1)^{\frac{n-r}{2}+r\nu} \frac{c\Gamma_\Omega\left(1+\frac{r}{n}\nu\right)}{(2\pi)^{n-r}} \in \mathbf{Q} \quad . \tag{3.8}$$

On the other hand, since

$$\frac{\mathbf{e}\left(\frac{n}{r}s\right)-1}{\mathbf{e}\left(\frac{r-i+1}{r}ns\right)-1} \to \frac{1}{r-i+1} \qquad \text{when} \qquad s \to -\frac{r}{n}\nu \quad ,$$

we see that $R(-(r/n)\nu)$ is equal to the coefficient of

$$t_1^{r\nu} \prod_{i=2}^{r} \tau_i^{(r-i+1)\left\{\nu+\frac{d}{2}(i-1)\right\}}$$

in the Laurent expansion of $\tilde{F}(t_1,\tau)$, i.e.,

$$R\left(-\frac{r}{n}\nu\right) = \tilde{c}_{(m_i)} \qquad \text{with} \quad m_i = (r-i+1)\left\{\nu + \frac{d}{2}(i-1)\right\}. \tag{3.9}$$

In this case,

$$\mu_i = m_i - m_{i+1} = \nu - \frac{d}{2}(r-2i+1) \quad .$$

Hence, applying Corollary to Proposition 2, we obtain the following

Theorem 1. *Assume that* Ω *is irreducible,* d *is even, and all* v_j*'s* $(1 \leqslant j \leqslant \ell)$ *are contained in* Ω. *Then* $\zeta_\Omega(-(r/n)\nu; (v_j), (\alpha_j))$ $(\nu = 0,1,\ldots)$ *is a* $\mathbb{Q}$*-linear combination of* $I((\nu_{ij}))$ *for* $(\nu_{ij}) \in M_{r,\ell}(\mathbb{Z})$ *satisfying the conditions* (2.12a,b) *and*

$$\nu - \frac{d}{2}(r-1) \leqslant \sum_{j=1}^{\ell} \nu_{ij} \leqslant \nu + \frac{d}{2}(r-1) \quad , \tag{3.10}$$

$$\sum_{\substack{1\leqslant i\leqslant r\\ 1\leqslant j\leqslant \ell}} \nu_{ij} = r\nu \quad . \tag{3.11}$$

When d is odd, our method does not work well, unless more informations on $I((\nu_{ij}))$ become available. In this case, according to the classification theory of self-dual homogeneous cones (cf. e.g., [2], [7]), one has either $d = 1$ (the case of $\mathcal{P}_r(\mathbb{R})$) or $r = 2$ (the case of quadratic cones of odd dimension). For the latter case, it seems more convenient to use the Laurent expansion of F in another set of variables, which we hope to have an opportunity to discuss elsewhere.

4. References

[1] A. Ash *et al.*, *Smooth Compactification of Locally Symmetric Varieties*, Math. Sci. Press, Brookline, 1975.

[2] H. Braun and M. Koecher, *Jordan-Algebren*, Springer-Verlag, 1966.

[3] S.G. Gindikin, "Analysis in homogeneous domains," *Uspehi Mat. Nauk.* 19 (1964), 3-92; = *Russian Math. Survey* 19 (1964), 1-89.

[4] S. Helgason, *Differential Geometry and Symmetric Spaces*, Academic Press, 1962.

[5] M. Koecher, "Positivitätsbereiche im $\mathbb{R}^n$," *Amer. J. Math.* 79 (1957), 575-596.

[6] H.L. Resnikoff, "On a class of linear differential equations for automorphic forms in several complex variables," *Amer. J. Math.* 95 (1973), 321-331.

[7] I. Satake, *Algebraic Structures of Symmetric Domains*, Iwanami-Shoten and Princeton Univ. Press, 1980.

[8] M. Sato and T. Shintani, "On zeta functions associated with prehomogeneous vector spaces," *Ann. of Math.* 100 (1974), 131-170.

[9] T. Shintani, "On Dirichlet series whose coefficients are class-numbers of integral binary cubic forms," *J. Math. Soc. Japan* 24 (1972), 132-188.

[10] T. Shintani, "On zeta-functions associated with the vector space of quadratic forms," *J. Fac. Sci. Univ. Tokyo* 22 (1975), 25-65.

[11] T. Shintani, "On evaluation of zeta functions of totally real algebraic number fields at non-positive integers," *J. Fac. Sci. Univ. Tokyo* 23 (1976), 393-417.

[12] C.L. Siegel, "Über die Zetafunktionen indefiniter quadratischer Formen," *Math. Z.* 43 (1938), 682-708.

[13] C.L. Siegel, "Berechnung von Zetafunktionen an ganzzähligen Stellen," *Nachr. Akad. Wiss. Göttingen*, Math.-Phys. Kl. 1968, 7-38.

[14] A. Kurihara, "On the values at non-negative integers of Siegel's zeta functions of Q-anisotropic quadratic forms with signature (1,n-1)," to appear in *J. Fac. Sci.*, Univ. Tokyo.

University of California
Berkeley, California 94720

and

Tôhoku University
Sendai 980
Japan

(Received February 6, 1981)

HESSIAN MANIFOLDS AND CONVEXITY

Hirohiko Shima

Let M be a *flat affine manifold*, that is, M admits open charts $(U_i, \{x_i^1, \ldots, x_i^n\})$ such that $M = \cup U_i$ and whose coordinate changes are all affine functions. Such local coordinate systems $\{x_i^1, \ldots, x_i^n\}$ will be called *affine local coordinate systems*. Throughout this note the local expressions for geometric concepts on M will be given in terms of affine local coordinate systems. A Riemannian metric g on M is said to be *Hessian* if for each point $p \in M$ there exists a C^∞-function ϕ defined on a neighborhood of p such that $g_{ij} = \partial^2\phi/\partial x^i \partial x^j$. Such a function ϕ is called a *primitive* of g on a neighborhood of p. Using the flat affine structure we define the exterior differentiation $\mathbb{d}$ for tensor bundle valued forms on M. Let g^0 be the cotangent bundle valued 1-form on M corresponding to a Riemannian metric g on M. Then we know that g is Hessian if and only if g^0 is $\mathbb{d}$-closed. A flat affine manifold provided with a Hessian metric is called a *Hessian manifold* [4], [5]. Koszul dealt with the case where g^0 is $\mathbb{d}$-exact [1], [2], [3].

<u>Theorem A</u> (Koszul [2]). *Let M be a connected Hessian manifold with Hessian metric g. Suppose g^0 is $\mathbb{d}$-exact, that is, M admits a closed 1-form α such that $g^0 = \mathbb{d}\alpha$. If there exists a subgroup G of affine transformations of M such that $G\backslash M$ is quasi-compact and that G leaves α invariant, then the universal covering manifold of M is a convex domain in a real affine space not containing any full straight line.*

In Section 1 we prove the convexity of a Hessian manifold and in Section 2 we discuss certain cohomology groups of the fundamental group of a Hessian manifold.

1. Let M be a Hessian manifold. A diffeomorphism of M onto itself is called an *automorphism* of M if it preserves both the flat affine structure and the Hessian metric. In this section we prove the following theorem along the same line as Koszul [2].

Theorem B. *Let* M *be a connected Hessian manifold. If there exists a subgroup* G *of automorphisms of* M *such that* $G\backslash M$ *is quasi-compact, then the universal covering manifold of* M *is a convex domain in a real affine space.*

Corollary. *Let* M *be a connected Hessian manifold. If* M *admits a transitive automorphism group or is compact, then the universal covering manifold of* M *is a convex domain in a real affine space.*

Remark. In [6] Yagi showed that a flat affine manifold whose universal covering manifold is a convex domain in a real affine space does not necessarily admit a Hessian metric.

Let M be a connected Hessian manifold with Hessian metric g. We denote by TM the tangent bundle over M and by E the domain of definition for the exponential mapping $\exp$ defined by the flat affine structure. For $y \in TM$ we set $\lambda(y) = \sup\{t \in \mathbb{R};\ ty \in E\}$ and $I(y) = (-\lambda(-y), \lambda(y))$. Let $\gamma_y(t) = \exp ty$. We can find a family $\{\phi_{(y,t)};\ t \in I(y)\}$ in such a way that $\phi_{(y,t)}$ is a primitive of g around $\gamma_y(t)$ and that if t' is sufficiently close to t then $\phi_{(y,t')} = \phi_{(y,t)}$ around $\gamma_y(t')$. Such a family is called a *primitive of* g *along* γ_y. Let $y \in TM$ and let $t \in (0, \lambda(y))$. We choose a subdivision $u_0 = 0 < u_1 < \cdots u_{k-1} < u_k < \cdots < u_m = t$ of $[0,t]$ in such a way that for each $k = 0,\ldots,m$ there exists an affine local coordinate system $\{x^1_{(k)},\ldots,x^n_{(k)}\}$ such that $\gamma_y(u)$ is contained in the domain of definition $\{x^1_{(k)},\ldots,x^n_{(k)}\}$ for all $u_{k-1} \leqq u \leqq u_k$. We put

$$\alpha_{(k)} = \sum_{i=1}^{n}\left(\sum_{p=1}^{n}\int_{u_{k-1}}^{u_k} g_{(k)ip}(\gamma_y(u))\,\frac{d}{du}x^p_{(k)}(\gamma_y(u))\,du\right)(dx^i_{(k)})_{\gamma_y(u_k)},$$

where $g_{(k)ij}$ are the component of g with respect to $\{x^1_{(k)},\ldots,x^n_{(k)}\}$. Let $P(y)^{t_2}_{t_1}$ be the parallel translation (with respect to the flat affine structure) from $\gamma_y(t_1)$ to $\gamma_y(t_2)$ along γ_y. We define a cotangent vector $\alpha_{\gamma_y(t)}$ at $\gamma_y(t)$ by

$$\alpha_{\gamma_y(t)} = \sum_{k=1}^{m} P(y)^{t}_{u_k}\,\alpha_{(k)} \qquad .$$

Note that the definition of $\alpha_{\gamma_y(t)}$ is independent of the choice of a subdivision of $[0,t]$ and affine local coordinate systems $\{x^1_{(k)},\dots,x^n_{(k)}\}$. For $t \in [0,\lambda(y))$ we put

$$h_y(t) = \int_0^t \alpha_{\gamma_y(u)}(\dot{\gamma}_y(u))\, du \qquad ,$$

where $\dot{\gamma}_y(u)$ is the tangent vector of γ_y at $\gamma_y(u)$. Let $\{\phi_{(y,t)}\}$ be a primitive of g along γ_y. Then we have

$$\alpha_{\gamma_y(t)} = (d\phi_{(y,t)})_{\gamma_y(t)} - P(y)_0^t(d\phi_{(y,0)})_{\gamma_y(0)} \quad ,$$

$$h_y(t) = \phi_{(y,t)}(\gamma_y(t)) - \phi_{(y,0)}(\gamma_y(0)) - (y\phi_{(y,0)})t \quad .$$

The function $h_y(t)$ is C^∞ in t and we have

$$\frac{d^2}{dt^2} h_y(t) = g(\dot{\gamma}_y(t),\dot{\gamma}_y(t)) > 0 \quad ,$$

$$\frac{d}{dt} h_y(t) = \dot{\gamma}_y(t)\phi_{(y,t)} - \dot{\gamma}_y(0)\phi_{(y,0)} > 0 \quad ,$$

$$h_y(t) > 0 \ ,$$

on $(0,\lambda(y))$. Let s be an automorphism of M. Then $\phi_{(y,t)}\circ s^{-1}$ is a primitive of g along $s\gamma_y = \gamma_{sy}$ and so

$$\begin{aligned} h_{sy}(t) &= (\phi_{(y,t)}\circ s^{-1})(\gamma_{sy}(t)) - (\phi_{(y,0)}\circ s^{-1})(\gamma_{sy}(0)) - (sy(\phi_{(y,0)}\circ s^{-1}))t \\ &= \phi_{(y,t)}(\gamma_y(t)) - \phi_{(y,0)}(\gamma_y(0)) - (y\phi_{(y,0)})t \\ &= h_y(t) \quad . \end{aligned}$$

$G\backslash M$ being quasi-compact there exists a compact subset K of M such that $M = GK$. For $y \in TM$ we denote by $\|y\|$ the norm $\sqrt{g(y,y)}$ of y. Since E is an open set in TM containing zero section and since K is compact, there exists $\varepsilon > 0$ such that if the origin of $y \in TM$ is contained in K and $\|y\| \leqq \varepsilon$ then $y \in E$. From this and $M = GK$ we obtain that if $y \in TM$ and $\|y\| \leqq \varepsilon$ then $y \in E$. Let A denote the set of $y \in TM$ such that $\|y\| = \varepsilon$ and A_K the set of $y \in A$ such that the origin of y is contained in K. We have then $A = GA_K$. The function defined on E by $y \to h_y(1)$ is positive, continuous and invariant by G. A_K being compact the infinimum r of $h_y(1)$ on A_K is positive and so

$$h_y(1) \geqq r > 0 \qquad \text{for all} \quad y \in A .$$

<u>Lemma 1</u>. $\lim_{t \to \lambda(y)} h_y(t) = +\infty$.

<u>Proof</u>. For a fixed $t_0 \in (0,\lambda(y))$ we put $z = \dot{\gamma}_y(t_0)$. Since $(\varepsilon/\|z\|)z \in A \subset E$, we have $t_0 + \varepsilon/\|z\| \in I(y)$ and

$$\gamma_z(t) = \exp tz = \exp(t_0 + t)y = \gamma_y(t_0 + t)$$

for $t \in [0,\varepsilon/\|z\|]$. Let $\{\phi_{(y,t)}\}$ be a primitive of g along γ_y. Then $\phi_{(z,t)} = \phi_{(y,t_0+t)}$ is a primitive of g along γ_z and so

$$\begin{aligned}
h_y(t_0 + \varepsilon/\|z\|) &= \phi_{(y,t_0+\varepsilon/\|z\|)}(\gamma_y(t_0+\varepsilon/\|z\|)) - \phi_{(y,0)}(\gamma_y(0)) \\
&\quad - (y\phi_{(y,0)})(t_0+\varepsilon/\|z\|) \\
&= \phi_{(z,\varepsilon/\|z\|)}(\gamma_z(\varepsilon/\|z\|)) - \phi_{(z,0)}(\gamma_z(0)) - (z\phi_{(z,0)})\varepsilon/\|z\| \\
&\quad + \phi_{(y,t_0)}(\gamma_y(t_0)) - \phi_{(y,0)}(\gamma_y(0)) - (y\phi_{(y,0)})t_0 \\
&\quad + (\dot{\gamma}_y(t_0)\phi_{(y,t_0)} - \dot{\gamma}_y(0)\phi_{(y,0)})\varepsilon/\|z\| \\
&= h_z(\varepsilon/\|z\|) + h_y(t_0) + \left(\frac{d}{dt}\Big|_{t=t_0} h_y(t)\right) \varepsilon/\|z\| \\
&= h_y(t_0) + h_{(\varepsilon/\|z\|)z}(1) + \left(\frac{d}{dt}\Big|_{t=t_0} h_y(t)\right) \varepsilon/\|z\| .
\end{aligned}$$

Therefore we have

$$h_y(t_0+\varepsilon/\|z\|) > h_y(t_0) + r$$

because

$$h_{(\varepsilon/\|z\|)z}(1) \geqq r \qquad \text{and} \qquad \left.\frac{d}{dt}\right|_{t=t_0} h_y(t) > 0 \quad .$$

This implies

$$\lim_{t\to\lambda(y)} h_y(t) = +\infty \quad .$$

Let T_xM be the tangent space of M at x and let $E_x = E \cap T_xM$. Then E_x is a domain in T_xM and admits a natural flat affine connection.

Lemma 2 (Koszul [2]). *Let* $\exp_x$ *denote the restriction of the exponential mapping* $\exp$ *to* E_x. *Then* $\exp_x$ *is an affine mapping from* E_x *to* M *and its rank is maximum at every point in* E_x *and equal to* $\dim M$. *Moreover if* E_x *is convex it is the universal covering manifold of* M *with covering projection* $\exp_x$.

It follows from Lemma 2 that the induced Riemannian metric $\tilde{g}$ on E_x is a Hessian metric.

Lemma 3 ([5]). *There exists a primitive* ψ *of* $\tilde{g}$ *globally defined on* E_x.

Lemma 4. *Let* $y \in T_xM$. *If* $\lambda(y) < +\infty$, *then we have*

$$\lim_{t\to\lambda(y)} \psi(ty) = +\infty \quad .$$

Proof. Let $\{\phi_{(y,t)}\}$ be a primitive of $\tilde{g}$ along γ_y. Then $\psi_{(y,t)} = \phi_{(y,t)} \circ \exp_x$ is a primitive of $\tilde{g}$ along the geodesic ty $(0 \leqq t < \lambda(y))$ and so $a(t) = \psi(ty) - \psi_{(y,t)}(ty)$ is an affine function in t. From this and Lemma 1 we have

$$\begin{aligned}\lim_{t\to\lambda(y)} \psi(ty) &= \lim_{t\to\lambda(y)} (\psi_{(y,t)}(ty) + a(t)) \\ &= \lim_{t\to\lambda(y)} (\phi_{(y,t)}(\gamma_y(t)) + a(t)) \\ &= \lim_{t\to\lambda(y)} (h_y(t) + \phi_{(y,0)}(\gamma_y(0)) + (y\phi_{(y,0)})t + a(t)) \\ &= +\infty \quad .\end{aligned}$$

According to Lemma 4, Lemma 4.2 in [1] and the fact that E_x is star-shaped with respect to the origin 0, E_x is a convex domain. Therefore, by Lemma 2, E_x is the universal covering manifold with covering projection $\exp_x : E_x \to M$.

2. In this section we consider certain cohomology groups of the fundamental group of a Hessian manifold. Let Ω be a convex domain in a real affine space $\mathbb{R}^n$ and Γ a properly discontinuous subgroup of affine transformation group of $\mathbb{R}^n$ acting freely on Ω. Then the quotient space $\Gamma\backslash\Omega$ admits in a natural way a flat affine structure induced from that of $\mathbb{R}^n$. Let A denote the vector space of all affine functions on $\mathbb{R}^n$. Then Γ acts on A by $sf = f\circ s^{-1}$ where $s \in \Gamma$, $f \in A$ and this action makes A into a Γ-module. Let $H^r(\Gamma,A)$ be the r-th cohomology group of Γ with coefficients in A. Then we have

Theorem C. *Suppose that* $\Gamma\backslash\Omega$ *admits a Hessian metric* g.

(i) *If* $H^1(\Gamma,A) = 0$, *there exists a primitive of* g *globally defined on* $\Gamma\backslash\Omega$.

(ii) *If* $\Gamma\backslash\Omega$ *is compact, then we have* $H^1(\Gamma,A) \neq 0$.

Proof. Let π be the projection from Ω onto $\Gamma\backslash\Omega$. Since the induced metric π^*g on Ω is Hessian and since Ω is convex, by Lemma 3 Ω has a primitive ψ of π^*g globally defined on Ω. Because $\psi\circ s$ is a primitive of π^*g for $s \in \Gamma$, we know $\psi\circ s^{-1} - \psi \in A$. Define a mapping

$$u_\psi : \Gamma \to A$$

by $u_\psi(s) = \psi\circ s^{-1} - \psi$ where $s \in \Gamma$. We have then

$$u_\psi(st) = su_\psi(t) + u_\psi(s)$$

for $s,t \in \Gamma$. This means that u_ψ is a 1-cocycle on Γ with coefficients in A. Being $H^1(\Gamma,A) = 0$ there exists an element f in A such that $u_\psi(s) = f - f\circ s^{-1}$ and so

$$(\psi + f)\circ s^{-1} = \psi + f$$

for all $s \in \Gamma$. Therefore there exists a C^∞-function ϕ globally defined on $\Gamma\backslash\Omega$ such that $\psi + f = \phi\circ\pi$. This function ϕ is a primitive of g and (i) is proved. For the proof of (ii) it suffices to show that if $\Gamma\backslash\Omega$ is compact, $\Gamma\backslash\Omega$ does not have a primitive of g globally defined on $\Gamma\backslash\Omega$. Suppose that $\Gamma\backslash\Omega$ has a globally defined primitive ϕ_0 of g. Let L be an elliptic differential operator given by

$$L = \sum_{j,k} g^{jk} \frac{\partial^2}{\partial x^j \partial x^k} .$$

Then we have

$$L(\phi_0) = \sum_{j,k} g^{jk} g_{jk} = n > 0 .$$

Therefore by a Theorem of E. Hopf ϕ_0 is a constant function on $\Gamma\backslash\Omega$ and so $g_{ij} = \partial^2\phi_0/\partial x^i \partial x^j = 0$. This is a contradiction.

Let P denote the vector space of all parallel (with respect to the natural flat affine connection on $\mathbb{R}^n$) 1-forms on $\mathbb{R}^n$. Then Γ acts on P by $s\omega = (s^{-1})^*\omega$ where $s \in \Gamma$, $\omega \in P$ and this action makes P into a Γ-module. Let $H^r(\Gamma,P)$ be the r-th cohomology group of Γ with coefficients in P. Then we have

Theorem D. *Suppose that* $\Gamma\backslash\Omega$ *admits a Hessian metric* g. *If* $H^1(\Gamma,P) = 0$, *the cotangent bundle valued* 1-*form* g^0 *corresponding to* g *is* d-*exact. Assuming further that* $\Gamma\backslash\Omega$ *is compact* Ω *is a convex domain not containing any full straight line.*

Proof. As in the proof of Theorem C we denote by ψ a primitive of π^*g globally defined on Ω. Define a mapping

$$v_\psi : \Gamma \to P$$

by $v_\psi(s) = d(u_\psi(s)) = sd\psi - d\psi$ where $s \in \Gamma$. Then we have

$$v_\psi(st) = sv_\psi(t) + v_\psi(s)$$

for all $s, t \in \Gamma$. This implies that v_ψ is a 1-cocycle on Γ with coefficients in P. Being $H^1(\Gamma, P) = 0$ there exists an element ω in P such that $v_\psi(s) = \omega - s\omega$ and so

$$s(d\psi + \omega) = d\psi + \omega$$

for all $s \in \Gamma$. Thus $\Gamma\backslash\Omega$ has a closed 1-form α such that $\pi^*\alpha = d\psi + \omega$. Therefore we have $g^0 = d\alpha$. The last assertion follows from this fact and Theorem A.

3. References

[1] J.L. Koszul, "Domaines bornés homogenès et orbites de groupes de transformations affines," *Bull. Soc. Math. France* 89 (1961), 515-533.

[2] J.L. Koszul, "Variétés localement plates et convexité," *Osaka J. Math.* 2 (1965), 285-290.

[3] J.L. Koszul, "Déformations de connexions localement plates," *Ann. Inst. Fourier, Grenoble* 18, 1 (1968), 103-114.

[4] H. Shima, "On certain locally flat homogeneous manifolds of solvable Lie groups," *Osaka J. Math.* 13 (1976), 213-229.

[5] H. Shima, "Homogeneous Hessian manifolds," *Ann. Inst. Fourier, Grenoble* 30, 3 (1980).

[6] K. Yagi, "On Hessian structures on an affine manifold," in *Manifolds and Lie Groups, Papers in Honor of Yozô Matsushima*, Progress in Mathematics, Vol. 14, Birkhäuser, Boston, Basel, Stuttgart, 1981, 449-459.

Yamaguchi University
Yamaguchi 753
Japan

(Received December 12, 1980)

INTRINSIC CHARACTERIZATION OF AFFINE ALGEBRAIC CONES

Wilhelm Stoll

I am deeply touched that I was asked to give an address at this conference in honor of my distinguished colleague Yozo Matsushima. We have been here together at Notre Dame for a long time. With deep regret we see you return to your native land. I am sure I speak in the name of all, when I say "Yozo we wish you well". We will always remember you and we hope we can have you back here occasionally or for longer. Mathematics knows no bounds; it spans over all nationalities and continents and within this community of mathematicians we are still united.

Your field, differential geometry, and my field, complex analysis, have many crossroads. We both worked on them. Let me discuss here a problem in complex analysis which is solved by elementary differential geometric means.

A simply connected Rieman surface is biholomorphic equivalent either to the Riemann sphere or the Gauss plane or the unit disc. No analogon to this theorem exists for higher dimensional complex manifolds. Already Poincaré observed that the ball and the polydisc are not biholomorphically equivalent.

Additional assumptions are needed to characterize complex manifolds of higher dimensions. Here differential geometric means are employed. One such tool is an *exhaustion function.*

Let M be an irreducible complex space of dimension m. A non-negative function τ of class C^∞ on M is said to be an *exhaustion of* M with maximal radius Δ if $\Delta = \sup \sqrt{\tau}$, if $0 \leqslant \tau < \Delta^2$ on M, and if the *closed pseudo-ball*

$$M[r] \quad = \quad \{x \in M \mid \tau(x) \leqslant r^2\}$$

is compact for every $r \in \mathbb{R}[0,\Delta)$. Obviously, M is not compact. Also we consider the *open pseudo-ball*

$$M(r) = \{x \in M \mid \tau(x) < r^2\}$$

and the *pseudo-sphere*

$$M\langle r\rangle = \{x \in M \mid \tau(x) = r^2\} \quad .$$

We call $M[0] = \tau^{-1}(0)$ the *center* of τ and $M_* = M - M[0]$ the *positivity* of τ.

In general $\mathbb{R}(M)$ will denote the set of regular points of M and $\Sigma(M)$ the set of singular points. Also we defined the twisted derivative $d^c = (i/4\pi)(\bar{\partial} - \partial)$. If u is a function of class C^2, then $dd^c u$ is called the *Levy form* of u and u is said to be *plurisubharmonic* if $dd^c u \geqslant 0$ and *strictly plurisubharmonic* if $dd^c u > 0$. We know that M is Stein if and only if there exists a strictly plurisubharmonic exhaustion.

Since the singularities pose difficulties, the manifold case will be considered first. Thus let M be a connected complex manifold of dimension m. An exhaustion τ of maximal radius Δ with $0 < \Delta \leqslant \infty$ is given on M. Then τ is said to be a strictly parabolic exhaustion of M and (M,τ) is said to be a strictly parabolic manifold if and only if $dd^c\tau > 0$ on M and $dd^c \log \tau \geqslant 0$ and $(dd^c \log \tau)^m \equiv 0$ on M_*. Thus τ is a strictly plurisubharmonic exhaustion of M and $\log \tau$ is plurisubharmonic and satisfies the Monge-Ampère equation on M_*.

For instance, define $\tau_0 : \mathbb{C}^m \to \mathbb{R}$ by $\tau_0(z) = \|z\|^2 = |z_1|^2 + \cdots + |z_m|^2$ for all $z \in \mathbb{C}^m$. Then τ_0 is a strictly parabolic exhaustion of $\mathbb{C}^m$. If $0 < \Delta \leqslant \infty$, then $(\mathbb{C}^m(\Delta), \tau_0)$ is a strictly parabolic manifold of dimension m with maximal radius Δ, and in fact the only one:

<u>Theorem 1</u>. *Let* (M,τ) *be a strictly parabolic manifold of dimension* m *and maximal radius* Δ. *Then there exists a biholomorphic map* $h : \mathbb{C}^m(\Delta) \to M$ *such that* $\tau \circ h = \tau_0$.

The map is an isometry of exhaustions and of Kähler metrics. In some sense, the theorem resembles the Riemann mapping theorem. Originally, I needed an additional assumption [3]. Dan Burns was able to eliminate the additional assumption.

A proof of Theorem 1 was published by Stoll [4]. Additional proofs were given by Burns [1] and P. Wong [6].

The manifold case is settled. If singularities are permitted, are there other strictly parabolic spaces? (M,τ) is said to be a *strictly parabolic space* of dimension m with maximal radius Δ if these four conditions are satisfied:

(1) M is an irreducible complex space of dimension $m \geqslant 1$.

(2) τ is an exhaustion of M with maximal radius Δ where $0 < \Delta \leqslant \infty$.

(3) On $\mathfrak{R}(M)$ we have $dd^c\tau > 0$ and on $\mathfrak{R}(M_*)$ we have $dd^c \log \tau \geqslant 0$ and $(dd^c \log \tau)^m \equiv 0$.

(4) At each singular point of M there exists a strictly parabolic extension.

Concerning (4), we adopt the following terminology. Let U be an open subset of M. Let G be an open subset of $\mathbb{C}^n$. Let $j:U \to G$ be a proper, injective holomorphic map. Then $U' = j(U)$ is an analytic subset of G. Assume that $j:U \to U'$ is biholomorphic. Then $j:U \to G$ is called a *chart* and a *chart at* $a \in M$ if $a \in U$. Usually we identify $U = U'$ such that $j:U \to G$ is the inclusion map. Then the chart $j:U \to G$ is called *embedded*. Let $J:U \to G$ be an embedded chart. A nonnegative function θ of class C^∞ on G is said to be a *strictly parabolic extension* of τ onto G if and only if $\theta|U = \tau|U$, if $dd^c\theta > 0$ on G and if for every point $p \in U \cap M_*$ there exists an open neighborhood V_p of p in G such that $\theta > 0$ and $dd^c \log \theta \geqslant 0$ on V_p. These conditions are quite natural and *imply* that

$$(dd^c \log \theta)^n = 0 \qquad \text{on} \quad U \cap M_*$$

but not necessarily on any open subset of G.

Define $\theta_0 : \mathbb{C}^n \to \mathbb{R}$ by $\theta_0(z) = \|z\|^2 = |z_1|^2 + \cdots + |z_n|^2$. Let K be an irreducible analytic cone of dimension m with vertex 0 in $\mathbb{C}^n$. Then $\tau_0 = \theta_0|K$ is a strictly parabolic exhaustion on K. Take Δ with $0 < \Delta \leqslant \infty$ and define

$$K(\Delta) = \{z \in K \,|\, \tau_0(z) < \Delta^2\} .$$

Then $(K(\Delta),\tau_0)$ is a strictly parabolic space of dimension m and of maximal radius Δ. Usually, strict parabolicity is not inherited by a subspace. However, that this is the case for the cone K, can be shown by the following considerations. Clearly $dd^c\tau_0 > 0$ on $\mathfrak{R}(K)$ and $dd^c \log \tau_0 \geqslant 0$ on $\mathfrak{R}(K_*)$. Let $\mathbb{P}:\mathbb{C}^n_* \to \mathbb{P}_{n-1}$ be the standard projection.

Then $\mathbb{P}(K)=\mathbb{P}(K_*)$ is an irreducible analytic subset of dimension $m-1$ of $\mathbb{P}_{n-1}$. Let $\iota:\mathbb{P}(K)\to\mathbb{P}_{n-1}$ be the inclusion map. Also the inclusion $j:K\to\mathbb{C}^n$ is an embedded chart. Let Ω be the Fubini-Study Kähler form on $\mathbb{P}_{n-1}$. Then $\mathbb{P}^*(\Omega)=dd^c\log\theta_0$ on $\mathbb{C}^n_*$. For dimension reasons, we have $\iota^*(\Omega)^m=0$ on $\mathbb{P}(K)$. Now $\mathbb{P}\circ j=\iota\circ\mathbb{P}$ implies

$$(dd^c\log\tau_0)^m = j^*(dd^c\log\theta_0)^m = j^*\mathbb{P}^*(\Omega^m)$$

$$= \mathbb{P}^*\iota^*(\Omega^m)) = \mathbb{P}^*(\iota^*(\Omega)^m) = \mathbb{P}^*(0) = 0 \quad .$$

The Monge-Ampère equation is satisfied. Obviously, θ_0 is a strictly parabolic extension of τ_0 onto $\mathbb{C}^n$. Hence we have shown, that τ_0 is a strictly parabolic exhaustion of K.

Theorem 2. *Let* (M,τ) *be a strictly parabolic space of dimension* m *with maximal radius* Δ. *Then there exists an irreducible, affine algebraic cone* K *of dimension* m *embedded into some* $\mathbb{C}^n$ *with* $n\geq m$, *and a biholomorphic map* $h:K(\Delta)\to M$ *such that* $\tau_0=\tau\circ h$.

Hence h is an isometry of exhaustions and Kähler metrics. The affine algebraic cones and truncated cones are the only strictly parabolic spaces up to biholomorphic isometries. The cone K in Theorem 2 can be determined the following way. First it is shown that the center $M[0]$ consists of one and only one point 0_M. Let K be the Whitney tangent cone at 0_M. Let n be the embedding dimension of M at 0_M. The Whitney cone K is naturally embedded as an affine algebraic cone in the complex tangent space $\mathfrak{T}$ of M at 0_M. Identify $\mathfrak{T}=\mathbb{C}^n$ via some linear isomorphism. The natural base in $\mathbb{C}^n$ defines the strictly parabolic exhaustions θ_0 of $\mathbb{C}^n$ and $\tau_0=\theta_0|K$ of K.

For a proof of Theorem 2 see Stoll [5]. The proof of Theorem 2 is not a simple minded extension of the proof of Theorem 1. The singularities of M provide considerable difficulties. The differential geometric machinery is available on manifolds but not on spaces. In fact, it was necessary to develop the concept of a vector field and its integral curves on complex spaces to obtain a proof of Theorem 2.

Let X be a vector field of class C^∞ on $\mathfrak{R}(M)$. Let $j:U\to G$ be an embedded chart of M. A vector field $\tilde{X}$ of class C^∞ on G is said to be an *extension of* X *onto* G if and only if $j_*(X)(p)=\tilde{X}(p)$ at every point $p\in U\cap\mathfrak{R}(M)$. Then X is said to be *a vector field of class* C^∞ *on* M if and only if for every point $a\in\Sigma(M)$ there exists an embedded chart $j:U\to G$ at a and extension $\tilde{X}$ of X onto G. Let $\varphi:\mathbb{R}(\alpha,\beta)\to G$ be an integral curve of the extension $\tilde{X}$ such that there exists $t_0\in\mathbb{R}(\alpha,\beta)$ with $\varphi(t_0)\in U$, then $\varphi(t)\in U$ for all $t\in\mathbb{R}(\alpha,\beta)$. This Lemma enables us to define *integral curves* and *one parameter groups of diffeomorphisms* associated to vector fields on complex spaces. The classical properties can be extended to vector fields on spaces. Observe, that the vector $X(p)$ is defined only for $p\in\mathfrak{R}(M)$ but not for $p\in\Sigma(M)$. Nevertheless we write $\dot{\varphi}(t)=X(\varphi(t))$ or $\dot{\varphi}=X\circ\varphi$ if φ is an integral curve of X even if $\varphi(t)\in\Sigma(M)$ and consequently $X(\varphi(t))$ has no meaning.

Some more notations are needed. We use Einstein summation. Greek indices run from 1 to m and Latin indices run from 1 to n. Let V be an open subset of $\mathfrak{R}(M)$. A biholomorphic map $\alpha:V\to H$ onto an open subset H of $\mathbb{C}^m$ is called a patch. Then $\alpha=(z^1,\dots,z^m)$ and $z^1,\dots,z^m$ are *local coordinates* on V. The vector field X is represented on V by

$$X = X^\mu \frac{\partial}{\partial z^\mu} + Y^{\bar{\mu}} \frac{\partial}{\partial \bar{z}^\mu}$$

where $X^1,\dots,X^m$ are functions of class C^∞ on V. If U is an open subset of M and if $j:U\to G$ is an embedded chart, then $j=(w^1,\dots,w^n)$ where $w^1,\dots,w^n$ are called *embedding coordinates*. An extension $\tilde{X}$ of X onto G is represented by

$$\tilde{X} = \tilde{X}^j \frac{\partial}{\partial w^j} + \tilde{Y}^j \frac{\partial}{\partial \bar{w}^j} \quad .$$

If $V\subseteq U$, then

$$\tilde{X}^j = X^\mu \frac{\partial w^j}{\partial z^\mu} \qquad \tilde{Y}^{\bar{j}} = Y^{\bar{\mu}} \overline{\frac{\partial w^j}{\partial z^\mu}}$$

on V.

Now an outline of the proof of Theorem 2 will be given.

On $\mathfrak{R}(M_*)$ real vector fields

$$F = f+\bar{f} = \frac{1}{2}\operatorname{grad}\tau \qquad Y = F/\sqrt{\tau} = \operatorname{grad}\sqrt{\tau}$$

are defined. When $z^1,\ldots,z^m$ are local coordinates on an open subset V of $\mathfrak{R}(M_*)$, then

$$\partial\tau = \tau_\mu dz^\mu \qquad \bar\partial\tau = \tau_{\bar\nu} d\bar z^\nu \qquad 2\pi dd^c\tau = i\tau_{\mu\bar\nu} dz^\mu \wedge d\bar z^\nu .$$

The hermitian matrix $(\tau_{\mu\bar\nu})$ is positive definite. Let $(\tau^{\bar\nu\mu})$ be the inverse matrix. Then

$$f = f^\mu \frac{\partial}{\partial z^\mu} \qquad f^\mu = \tau_{\bar\nu}\tau^{\bar\nu\mu} \qquad f^\mu\tau_\mu = \tau .$$

In particular, F, f, $\partial\tau$, $\bar\partial\tau$ and $d\tau$ are not zero on $\mathfrak{R}(M_*)$. Let $j:U\to G$ be a chart and let $j=(w^1,\ldots,w^n)$. Let θ be a strictly parabolic extension of τ onto G. Then

$$\partial\theta = \theta_j dw^j \qquad \bar\partial\theta = \theta_{\bar k} d\bar w^k \qquad 2\pi dd^c\theta = i\theta_{j\bar k} dw^j \wedge d\bar w^k .$$

The hermitian matrix $(\theta_{j\bar k})$ is positive definite. Let $(\theta^{\bar k j})$ be the inverse matrix. Then F, f extend to vector fields $\tilde F$ and $\tilde f$ on G by setting

$$\tilde F = \tilde f + \bar{\tilde f} \qquad \tilde f = \tilde f^j \frac{\partial}{\partial w^j} \qquad \tilde f^j = \theta_{\bar k}\theta^{\bar k j} .$$

If $U\subseteq M_*$ and if G is taken so small that $\theta > 0$ on U, then Y extends to a vector field $\tilde Y = (1/\sqrt{\theta})\tilde F$ onto G.

The vector field Y on M_* defines a flow

$$\psi : \mathbb{R}(0,\Delta)\times M_* \to M_*$$

of class C^∞ called the gradient flow such that

$$\dot\psi(t,p) = Y(\psi(t,p)) \qquad \psi(\sqrt{\tau(p)},p) = p$$

for all $t\in\mathbb{R}(0,\Delta)$ and $p\in M_*$. Then

$$\tau(\psi(t,p)) = t^2$$

for all $t\in\mathbb{R}(0,\Delta)$ and $p\in M_*$. For each fixed $t\in\mathbb{R}(0,\Delta)$, the map

$$\psi(t,\square):M_*\to M_*$$

is a diffeomorphism of class C^∞.

Now, we will show that the center consists of one and only one point.

If $M[0]=\emptyset$, then $\tau>0$ on M. Take $r\in\mathbb{R}(0,\Delta)$ such that Stokes Theorem holds for $M(r)\neq\emptyset$. Then

$$0 = \int_{M(r)} (dd^c \log\tau)^m = \int_{M\langle r\rangle} d^c\log\tau\wedge(dd^c\log\tau)^{m-1}$$

$$= \int_{M\langle r\rangle} \frac{d^c\tau}{\tau}\wedge\left(\frac{dd^c\tau}{\tau}-\frac{d\tau\wedge d^c\tau}{\tau^2}\right)^{m-1}$$

$$= r^{-2m}\int_{M\langle r\rangle} d^c\tau\wedge(dd^c\tau)^{m-1}$$

$$= r^{-2n}\int_{M(r)} (dd^c\tau)^m > 0$$

Contradiction! Hence $M[0]\neq\emptyset$. For another proof using ψ see [5] Theorem 4.4. Take any $a\in M$. Let $j:U\to G$ be a chart at a with a strictly parabolic extension θ on G. Since $\theta\geqslant 0$ on G and since $\theta(a)=0$, the function θ assumes a minimum at a. The Monge-Ampère equation shows that the Hessian at a is positive on the non-zero vectors in the Whitney tangent cone of M at a. Hence $\tau=\theta|U$ has a strict minimum at a, which means that a is an isolated point of the center $M[0]$. Since $M[0]$ is compact, $M[0]$ is finite.

For every $a\in M$, take an open neighborhood U_a such that $\bar{U}_a$ is compact and such that $\bar{U}_a\cap\bar{U}_b=\emptyset$ if $a\in M[0]$ and $b\in M[0]$ and $a\neq b$. Then $U=\bigcup_{a\in M[0]}U_a$ is open and $\bar{U}=\bigcup_{a\in M[0]}\bar{U}_a$ is compact. A number $r\in\mathbb{R}(0,\Delta)$ exists such that $\bar{U}\subset M(r)$. Then $V=M[r]-U\neq\emptyset$ is compact with $V\cap M[0]=\emptyset$. A number $s\in\mathbb{R}(0,r)$ exists such that $\tau>s^2$ on V. Then $M[s]\subset U$. Take any $p\in M_*$, then $\psi(s,p)\in U$. One and only one $\alpha(p)\in M[a]$ exists such that $\psi(s,p)\in U_{\alpha(p)}$. Then $\psi(t,p)\in U$ for all $0<t\leqslant s$. Since U is a disjoint union, $\psi(t,p)\in U_{\alpha(p)}$ for all $t\in\mathbb{R}(0,s)$. Take $a\in M[0]$. Take $p\in M_*\cap U_a\cap M[s]$. Then $\tau(p)\leqslant s^2$.

Hence $\psi(\sqrt{\tau(p)},p) = p \in U_a \cap U_{\alpha(p)}$. Consequently, $a = \alpha(p)$. The map $\alpha : M_* \to M[0]$ is surjective. Take $p_0 \in M_*$. Then $\psi(s,p_0) \in U_{\alpha(p_0)}$. By continuity an open neighborhood N of p_0 exists such that $\psi(s,p) \in U_{\alpha(p_0)}$ for all $p \in N$. Hence $\alpha(p) = \alpha(p_0)$ for all $p \in V$. The map α is locally constant. Since $M[0]$ is finite and M is irreducible, $M_* = M - M[0]$ is connected. Therefore α is constant. Hence $M[0] = \alpha(M_*)$ consists of one and only one point 0_M which shall be called the *center point*.

A switch of parameter $t = e^x$ is opportune. Define $\Delta_0 = \log \Delta$. A map

$$\chi : \mathbb{R}(-\infty,\Delta_0) \times M_* \to M_*$$

is defined by

$$\chi(x,p) = \psi(e^x,p)$$

for all $x \in \mathbb{R}(-\infty,\Delta_0)$ and $p \in M_*$. Then

$$\dot{\chi}(x,p) = F(\chi(x,p)) \qquad \chi\left(\frac{1}{2}\log \tau(p),p\right) = p$$

$$\tau(\chi(x,p)) = e^{2x}$$

for all $x \in \mathbb{R}(-\infty,\Delta_0)$ and $p \in M_*$.

Consider the rotated vector field $\mathfrak{J}F$ on M_*. Each integral curve of $\mathfrak{J}F$ remains in a fixed pseudo-sphere which is compact. Hence $\mathfrak{J}F$ is complete. A one parameter group of diffeomorphisms $\sigma : \mathbb{R} \times M_* \to M_*$ is associated to $\mathfrak{J}F$. Now $[F,\mathfrak{J}F] = 0$ can be shown and implies the fundamental relation

$$\chi(x,\sigma(y,p)) = \sigma(y,\chi(x,p))$$

for all $x \in \mathbb{R}(-\infty,\Delta_0)$, all $y \in \mathbb{R}$ and all $p \in M_*$. Define $D = \mathbb{R}(-\infty,\Delta_0) \times \mathbb{R}$. A map

$$\mathfrak{w} : D \times M_* \to M_*$$

of class C^∞, called the *complex flow* of τ, is defined by

$$\mathfrak{w}(x+iy,p) = \chi(x,\sigma(y,p)) = \sigma(y,\chi(x,p))$$

for all $x \in \mathbb{R}(-\infty,\Delta_0)$, all $y \in \mathbb{R}$ and all $p \in M_*$. Since

$$\mathfrak{w}_x + \mathfrak{J}\mathfrak{w}_y = F + \mathfrak{J}\mathfrak{J}F = F - F = 0$$

the map $\mathfrak{w}(\square, p): D \to M_*$ is holomorphic for each $p \in M_*$.

Further progress is obtained by a close look at the situation near the center point 0_M. Let n be the embedding dimension of M at 0_M and identify the embedding space with $\mathbb{C}^n$. Let $j: U \to G$ be an embedded chart of M at 0_M where G is open in $\mathbb{C}^n$ and where $0_M = j(0_M) = 0$ is the origin of $\mathbb{C}^n$. Here G is taken so small that there exists a strictly parabolic extension θ of τ on G. Also the base of $\mathbb{C}^n$ is chosen such that $\theta_{j\bar{k}}(0) = 0$ if $j \neq k$ and $\theta_{j\bar{k}}(0) = 1$ if $j = k$. Then $\|w\|^2 = \theta_{j\bar{k}}(0) w^j w^{-k}$ defines the standard norm on $\mathbb{C}^n$. Let K be the Whitney tangent cone at 0_M embedded in $\mathbb{C}^n$. A strictly parabolic exhaustion $\tau_0: K \to \mathbb{R}$ is defined by $\tau_0(w) = \|w\|^2$ for all $w \in K$. Here K has its vertex at 0. As usual define

$$K(r) = \{w \in K \mid \tau_0(w) < r\} \qquad K[r] = \{w \in K \mid \tau_0(w) \leqslant r\}$$

$$K\langle r\rangle = \{w \in K \mid \tau_0(w) = r\}$$

for all $r \geqslant 0$. Also G can be taken so small, that there is an open, convex neighborhood of 0 in $\mathbb{C}^n$ such that $\exp_0: G_0 \to G$ is a diffeomorphism where $\exp_0$ is the exponential map at 0 associated to the Kähler metric $dd^c\theta > 0$ on G. A number $t_0 \in \mathbb{R}(0, \Delta)$ exists such that $M[t_0] \subset U \subseteq G$.

It can be shown, that for each $p \in M_*$ the curve $\psi(\square; p): \mathbb{R}[0, t_0) \to G$ is a geodesic if we define $\psi(0, p) = 0_M$. Hence

$$\psi: \mathbb{R}[0, \Delta) \times M_* \to M$$

is a map of class C^∞. In fact $\dot{\psi}(0, p) \in K\langle 1\rangle$. A diffeomorphism

$$\mathfrak{y}: K\langle 1\rangle \to M\langle t_0\rangle$$

is uniquely defined by $\mathfrak{y}(\dot{\psi}(0, p)) = p$ for all $p \in M\langle t_0\rangle$. Maps of class C^∞ are defined by

$$\psi: \mathbb{R}[0, \Delta) \times K\langle 1\rangle \to M \qquad \text{by} \qquad \psi(t, \xi) = \psi(t, \mathfrak{y}(\xi))$$

$$\sigma: \mathbb{R} \times K\langle 1\rangle \to M_* \qquad \text{by} \qquad \sigma(y, \xi) = \sigma(y, \mathfrak{y}(\xi))$$

$$\chi:\mathbb{R}(-\infty,\Delta_0)\times K\langle 1\rangle\to M_* \qquad \text{by} \qquad \chi(1,\xi) = \chi(x,\xi) = \chi(x,\mathfrak{y}(\xi))$$

$$\mathfrak{w}:D\times K\langle 1\rangle\to M_* \qquad \text{by} \qquad \mathfrak{w}(z,\xi) = \mathfrak{w}(x,\mathfrak{y}(\xi)) \quad .$$

Then

$$\mathfrak{w}(x+iy,\xi) = \chi(x,\sigma(y,\xi)) = \sigma(y,\chi(x,\xi))$$

$$\chi(x,\xi) = \psi(e^x,\xi)$$

$$\tau(\psi(t,\xi)) = t^2 \qquad\qquad \tau(\chi(x,\xi)) = e^{2x}$$

$$\sigma(y,\xi)\in M\langle t_0\rangle$$

for all $x\in\mathbb{R}(-\infty,\Delta_0)$, all $y\in\mathbb{R}$, all $t\in\mathbb{R}[0,\Delta)$ and all $\xi\in K\langle 1\rangle$. Also if $t\in\mathbb{R}[0,\Delta_0)$ and $\xi\in K\langle 1\rangle$, then

$$\psi(t,\xi) = \exp_0(t\xi) \quad .$$

A flow $\zeta:\mathbb{R}\times K\langle 1\rangle\to K\langle 1\rangle$ is defined by

$$\zeta(y,\xi) = \mathfrak{y}^{-1}(\sigma(y,\xi)) \qquad \text{for all} \quad y\in\mathbb{R} \quad \text{and} \quad \xi\in K\langle 1\rangle \quad .$$

Obviously $\zeta(0,\xi)=\xi$. Local considerations at the center establish the differential equation

$$\zeta_y(y,\xi) = i\zeta(y,\xi) \qquad \text{for all} \quad y\in\mathbb{R} \quad \text{and} \quad \xi\in K\langle 1\rangle \quad .$$

Therefore

$$\zeta(y,\xi) = e^{iy}\xi \qquad \text{for all} \quad y\in\mathbb{R} \quad \text{and} \quad \xi\in K\langle 1\rangle \quad .$$

Consequently

$$\mathfrak{w}(x+iy,\xi) = \chi(x,\sigma(y,\xi)) = \mathfrak{w}(x,\zeta(y,\xi)) = \mathfrak{w}(x,e^{iy}\xi) \quad .$$

Hence there exists uniquely a map

$$\mathfrak{x}:\mathbb{C}(\Delta)\times K\langle 1\rangle\to M$$

such that

$$\mathfrak{x}(e^z,\xi) = \mathfrak{w}(z,\xi) \qquad \mathfrak{x}(0,\xi) = 0_M$$

for all $z \in D$ and $\xi \in K\langle 1\rangle$. It can be shown, that $\mathfrak{x}$ is of class C^∞ and that for each $\xi \in K\langle 1\rangle$ the map $\mathfrak{x}(\square,\xi):\mathbb{C}(\Delta) \to M$ is holomorphic.

A map $h:K(\Delta) \to M$ is defined by setting $h(w) = \psi(w,w/\|w\|)$ if $0 \neq w \in K(\Delta)$ and by $h(0) = 0_M$. If $w \in K(t_0)$, then $h(w) = \exp_0 w$. Therefore $h:K(\Delta) \to M$ is a diffeomorphism of class C^∞. For each $\xi \in K\langle 1\rangle$ define $j_\xi:\mathbb{C}(\Delta) \to K(\Delta)$ by $\dot{f}_\xi(z) = z\xi$. Then $h \circ j_\xi = \mathfrak{x}(\square,\xi)$ is holomorphic on $\mathbb{C}(\Delta)$. Since M is Stein, the map $h:K(\Delta) \to M$ is holomorphic. By Malgrange [2] a holomorphic diffeomorphism of class C^∞ is biholomorphic. Also $\tau(h(w)) = \tau(\psi(w,w/\|w\|)) = \|w\|^2 = \tau_0(w)$. Thus a proof of Theorem 2 has been sketched. For details see [5].

This is a lecture given at the Conference in honor of Yozo Matsushima on May 15, 1979 at the University of Notre Dame. The research was partially supported by a grant of the National Science Foundation, Grant M.C.S. 8003257.

References

[1] Burns, D., "Curvature of Monge-Ampère foliation and parabolic manifolds," Preprint, 47 pp. of ms.

[2] Malgrange, B., "Sur les fonctions differentiables et les ensembles analytiques," *Bull. Soc. Math. France* 91 (1963), 113-127.

[3] Stoll, W., "Variétés strictement parabolique," *C.R. Acad. Sc. Paris*, 285 (1977), Serie A, 757-759.

[4] Stoll, W., "The characterization of strictly parabolic manifolds," *Ann. Scuola. Norm. Sup. Pisa* 7 (1980), 87-154.

[5] Stoll, W., "The characterization of strictly parabolic spaces," Preprint, 89 pp. of ms.

[6] Wong, P., "Geometry of the equation $(\partial\bar{\partial}u)^n = 0$," in preparation.

University of Notre Dame
Notre Dame, Indiana 46544

(Received December 26, 1980)

THE TANNAKA DUALITY THEOREM FOR SEMISIMPLE LIE GROUPS AND THE UNITARIAN TRICK

Mitsuo Sugiura

0. Introduction

Harish-Chandra [6] found that the Tannaka duality theorem almost holds for connected semisimple Lie groups but it does not hold exactly. It remained an open question when the Tannaka duality theorem holds for semisimple Lie groups. In this paper, we answer the question in the following way.

Theorem 4. *Let* G *be a semisimple Lie group with a finite number of connected components. Then the Tannaka duality theorem holds for* G *if and only if* G *is a real affine algebraic group.*

This theorem is proved immediately by our previous result [9] linked with the following theorem.

Theorem 3. *Every continuous representation of a real semisimple algebraic group is a rational representation.*

However, in this paper, we shall prove the above two theorems together with the other theorems independently of [9] and consistently using the unitarian trick.

The principle of the unitarian trick was found by H. Weyl [11]. He used it successfully to reduce the theory of connected complex semisimple Lie groups to that of compact Lie groups. Afterward the principle was algebraically formulated for semisimple Lie algebras by Chevalley-Eilenberg [4]. But their formulation of the principle cannot be transferred directly to semisimple Lie groups because a semisimple Lie group generally has no complexification unless it has a faithful representation (cf. Proposition 12). Nevertheless we can establish the principle of the unitarian trick for an arbitrary semisimple Lie group with a finite number of components by combining a theorem of Matsushima [8] (Appendix) with Hochschild-Mostow's notion of the universal complexification [7]. Matsushima's theorem is a converse of Chevalley's theory [2] which associates each compact Lie group G with a complex affine algebraic group G^{*c} (the associated algebraic group of G) defined over $\mathbb{R}$. Chevalley's version of the Tannaka duality theorem [10] implies that G is isomorphic to the subgroup G^* of G^{*c} consisting of all $\mathbb{R}$-rational points.

In this paper, we first extend Matsushima's theorem to nonconnected groups and prove the following theorem.

Theorem 1. *Every complex semisimple Lie group with a finite number of components is the associated algebraic group of its maximal compact subgroup.*
The existence of a maximal compact subgroup is proved in Propositon 8.

Let G be a semisimple Lie group with a finite number of components, G^+ the universal complexification of G (for the definition of G^+, cf. §3) and G_u a maximal compact subgroup of G.

Then the principle of the unitarian trick establishes the connection between two groups G and G_u. More precisely we have the following theorem.

Theorem 2. (1) *There exists a bijection* w *of the set* $R(G)$ *of all representations of* G *onto the set* $R(G_u)$ *of all representations of* G_u. *The bijection* w *preserves the direct sum, the tensor product and the equivalence. The associated algebraic group* G^{*c} *of* G *is isomorphic to the universal complexification* G^+ *and* G_u^{*c}. (2) *Every representation of* G *is completely reducible.* (3) *The representative algebra* $R(G)$ *(the algebra spanned by the coefficients of all representations) of* G *is isomorphic to the representative algebra* $R(G_u)$ *of* G_u. *The algebras* $R(G)$ *and* $R(G_u)$ *are finitely generated.*

Using the principle of the unitarian trick established in Theorem 2, we shall prove the above cited Theorems 3 and 4 together with the following Theorems 5 and 6.

The above Theorem 4 shows that the Tannaka duality does not hold for semisimple Lie groups which are not algebraic groups. The examples of such groups are the universal covering group of $SL(2,R)$ and the proper Lorentz group $SO_0(n,1)$. But the Tannaka duality holds infinitesimally for an arbitrary semisimple Lie group as the following theorem shows.

Theorem 5. *The Tannaka duality holds between the Lie algebra* $\mathfrak{g}$ *of an arbitrary semisimple Lie group* G *with a finite number of components and the dual object* $d\mathcal{R}(G) = \{dD \mid D \in \mathcal{R}(G)\}$.

Corollary to Theorem 5. *The Tannaka duality holds for any real semisimple Lie algebra* $\mathfrak{g}$ *and* $\mathcal{R}(\mathfrak{g})$ *(the set of all representations of* $\mathfrak{g}$*).*

By Theorem 5, we can conclude that the Tannaka duality for semisimple Lie groups nearly holds as the following theorem shows.

Theorem 6. *The canonical image of* G *in* G^* *is an open subgroup.* G^* *is the smallest real algebraic group containing the canonical image of* G.

Theorem 5 and the first half of Theorem 6 were proved by Harish-Chandra [6] in a quite different way.

0.1 Notations

G_0: the connected component of a topological group G containing the identity e.

$A \dot{+} B = \begin{pmatrix} A & 0 \\ 0 & B \end{pmatrix}$: the direct sum of A and B.

$A \otimes B = \begin{pmatrix} a_{11}B & \cdots & a_{1n}B \\ & \cdots\cdots\cdots & \\ a_{n1}B & \cdots & a_{nn}B \end{pmatrix}$: the tensor product of A and B.

$A \oplus B = A \otimes 1_m + 1_n \otimes B$: the tensor sum of A (degree n) and B (degree m).

$\mathcal{R}(G)$: the set of all continuous matricial representations of G.

$R(G)$: the algebra spanned by the coefficients of all D in $\mathcal{R}(G)$.

$R_r(G)$: the algebra spanned by the coefficients of all rational representations of an algebraic group G.

$R_h(G)$: the algebra spanned by the coefficients of all holomorphic representations of a complex Lie group G.

1. The Associated Algebraic Groups

In this section, we gather the fundamental definitions and the elementary properties concerning the Tannaka duality theorem and the associated algebraic groups.

Let G be a topological group. In this paper, a representation D of G is defined as a continuous homomorphism of G into $GL(n,\mathbb{C})$ for a certain integer n. n is called the degree of D and denoted by $d(D)$. The set $\mathcal{R}(G)$ of all representations of G is called the *dual object* of G and denoted by $\mathcal{R}=\mathcal{R}(G)$. A *complex representation* ζ of the dual object $\mathcal{R}$ is, by definition, a mapping ζ from $\mathcal{R}$ into $\bigcup_{n=1}^{\infty} GL(n,\mathbb{C})$ which satisfies the following four conditions:

(0) $\zeta(D)\in GL(d(D),\mathbb{C})$, (1) $\zeta(D\dotplus D') = \zeta(D)\dotplus\zeta(D')$,

(2) $\zeta(D\otimes D') = \zeta(D)\otimes\zeta(D')$, (3) $\zeta(\gamma D\gamma^{-1}) = \gamma\zeta(D)\gamma^{-1}$

for any $D,D'\in\mathcal{R}$ and any regular matrix γ of degree $d(D)$.

The set $G^{*c}(\mathcal{R})$ of all complex representations of $\mathcal{R}$ becomes a topological group under the weakest topology which makes the mapping $\zeta\mapsto\zeta(D)\in GL(d(D),\mathbb{C})$ continuous for every $D\in\mathcal{R}$. The group operation is defined by $(\zeta_1\zeta_2^{-1})(D)=\zeta_1(D)\zeta_2(D)^{-1}$. The topological group $G^{*c}(\mathcal{R})$ is called the *complex Tannaka group* of G. An element ζ of $G^{*c}(\mathcal{R})$ satisfying the following condition (4) is called a *representation* of $\mathcal{R}$.

(4) $\zeta(\bar{D}) = \overline{\zeta(D)}$ for every $D\in\mathcal{R}$.

The set of all representations of $\mathcal{R}$ is a subgroup $G^*(\mathcal{R})$ of $G^{*c}(\mathcal{R})$ and is called the *Tannaka group* of G. For any element g of G, the mapping $\zeta_g: D\mapsto D(g)$ belongs to the Tannaka group $G^*(\mathcal{R})$. Moreover, the mapping $\Phi: g\mapsto\zeta_g$ is a continuous homomorphism of G into $G^*(\mathcal{R})$. When this homomorphism Φ is a topological isomorphism of G onto $G^*(\mathcal{R})$, we say that the *Tannaka duality* holds for the group G.

The set $R(G)$ of all finite linear combinations of the matricial coefficients of all representations in $\mathcal{R}$ forms an algebra over $\mathbb{C}$ and is called the *representative algebra* of G. Any element g of G defines two automorphisms L_g and R_g of the algebra $R(G)$ defined by

$$(L_g f)(x) = f(gx) \quad \text{and} \quad (R_g f)(x) = f(xg) \quad .$$

A continuous function f on G belongs to $R(G)$ if and only if $\dim\{R_g f \mid g \in G\}_{\mathbb{C}}$ is finite. let $\bar{f}$ be the complex conjugate function of f in $R(G)$ and put

$$G^{*c} = \{\alpha \in \mathrm{Aut}\, R(G) \mid \alpha \circ L_g = L_g \circ \alpha \quad \text{for all} \quad g \in G\}$$

and

$$G^{*} = \{\alpha \in G^{*c} \mid \alpha(\bar{f}) = \overline{\alpha(f)} \quad \text{for all} \quad f \quad \text{in} \quad R(G)\}.$$

An element of G^* (or G^{*c}) is called a *proper automorphism* (or a *complex proper automorphism*) of $R(G)$. The group G^{*c} becomes a topological group under the weakest topology which makes the mapping $\alpha \mapsto \lambda(\alpha(f)) \in \mathbb{C}$ is continuous for every linear form λ on $R(G)$ and every f in $R(G)$. The canonical mapping $R: g \mapsto R_g$ is a continuous homomorphism of G into G^*. We say that the *duality* holds for the group G if the canonical mapping R is a topological isomorphism of G onto G^*.

The set of all homomorphisms ω of the algebra $R(G)$ into $\mathbb{C}$ satisfying $\omega(1) = 1$ is denoted by $\mathrm{Hom}(R(G), \mathbb{C}) = \Omega$. Let e be the identity element of G. Then the mapping $\Gamma: G^{*c} \to \mathrm{Hom}(R(G), \mathbb{C})$ is defined by $\Gamma(\alpha)(f) = (\alpha f)(e)$. The mapping Γ is a bijection of G^{*c} onto $\mathrm{Hom}(R(G), \mathbb{C}) = \Omega$ and maps G^* onto $\Omega_{\mathbb{R}} = \mathrm{Hom}_{\mathbb{R}}(R(G), \mathbb{C}) = \{\omega \in \mathrm{Hom}(R(G), \mathbb{C}) \mid \omega(\bar{f}) = \overline{\omega(f)}$ for all $f \in R(G)\}$. If the topology of Ω is defined as the weakest topology which makes the mapping $\omega \to \omega(f) \in \mathbb{C}$ continuous for every f in $R(G)$, then the mapping Γ is a homeomorphism (cf. Sugiura [9]).

We define the following mappings:

$$T: \Omega \to G^{*c}(\mathcal{R}), \qquad T(\omega) = \zeta_\omega,$$

where $\zeta_\omega : D = (D_{ij}) \mapsto (\omega(D_{ij})) \in GL(d(D), \mathbb{C})$,

$$F = T \circ \Gamma \qquad \text{and} \qquad S: G \to \Omega_R, \quad S(g) = \omega_g$$

where $\omega_g \in \Omega_{\mathbb{R}}$ is defined by $\omega_g(f) = f(g)$. Then the following diagram is commutative.

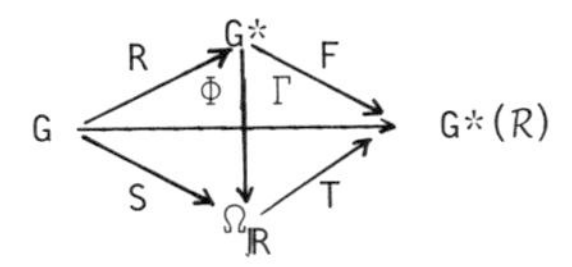

Since the mapping T is injective, the homomorphism F is a continuous isomorphism of G^{*c} into $G^{*c}(\mathcal{R})$ which maps G^* into $G^*(\mathcal{R})$. If the canonical mapping Φ is a bijection, then the homomorphism F and $R = F^{-1} \circ \Phi$ are also bijections. Hence if the Tannaka duality holds for G, then the duality holds for G. The converse of this assertion is true under the additional condition that all representations of G are completely reducible (Sugiura [9]).

Proposition 1. *If all representations of* G *are completely reducible, then the homomorphism* F *is a bijection and a homeomorphism. We have* $G^{*c} \cong G^{*c}(\mathcal{R})$ *and* $G^* \cong G^*(\mathcal{R})$. *In this case, the Tannaka duality for* G *is equivalent to the duality for* G.

Proof. Let $\mathcal{D} = \{D^\alpha | \alpha \in A\}$ be a complete set of representatives of the set A of all equivalence classes of irreducible representations of G. By the assumption of complete reducibility, the set $\mathcal{B} = \{D^\alpha_{ij} | \alpha \in A,\ 1 \leqq i,\ j \leqq d(D)\}$ spannes the vector space $R(G)$. B is a basis of $R(G)$, because $\mathcal{B}$ is linearly independent by Burnside's theorem and its generalization. Hence every element ζ in $G^{*c}(\mathcal{R})$ defines uniquely a linear form ω on $R(G)$ which maps D^α_{ij} to the (i,j)-element of the matrix $\zeta(D^\alpha)$. Now we shall prove that ω belongs to Ω and satisfies $\zeta = \zeta_\omega = T(\omega)$. Let D be an arbitrary representation of G. Then D is decomposed as $D = \gamma(D^{\alpha_1} + \cdots + D^{\alpha_n})\gamma^{-1}$ $(\alpha_i \in A)$. Since ω is a linear form and $\zeta(D^\alpha) = (\omega(D^\alpha_{ij}))$, we have $\zeta(D) = (\omega(D_{ij}))$. Let D and D' be in $\mathcal{R}$. Then we have

$$(\omega(D_{ij}D'_{kl})) = \zeta(D \otimes D') = \zeta(D) \otimes \zeta(D') = (\omega(D_{ij})\omega(D'_{kl})) \quad .$$

Hence ω belongs to $\Omega = \mathrm{Hom}(R(G), \mathbb{C})$ and $\zeta = \zeta_\omega$. Thus the mapping T is surjective and hence bijective. T is also a homeomorphism by the definition of topologies on Ω and $G^{*c}(\mathcal{R})$. Hence $F = T \circ \Gamma$ is also a homeomorphism. Therefore $G^{*c} \cong G^{*c}(\mathcal{R})$ as topological groups. Moreover if ζ belongs to $G^*(\mathcal{R})$, then the above defined element $\omega = T^{-1}(\zeta)$ belongs to $\Omega_{\mathbb{R}}$ and $\alpha = F^{-1}(\zeta)$ belongs to G^*. Hence $G^* \cong G^*(\mathcal{R})$. ∎

Every function f in $R(G)$ defines a function $\hat{f}$ on the set Ω by putting $\hat{f}(\omega) = \omega(f)$ for $\omega \in \Omega$. We identify G^{*c} with Ω and G^* with $\Omega_{\mathbb{R}}$ by the homeomorphism Γ. Then $\hat{f}$ is regarded as a function on G^{*c}. We note that the algebra $\hat{R}(G) = \{\hat{f} | f \in R(G)\}$ is canonically isomorphic to $R(G)$.

Proposition 2. *If the representative algebra* $R(G)$ *is finitely generated, then the group* G^{*c} *is a complex affine algebraic group with the affine algebra* $\hat{R}(G)$ $(\cong R(G))$. G^{*c} *is defined over* $\mathbb{R}$ *and* G^* *is a real form (the set of all* $\mathbb{R}$*-rational points) of* G^{*c}. *For any* D *in* $R(G)$, $\hat{D} = (\hat{D}_{ij})$ *is a rational representation of* G^{*c} *satisfying* $\hat{D}(R_g) = D(g)$.

Proof. If the algebra $R(G)$ $(\cong \hat{R}(G))$ is finitely generated, G^{*c} identified with Ω is a complex affine algebraic set with the affine algebra $\hat{R}(G)$. Since the real and imaginary parts of a function in $R(G)$ belong to $R_{\mathbb{R}}(G) = \{f \in R(G) \mid f = \bar{f}\}$, the algebra $R_{\mathbb{R}}(G)$ over $\mathbb{R}$ is finitely generated and $R(G) = R_{\mathbb{R}}(G) \otimes \mathbb{C}$. Hence G^* is a real affine algebraic set with the affine algebra $\hat{R}_{\mathbb{R}}(G)$ and a real form of G^{*c}.

Let D be an arbitrary representation of G, and α, β be in G^{*c}. Put $\Gamma(\alpha) = \omega$, $\Gamma(\beta) = \omega'$ and apply ω' to $L_g D_{ij} = \Sigma_k D_{ik}(g) D_{kj}$, we get $\beta(D_{ij}) = \Sigma_k D_{ik}\omega'(D_{kj})$. Applying α to the last equality, we get $(\alpha\beta)(D_{ij}) = \Sigma\, \alpha(D_{ik})\omega'(D_{kj})$ and

$$\hat{D}_{ij}(\alpha\beta) = (\alpha\beta)(D_{ij})(e) = \sum_k \hat{D}_{ik}(\alpha)\hat{D}_{kj}(\beta) .$$

Similarly we have

$$\hat{D}_{ij}(\alpha^{-1}) = [(i,j)\text{-cofactor of } \hat{D}(\alpha)][\det \hat{D}(\alpha)]^{-1} .$$

Hence the group operations $(\alpha,\beta) \mapsto \alpha\beta$ and $\alpha \mapsto \alpha^{-1}$ are everywhere defined rational mappings and G^{*c} is an affine algebraic group. Moreover $\hat{D}: \alpha \mapsto \hat{D}(\alpha)$ is a rational representation. We have $\hat{D}(R_g) = \hat{D}(\omega_g) = D(g)$. ∎

If $R(G)$ is finitely generated, the group G^{*c} and G^* are called the *associated algebraic group* and the *real associated algebraic group* of G respectively.

Corollary to Proposition 2. *Assume that* $R(G)$ *is finitely generated and the duality holds for* G. *Then we have the following.*

(1) *The mapping* $A: D \mapsto \hat{D}$ *is a bijection of* $R(G)$ *onto the set* $R_r(G^{*c})$ *of all rational representations of* G^{*c}. *The inverse mapping* A^{-1} *is the restriction to* G *identified with* G^*.

(2) *The mapping* $B: f \mapsto \hat{f}$ *is an isomorphism of the algebra* $R(G)$ *onto the algebra* $R_r(G^{*c})$ *spanned by the coefficients of rational*

representations of G^{*c}. *The inverse mapping* B^{-1} *is the restriction to* $G = G^*$.

(3) *Every holomorphic representation* D' *of a complex Lie group* G^{*c} *is a rational representation of the algebraic group* G^{*c}.

Proof. (1) (3) For every $D \in R(G)$, $\hat{D}$ belongs to $R_r(G^{*c})$ by Proposition 2. Since $\hat{D}(R_g) = D(g)$, the restirction to $G = G^*$ is the original D. Hence the mapping $A: D \mapsto \hat{D}$ is injective. Let D be the restriction to G of a holomorphic representation D' of G^{*c}. Then two holomorphic representations D' and $\hat{D}$ of G^{*c} coincide with each other, because they are identical on a real form G. Hence D' is a rational representation. Moreover the mapping A is surjective and hence bijective. (2) The mapping B is surjective by (1). B is injective, because $\hat{f}(R_g) = f(g)$.

Proposition 3. *Assume that* $R(G)$ *is finitely generated. Let* M *be a complex affine algebraic subgroup of* G^{*c} *containing* $G_1 = \{R_g | g \in G\}: G^{*c} \supset M \supset G_1$. *Then* M *coincides with* G^{*c}. *Similarly* G^* *is the smallest real algebraic group containing* G_1,

Proof. Let D be a representation of G whose coefficients generate $R(G)$. Let $n = d(D)$ and $A = \mathbb{C}[\ldots, X_{ij}, \ldots]$ be the polynomial algebra over $\mathbb{C}$ of n^2 variables X_{ij} $(1 \leqq i, j \leqq n)$. Let I be the ideal of A defined by

$$I = \{P \in A | P(.., D_{ij}, ..) = 0\} = \{P \in A | P(.., \hat{D}_{ij}, ..) = 0\} .$$

(Remark that $\hat{R}(G) = R(G)$.) Since $R(G) = A/I$, it is easily seen that I is the defining ideal of the algebraic set $\hat{D}(G^{*c})$, i.e.,

$$\hat{D}(G^{*c}) = \{a = (a_{ij}) \in GL(n, \mathbb{C}) | P(.., a_{ij}, ..) = 0 \quad \text{for all} \quad P \in I\} .$$

Let J be the ideal of A defined by

$$J = \{Q \in A | A(.., \hat{D}(m)_{ij}, ..) = \quad \text{for all} \quad m \in M\} .$$

Since $M \subset G^{*c}$, we get

$$J \supset I \tag{1}$$

On the other hand, since $M \supset G_1$ and $\hat{D}(R_g) = D(g)$, we get

$$Q(\ldots, D(g)_{ij}, \ldots) = 0 \quad \text{for all} \quad g \in G \quad \text{and} \quad Q \in J .$$

Hence we have $Q(\ldots, D_{ij}, \ldots) = 0$ for all $Q \in J$. Therefore every Q in J belongs to I:

$$I \supset J \quad . \tag{2}$$

(1) and (2) implies that $J = I$. Hence two algebraic sets $\hat{D}(M)$ and $\hat{D}(G^{*c})$ coincide with each other: $\hat{D}(M) = \hat{D}(G^{*c})$. Since the coefficients of D generate $R(G)$, $\hat{D}$ is a faithful representation of G^{*c}. Hence we get $M = G^{*c}$.

Replacing $R(G)$ with the real representative algebra $R_{\mathbb{R}}(G)$, the statement for G^* can be proved similarly. ■

The condition that the canonical mapping Φ is an injection can be expressed as follows.

Proposition 4. (1) $\operatorname{Ker} \Phi = \bigcap_{D \in \mathcal{R}(G)} \operatorname{Ker} D$. (2) *The following condition* (a) *is equivalent to* (b). *If the algebra* $R(G)$ *is finitely generated, then the condition* (c) *is equivalent to* (a) *and* (b).
(a) Φ *is an injection.* (b) G *has sufficiently many representations (i.e., for each* $g \neq e$, *there exists a representation* $D \in \mathcal{R}(G)$ *such that* $D(g) \neq 1$*).* (c) G *has a faithful representation.*

Proof. (1) $g \in \operatorname{Ker} \Phi \iff \zeta_g(D) = D(g) = 1$ for all $D \in \mathcal{R}(G) \iff g \in \bigcap_{D \in \mathcal{R}(G)} \operatorname{Ker} D$.

(2) (a) is equivalent to (b) by (1). (c) $\Rightarrow$ (b) is trivial.
(b) $\Rightarrow$ (c) Let D_0 be a representation of G whose coefficients generate the algebra $R(G)$. Suppose that two elements g and h in G satisfy $D_0(g) = D_0(h)$. Then we have $f(g) = f(h)$ for all f in $R(G)$. Hence $D(g) = D(h)$ for all D in $\mathcal{R}(G)$. Hence the condition (b) implies that $g = h$ and D_0 is a faithful representation of G.

Remark. If G is a connected Lie group, the condition (b) is equivalent to (c) without the assumption that $R(G)$ is finitely generated. Cf. Goto [5].

2. A Theorem of Matsushima

In this section we extend a thoerem of Matsushima to nonconnected groups.

Proposition 5 (Chevalley). *Every compact Lie group* G *has the following properties.*

(1) *Every representation of* G *is completely reducible.*

(2) *The representative algebra* R(G) *is finitely generated.*

(3) *The Tannaka duality holds for* G.

(4) *The Lie algebra* $\mathfrak{g}^c$ *of* G^{*c} *is the complexification of the Lie algebra* $\mathfrak{g}$ *of* G.

(5) G^{*c} *has the unique and continuous decomposition* $G^{*c} = G\exp(i\mathfrak{g})$.

(6) G *is a maximal compact subgroup of* G^{*c}.

Proof. Chevalley [2] Chapter VI.

Corollary 1 to Proposition 5. *A compact Lie group* G *has the unique real affine algebraic group structure compatible with its Lie group structure.*

Proof. G is a real affine algebraic group with the affine algebra $R_{\mathbb{R}}(G)$ (Propositions 2 and 5). Now we shall prove the uniqueness. Let G' be a real algebraic group with the affine algebra A and assume that the underlying Lie group of G' is the Lie group G. Since the affine algebraic structure is compatible with the analytic structure, A is contained in the space of real valued continuous functions on G. Since the mapping $(g,h) \to gh$ is a morphism of algebraic sets, for each f in A there exists a finite number of elements $u_1,\ldots,u_n;v_1,\ldots,v_n$ in A such that $f(gh) = \Sigma\, u_i(g)v_i(h)$. Hence $\dim\{R_h f \mid h \in G\}_{\mathbb{R}}$ is a finite and A is contained in $R_{\mathbb{R}}(G): A \subset R_{\mathbb{R}}(G)$. Let D be a real faithful rational representation of G'. Then we have $A \supset R[\ldots,D_{ij},\ldots]$. On the other hand the algebra $R_{\mathbb{R}}(G)$ is generated by $\{D_{ij} \mid 1 \leqq i,j \leqq d(D)\}$ (Chevalley [2], p. 190, Proposition 3). Hence $R_{\mathbb{R}}(G) \subset A$. We have proved that $A = R_{\mathbb{R}}(G)$ and $G' = G$ as algebraic groups.

Corollary 2 to Proposition 5. *Let* G *be a compact Lie group. Then every holomorphic representation* D' *of* G^{*c} *is completely reducible.*

Proof. By Proposition 5 and Corollary to Proposition 2, there exists a representation D of G such that $D' = \hat{D}$. Let V be the representation space of D and W be a subspace of V invariant under D. Then W is invariant under $D'(G^{*c}) = D(G)\exp i dD(\mathfrak{g})$. Hence W is invariant under D' if and only if W is invariant under D. Since a representation D of a compact group G is completely reducible, D' is also completely reducible. ∎

Let G be a connected complex semisimple Lie group, $\mathfrak{g}$ the Lie algebra of G and $\mathfrak{k}$ a compact real form of $\mathfrak{g}$. Then the analytic subgroup K with the Lie algebra $\mathfrak{k}$ is a maximal compact subgroup of G.

Proposition 6 (Matsushima [8]). *Let* K *and* K' *be maximal compact subgroups of connected complex semisimple Lie groups* G *and* G', *respectively. Then any continuous isomorphism* ϕ *of* K *onto* K' *is uniquely extended to a holomorphic isomorphism* ϕ^c *of* G *onto* G'.

Outline of Proof (Matsushima). Let $\mathfrak{g}$ and $\mathfrak{g}'$ be the Lie algebras of G and G', $S(\mathfrak{g})$ and $S(\mathfrak{g}')$ the simply connected Lie groups with $\mathfrak{g}$ and $\mathfrak{g}'$ as their Lie algebras, respectively. Let $\pi:S(\mathfrak{g}) \to G$ and $\pi':S(\mathfrak{g}') \to G'$ be the covering homomorphisms. Put $D = \mathrm{Ker}\,\pi$ and $D' = \mathrm{Ker}\,\pi'$. Since the Lie algebra isomorphism $d\phi$ of $\mathfrak{k}$ onto $\mathfrak{k}'$ can be extended complex linearly to an isomorphism $(d\phi)^c$ of $\mathfrak{g}$ onto $\mathfrak{g}'$, there exists a unique holomorphic isomorphism ψ of $S(\mathfrak{g})$ onto $S(\mathfrak{g}')$ satisfying $d\psi = (d\phi)^c$. Then a holomorphic isomorphism ϕ^c of $G = S(\mathfrak{g})/D$ onto $G' = S(\mathfrak{g}')/D'$ is defined by $\phi^c(xD) = \psi(x)D'$. Since $\phi^c(k\exp ik) = \phi(K)\exp i d\phi(k)$, ϕ^c is uniquely determined by ϕ.

Proposition 7 (Matsushima [8]). *Let* G *be a connected complex semisimple Lie group and* K *be a maximal compact subgroup of* G. *Then* G *is isomorphic to the associated algebraic group* K^{*c} *of* K. *In particular,* G *is uniquely and continously decomposed as* $G = K\cdot\exp i\mathfrak{k}$.

Proof. The canonical isomorphism $R:K \to K^*$ can be extended to an isomorphism of G onto K^{*c} by Proposition 6. Since $G \cong K^{*c}$, the decomposition $G = K\cdot\exp i\mathfrak{k}$ follows from Proposition 5.

Proposition 8. *Let* G *be a complex semisimple Lie group with a finite number of components. Then* G *has a maximal compact subgroup* K. *The Lie algebra* $\mathfrak{k}$ *of* K *is a compact real form of the Lie algebra* $\mathfrak{g}$ *of* G. *The identity component* G_0 *of* G *satisfies* $G = KG_0 = G_0K$. *The identity component* $K_0 = K \cap G_0$ *of* K *is a maximal compact subgroup of* G_0. G *is uniquely and continuously decomposed as* $G = K\cdot\exp i\mathfrak{k}$.

Proof. Let $\mathfrak{k}$ be a compact real form of $\mathfrak{g}$. Put $H = \{h \in \mathrm{Aut}\,\mathfrak{g} | h(\mathfrak{k}) = \mathfrak{k}\}$ and $H_0 = H \cap \mathrm{Int}\,\mathfrak{g}$. Then H and H_0 are maximal compact subgroups of $\mathrm{Aut}\,\mathfrak{g}$ (the automorphism group of $\mathfrak{g}$) and $\mathrm{Int}\,\mathfrak{g}$ (the identity component of $\mathrm{Aut}\,\mathfrak{g}$), respectively. We shall prove that $K = (\mathrm{Ad})^{-1}(H) = \{k \in G | \mathrm{Ad}\,k \in H\}$ is a maximal compact subgroup of G. Since compact real forms of $\mathfrak{g}$ are conjugate with each other under the group $\mathrm{Int}\,\mathfrak{g} = \mathrm{Ad}\,G_0$, for each g in G there exists an element g_0 in G_0 satisfying $g_0^{-1}g = k \in K$. Hence we have $G = G_0K = KG_0$. Both K and $K \cap G_0$ are closed subgroups of G with the Lie algebra $\mathfrak{k}$. Hence the identity component K_0 of K is the analytic subgroup of G_0 with the Lie algebra $\mathfrak{k}$ and a maximal compact subgroup of G_0. The center Z of G_0 is contained in K_0. Since

$$H_0 = \mathrm{Ad}\,K_0 \subset \mathrm{Ad}(K \cap G_0) \subset H \cap \mathrm{Int}\,\mathfrak{g} = H_0 .$$

We have $\mathrm{Ad}\,K_0 = \mathrm{Ad}(K \cap G_0)$ and $K \cap G_0 = K_0Z = K_0$. Moreover we have

$$K/K_0 \cong G_0K/G_0 = G/G_0 .$$

Therefore K/K_0 is a finite group and K is a compact subgroup of G. Let K' be a compact subgroup of G containing K. Then $\mathrm{Ad}\,K'$ is a compact subgroup of $\mathrm{Aut}\,\mathfrak{g}$ containing $\mathrm{Ad}\,K = H$ which is a maximal compact subgroup of $\mathrm{Aut}\,\mathfrak{g}$. Hence $\mathrm{Ad}\,K' = H$, $K' \subset K$ and hence $K' = K$. Therefore K is a maximal compact subgroup of G. By Proposition 7, we have $G = KG_0 = KK_0\exp i\mathfrak{k} = K\exp i\mathfrak{k}$. The decomposition is unique. Assume that $k\exp X = k'\exp Y$ where $k, k' \in K$ and $X, Y \in i\mathfrak{k}$. Then $k_0 = k'^{-1}k = (\exp Y)(\exp X)^{-1}$ belongs to $K \cap G_0 = K_0$ and we have $k_0\exp X = \exp Y$. Then the uniqueness of the decomposition in G_0 (Proposition 7) proves that $k_0 = e$ and $X = Y$. The uniqueness is proved. The decomposition is continuous on each connected component of G by Proposition 7 and hence continuous on the whole G.

Proposition 9. *Let G and G' be two complex semisimple Lie groups with a finite number of components and K and K' be maximal compact subgroups of G and G', respectively. Then any continuous isomorphism ϕ of K onto K' can be extended uniquely to a holomorphic isomorphism ϕ^c of G onto G'.*

Proof. Let $\psi = \phi|K_0$ be the restriction of ϕ to the identity component K_0 of K. Then $\psi(K_0) = K_0'$. By Proposition 6, ψ can be extended uniquely to a holomorphic isomorphism ψ^c of G_0 onto G_0'. For each $k \in K$ and $X \in \mathfrak{k}$, we have $\phi(k\exp Xk^{-1}) = \exp\{(\mathrm{Ad}\phi(k))d\phi(X)\}$ and hence $(d\phi)((\mathrm{Ad}\,k)X) = (\mathrm{Ad}\phi(k))(d\phi(X))$. Let $(d\phi)^c$ be the isomorphism of $\mathfrak{g}$ onto $\mathfrak{g}'$ which is the $\mathbb{C}$-linear extension of $d\phi = d\psi$. Then $(d\phi)^c((\mathrm{Ad}\,k)X) = (\mathrm{Ad}\phi(k))((d\phi)^cX)$ for all $X \in \mathfrak{g}$ and $k \in K$. Since $(d\phi)^c$ is equal to the differential $d\psi^c$ of ψ^c, we get

$d\psi^c \circ \mathrm{Ad}\, k = \mathrm{Ad}\phi(k) \circ d\psi^c$ for all $k \in K$. Hence $\psi^c(k\exp Xk^{-1}) = \exp(d\psi^c \circ \mathrm{Ad}\, k)X$ $= \exp(\mathrm{Ad}\phi(k) d\psi^c(X)) = \phi(k)\psi^c(\exp X)\phi(k^{-1})$ for all $X \in \mathfrak{g}$ and $k \in K$. We have proved that

$$\psi^c(kg_0k^{-1}) = \phi(k)\psi^c(g_0)\phi(k)^{-1} \quad \text{for all} \quad k \in K \quad \text{and} \quad g_0 \in G_0. \quad (3)$$

Now we define the mapping $\phi^c : G \to G'$ by putting

$$\phi^c(kg_0) = \phi(k)\psi^c(g_0) \qquad \text{for} \quad k \in K \quad \text{and} \quad g_0 \in G_0 \quad .$$

Since the restriction $\psi^c|K_0$ of ψ^c to $K_0 = K \cap G_0$ is equal to $\phi|K_0 = \psi$, the mapping ϕ^c is well defined. By the equality (3), ϕ^c is a homomorphism of G into G'. Since ϕ^c is continuous and holomorphic on G_0, ϕ^c is holomorphic on G. Moreover $\phi^c(G) = \phi(K)\psi^c(G_0) = K'G_0' = G'$. Assume that $\phi^c(k\exp iX) = e$, $k \in K$ and $X \in \mathfrak{k}$. Then $\phi(k)\exp i d\phi(X) = e$, $\phi(k) \in K'$ and $d\phi(X) \in \mathfrak{k}'$. The uniqueness of the decomposition in G' (Proposition 8) proves that $\phi(k) = e$ and $d\phi(X) = 0$. Since ϕ is an isomorphism, $d\phi$ is injective and we have $k = e$ and $X = 0$. Hence ϕ^c is a holomorphic isomorphism of G onto G' which is an extension of ϕ. The equality $\phi^c(K\exp i\mathfrak{k}) = \phi(K)\exp i d\phi(\mathfrak{k})$ shows that ϕ^c is uniquely determined by ϕ.

Theorem 1. *Let* G *be a complex semisimple Lie group with a finite number of components and* K *be a maximal compact subgroup of* G. *Then* G *is isomorphic to the associated algebraic group* K^{*c} *of* K *as complex Lie groups.* G *has the unique structure of complex affine algebraic group defined over* $\mathbb{R}$ *compatible with its Lie group structure. Every holomorphic representation of* G *is rational.* G *has a faithful representation.*

Proof. By Proposition 5, $K = K^*$ is a maximal compact subgroup of a semisimple complex algebraic group K^{*c} and the canonical mapping R is an isomorphism of K onto K^*. By Proposition 9, the isomorphism R of K onto K^* can be extended uniquely to a holomorphic isomorphism of G onto K^{*c}. Hence G has a structure of a complex affine algebraic group defined over $\mathbb{R}$ which is compatible with its complex Lie group structure. As an affine algebraic group, G has a faithful rational representation. Every holomorphic representation of G is rational by Corollary to Proposition 2. It remains to prove the uniqueness of the algebraic group structure. We identify G with K^{*c}.

Let G' be a complex affine algebraic group whose underlying complex Lie group is G. Then every rational representation D of G' is a holomorphic representation of G. Hence D is a rational representation of $G = K^{*c}$ by Corollary to Proposition 2. Therefore

$$\mathcal{R}_r(G') \subset \mathcal{R}_r(G) \qquad \text{and} \qquad R_r(G') \subset R_r(G) .$$

Since $R_r(G')$ and $R_r(G)$ are the affine algebras of G' and G respectively, the identity mapping I of the set G is a morphism of affine algebraic groups from G onto G'. In particular, $I(K) = K$ is a real affine algebraic subgroup of G'. Let M be a complex algebraic subgroup of G' containing K. Then $M = I^{-1}(M)$ is a complex algebraic subgroup of $G = K^{*c}$ containing K. Hence $M = G$ by Proposition 3. Therefore $M = G'$ and G' is the smallest algebraic group containing K. Hence G' is the complexification of K and is equal to $G = K^{*c}$.

3. The Universal Complexification of a Lie Group

Let G be a Lie group with a finite number of connected components $\mathfrak{g}$ the Lie algebra of G, $\mathfrak{g}^c$ the complexification of $\mathfrak{g}$ and $S(\mathfrak{g}^c)$ the simply connected Lie group with the Lie algebra $\mathfrak{g}^c$.

For any element x in G, there exists a unique automorphism $h(x)$ of $S(\mathfrak{g}^c)$ satisfying $d(h(x)) = \mathrm{Ad}_{\mathfrak{g}^c}(x)$. Since h is an analytic homomorphism of G into $\mathrm{Aut}\, S(\mathfrak{g}^c)$, the semidirect product Lie group $H = S(\mathfrak{g}^c) \times_h G$ is defined. The product in H is defined by

$$(x,y)(x',y') \;=\; (x \cdot h(y)x', yy') .$$

The differential dD of any representation D in $\mathcal{R}(G)$ has the $\mathbb{C}$-linear extension $(dD)^c$ which is a representation of $\mathfrak{g}^c$. There exists a representation D_1 of $S(\mathfrak{g}^c)$ satisfying $dD_1 = (dD)^c$ and a representation D_2 of H defined by $D_2(x,y) = D_1(x)D(y)$. Put $P = \bigcap_{D\in\mathcal{R}(G)} \mathrm{Ker}\, D_2$. Then P is a closed normal subgroup of H. The quotient group $G^+ = H/P$ is called the *universal complexification* of G. Let p be the canonical projection of H onto G^+ and ψ be the restriction of dp to $\mathfrak{g}^c \times \{0\}$. Put $j(g) = (e,g)$ and $\phi = p \circ j$. Then the universal complexification G^+ has the following properties.

Proposition 10 (Hochschild-Mostow). (1) G^+ *is a complex Lie group.* (2) ϕ *is an analytic homomorphism of* G *into* G^+. (3) *For any representation* D *of* G, *there exists a unique holomorphic representation* D^+ *of* G^+ *satisfying* $D^+ \circ \phi = D$.

Proof. Hochschild-Mostow [7].

Corollary to Proposition 10. (1) $G^+ = (G^+)_0 \cdot \phi(G)$. G^+ *has a finite number of components.* G^+ *is connected if* G *is connected.* (2) *Let* M *be a complex Lie group satisfying* $G^+ \supset M \supset \phi(G)$. *Then* M *is equal to* G^+. (3) *The mapping* $v: D \to D^+$ *is a bijection of* $R(G)$ *onto the set* $R_h(G^+)$ *of all holomorphic representations of* G^+. (4) *If there exists a representation* D' *of* G *such that* dD' *is faithful (especially if* $\mathfrak{g}$ *is semisimple), then the homomorphism* $\psi = dp|\mathfrak{g}^c \times \{0\}$ *is an isomorphism of* $\mathfrak{g}^c$ *onto the Lie algebra* $\mathfrak{g}^+$ *of* G^+ *and* $d\phi$ *is injective. We identify* $\mathfrak{g}$ *with* $d\phi(\mathfrak{g})$ *and* $\mathfrak{g}^c$ *with* $\mathfrak{g}^+$ *by* ψ. *Then* dD^+ *satisfies*

$$dD^+(X + iY) = dD(X) + i dD(Y) \quad \textit{for any} \quad X, y \in \mathfrak{g} \quad \textit{and} \quad D \in R(G).$$

(5) *If* G *has a faithful representation* D_0, *then* ϕ *is injective.*

Proof. (1) Since the identity component H_0 of H satisfies $H_0 = S(\mathfrak{g}^c) \times_h G_0$, we have $H = H_0 \cdot j(G)$ and $G^+ = p(H) = p(H_0) \cdot \phi(G)$. Hence the index $[G^+ : (G^+)_0] \leqq [G : G_0] < +\infty$. In particular, G^+ is connected if G is connected. (2) ψ is a homomorphism of $\mathfrak{g}^c$ onto $\mathfrak{g}^+$ which introduce a complex structure on $\mathfrak{g}^+$ via that of $\mathfrak{g}^c$. We have the congruence $(X,0) \equiv (0,X)$ modulo an ideal $\mathfrak{d} = \{(X,-X) \mid X \in \mathfrak{g}^c\}$ of $\mathfrak{h}$. Hence we have the equalities

$$dp(X,0) + i dp(Y,0) = \psi(X) + i\psi(Y) = \psi(X + iY) = dp(X + iY, 0)$$

and

$$[d\phi(\mathfrak{g})]_{\mathbb{C}} = [dp(0,\mathfrak{g})]_{\mathbb{C}} = [dp(\mathfrak{g},0)]_{\mathbb{C}} = dp(\mathfrak{g}^c,0) = \psi(\mathfrak{g}^c) = \mathfrak{g}^+ .$$

Let $\mathfrak{m}$ be the Lie algebra of M. Then we have $\mathfrak{g}^+ \supset \mathfrak{m} \supset [d\phi(\mathfrak{g})]_{\mathbb{C}} = \mathfrak{g}^+$. Hence $\mathfrak{m} = \mathfrak{g}^+$ and $M_0 = (G^+)_0$. By (1) we get $G^+ \supset M = M \cdot \phi(G) \supset M_0 \cdot \phi(G) = (G^+)_0 \cdot \phi(G) = G^+$ and $G^+ = M$. (3) If $D_1^+ = D_2^+$, then $D_1 = D_1^+ \circ \phi = D_2^+ \circ \phi = D_2$. Hence the mapping $v: D \mapsto D^+$ is injective. Let D_0 be a holomorphic representation of G^+. Then $D = D_0 \circ \phi$ belongs to $R(G)$ and $D^+ \circ \phi = D = D_0 \circ \phi$. Two holomorphic representations D_0 and D^+ are identical on $\phi(G)$. Hence $D_0 = D^+$ by (2) and v is surjective.

(4) By the definition of ψ, we have

$$\mathrm{Ker}\ \psi = \{X \in \mathfrak{g}^c \mid (X,0) \in \mathfrak{p}\} = \{X \in \mathfrak{g}^c \mid dD_1(X) = 0 \quad \text{for all} \quad D \in R(G)\}$$

$$= \bigcap_{D \in R(G)} \mathrm{Ker}\ dD_1 \subset \mathrm{Ker}\ dD_1' = \{0\}.$$

Hence ψ is an isomorphism of $\mathfrak{g}^c$ onto $\mathfrak{g}^+$. (If $\mathfrak{g}$ is semisimple, then we can take the adjoint representation Ad as D'.) If an element $X \in \mathfrak{g}$ satisfies $d\phi(X) = 0$, then we have $dD'(X) = (dD'^+ \circ d\phi)(X) = 0$ and $X = 0$. Therefore $d\phi$ is injective. Since D^+ is holomorphic, dD^+ is $\mathbb{C}$-linear. After the above stated identifications, we have $dD^+(X) = dD(X)$ and $dD^+(X + iY) = dD(X) + idD(Y)$ for any X, Y in $\mathfrak{g}$ and any D in $R(G)$. (5) Suppose that $\phi(x) = e$. Then $D_0(x) = D_0^+(\phi(x)) = D_0^+(e) = 1$ and $x = e$. Hence ϕ is injective.

4. The Tannaka Duality Theorem for Semisimple Lie Groups

Theorem 2. *Let G be a semisimple Lie group with a finite number of components, G^+ the universal complexification of G and G_u a maximal compact subgroup of G^+. Then we have the following.*

(0) *G^+ is a complex semisimple Lie group with a finite number of components. The Lie algebra $\mathfrak{g}^+$ of G^+ is the complexification of the Lie algebra $\mathfrak{g}$ of G.*

(1) *The mapping $v: D \mapsto D^+$ is a bijection of $R(G)$ onto $R_h(G^+)$. The unitary restriction mapping $u: D^+ \mapsto D_u = D^+|G_u$ is a bijection of $R_h(G^+)$ onto $R(G_u)$. The mapping $w = u \circ v$ is a bijection of $R(G)$ onto $R(G_u)$.*

(2) *w preserves the direct sum, the tensor product and the equivalence. We have $G^{*c}(R) \cong G_u^{*c}(R) \cong G^+$.*

(3) *Every representation of G is completely reducible. A representation D of G is irreducible if and only if D^+ is irreducible.*

(4) *The mapping $M: f^+ \mapsto f = f^+ \circ \phi$ is an isomorphism of the algebra $R_h(G^+)$ onto $R(G_u)$. Three mutually isomorphic algebras $R(G)$, $R_h(G^+)$ and $R(G_u)$ are finitely generated.*

Proof. (0) is proved in Corollary to Proposition 10.

(1) v is a bijection by Corollary to Proposition 10. u is a bijection by Corollary to Proposition 2, Theorem 1 and the above (o).

(2) Let D, D' and D'' be three representations of G. If

$D = D' \dot{+} D''$, then we have $dD = dD' \dot{+} dD''$, $D_1 = D_1' \dot{+} D_1''$, $D_2 = D_2' \dot{+} D_2''$ and $D^+ = D'^+ \dot{+} D''^+$. Conversely since $D_2(e,y) = D(y)$, if $D^+ = D'^+ \dot{+} D''^+$ then $D_2 = D_2' \dot{+} D_2''$ and $D = D' \dot{+} D''$. Similarly $D = D' \otimes D''$ if and only if $D^+ = D'^+ \otimes D''^+$ and $D = \gamma D' \gamma^{-1}$ if and only if $D^+ = \gamma D'^+ \gamma^{-1}$. The mapping u also preserves the three kinds of operations. Hence the structures of the dual objects $\mathcal{R}(G)$ and $\mathcal{R}(G_u)$ are isomorphic. The natural mapping $\zeta \mapsto \zeta_u$ defined by $\zeta(D) = \zeta_u(D_u)$ defines an isomorphism of $G^{*c}(\mathcal{R})$ onto $G_u^{*c}(\mathcal{R})$. $G_u^{*c}(\mathcal{R})$ is isomorphic to G^+ by Theorem 1.

(3) Let D be an arbitrary representation of G, W a subspace of the representation space of D. W is stable under D^+ if and only if it is stable under D_2. Since $D_2(e,y) = D(y)$, W is stable under D if it is stable under D_2 and D^+. Conversely if W is stable under D, then W is stable under dD, $(dD)^c = dD_1$, D_1, D_2 and D^+. Hence W is stable under D if and only if W is stable under D^+. On the other hand, by Corollary 2 to Proposition 5 and Theorem 1, every holomorphic representation D^+ of G^+ is completely reducible. Hence D is completely reducible. Moreover D is irreducible if and only if D^+ is irreducible.

(4) The restriction mapping N is an isomorphism of $R_h(G^+)$ onto $R(G_u)$ by Corollary to Proposition 2. Let $\mathcal{D} = \{D^\alpha | \alpha \in A\}$ be a set of the complete representatives of the set A of equivalence classes of all irreducible representations of G. Then $\mathcal{B} = \{D^\alpha_{ij} | \alpha \in A,\ 1 \leqq i,\ j \leqq d(D^\alpha)\}$ is a basis of $R(G)$. By the above (1), (2) and (3) $\mathcal{D}^+ = \{D^{\alpha+} | \alpha \in A\}$ is a set of the complete representatives of the set of equivalence classes of all holomorphic irreducible representations of G^+. Then $\mathcal{B}^+ = \{D^{\alpha+}_{ij} | \alpha \in A,\ 1 \leqq i,\ j \leqq d(D^\alpha)\}$ is a basis of $R_h(G^+)$. Hence the mapping $M: f^+ \mapsto f$ is a linear bijection of $R_h(G^+)$ onto $R(G)$. Since M is an algebra homomorphism, it is an isomorphism. We have proved that $R(G) \cong R_h(G^+) \cong R(G_u)$. Since $R(G_u)$ is finitely generated by Proposition 5, $R(G)$ and $R_h(G^+)$ are also finitely generated.

<u>Theorem 3</u>. *Let G be a real semisimple algebraic group.*
(1) *Then the canonical homomorphisms $\phi: G \to G^+$ and $\Phi: G \to G^*(\mathcal{R})$ are morphisms of real algebraic groups.* (2) *Every continuous representation of the underlying Lie group of G is a rational representation of G.*

<u>Proof</u>. (1) Since G^+ is a complex semisimple Lie group with a finite number of components, G^+ has the unique complex affine algebraic

structure compatible with its analytic structure. Hence G^+ has the natural structure of a real affine algebraic group. We shall show that such a structure is naturally defined by the construction of G^+ itself. We use the notations in Section 3 freely. The holomorphic automorphism group $\mathrm{Aut}\, S(\mathfrak{g}^c)$ of a simply connected group $S(\mathfrak{g}^c)$ is canonically isomorphic to $\mathrm{Aut}\, \mathfrak{g}^c$. This isomorphism gives a structure of an affine algebraic group to the group $\mathrm{Aut}\, S(\mathfrak{g}^c)$. Since the adjoint representation Ad on $\mathfrak{g}^c$ is a rational representation of G, we have a rational homomorphism $h: G \to \mathrm{Aut}\, S(\mathfrak{g}^c)$ by the above isomorphism. Hence the semidirect product $H = S(\mathfrak{g}^c) \times_h G$ is a real affine algebraic group. Since G is a real affine algebraic group, G has a faithful real representation F and is identified with $F(G)$ and regarded as an algebraic subgroup of $GL(n,\mathbb{R})$. Let F_1 be the holomorphic hence rational representation of $S(\mathfrak{g}^c)$ such that $dF_1 = (dF)^c$ and put $G' = F_1^{-1}(G)$. Since dF and dF_1 are faithful, $\mathrm{Ker}\, F_1$ is discrete. An abstract homomorphism f of an algebraic group G into an algebraic group is a morphism if the restriction f° of f to the irreducible component G° of G is a morphism. Hence we may assume that G is irreducible. In this case, the complexification G^c of G is irreducible and hence connected in Lie group topology (Chevalley [3], p. 176). Hence F_1 maps G' onto G and the restriction $F_1|G'$ of F_1 to G' is a covering homomorphism of G' onto G. Let D be an arbitrary representation of G and define D_1 and D_2 as in Section 3. Then we have $D_1(x) = D(F_1(x))$ for all $x \in G'$. Hence if (x,y) belongs to $\mathrm{Ker}\, F_2$, then $F_1(x) = F(y)^{-1} = y^{-1}$, $x \in G'$, $D_1(x) = D(F_1(x)) = D(y^{-1})$ and (x,y) belongs to $\mathrm{Ker}\, D_2$. We have proved that $\mathrm{Ker}\, F_2 \subset \mathrm{Ker}\, D_2$ for all $D \in \mathcal{R}(G)$ and $P = \cap_{D \in \mathcal{R}(G)} \mathrm{Ker}\, D_2$ is equal to the kernel $\mathrm{Ker}\, F_2$ of a rational representation F_2. Therefore the normal subgroup P is an algebraic subgroup of H and G^+ is a real algebraic group. Since the injection $j: g \mapsto (e,g)$ and the projection $p: H \to H/P = G^+$ are morphisms of real algebraic groups, the canonical homomorphism $\phi = p \circ j$ is also a morphism. Since $G^{*c}(\mathcal{R})$ is an affine algebraic group, it is identified with an algebraic subgroup of $GL(n,\mathbb{C})$. Hence the canonical mapping Φ can be regarded a representation for G. Therefore we can define a holomorphic and hence rational representation Φ^+ of G^+. Therefore $\Phi = \Phi^+ \circ \phi$ is a morphism.

(2) Every continuous representation D of G is rational, because $D = D^+ \circ \phi = \hat{D} \circ \Phi$.

Theorem 4. *Let* G *be a semisimple Lie group with a finite number of components. Then the following three conditions are mutually equivalent.* (a) *The Tannaka duality holds for* G. (b) *The duality holds for* G. (c) G *is the underlying Lie group of a real affine algebraic group.*

Proof. (a) ⇔ (b) Since every representation of G is completely reducible (Theorem 2), the canonical homomorphism F is an isomorphism of G^* onto $G^*(\mathcal{R})$ (Proposition 1). Hence (a) is equivalent to (b). (b) ⇒ (c) Since the algebra $R(G)$ is finitely generated and the real and imaginary parts of a function in $R(G)$ belong to $R_{\mathbb{R}}(G)$, $R_{\mathbb{R}}(G)$ is finitely generated. Hence the duality implies that the group G is a real affine algebraic group with the affine algebra $R_{\mathbb{R}}(G)$ (Proposition 2). (c) ⇒ (a) Since Φ is a rational homomorphism, $\Phi(G)$ is a real algebraic subgroup of $G^{*c}(\mathcal{R})$ and contained in $G^*(\mathcal{R})$. By Proposition 3 and Proposition 1, $G^*(\mathcal{R})$ is the smallest real algebraic group containing $\Phi(G)$. Hence we get $\Phi(G) = G^*(\mathcal{R})$ and Φ is surjective. Since G is an affine algebraic group, G has a faithful representation and Φ is injective (Proposition 4). Hence Φ is an isomorphism of G onto $G^*(\mathcal{R})$. Since G is σ-compact, a continuous isomorphism Φ is an open mapping and hence a homeomorphism. Therefore Φ is a topological isomorphism of G onto $G^*(\mathcal{R})$ and the Tannaka duality holds for G.

5. Infinitesimal Tannaka Duality Theorem

In this section we shall show that the Tannaka duality holds infinitesimally for every semisimple Lie group. One of the consequences of this fact is that the canonical image $\Phi(G)$ of G is "large" in $G^*(\mathcal{R})$, that is, open in $G^*(\mathcal{R})$.

Now we formulate the Tannaka duality for real Lie algebras. Since our main concern is Lie group, it is convenient to choose the set $dR(G) = \{dD \mid D \in R(G)\}$ as the dual object of the Lie algebra $\mathfrak{g}$ of a Lie group G. More generally, we define as follows. Let $\mathfrak{g}$ be a real Lie algebra and $\mathfrak{M}$ be a set of (finite dimensional) representations of $\mathfrak{g}$ by complex matrices. $\mathfrak{M}$ is assumed to be closed under the operations of the direct sum, the tensor product, the equivalence and the complex conjugation. Let $\hat{\mathfrak{g}}^c(\mathfrak{M})$ be the set of all mappings $\eta: M \to \bigcup_{n=1}^{\infty} \mathfrak{gl}(n,\mathbb{C})$ which preserves the direct sum, the tensor product and the equivalence and satisfies $\eta(D) \in \mathfrak{gl}(d(D),\mathbb{C})$. Then the set $\hat{\mathfrak{g}}^c(\mathfrak{M})$ becomes a Lie algebra over $\mathbb{C}$ by defining $(a\eta_1 + b\eta_2)(D) = a\eta_1(D) + b\eta_2(D)$ and $[\eta_1,\eta_2](D) = [\eta_1(D),\eta_2(D)]$ for $a, b \in \mathbb{C}$, $D \in \mathfrak{M}$ and $\eta_1,\eta_2 \in \hat{\mathfrak{g}}^c(\mathfrak{M})$. Let $\mathfrak{g}(\mathfrak{M})$ be the set of all η in $\hat{\mathfrak{g}}^c(\mathfrak{M})$ satisfying $\eta(\bar{D}) = \overline{\eta(D)}$ for all $D \in \mathfrak{M}$. Then $\hat{\mathfrak{g}}(\mathfrak{M})$ is a Lie algebra over $\mathbb{R}$. Every element X in $\mathfrak{g}$ defines an element η_X of $\hat{\mathfrak{g}}(\mathfrak{M})$ which maps $D \in \mathfrak{M}$ to $D(X)$. The mapping $\Psi: X \mapsto \eta_X$ is a homomorphism of $\mathfrak{g}$ into $\hat{\mathfrak{g}}(\mathfrak{M})$. We say that the Tannaka duality between $\mathfrak{g}$ and $\mathfrak{M}$ holds if Ψ is an isomorphism of $\mathfrak{g}$ onto $\hat{\mathfrak{g}}(\mathfrak{M})$. Taking the real and imaginary parts of $\zeta(D)$, any ζ in $\hat{\mathfrak{g}}^c(\mathfrak{M})$ can be written uniquely as $\zeta = \xi + i\eta, \xi,\eta \in \hat{\mathfrak{g}}(\mathfrak{M})$. Hence $\hat{\mathfrak{g}}(\mathfrak{M})$ is a real form of $\hat{\mathfrak{g}}^c(\mathfrak{M})$.

<u>Proposition 11</u>. *Let* G *be a Lie group and assume that the algebra* $R(G)$ *is finitely generated and every representation of* G *is completely reducible. We identify the real algebraic group* G^* *with* $G^*(R)$ *by the canonical mapping* F *(Proposition* 1*). For each element* Z *in the Lie algebra* $\mathfrak{g}^*(R)$ *of* $G^*(R)$*, we define an element* $\hat{\eta}_Z$ *in* $\hat{\mathfrak{g}}(dR)$ *by* $\hat{\eta}_Z(dD) = d\hat{D}(Z)$ *for* $D \in R(G)$. (1) *Then the mapping* $H: Z \mapsto \hat{\eta}_Z$ *is a Lie algebra isomorphism of* $\mathfrak{g}^*(R)$ *onto* $\hat{\mathfrak{g}}(dR)$. (2) $H \circ d\Phi = \Psi$. (3) *If the Tannaka duality holds for* G*, then the Tannaka duality holds between the Lie algebra* $\mathfrak{g}$ *and* $\hat{\mathfrak{g}}(dR)$.

<u>Proof</u>. As is easily seen, the mapping $\hat{\eta}_Z: dD \mapsto d\hat{D}(Z)$ belongs to $\hat{\mathfrak{g}}(dR)$ for any Z in $\mathfrak{g}^*(R)$ and the mapping $H: Z \mapsto \hat{\eta}_Z$ is a Lie algebra homomorphism of $\mathfrak{g}^*(R)$ into $\hat{\mathfrak{g}}(dR)$. Now we shall prove that the mapping H is injective. Assume that $\hat{\eta}_Z = 0$. Then $d\hat{D}(Z) = 0$ for all $D \in R(G)$. Let $\zeta^t = \exp tZ \in G^*(R)$ $(t \in \mathbb{R})$ be the one-parameter subgroup of $G^*(R)$ generated by $Z \in \mathfrak{g}^*(R)$. Then we have

$$\zeta^t(D) = \hat{D}(\zeta^t) = \exp td\hat{D}(Z) = 1 \quad \text{for all } D \in R(G) \text{ and } t \in \mathbb{R}.$$

Hence $\zeta^t = e$ for all $t \in \mathbb{R}$ and $Z = 0$. The injectivity of H is proved. Now we prove the surjectivity. Choose an arbitrary element η in $\hat{\mathfrak{g}}(dR)$. Put $\zeta^t(D) = \exp t\eta(dD)$ for $t \in \mathbb{R}$ and $D \in R(G)$. Then ζ^t is a one-parameter subgroup of $G^*(R)$. Hence there exists an element Z in the Lie algebra $\mathfrak{g}^*(R)$ of $G^*(R)$ such that $\zeta^t = \exp tZ$. Then we have $\exp t\eta(dD) = \zeta^t(D) = \hat{D}(\zeta^t) = \exp td\hat{D}(Z)$ for all $t \in \mathbb{R}$. Hence $\eta(dD) = d\hat{D}(Z) = \hat{\eta}_Z(dD)$ for all $D \in R(G)$. Therefore we get $\eta = \hat{\eta}_Z$. The bijectivity of H is proved. (2) We have

$$[(H\circ d\Phi)(X)](dD) = d\hat{D}(d\Phi(X)) = d(\hat{D}\circ\Phi)(X) = dD(X) = \eta_X(dD)$$
$$= \Psi(X)(dD) \qquad \text{for all} \quad D\in R(G) \quad \text{and} \quad X\in \mathfrak{g}.$$

Hence we get $H\circ d\Phi = \Psi$. (3) Assume that the Tannaka duality holds for G. Then Φ is a continuous and hence analytic isomorphism of G onto $G^*(R)$. Hence the differential $d\Phi$ is an isomorphism of the Lie algebra $\mathfrak{g}$ onto $\mathfrak{g}^*(R)$. By the above (1) and (2). The canonical mapping $\Psi = H\circ d\Phi$ is an isomorphism of $\mathfrak{g}$ onto $\hat{\mathfrak{g}}(dR)$.

Theorem 5. *Let* G *be a semisimple Lie group with a finite number of components. Then the Tannaka duality holds between the Lie algebra* $\mathfrak{g}$ *of* G *and* $dR = dR(G)$.

Proof. Let $D' = \mathrm{Ad}$ be the adjoint representation of G. Then $dD' = \mathrm{ad}$ is a faithful representation of a semisimple Lie algebra $\mathfrak{g}$. Hence if $\eta_X = 0$, then $\mathrm{ad}\, X = dD'(X) = \eta_X(dD') = 0$ and $X = 0$. Therefore the canonical homomorphism $\Psi: X\mapsto \eta_X$ is injective.

Let G^+ be the universal complexification of G and G_u be a maximal compact subgroup of G^+. Then the Lie algebra $\mathfrak{g}^+$ of G^+ is the complexification of $\mathfrak{g}$ (Corollary to Proposition 10). By Theorem 2 we can identify two sets $R(G)$ and $R(G_u)$ by the mapping w. Since w preserves the direct sum, the tensor product and the equivalence, we can identify two complex Lie algebras $\hat{\mathfrak{g}}^c(dR(G))$ and $\hat{\mathfrak{g}}^c(dR(G_u))$. For each element Z of $\mathfrak{g}^+ = \mathfrak{g}_u^c$, the mapping $\eta_Z^+: Z\mapsto dD^+(Z)$ belongs to $\hat{\mathfrak{g}}^c(dR(G_u))$. The mapping $\Psi^c: Z\mapsto \eta_Z^+$ is the $\mathbb{C}$-linear extension of $\Psi_u: X\mapsto \eta_X$. Since the Tannaka duality holds for compact Lie group G_u, Ψ_u is an isomorphism of $\mathfrak{g}_u$ onto $\hat{\mathfrak{g}}(dR(G_u))$ (Proposition 11). Hence the mapping Ψ^c is an isomorphism of $\mathfrak{g}^+ = \mathfrak{g}_u^c$ onto $\hat{\mathfrak{g}}^c(dR(G_u))$ which is the complexification of $\hat{g}(dR(G))$. Now we shall prove that the canonical mapping $\Psi: \mathfrak{g}\to \hat{\mathfrak{g}}(dR(G))$ is surjective. Choose η in $\hat{\mathfrak{g}}(dR)$. Since η is an element of $\hat{\mathfrak{g}}^c(dR(G)) = \hat{\mathfrak{g}}^c(dR(G_u))$ and the mapping Ψ^c is surjective, there exists an element Z in $\mathfrak{g}^+$ such that

$$\eta(dD) = dD^+(Z) \qquad \text{for all} \quad D\in R(G) \quad . \tag{5.1}$$

Let σ be the conjugation of $\mathfrak{g}^+$ with respect to a real form $\mathfrak{g}$. Then we have

$$d(\bar{D})^+(Z) = \overline{dD^+(\sigma Z)} \qquad \text{for all} \quad Z\in \mathfrak{g}^+ \quad \text{and} \quad D\in R(G) \ . \tag{5.2}$$

In fact, by the identification $\mathfrak{g}^+ = \mathfrak{g}^c$, $d(D^+)$ is equal to dD on $\mathfrak{g}$, hence any element $Z = X + iY$ $(X, Y \in \mathfrak{g})$ of $\mathfrak{g}^+$ satisfies

$$d(\bar{D})^+(Z) = d\bar{D}(X) + id\bar{D}(Y) = \overline{dD(X) - idD(Y)} = \overline{dD^+(X - iY)} = \overline{dD^+(\sigma Z)} .$$

Since $\eta \in \hat{\mathfrak{g}}(dR)$ preserves the complex conjugation, we have

$$\eta(d\bar{D}) = \overline{\eta(dD)} \qquad \text{for all} \quad D \in R(G) . \tag{5.3}$$

By the equalities (5.1), (5.2) and (5.3), we get

$$dD^+(\sigma Z) = dD^+(Z) \qquad \text{for all} \quad D \in R(G) . \tag{5.4}$$

Put $D = Ad$ in (5.4), we get $\sigma Z = Z$, because $d(Ad)^+ = ad_{g^+}$ is faithful. Hence Z belongs to g and $\eta = \eta_Z = \Psi(Z)$. The surjectivity of Ψ is proved.

Corollary to Theorem 5. *The Tannaka duality holds between any real semisimple Lie algebra* $\mathfrak{g}$ *and* $R(\mathfrak{g})$ *(the set of all representations of* $\mathfrak{g}$*).*

Proof. Let G be a simply connected Lie group with the Lie algebra $\mathfrak{g}$. Then we have $dR(G) = R(\mathfrak{g})$. Hence Corollary is a special case of Theorem 5.

Theorem 6. *Let* G *be a semisimple Lie group with a finite number of components and* $\Phi: g \mapsto \zeta_g$ *be the canonical homomorphism of* G *into* $G^*(R)$. *Then* $\Phi(G)$ *is an open subgroup of* $G^*(R)$ *and the identity component* $\Phi(G)_0$ *of* $\Phi(G)$ *is equal to* $G^*(R)_0$. $G^*(R)$ *is the smallest real affine algebraic group containing* $\Phi(G)$.

Proof. By Theorem 5 the canonical mapping Ψ is an isomorphism of $\mathfrak{g}$ onto $\hat{\mathfrak{g}}(dR)$. Hence by Proposition 11, we have

$$d\Phi(\mathfrak{g}) = H^{-1}(d\Psi(\mathfrak{g})) = H^{-1}(\hat{\mathfrak{g}}(dR)) = \mathfrak{g}^*(R) .$$

Hence $\Phi(G)$ is an open subgroup of the Lie group $G^*(R)$. Therefore their identity components coincide with each other: $\Phi(G)_0 = G^*(R)_0$. $G^*(R)$ is the smallest real algebraic group containing $\Phi(G)$ by Proposition 3.

Corollary to Theorem 6. *A connected semisimple Lie group* G *admitting a faithful representation has the following properties.*

(1) G *is isomorphic to the topological identity component of a real affine algebraic group* $G^*(R)$.

(2) *If* G *is an analytic subgroup of* $GL(n,\mathbb{C})$, *then* G *is closed in* $GL(n,\mathbb{C})$.

Proof. (1) Φ is injective by Proposition 4. Hence $G = G_0$ is isomorphic to $\Phi(G)_0 = G^*(R)_0$. (2) Let D be a representation of G defined by $D(g) = g$. Then $\hat{D}$ is a rational representation of $G^*(R)$. Hence $\hat{D}(G) = D(G) = G$ is the identity component of a linear algebraic group $\hat{D}(G^*(R))$ and is closed in $GL(n,\mathbb{C})$. ■

Definition. A Lie group H is called a complexification of a Lie group G if H satisfies the following three conditions: (1) The Lie algebra $\mathfrak{h}$ of H is the complexification of $\mathfrak{g}$ (the Lie algebra of G). (2) G is a Lie subgroup of H. (3) G intersects every connected component of H.

Proposition 12. *Let* G *be a semisimple Lie group with a finite number of components. Then* G *has a complexification if and only if* G *has a faithful representation.*

Proof. If G has a complexification H, then H is a complex semisimple Lie group with a finite number of components. Hence H has a faithful representation D (Theorem 1). Then the restriction of D to the subgroup G is a faithful representation of G.

Conversely assume that G has a faithful representation D. We can assume that D is a real representation. By D, G is identified with a Lie subgroup of $GL(n,\mathbb{R})$. Let H_0 be the analytic subgroup of $GL(n,\mathbb{C})$ whose Lie algebra $\mathfrak{h}$ is the complexification of $\mathfrak{g}$. Then H_0 is a closed subgroup of $GL(n,\mathbb{C})$ by Corollary to Theorem 6. Put $H = H_0 G$. Since $\mathrm{Ad}_{\mathfrak{gl}(n,\mathbb{C})}(G)$ leaves $\mathfrak{h}$ stable, G normalize H_0 and H is a subgroup of $GL(n,\mathbb{C})$. Since $[H:H_0] \leqq [G;G_0] < +\infty$, H is closed in $GL(n,\mathbb{C})$ and G is closed in H. Hence G is a Lie subgroup of H. Therefore H is a complexification of G.

6. References

[1] P. Cartier, "Dualité de Tannaka des groupes et algèbres de Lie," *C.R., Paris* 242 (1956), 322-325.

[2] C. Chevalley, *Theory of Lie Groups I*, Princeton Univ. Press.

[3] C. Chevalley, *Théorie des Groupes de Lie, t II*, Hermann, 1948.

[4] C. Chevalley and S. Eilenberg, "Cohomology theory of Lie groups and Lie algebras," *Trans. A.M.S.* 63 (1948), 85-124.

[5] M. Goto, "Faithful representations of Lie groups," *Math. Japonica* 1 (1948), 107-119.

[6] Harish-Chandra, "Lie algebras and the Tannaka duality theorem," *Ann. of Math.* 51 (1950), 299-330.

[7] G. Hochschild and G.D. Mostow, "Representations and representative functions of Lie groups," *Ann. of Math.* 66 (1957), 495-542.

[8] Y. Matsushima, "Espaces homogènes de Stein des groupes de Lie complexes," *Nagoya Math. J.* 16 (1960), 205-218.

[9] M. Sugiura, "Some remarks on duality theorems of Lie groups," Proc. Japan Academy 43 (1967), 927-931.

[10] T. Tannaka, "Dualität der nicht-kommutativen Gruppen," *Tôhoku Math. J.* 53 (1938), 1-12.

[11] H. Weyl, "Theorie der Darstellung kontinuierlicher halbeinfacher Gruppen durch lineare Transformationen I, II, III," *Math. Z.* 23 (1925), 271-309, 24 (1926), 328-395.

College of General Education
University of Tokyo
Komaba, Meguroku, Tokyo
153 Japan

(Received January 12, 1981)

PARALLEL SUBMANIFOLDS OF SPACE FORMS

Masaru Takeuchi

0. Introduction

A submanifold M of a Riemannian manifold $\bar{M}$ is said to be parallel if the second fundamental form of M is parallel. For example, an affine subspace M of $\mathbb{R}^m$ or a symmetric R-space $M \subset \mathbb{R}^m$, which is minimally imbedded in a hypersphere of $\mathbb{R}^m$ (cf. Takeuchi-Kobayashi [12]), is a parallel submanifold of $\mathbb{R}^m$. Ferus ([3],[4]) showed that essentially these submanifolds exhaust all parallel submanifolds of $\mathbb{R}^m$ in the following sense: A complete full parallel submanifold of the Euclidean space $\mathbb{R}^m = M^m(0)$ is congruent to

(a) $M = \mathbb{R}^{m_0} \times M_1 \times \cdots \times M_s \subset \mathbb{R}^{m_0} \oplus \mathbb{R}^{m_1} \oplus \cdots \oplus \mathbb{R}^{m_s} = \mathbb{R}^m$, $m = m_0 + \Sigma m_i$, $s \geqslant 0$, or to

(b) $M = M_1 \times \cdots \times M_s \subset \mathbb{R}^{m_1} \oplus \cdots \oplus \mathbb{R}^{m_s} = \mathbb{R}^m$, $m = \Sigma m_i$, $s \geqslant 1$,

where each $M_i \subset \mathbb{R}^{m_i}$ is an irreducible symmetric R-space.

Note that in case (a) M is not contained in any complete totally umbilic hypersurface of $\mathbb{R}^m$, i.e., in any hypersphere of $\mathbb{R}^m$, but in case (b) M is contained in a hypersphere of $\mathbb{R}^m$.

In this note we classify parallel submanifolds in spheres:

$$M^m(\bar{c}) = \left\{(x_i) \in \mathbb{R}^{m+1};\ \Sigma x_i^2 = 1/\bar{c}\right\}, \quad \bar{c} > 0,$$

and those in hyperbolic spaces:

$$M^m(\bar{c}) = \left\{(x_i) \in \mathbb{R}^{m+1};\ -x_1^2 + x_2^2 + \cdots + x_{m+1}^2 = 1/\bar{c},\ x_1 > 0\right\}, \quad \bar{c} < 0.$$

Together with the result of Ferus, we get the classification of parallel submanifolds of space forms $M^m(\bar{c})$ as follows.

Let M *be a complete full parallel submanifold of a space form* $M^m(\bar{c})$. *Then:*

(A) *Case where* M *is not contained in any complete totally umbilic hypersurface of* $M^m(\bar{c})$; *In case* $\bar{c} > 0$, M *is congruent to the product* $M_1 \times \cdots \times M_s$ *of irreducible symmetric* R-*spaces as* (b), *replacing* m *by* $m+1$, *which is regarded as a submanifold of a hypersphere of* $\mathbb{R}^{m+1}$. *In case* $\bar{c} = 0$, M *is congruent to the product* $\mathbb{R}^{m_0} \times M_1 \times \cdots \times M_s \subset \mathbb{R}^m$ *as* (a). *In case* $\bar{c} < 0$, M *is congruent to the product* $M^{m_0}(c_0) \times M_1 \times \cdots \times M_s \subset M^{m_0}(c_0) \times M^{m-m_0-1}(c') \subset M^m(\bar{c})$ *with* $c_0 < 0$, $c' > 0$, $1/c_0 + 1/c' = 1/\bar{c}$, $s \geqslant 0$, *where* $M_1 \times \cdots \times M_s \subset M^{m-m_0-1}(c')$ *is a submanifold as the one in case* $\bar{c} > 0$ *and the second inclusion is the natural one.*

(B) *Case where* M *is contained in a complete totally umbilic hypersurface* N *of* $M^m(\bar{c})$; *The hypersurface* N, *endowed with the induced metric, is isometric to* $M^{m-1}(c)$ *with* $c = \bar{c} + h^2$, *where* h *denotes the length of the mean curvature of* N, *and* M *is a submanifold of* $N = M^{m-1}(c)$ *as the one in* (A), *replacing* m *and* $\bar{c}$ *by* $m-1$ *and* c, *respectively.*

As an application, we give an alternative proof of Sakamoto's classification theorem (Sakamoto [8]) for planar geodesic submanifolds of space forms.

1. Preliminaries

In this section we recall some basic facts on isometric immersions and prepare some lemmas.

Let $f:(M,g) \to (\bar{M},\bar{g})$ be an isometric immersion of a pseudo-Riemannian manifold (M,g) into a pseudo-Riemannian manifold $(\bar{M},\bar{g})$. The metrics on the tangent bundles TM, $T\bar{M}$ are denoted by $\langle\ ,\ \rangle$. Let ∇, $\bar{\nabla}$ be the Levi-Civita connections on (M,g), $(\bar{M},\bar{g})$. The metric and the connection on the pull back $f^{-1}T\bar{M}$ induced from $\langle\ ,\ \rangle$ and $\bar{\nabla}$ are also denoted by $\langle\ ,\ \rangle$ and $\bar{\nabla}$. We have an orthogonal sum:

$$f^{-1}T\bar{M} = TM \oplus NM \qquad ,$$

where NM denotes the normal bundle for f. Let $\nabla^{\perp}$ denote the normal connection on NM induced from $\bar{\nabla}$. Then we have Gauss-Weingarten formulas:

$$\bar{\nabla}_X Y = \nabla_X Y + \alpha(X,Y) \qquad , \tag{1.1}$$

$$\bar{\nabla}_X \xi \quad = \quad -A_\xi X + \nabla^\perp_X \xi \tag{1.2}$$

for vector fields X, Y on M and a normal vector field ξ. The tensors α and A_ξ are called the *second fundamental form* and the *shape operator* of f, respectively. $\alpha(X,Y)$ is symmetric in X, Y and related to A_ξ by $\langle A_\xi X, Y\rangle = \langle \alpha(X,Y), \xi\rangle$. We define

$$(\nabla^*\alpha)(X,Y,Z) \quad = \quad (\nabla^*_X\alpha)(Y,Z) \quad = \quad \nabla^\perp_X \alpha(Y,Z) - \alpha(\nabla_X Y, Z) - \alpha(Y, \nabla_X Z)$$

for vector fields X, Y, Z on M. The immersion f is said to be *totally geodesic* (resp. *parallel*) if $\alpha = 0$ (resp. $\nabla^*\alpha = 0$). The image $f(M)$ of a totally geodesic (resp. parallel) isometric imbedding f is called a *totally geodesic* (resp. *parallel*) *submanifold* of $(\bar{M},\bar{g})$.

We shall use the following Gauss-Codazzi-Ricci equations in the case where $(\bar{M},\bar{g})$ is of constant sectional curvature $\bar{c}$ (cf. Kobayashi-Nomizu [5]).

$$\begin{aligned}\langle R(X,Y)Z,W\rangle \quad &= \quad \bar{c}(\langle X,W\rangle\langle Y,Z\rangle - \langle X,Z\rangle\langle Y,W\rangle) \\ &\quad + \langle\alpha(X,W),\alpha(Y,Z)\rangle - \langle\alpha(X,Z),\alpha(Y,W)\rangle \quad ,\end{aligned} \tag{1.3}$$

$$(\nabla^*_X\alpha)(Y,Z) \quad = \quad (\nabla^*_Y\alpha)(X,Z) \quad , \tag{1.4}$$

$$\langle R^\perp(X,Y)\xi,\zeta\rangle = \langle [A_\xi, A_\zeta]X, Y\rangle \tag{1.5}$$

for vector fields X, Y, Z, W on M and normal vector fields ξ, ζ. Here, R and $R^\perp$ denote the curvatures of ∇ and $\nabla^\perp$, respectively.

Now we assume that the source (M,g) of f is a Riemannian manifold of dimension n. The normal vector field

$$\eta \quad = \quad \frac{1}{n}\,\mathrm{Tr}_g \alpha$$

is called the *mean curvature* of f. If $\eta = 0$, f is said to be *minimal*. It is said to be *totally umbilic* if (1) the mean curvature η is nowhere zero and $\alpha(X,Y) = \langle X,Y\rangle\eta$ for any vector fields X,Y on M, which is equivalent to that $A_\xi = \langle\eta,\xi\rangle I$ for any normal vector field ξ; and (2) η is parallel: $\nabla^\perp\eta = 0$. The image $f(M)$ of a totally umbilic isometric imbedding f is called a *totally umbilic submanifold* of $(\bar{M},\bar{g})$.

Remark 1. A totally umbilic isometric immersion is always parallel.

Remark 2. Usually an isometric immersion f satisfying (1) is said to be totally umbilic. In the case where $n \geqslant 2$ and $(\bar{M},\bar{g})$ has constant curvature, our definition coincides with the usual one, since (1) implies (2) in this case, in virtue of the Codazzi equation (1.4).

An isometric immersion f of a Riemannian manifold (M,g) into a pseudo-Riemannian manifold $(\bar{M},\bar{g})$ is said to be *isotropic* or *λ-isotropic* if there is a constant λ such that $\langle\alpha(x,x),\alpha(x,x)\rangle = \lambda$ for any unit vector $x \in TM$.

Let $f_i:(M_i,g_i) \to (\bar{M}_i,\bar{g}_i)$, $1 \leqslant i \leqslant s$, be isometric immersions of Riemannian manifolds (M_i,g_i) with $\dim M_i = n_i$ into pseudo-Riemannian manifolds $(\bar{M}_i,\bar{g}_i)$. Let $(M,g) = (M_1,g_1) \times \cdots \times (M_s,g_s)$, $(\bar{M},\bar{g}) = (\bar{M}_1,\bar{g}_1) \times \cdots \times (\bar{M}_s,\bar{g}_s)$ and denote by $\pi_i:M \to M_i$, $1 \leqslant i \leqslant s$, the projection to the i-th factor. We define an isometric immersion $f = f_1 \times \cdots \times f_s$: $(M,g) \to (\bar{M},\bar{g})$, called the *product immersion*, by $f(p) = (f_1(\pi_1 p),\ldots, f_s(\pi_s p))$ for $p \in M$. Then, denoting by α_i, η_i the second fundamental form and the mean curvature of f_i, $1 \leqslant i \leqslant s$, we have

$$\alpha(X_1 + \cdots + X_s, Y_1 + \cdots + Y_s) = \alpha_1(X_1,Y_1) + \cdots + \alpha_s(X_s,Y_s)$$

for vector fields X_i, Y_i on M_i, $1 \leqslant i \leqslant s$, and hence

$$n\eta = n_1\eta_1\circ\pi_1 + \cdots + n_s\eta_s\circ\pi_s \quad ,$$

where $n = \Sigma n_i$. Therefore, f is parallel (resp. minimal) if and only if each f_i is parallel (resp. minimal). Moreover, η is parallel if and only if each η_i is parallel.

Lemma 1.1. *Let* (M,g), (M',g') *be Riemannain manifolds and* $(\bar{M},\bar{g})$ *a pseudo-Riemannian manifold. Let* $f':(M,g) \to (M',g')$ *be an isometric immersion,* $f'':(M',g') \to (\bar{M},\bar{g})$ *a totally geodesic or totally umbilic isometric immersion and let*

$$f:(M,g) \xrightarrow{f'} (M',g') \xrightarrow{f''} (\bar{M},\bar{g})$$

be the composite of f' *and* f''. *Then:*

(1) *We have*

$$\alpha(X,Y) = \alpha'(X,Y) + \langle X,Y\rangle(\eta''\circ f') \tag{1.6}$$

for vector fields X,Y *on* M;

$$\eta = \eta' + \eta''\circ f' \quad ; \tag{1.7}$$

$$\nabla^{\perp}_{X}\eta = \nabla'^{\perp}_{X}\eta' \tag{1.8}$$

for a vector field X *on* M

$$A_{\eta} = A'_{\eta'} + (\langle\eta'',\eta''\rangle\circ f')I \quad . \tag{1.9}$$

Here, α, α' *denote the second fundamental forms of* f, f', *etc.*

(2) η *is parallel if and only if* η' *is parallel.*

(3) f' *is* λ'*-isotropic if and only if* f *is* λ*-isotropic, where* $\lambda' + \langle\eta'',\eta''\rangle = \lambda$.

(4) f *is parallel if and only if* f' *is parallel.*

<u>Proof</u>. The proof of (1.6), (1.7), (1.9) is immediate from definitions. The equality (1.6) implies the assertion (3). We prove (1.8) and the assertion (4). Let X be a vector field on M. Then, by (1.2), $\bar\nabla_X\eta'' = -A''_{\eta''}X + \nabla''^{\perp}_X\eta'' = -\langle\eta'',\eta''\rangle X$, which is tangent to M, and hence

$$\nabla^{\perp}_{X}(\eta''\circ f') = 0 \quad . \tag{1.10}$$

Moreover, for a normal vector field ξ' for the immersion f', by (1.1), $\bar\nabla_X\xi' = \nabla'_X\xi' + \alpha''(X,\xi') = \nabla'_X\xi' + \langle X,\xi'\rangle\eta'' = \nabla'_X\xi'$, which is tangent to M', and hence

$$\nabla^{\perp}_{X}\xi' = \nabla'^{\perp}_{X}\xi' \quad . \tag{1.11}$$

By (1.7), (1.10), (1.11) we get (1.8), which implies also the assertion (2).

Now, for vector fields X, Y, Z on M, by (1.6), (1.10), (1.11) we have

$$\begin{aligned}(\nabla^*_X\alpha)(Y,Z) &= \nabla^\perp_X\alpha'(Y,Z) + \nabla^\perp_X(\langle Y,Z\rangle\eta''\circ f') - \alpha'(\nabla_X Y,Z) \\ &\quad - \langle\nabla_X Y,Z\rangle\eta''\circ f' - \alpha'(Y,\nabla_X Z) - \langle Y,\nabla_X Z\rangle\eta''\circ f' \\ &= (X\langle Y,Z\rangle - \langle\nabla_X Y,Z\rangle - \langle Y,\nabla_X Z\rangle)\eta''\circ f' \\ &\quad + \nabla'^\perp_X\alpha'(Y,Z) - \alpha'(\nabla_X Y,Z) - \alpha'(Y,\nabla_X Z) \\ &= (\nabla'^*_X\alpha')(Y,Z) \quad .\end{aligned}$$

Thus we get a proof of the assertion (4). ∎

The real vector space $\mathbb{R}^m$, $m \geq 1$, endowed with the Euclidean metric:

$$\langle x,y\rangle = \Sigma\, x_i y_i \qquad \text{for} \quad x = (x_i), \quad y = (y_i) \in \mathbb{R}^m ,$$

is denoted by E^m and called the *Euclidean space*. Then E^m is simply connected and complete, and for $m \geq 2$ E^m has constant sectional curvature 0. We write $E^m = M^m(0)$. The complete totally geodesic submanifolds of $M^m(0)$ are just the affine subspaces of $\mathbb{R}^m$.

For $\bar{c} > 0$, $m \geq 1$ we define

$$M^m(\bar{c}) = \{x \in \mathbb{R}^{m+1};\ \langle x,x\rangle = 1/\bar{c}\} ,$$

and endow it with the Riemannian metric induced from that of E^{m+1}. Then $M^m(\bar{c})$ is complete, and the inclusion $f: M^m(\bar{c}) \to E^{m+1}$ is a totally umbilic isometric imbedding with $\langle\eta,\eta\rangle = \bar{c}$. Actually we have $\eta = -\bar{c}f$, identifying each tangent space T_pE^{m+1} with $\mathbb{R}^{m+1}$. For $m \geq 2$ $M^m(\bar{c})$ is simply connected and has positive constant sectional curvature $\bar{c}$, which follows from the Gauss equation (1.3). The complete totally geodesic submanifolds of $M^m(\bar{c})$ are just the intersections of $M^m(\bar{c})$ with linear subspaces of $\mathbb{R}^{m+1}$.

The real vector space $\mathbb{R}^m$, $m \geq 1$, endowed with the Lorentz metric:

$$\langle x,y\rangle = -x_1y_1 + x_2y_2 + \cdots + x_my_m \qquad \text{for} \quad x = (x_i), \quad y = (y_i) \in \mathbb{R}^m ,$$

is denoted by L^m and called the *Lorentz space*. For $m \geq 2$ L^m has constant sectional curvature 0. For $\bar{c} < 0$, $m \geq 1$ we define

$$M^m(\bar{c}) = \{x = (x_i) \in \mathbb{R}^{m+1};\ \langle x,x\rangle = 1/\bar{c},\ x_1 > 0\} \quad ,$$

and endow it with the Riemannian metric induced from the pseudo-Riemannian metric of L^{m+1}. Then $M^m(\bar{c})$ is complete and simply connected, and the inclusion $f: M^m(\bar{c}) \to L^{m+1}$ is a totally umbilic isometric imbedding with $\eta = -\bar{c}f$ and $\langle\eta,\eta\rangle = \bar{c}$. For $m \geq 2$ $M^m(\bar{c})$ has negative constant sectional curvature $\bar{c}$, which also follows from (1.3). The complete totally geodesic submanifolds of $M^m(\bar{c})$ are just the intersections of $M^m(\bar{c})$ with linear subspaces of $\mathbb{R}^{m+1}$.

For an arbitrary $\bar{c}$ and $m \geq 1$, the complete totally geodesic submanifolds of $M^m(\bar{c})$ of the same dimension are mutually congruent under the group $I(M^m(\bar{c}))$ of isometries of $M^m(\bar{c})$. Any totally geodesic immersion of an n-dimensional Riemannian manifolds (resp. complete Riemannian manifold) into $M^m(\bar{c})$ is an isometric immersion (resp. an isometric covering) to an n-dimensional complete totally geodesic submanifold of $M^m(\bar{c})$. The Riemannian manifolds $M^m(\bar{c})$ with $m \geq 2$ are called the *space forms*. The following theorem describes completely all totally umbilic submanifolds of space forms (see Takahashi [10], for example).

Theorem 1.2. *Let* $m \geq 2$ *and* $1 \leq n \leq m-1$.

(1) *Any totally umbilic immersion* f *of an* n*-dimensional Riemannian manifold (resp. complete Riemannian manifold)* (M,g) *into* $M^m(\bar{c})$ *is an isometric immersion (resp. an isometric covering) to an* n*-dimensional complete totally umbilic submanifold of* $M^m(\bar{c})$. *And* f *is equivariant in the following sense: There exists a homomorphism* ρ *from the group* $I(M,g)$ *of isometries of* (M,g) *to the group* $I(M^m(\bar{c}))$ *such that* $f(ap) = \rho(a)f(p)$ *for any* $a \in I(M,g)$, $p \in M$.

(2) *For each* $h > 0$ *there exists an* n*-dimensional complete totally umbilic submanifold* M *of* $M^m(\bar{c})$ *whose mean curvature has the length* h*, and such an* M *is unique up to the congruence by an element of* $I(M^m(\bar{c}))$. *Moreover,* (M,g), g *being the Riemannian metric induced from that of* $M^m(\bar{c})$, *is isometric to* $M^n(c)$ *with* $c = \bar{c} + h^2$.

(3) *The* n*-dimensional complete totally umbilic submanifolds of* $M^m(0)$ *are just the hyperspheres in* $(n+1)$*-dimensional affine subspaces of* $\mathbb{R}^m$. *In case* $\bar{c} \neq 0$, *the* n*-dimensional complete totally umbilic submanifolds of* $M^m(\bar{c})$ *are just the intersections of* $M^m(\bar{c})$ *with* $(n+1)$*-dimensional affine subspaces of* $\mathbb{R}^{m+1}$ *not passing through the origin* 0.

An immersion $f:M\to \mathbb{R}^m$ is said to be *substantial*, if $f(M)$ is not contained in any affine hyperplane of $\mathbb{R}^m$. An isometric immersion $f:(M,g)\to M^m(\bar{c})$, $m\geq 1$, is said to be *full*, if $f(M)$ is not contained in any totally geodesic hypersurface of $M^m(\bar{c})$. Thus, for an isometric immersion $f:(M,g)\to M^m(0)$, f is full if and only if f is substantial. An isometric immersion $f:(M,g)\to M^m(\bar{c})$, $m\geq 1$, is said to be *strongly full*, if f is full and further f is not contained in any totally umbilic hypersurface of $M^m(\bar{c})$. Now Theorem 1.2, (3) implies the following

Corollary. *Let* $f:(M,g)\to M^m(\bar{c})$, $m\geq 1$, $\bar{c}\neq 0$, *be an isometric immersion, and* $\iota:M^m(\bar{c})\to \mathbb{R}^{m+1}$ *denote the inclusion map. Then* f *is strongly full if and only if the composite* $\iota\circ f$ *is substantial.*

Lemma 1.3. *Let* $m\geq 2$ *and* $c>\bar{c}$. *Let* $f':(M,g)\to M^{m-1}(c)$ *be an isometric immersion,* $f'':M^{m-1}(c)\to M^m(\bar{c})$ *a totally umbilic isometric imbedding and let* $f=f''\circ f':(M,g)\to M^m(\bar{c})$. *Then* f *is full if and only if* f' *is strongly full.*

Proof. If $m=2$, then $\dim M=1$ and hence always f is full and f' is strongly full. Thus we may assume $m\geq 3$. In case $\bar{c}=0$, the lemma is a restatement of the above Corollary for $\bar{c}>0$. Therefore, we may further assume $\bar{c}\neq 0$. But in this case the assertion follows easily from Theorem 1.2, (3) and Lemma 1.1. ■

The following lemma is proved in the same way as Moore [7].

Lemma 1.4. *Let* $(M,g)=(M_1,g_1)\times\cdots\times(M_s,g_s)$ *be the product of Riemannian manifolds* (M_i,g_i), $1\leq i\leq s$, *and denote by* $\pi_i:M\to M_i$, $1\leq i\leq s$, *the projection to the* i*-th factor, and thus* TM *is the orthogonal sum:* $TM=T_1\oplus\cdots\oplus T_s$ *of the pullback bundles* $T_i=\pi_i^{-1}(TM_i)$, $1\leq i\leq s$. *Let* F *be a pseudo-Euclidean space, i.e., the real vector space* $\mathbb{R}^m$ *endowed with a nondegenerate metric* $\langle\,,\rangle$, *and let* $f:(M,g)\to F$ *be a substantial isometric immersion such that the second fundamental form* α *satisfies* $\alpha(T_i,T_j)=0$ *for* $i\neq j$. *Then there exist substantial isometric immersions* $f_i:(M_i,g_i)\to F_i$ *into pseudo-Euclidean spaces* F_i, $1\leq i\leq s$, *and an isometry* $\varphi:F_1\times\cdots\times F_s\to F$ *such that* $f=\varphi\circ(f_1\times\cdots\times f_s)$.

Lemma 1.5. *Let* $f:(M,g)\to M^m(\bar{c})$, $m\geq 1$, $\bar{c}\neq 0$, *be an isometric immersion of a Riemannian manifold* (M,g), $\iota:M^m(\bar{c})\to F^{m+1}$ *be the inclusion, where* $F^{m+1}=E^{m+1}$ *if* $\bar{c}>0$ *and* $F^{m+1}=L^{m+1}$ *if* $\bar{c}<0$, *and let* $f'=\iota\circ f:(M,g)\to F^{m+1}$, *which is also an isometric immersion. Suppose*

(1) f' *is substantial;*

(2) *the mean curvature* η' *of* f' *is parallel;*

(3) $A'_{\eta'} = \lambda I$ *for some constant* λ.

Then f *is minimal and* $\lambda = \bar{c}$.

Proof. Let $\bar{\nabla}$ and $\nabla'^{\perp}$ denote the flat connection of F^{m+1} and the normal connection of f', respectively. We identify tangent spaces T_pF^{m+1} with $\mathbb{R}^{m+1}$, and hence tangent spaces T_pM with subspaces of $\mathbb{R}^{m+1}$. Recall that the mean curvature of ι is given by $-\bar{c}\iota$. It follows from (1.7)

$$\eta' = \eta - \bar{c}f' \qquad , \tag{1.12}$$

where η denotes the mean curvature of f. Now, for each vector field X on M, we have $\nabla'^{\perp}_X\eta' = 0$ and $\bar{\nabla}_X f' = X$, and hence by (3)

$$\begin{aligned}\bar{\nabla}_X(\eta' + \lambda f') &= -A'_{\eta'}X + \nabla'^{\perp}_X\eta' + \lambda\bar{\nabla}_X f' \\ &= -\lambda X + \lambda X = 0 \qquad .\end{aligned}$$

There exists therefore a vector $a \in \mathbb{R}^{m+1}$ such that $\eta' + \lambda f' = a$. We show first that $\lambda \neq 0$. Suppose $\lambda = 0$. Then $\eta' = a$, and hence $\langle f'_*TM, a\rangle = \langle f'_*TM, \eta'\rangle = 0$. Therefore $\langle f', a\rangle$ is constant on M. But $a \neq 0$ in virtue of (1.12). This contradicts the assumption (1). We show next that $a = 0$. The function:

$$\begin{aligned}\langle\eta', \eta'\rangle &= \langle a - \lambda f', a - \lambda f'\rangle \\ &= \langle a, a\rangle - 2\lambda\langle f', a\rangle + \lambda^2\langle f', f'\rangle \\ &= (\langle a, a\rangle + \lambda^2/\bar{c}) - 2\lambda\langle f', a\rangle\end{aligned}$$

is constant on M by (2), and hence $\langle f', a\rangle$ is constant on M. This together with (1) implies $a = 0$.

Therefore we get $\eta' = -\lambda f'$. Comparing this with (1.12) we obtain $\eta = 0$ and $\lambda = \bar{c}$. ∎

Lemma 1.6. *Any minimal parallel isometric immersion* $f: (M,g) \to M^m(\bar{c})$, $m \geqslant 1$, $\bar{c} \leqslant 0$, *is totally geodesic.*

Proof. We may assume $m \geqslant 2$. Chern-do Carmo-Kobayashi [1] proved the following equality for the second fundamental form α of a minimal isometric immersion f of (M,g) into a space form $M^m(\bar{c})$:

$$\langle\Delta\alpha,\alpha\rangle = n\bar{c}\|\alpha\|^2 - \|\tilde{\alpha}\|^2 - \|R^{\perp}\|^2 .$$

Here, Δ is the Laplacian defined by $\Delta = \mathrm{Tr}_g \nabla^{*2}$; $n = \dim M$; $\tilde{\alpha}$ is the endomorphism of the normal bundle NM for f defined by $\tilde{\alpha} = \alpha \circ {}^t\alpha$, regarding α as a homomorphism from $TM \otimes TM$ to NM; $\langle,\rangle$ and $\| \|$ are the metric and the norm induced from those on tangent bundles. Therefore, in our case we have

$$n\bar{c}\|\alpha\|^2 = \|\tilde{\alpha}\|^2 + \|R^{\perp}\|^2 , \qquad \bar{c} \leqslant 0.$$

This implies the lemma. ■

Remark. Ferus [3] proved this lemma for $\bar{c} = 0$ by making use of the Gauss map of f.

2. Symmetric R-Spaces

In this section we recall the notion of symmetric R-spaces (cf. Takeuchi [11]) and their standard isometric imbeddings.

Let

$$\mathfrak{g} = \mathfrak{g}_{-1} + \mathfrak{g}_0 + \mathfrak{g}_1, \quad [\mathfrak{g}_p, \mathfrak{g}_q] \subset \mathfrak{g}_{p+q}$$

be a real semi-simple graded Lie algebra and σ an involutive automorphism of $\mathfrak{g}$. The pair $(\mathfrak{g},\sigma)$ is called a *symmetric graded Lie algebra* if (1) $\mathfrak{g}_{-1} \neq 0$; (2) $\mathfrak{g}_0$ acts effectively on $\mathfrak{g}_{-1}$; and (3) $\sigma\mathfrak{g}_\nu = \mathfrak{g}_{-\nu}$ for $\nu = -1, 0, 1$. It is said to be *positive definite* if σ is a Cartan involution of $\mathfrak{g}$. The positive definite graded Lie algebras are completely classified (Kobayashi-Nagano [5], Takeuchi [11]).

Let $(\mathfrak{g},\sigma)$ be a positive definite symmetric graded Lie algebra. It follows from the semi-simplicity of $\mathfrak{g}$ that there exists uniquely $e \in \mathfrak{g}_0$ such that $[e,x_\nu] = \nu x_\nu$ for each $x_\nu \in \mathfrak{g}_\nu$, $\nu = -1, 0, 1$. Let $\mathfrak{g} = \mathfrak{k} + \mathfrak{p}$, where

$$\mathfrak{k} = \{x \in \mathfrak{g};\ \sigma x = x\} ,$$

$$\mathfrak{p} = \{x \in \mathfrak{g};\ \sigma x = -x\} ,$$

be the Cartan decomposition associated to σ. The condition (3) implies $e \in \mathfrak{p}$. We define an involutive automorphism θ of the complexification of $\mathfrak{g}$ by $\theta = \exp \operatorname{ad} \pi\sqrt{-1}e$. Then θ leaves $\mathfrak{g}$ invariant, and hence it induces an automorphism θ of $\mathfrak{g}$. Let G be the adjoint group of $\mathfrak{g}$, and K the connected (compact) Lie subgroup of G generated by $\mathfrak{k}$. We define

$$K_0 = \{k \in K;\ ke = e\} .$$

Then $\theta K \theta^{-1} = K$ and $\theta k \theta^{-1} = k$ for each $k \in K_0$. We set

$$\mathfrak{k}_0 = \{x \in \mathfrak{k};\ \theta x = x\} ,$$

$$\mathfrak{m} = \{x \in \mathfrak{k};\ \theta x = -x\} .$$

Since $\mathfrak{k}_0$ is the Lie algebra of K_0, the pair (K,K_0) is a compact symmetric pair. We set

$$M = K/K_0 ,$$

and identify the tangent space of M at the origin $o = K_0 \in M$ with the subspace $\mathfrak{m}$. Let B denote the Killing form of $\mathfrak{g}$. Now M has a unique K-invariant Riemannian metric g_0 which coincides with $-B|\mathfrak{m}\times\mathfrak{m}$ at the origin o. Take an arbitrary Riemannian metric g of M which is homethetic to g_0. The Riemannian manifold (M,g) is called a *symmetric R-space* associated to $(\mathfrak{g},\sigma)$. It is said to be *irreducible* if $\mathfrak{g}$ is simple.

Let $n = \dim M$ and $m = \dim \mathfrak{p} - 1$. Then we see $B(e,e) = 2n$. We identify $\mathfrak{p}$ with the Euclidean space $\mathbb{R}^{m+1}$ by an orthonormal basis of $\mathfrak{p}$ with respect to $B|\mathfrak{p}\times\mathfrak{p}$. We define a K-equivariant imbedding $f_0 : (M,g_0) \to M^m(1/2n)$ by

$$f_0(ko) = ke \qquad \text{for} \quad k \in K .$$

It is known (Takeuchi-Kobayashi [12]) that f_0 is a minimal isometric imbedding such that the composite $f_0' = \iota \circ f_0$ is substantial, where $\iota : M^m(1/2n) \to \mathbb{R}^{m+1}$ denotes the inclusion. Moreover, f_0' is parallel (Ferus [2]), and hence f_0 is also parallel by Lemma 1.1,(4). Therefore, for an arbitrary $c > 0$, putting $f = (1/\sqrt{2nc})f_0$ and $g = (1/2nc)g_0$, we obtain a full minimal parallel isometric imbedding $f:(M,g) \to M^m(c)$ of a symmetric R-space (M,g). Such an imbedding f is called a *standard isometric imbedding* of a symmetric R-space associated to $(\mathfrak{g},\sigma)$.

Remark. Assume that $(\mathfrak{g},\sigma)$ is the direct sum:

$$(\mathfrak{g},\sigma) = (\mathfrak{g}_1,\sigma_1) \oplus \cdots \oplus (\mathfrak{g}_s,\sigma_s)$$

of simple positive definite symmetric graded Lie algebras $(\mathfrak{g}_i,\sigma_i)$, $1 \leqslant i \leqslant s$. Then a symmetric R-space (M,g) associated to $(\mathfrak{g},\sigma)$ is the product $(M,g) = (M_1,g_1) \times \cdots \times (M_s,g_s)$ of irreducible symmetric R-spaces (M_i,g_i) associated to $(\mathfrak{g}_i,\sigma_i)$, $1 \leqslant i \leqslant s$. And a standard isometric imbedding $f:(M,g) \to M^m(c)$ is factored to the composite $j \circ (f_1 \times \cdots \times f_s)$ of the product imbedding

$$f_1 \times \cdots \times f_s : (M_1,g_1) \times \cdots \times (M_s,g_s) \to M^{m_1}(c_1) \times \cdots \times M^{m_s}(c_s)$$

of standard isometric imbeddings $f_i:(M_i,g_i) \to M^{m_i}(c_i)$ associated to $(\mathfrak{g}_i,\sigma_i)$, $1 \leqslant i \leqslant s$, and the natural isometric imbedding

$$j: M^{m_1}(c_1) \times \cdots \times M^{m_s}(c_s) \to M^m(c) \qquad ,$$

where $m = \Sigma m_i + s - 1$, $\Sigma 1/c_i = 1/c$ and $c \dim M = c_i \dim M_i$, $1 \leqslant i \leqslant s$.

It is known that our standard isometric imbeddings of symmetric R-spaces exhaust all full minimal parallel isometric immersions into spheres in the following sense.

Theorem 2.1 (Ferus [4]). *Let* $f:(M,g) \to M^m(c)$, $m \geqslant 1$, $c > 0$, *be a full minimal parallel isometric immersion of a Riemannian manifold (resp. a complete Riemannian manifold)* (M,g) *with* $\dim M = n$. *Then there exist a standard isometric imbedding* $\bar{\bar{f}}:(M',g') \to M^m(c)$ *of a symmetric* R*-space* (M',g') *with* $\dim M' = n$ *and an isometric immersion (resp. an isometric covering)* $\bar{f}:(M,g) \to (M',g')$ *such that* $f = \bar{\bar{f}} \circ \bar{f}$.

3. Models for Parallel Submanifolds of Space Forms

In this section we give some examples of full parallel submanifolds of $M^m(\bar{c})$.

In the following, let $f_i:(M_i,g_i) \to M^{m_i}(c_i)$, $m_i \geq 1$, $c_i > 0$, denote a standard isometric imbedding of an irreducible symmetric R-space (M_i,g_i), and $f_i' = \iota_i \circ f_i:(M_i,g_i) \to \mathbb{R}^{m_i+1}$ be the composite of f_i and the inclusion $\iota_i:M^{m_i}(c_i) \to \mathbb{R}^{m_i+1}$, $1 \leq i \leq s$.

(i) Case $\bar{c} > 0$.

(a) $s \geq 1$, $m = \Sigma m_i + s - 1$, $\Sigma 1/c_i = 1/\bar{c}$. Let $(M,g) = (M_1,g_1) \times \cdots \times (M_s,g_s)$ and $j:M^{m_1}(c_1) \times \cdots \times M^{m_s}(c_s) \to M^m(\bar{c})$ be the natural isometric imbedding. We define an isometric imbedding $f:(M,g) \to M^m(\bar{c})$ by the composite $f = j \circ (f_1 \times \cdots \times f_s)$. Then the composite $\iota \circ f:(M,g) \to \mathbb{R}^{m+1}$ of f and the inclusion $\iota:M^m(\bar{c}) \to \mathbb{R}^{m+1}$ coincides with the product $f_1' \times \cdots \times f_s'$ which is substantial. Therefore, by Corollary of Theorem 1.2 f is a strongly full isometric imbedding. Moreover, since $\iota \circ f = f_1' \times \cdots \times f_s'$ is parallel, f is also parallel by Lemma 1.1,(4).

(b) $s \geq 1$, $m = \Sigma m_i + s$, $c > \bar{c}$, $\Sigma 1/c_i = 1/c$. Let $(M,g) = (M_1,g_1) \times \cdots \times (M_s,g_s)$ and $f':(M,g) \to M^{m-1}(c)$ a strongly full parallel isometric imbedding as f in (a). Let $f'':M^{m-1}(c) \to M^m(\bar{c})$ be a totally umbilic isometric imbedding. We define an isometric imbedding $f:(M,g) \to M^m(\bar{c})$ by the composite $f = f'' \circ f'$. Then, by Lemma 1.3 and Lemma 1.1,(4), f is a full parallel isometric imbedding.

(ii) Case $\bar{c} = 0$.

(a) $m_0 \geq 1$, $s \geq 0$, $m = m_0 + \Sigma m_i + s$. Let $(M,g) = M^{m_0}(0) \times (M_1,g_1) \times \cdots \times (M_s,g_s)$ and identify $M^m(0)$ with $M^{m_0}(0) \times \mathbb{R}^{m_1+1} \times \cdots \times \mathbb{R}^{m_s+1}$. We define a parallel isometric imbedding $f:(M,g) \to M^m(0)$ by the product imbedding $f = id \times f_1' \times \cdots \times f_s'$, which is strongly full because of $m_0 \geq 1$.

(b) $s \geq 1$, $m = \Sigma m_i + s$, $c > 0$, $\Sigma 1/c_i = 1/c$. Let $(M,g) = (M_1,g_1) \times \cdots \times (M_s,g_s)$, $f':(M,g) \to M^{m-1}(c)$ a strongly full parallel isometric imbedding as f in (i)(a), and $f'':M^{m-1}(c) \to M^m(0)$ be a totally umbilic isometric imbedding. We define an isometric imbedding $f:(M,g) \to M^m(0)$ by $f = f'' \circ f'$. It is seen as in (i)(b) that f is full and parallel.

(iii) Case $\bar{c} < 0$.

(a) $m_0 \geq 1$, $s \geq 0$, $m = m_0 + \Sigma m_i + s$, $\bar{c} \leq c_0 < 0$, $1/c_0 + \Sigma 1/c_i = 1/\bar{c}$. Let $(M,g) = M^{m_0}(c_0) \times (M_1,g_1) \times \cdots \times (M_s,g_s)$ and

$j: M^{m_0}(c_0) \times M^{m_1}(c_1) \times \cdots \times M^{m_s}(c_s) \to M^m(\bar{c})$ be the natural isometric imbedding. We define an isometric imbedding $f:(M,g) \to M^m(\bar{c})$ by $f = j \circ (id \times f_1 \times \cdots \times f_s)$. It is seen as in (i)(a) that f is strongly full and parallel.

The following three examples of isometric imbeddings $f:(M,g) \to M^m(\bar{c})$ are seen to be full and parallel in the same way as (i)(b).

(b) $s \geqslant 1$, $m = \Sigma m_i + s$, $c > 0$, $\Sigma 1/c_i = 1/c$. Let $(M,g) = (M_1,g_1) \times \cdots \times (M_s,g_s)$, $f':(M,g) \to M^{m-1}(c)$ a strongly full parallel isometric imbedding as f in (i)(a), and $f'':M^{m-1}(c) \to M^m(\bar{c})$ be a totally umbilic isometric imbedding. We define $f = f'' \circ f'$.

(c) $m_0 \geqslant 1$, $s \geqslant 0$, $m = m_0 + \Sigma m_i + s + 1$. Let $(M,g) = M^{m_0}(0) \times (M_1,g_1) \times \cdots \times (M_s,g_s)$, $f':(M,g) \to M^{m-1}(0)$ a strongly full parallel isometric imbedding as f in (ii)(a), and $f'':M^{m-1}(0) \to M^m(\bar{c})$ be a totally umbilic isometric imbedding. We define $f = f'' \circ f'$.

(d) $m_0 \geqslant 1$, $s \geqslant 0$, $m = m_0 + \Sigma m_i + s + 1$, $\bar{c} < c \leqslant c_0 < 0$, $1/c_0 + \Sigma 1/c_i = 1/c$. Let $(M,g) = M^{m_0}(c_0) \times (M_1,g_1) \times \cdots \times (M_s,g_s)$, $f':(M,g) \to M^{m-1}(c)$ a strongly full parallel isometric imbedding as f in (iii)(a), and $f'':M^{m-1}(c) \to M^m(\bar{c})$ be a totally umbilic isometric imbedding. We define $f = f'' \circ f'$.

4. Classification of Parallel Submanifolds of Space Forms

In this section we show that the examples in Section 3 exhaust all full parallel submanifolds of space forms.

Theorem 4.1. *Let* $f:(M,g) \to M^m(\bar{c})$ *be a full parallel isometric immersion of a complete Riemannian manifold* (M,g) *into a space form* $M^m(\bar{c})$. *Then* f *is congruent to the composite* $\bar{\bar{f}} \circ \bar{f}$ *of a parallel isometric imbedding* $\bar{\bar{f}}:(M',g') \to M^m(\bar{c})$ *among the models in Section 3 and an isometric covering* $\bar{f}:(M,g) \to (M',g')$.

Proof. In case $\bar{c} = 0$, this was proved by Ferus [3], [4]. Hence we may assume $\bar{c} \neq 0$. Moreover we may assume that M is simply connected. Let $\iota: M^m(\bar{c}) \to F^{m+1}$ be the inclusion, where $F^{m+1} = E^{m+1}$ if $\bar{c} > 0$, $F^{m+1} = L^{m+1}$ if $\bar{c} < 0$, and let $f' = \iota \circ f:(M,g) \to F^{m+1}$ be the composite, which is parallel isometric immersion by Lemma 1.1,(4).

Case I. f' *is substantial.*

Recall that this condition is equivalent to that f is strongly full. We follow the argument of Ferus [3]. Since f is parallel, the shape operator A_η, η being the mean curvature of f, is a parallel symmetric endomorphism of TM. Therefore eigenvalues $\lambda_1,\dots,\lambda_s$ of A_η are constant on M, and eigenspaces of A_η corresponding to λ_i constitute a parallel and integrable subbundle T_i of TM for each i. Hence we get an orthogonal decomposition: $TM = T_1 \oplus \cdots \oplus T_s$. We can show by the Ricci equation (1.5) that the second fundamental form α of f satisfies $\alpha(T_i,T_j) = 0$ for $i \neq j$. Let $(M,g) = (M_1,g_1) \times \cdots \times (M_s,g_s)$ be the de Rham decomposition into the product of simply connected complete Riemannian submanifolds (M_i,g_i), $1 \leqslant i \leqslant s$, such that $\pi_i^{-1}TM_i = T_i$, $\pi_i : M \to M_i$ being the projection to the i-th factor.

Now, for the mean curvature η' of f', by (1.9) we have $A'_{\eta'} = A_\eta + \bar{c}I$, and hence $A'_{\eta'}$ is the scalar operator $\lambda'_i I$ on T_i, $1 \leqslant i \leqslant s$, with $\lambda'_i = \lambda_i + \bar{c}$. Moreover, by (1.6) the second fundamental form α' of f' satisfies $\alpha'(T_i,T_j) = 0$ for $i \neq j$. Therefore, by Lemma 1.4 there exist linear subspaces F_i of $\mathbb{R}^{m+1}$ and substantial isometric immersions $f'_i : (M_i,g_i) \to F_i$, $1 \leqslant i \leqslant s$, such that

$$\mathbb{R}^{m+1} = F_1 \oplus \cdots \oplus F_s \tag{4.1}$$

(an orthogonal direct sum with respect to the pseudo-Riemannian metric $\langle , \rangle$) and

$$f' = f'_1 \times \cdots \times f'_s \quad .$$

Here, each f'_i is parallel since f' is parallel, which implies that the mean curvature η'_i of f'_i is parallel. Moreover the shape operator $A'^{(i)}_{\eta'_i}$ of f'_i for the direction η'_i satisfies

$$A'^{(i)}_{\eta'_i} = (n\lambda'_i/n_i)I \quad , \tag{4.2}$$

where $n_i = \dim M_i$, $n = \dim M = \Sigma n_i$. In fact, we have $\eta' = (n_1/n)\eta'_1 \circ \pi_1 + \cdots + (n_s/n)\eta'_s \circ \pi_s$, and hence $A'_{\eta'} = (n_1/n)A'^{(1)}_{\eta'_1} \oplus \cdots \oplus (n_s/n)A'^{(s)}_{\eta'_s}$. Recalling that $A'_{\eta'} = \lambda'_i I$ on T_i, we get (4.2).

In case $\bar{c} > 0$, the metric $\langle , \rangle$ restricted to F_i is Euclidean for each i. In case $\bar{c} < 0$, it is Lorentz on one of F_i, say F_1, and Euclidean on the other F_j, $2 \leqslant j \leqslant s$. Let $\dim F_i = m_i + 1$, $1 \leqslant i \leqslant s$, and so $m=\Sigma m_i + s - 1$. Choose an orthogonal basis $\{e_\alpha\}$ of $\mathbb{R}^{m+1}$ such that $\langle e_1,e_1\rangle = \operatorname{sgn}\bar{c}$, $\langle e_\alpha,e_\alpha\rangle = 1$ for $2\leqslant\alpha\leqslant m+1$ and that $\{e_\alpha; m_1+\cdots+m_{i-1}+i\leqslant\alpha\leqslant m_1+\cdots+m_i+i\}$ spans F_i, $1\leqslant i\leqslant s$. Moreover, in case $\bar{c}<0$ we choose e_1 in such a way that $\langle f_1'(M_1),e_1\rangle<0$. This is possible because f_1' is substantial. Then the linear map φ of $\mathbb{R}^{m+1}$ which sends e_α to the standard α-th unit vector ${}^t(0,\dots,\overset{\alpha}{1},\dots 0)$ is an isometry of $\langle\, ,\rangle$ such that $\varphi M^m(\bar{c}) = M^m(\bar{c})$. We may therefore assume that the decomposition (4.1) is the natural one:

$$\mathbb{R}^{m+1} = \mathbb{R}^{m_1+1} \oplus \cdots \oplus \mathbb{R}^{m_s+1} .$$

Let b_i, $1\leqslant i\leqslant s$, be the constants with $\Sigma b_i = 1/\bar{c}$ determined by $\langle f_i',f_i'\rangle = b_i$.

Case $\bar{c}>0$. In this case each b_i is positive. Put $c_i = 1/b_i$, $1\leqslant i\leqslant s$. Then $\Sigma 1/c_i = 1/\bar{c}$ and $f_i'(M_i)\subset M^{m_i}(c_i)$ for each i. Let $f_i:(M_i,g_i)\to M^{m_i}(c_i)$ be the isometric immersion induced from f_i'. It follows from (4.2) and Lemma 1.5 that f_i is minimal. Moreover f_i is parallel and full, since f_i' is parallel and substantial. Therefore Theorem 2.1 implies that there exist a standard isometric imbedding $\bar{\bar{f}}_i:(M_i',g_i')\to M^{m_i}(c_i)$ of a symmetric R-space (M_i',g_i') and an isometric covering $\bar{f}_i:(M_i,g_i)\to(M_i',g_i')$ such that $f_i = \bar{\bar{f}}_i\circ\bar{f}_i$. Let $(M',g') = (M_1',g_1')\times\cdots\times(M_s',g_s')$, $\bar{f} = \bar{f}_1\times\cdots\times\bar{f}_s:(M,g)\to(M',g')$, and $\bar{\bar{f}}:(M',g')\to M^m(\bar{c})$ be the composite of $\bar{\bar{f}}_1\times\cdots\times\bar{\bar{f}}_s$ and the natural inclusion $j:M^{m_1}(c_1)\times\cdots\times M^{m_s}(c_s)\to M^m(\bar{c})$. We have then $f=\bar{\bar{f}}\circ\bar{f}$, where $\bar{f}$ is an isometric covering and $\bar{\bar{f}}$ is seen by Remark in Section 2 to be an isometric imbedding of the model (i)(a).

Case $\bar{c}<0$. In this case $b_j>0$ for $2\leqslant j\leqslant s$. Therefore $b_1 = 1/\bar{c} - \Sigma b_j < 0$. Put $c_1 = 1/b_1$ and $c_j = 1/b_j$ for $2\leqslant j\leqslant s$. Then $\bar{c}\leqslant c_1<0$, $c_j>0$ for $2\leqslant j\leqslant s$, $1/c_1+\Sigma 1/c_j = 1/\bar{c}$ and $f_i'(M_i)\subset M^{m_i}(c_i)$ for each i. By the same reason as the case $\bar{c}>0$, the induced immersion $f_i:(M_i,g_i)\to M^{m_i}(c_i)$ is a full minimal parallel isometric immersion. In particular, by Lemma 1.6 the first immersion $f_1:(M_1,g_1)\to M^{m_1}(c_1)$ is totally geodesic, and hence, by the fullness of f, f_1 is an isometry. Now in the same way as the case $\bar{c}>0$, we obtain an isometric imbedding $\bar{\bar{f}}:(M',g')\to M^m(\bar{c})$ of the model (iii)(a) and an isometric covering $\bar{f}:(M,g)\to(M',g')$ such that $f=\bar{\bar{f}}\circ\bar{f}$.

Case II. f' *is not substantial.*

By Theorem 1.2 there exist a totally umbilic isometric imbedding $f'':M^{m-1}(c) \to M^m(\bar{c})$ with $c > \bar{c}$ and an isometric immersion $f':(M,g) \to M^{m-1}(c)$ such that $f = f'' \circ f'$. By Lemma 1.3 f' is strongly full. Note that the argument in Case I and the theorem of Ferus for $\bar{c} = 0$ are valid also for $m \geqslant 1$. Thus we know already all strongly full parallel isometric immersions $f':(M,g) \to M^{m-1}(c)$. Now, together with the uniqueness and equivariantness of totally umbilic immersions, we conclude that f is congruent to the composite $\bar{\bar{f}} \circ \bar{f}$ of an isometric imbedding $\bar{\bar{f}}:(M',g') \to M^m(\bar{c})$ in the models (i)(b), (iii)(b), (iii)(c) or (iii)(d) and an isometric covering $\bar{f}:(M,g) \to (M',g')$. ∎

5. Planar Geodesic Submanifolds

In this section we give an alternative proof of Sakamoto's classification theorem for planar geodesic submanifolds.

An isometric immersion $f:(M,g) \to M^m(\bar{c})$ of a Riemannian manifold (M,g) with $\dim M \geqslant 2$ is said to be *planar geodesic* if (1) it is not totally geodesic; and (2) for any geodesic $c(t)$, $|t| < \ell$, of (M,g), there exist $\varepsilon > 0$ and a 2-dimensional totally geodesic submanifold P of $M^m(\bar{c})$ such that $f(c(t)) \in P$ for each t with $|t| < \varepsilon$. These conditions are known (Sakamoto [8]) to be equivalent to that f is parallel and λ-isotropic for some $\lambda > 0$. The following are examples of full planar geodesic isometric imbeddings.

(α) Let $f:M^{m-1}(c) \to M^m(\bar{c})$, $m \geqslant 3$, $c > \bar{c}$, be a totally umbilic isometric imbedding. Then f is λ-isotropic with $\lambda = c - \bar{c}$.

(β) Let $f:(M,g) \to M^m(\bar{c})$, $\bar{c} > 0$, be a proper standard isometric imbedding of a rank one symmetric R-space, i.e., the generalized Veronese imbedding of the projective space $M = P_n(\mathbb{F})$ over $\mathbb{F} = \mathbb{R}$, $\mathbb{C}$ or real quaternions with $n \geqslant 2$ or of the Cayley projective plane M (cf. Tai [9]). Since the isometry group $I(M,g)$ of a rank one symmetric space (M,g) acts transitively on the unit tangent bundle, our imbedding f is λ-isotropic for some $\lambda > 0$.

(γ) Let $f':(M,g) \to M^{m-1}(c)$, $c > 0$, be a generalized Veronese imbedding, $f'':M^{m-1}(c) \to M^m(\bar{c})$, $c > \bar{c}$, a totally umbilic isometric imbedding, and let $f = f'' \circ f':(M,g) \to M^m(\bar{c})$. Then, by Lemma 1.1,(3) f is λ-isotropic for some $\lambda > 0$.

Theorem 5.1 (Sakamoto[8]). *Let* $f:(M,g) \to M^m(\bar{c})$ *be a full planar geodesic immersion of a complete Riemannian manifold* (M,g) *with* dim $M \geq 2$. *Then* f *is congruent to the composite* $f'' \circ f'$ *of an isometric imbedding* $f'':(M',g') \to M^m(\bar{c})$ *in the above* (α), (β), (γ) *and an isometric covering* $f':(M,g) \to (M',g')$.

Proof. We want to pick up positive-isotropic ones among the models in Section 3. Observe first that if a product immersion $f_1 \times \cdots \times f_s$ of a Riemannian product into a pseudo-Riemannian product is λ-isotropic, then each f_i is λ-isotropic. It follows that if $f_1 \times \cdots \times f_s$ is λ-isotropic with $\lambda \neq 0$ then $s = 1$.

Now let $f:(M,g) \to M^m(\bar{c})$ be an isometric imbedding of the models in Section 3 and suppose that it is λ-isotropic with $\lambda > 0$. Then the above observation together with Lemma 1.1,(3) yields that $s = 1$ in cases (i)(a), (i)(b), (ii)(b), (iii)(b), and $s = 0$ in cases (iii)(c), (iii)(d), and that the cases (ii)(a), (iii)(a) do not occur. It remains therefore to show that if $f:(M,g) \to M^m(\bar{c})$, $\bar{c} > 0$, is an isotropic standard isometric imbedding of a symmetric R-space (M,g), then $r = \text{rank}(M,g)$ equals 1. Let A be an r-dimensional complete flat totally geodesic submanifold of (M,g). Let $M^n(\bar{c})$ be the smallest complete totally geodesic submanifold of $M^m(\bar{c})$ containing $f(A)$. Then, by Lemma 1.1 the induced isometric imbedding $f':(A,g) \to M^n(\bar{c})$ is an isotropic full parallel minimal isometric imbedding of the flat torus (A,g). Now Theorem 4.1 implies that f' is congruent to the isometric imbedding of (i)(a) with $s = r$ and each (M_i,g_i) is a flat circle. But this is isotropic only if $r = 1$. ∎

6. References

[1] Chern, S.S., do Carmo, M., Kobayashi, S., "Minimal submanifolds of a sphere with second fundamental form of constant length," *Functional Analysis and Related Fields*, ed. by F.E. Browder, Springer, 1970, 59-75.

[2] Ferus, D., "Immersionen mit paralleler zweiter Fundamentalform: Beispiele and Nicht-Beispiele," *Manus. Math.* 12 (1974), 153-162.

[3] Ferus, D., "Produkt-Zerlegung von Immersionen mit paralleler zweiter Fundamentalform," *Math. Ann.* 211 (1974), 1-5.

[4] Ferus, D., "Immersions with parallel second fundamental form," *Math. Z.* 140 (1974), 87-93.

[5] Kobayashi, S., Nagano, T., "On filtered Lie algebras and geometric structures I," *J. Math. Mech.* 13 (1964), 875-908.

[6] Kobayashi, S., Nomizu, K., *Foundations of Differential Geometry II*, Interscience, New York, 1969.

[7] Moore, J.D., "Isometric immersions of Riemannian products," *J. of Diff. Geom.* 5 (1971), 159-168.

[8] Sakamoto, K., "Planar geodesic immersions," *Tôhoku Math. J.* 29 (1977), 25-56.

[9] Tai, S.S., "On minimum imbeddings of compact symmetric spaces of rank one," *J. Diff. Geom.* 2 (1968), 55-66.

[10] Takahashi, T., "Homogeneous hypersurfaces in spaces of constant curvature," *J. Math. Soc. Japan* 22 (1970), 395-410.

[11] Takeuchi, M., "Cell decomposition and Morse equalities on certain symmetric spaces," *J. Fac. Sci. Univ. Tokyo* 12 (1965), 81-192.

[12] Takeuchi, M., Kobayashi, S., "Minimal imbeddings of R-spaces," *J. Diff. Geom.* 2 (1968), 203-215.

Osaka University
Toyonaka, Osaka 560
Japan

(Received December 2, 1980).

ON HESSIAN STRUCTURES ON AN AFFINE MANIFOLD

Katsumi Yagi

On a smooth manifold, an affine connection whose torsion and curvature vanish identically is called an *affine structure*. A smooth manifold provided with an affine structure is called an *affine manifold*. Let M be an affine manifold with an affine structure D. The covariant differentiation by D will be also denoted by D. A Riemannian metric h on M is called a *hessian* metric if for each point $x \in M$ there exist a neighborhood U of x and a smooth function ϕ on U such that $g = D^2\phi$ on U [5]. In this note we shall give an example of an affine manifold which does not admit any hessian metric and then determine the structure of A-Lie algebras which admit hessian metrics. For these purposes, we shall also establish a vanishing theorem of a certain cohomology group. The author would like to thank Professor H. Shima who introduced him to the problem discussed here.

1. Let M be a smooth manifold and E a vector bundle over M with a connection D whose curvature vanishes. We denote by $\Omega^p(M;E)$ the space of E-valued p-forms on M. In particular, $\Omega^0(M;E)$ is the space of all sections of E and is denoted by $\Gamma(M;E)$ as is customary. Then D defines a differential $\mathbf{d}$ on $\Omega^*(M;E)$; for $\Phi \in \Omega^p(M;E)$ and vector fields $X_0,\dots,X_p$ on M,

$$\mathbf{d}\,\Phi(X_0,\dots,X_p) = \sum_{i=0}^{p} (-1)^i D_{X_i}(\Phi(X_0,\dots,\hat{X}_i,\dots,X_p)) + \sum_{i<j} (-1)^{i+j}\Phi([X_i,X_j],X_o,\dots,\hat{X}_i,\dots,\hat{X}_j,\dots,X_p) .$$

Since the curvature of D vanishes, we have $\mathbf{d}\,\mathbf{d} = 0$. The p-th cohomology group of the complex $\{\Omega^*(M;E),\mathbf{d}\}$ will be denoted by $H^p(M;E)$. Let G be a Lie group acting on the vector bundle E preserving the flat connection D; that is, for a vector field X on M, a section s of E and $a \in G$, there holds

$$T_a(D_X s) = D_{T_{a_*}(X)} T_a(s)$$

where T_a denotes the action of $a \in G$ on E and M, and $(T_a(s))(x) = T_a(s(T_a^{-1}(x))$ for $x \in M$.

The action T of G on E and M induces the action $T^{\#}$ of G on $\Omega^*(M;E)$ naturally; for $a \in G$ and $\Phi \in \Omega^*(M;E)$

$$(T_a^{\#}\Phi)(X_1,\ldots,X_p) = T_a^{-1}(\Phi(T_{a_*}X_1,\ldots,T_{a_*}X_p)) \quad .$$

Moreover since the action of G on E preserves the connection D, we can show the following: for $a \in G$

$$T_a^{\#}\mathbf{d} = \mathbf{d}\,T_a^{\#} \qquad \text{on} \quad \Omega^*(M;E) \quad .$$

Therefore the invariant elements of $\Omega^*(M;E)$ form a subcomplex $\{\Omega^*(M;E)_G, \mathbf{d}\}$, whose p-th cohomology group will be denoted by $H^p(M;E)_G$. Suppose that G is compact and connected. Then we have the following homomorphism I of $\Omega^p(M;E)$ into $\Omega^p(M;E)_G$: for $\Phi \in \Omega^p(M;E)$,

$$I(\Phi) = \int_{a \in G} T_a^{\#}(\Phi)\, da$$

where da denotes the invariant volume element of G with the total volume one. Since $T_a^{\#}\mathbf{d} = \mathbf{d}\,T_a^{\#}$ for each $a \in G$, we can prove that $\mathbf{d}I = I\mathbf{d}$ and I induces a homomorphism of $H^p(M;E)$ into $H^p(M;E)_G$. Moreover, applying the usual procedure [1], we obtain the following isomorphism: if G is compact and connected, then I induces an isomorphism of $H^p(M;E)$ onto $H^p(M;E)_G$.

Let M be an affine manifold with an affine structure D. Then the vector bundle $\otimes_s^t T(M) = (\otimes^s T(M)) \otimes (\otimes^t T^*(M))$ of (s,t)-tensor fields on M is provided with a flat connection D in the usual way. Thus we can define the cohomology group $H^p(M;\otimes_t^s T(M))$. This cohomology group has been investigated and a vanishing theorem has been obtained by Koszul [3]. If a Lie group G acts on M affinely, then the invariant cohomology group $H^p(M;\otimes_t^s T(M))_G$ is defined as above, and furthermore if G is compact and connected then $H^p(M;\otimes_t^s T(M))_G$ is isomorphic to $H^p(M;\otimes_t^s T(M))$.

2. Let M be an affine manifold with an affine structure D. We denote by $\Omega^p(M)$ the space of all p-forms on M. Then we have the following commutative diagram:

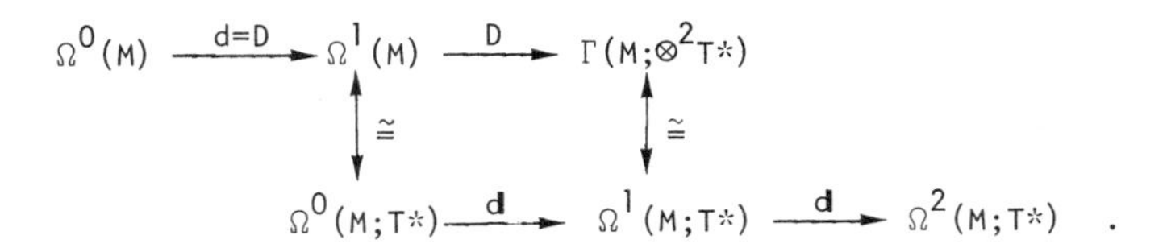

$$\begin{array}{ccccc} \Omega^0(M) & \xrightarrow{d=D} & \Omega^1(M) & \xrightarrow{D} & \Gamma(M;\otimes^2T^*) \\ & & \updownarrow\cong & & \updownarrow\cong \\ & & \Omega^0(M;T^*) \xrightarrow{\mathbf{d}} & \Omega^1(M;T^*) & \xrightarrow{\mathbf{d}} \Omega^2(M;T^*) \end{array} .$$

where each vertical isomorphism is a natural identification. In what follows, $\gamma \in \Omega^1(M;T^*)$ is called a *Riemannian* (resp. *hessian*) metric if the corresponding element in $\Gamma(M;\otimes^2T^*)$ is a Riemannian (resp. hessian) metric on M. Then we have the following [5]: $\gamma \in \Omega^1(M;T^*)$ is a hessian metric on M if and only if γ is $\mathbf{d}$-closed. Thus a hessian metric γ on M defines a cohomology class $[\gamma]$ in $H^1(M;T^*)$. A hessian metric $\gamma \in \Omega^1(M;T^*)$ is said to be *exact* if γ is $\mathbf{d}$-exact.

An affine manifold is said to be *convex* (resp. *hyperbolic*) if the universal covering space of the manifold with the induced affine structure is affinely diffeomorphic to a convex domain in an affine space (resp. a convex domain in an affine space which does not contain any full straight line [3]. The theorems due to Koszul [3] give conditions for an affine manifold to be hyperbolic.

For our purposes we reformulate Koszul's theorems in terms of hessian metrics as follows:

Theorem (Koszul). *Let* M *be a connected affine manifold. Then*

(1) (homogeneous case). *Let* G *be a Lie group acting on* M *affinely and transitively. Then* M *is hyperbolic if and only if there exists a* G*-invariant* 1*-form* $\eta \in \Omega^0(M;T^*)$ *such that* $\mathbf{d}\eta \in \Omega^1(M;T^*)$ *is a Riemannian metric.*

(2) (compact case). *When* M *is compact,* M *is hyperbolic if and only if* M *admits an exact hessian metric.*

Remark 1. The condition in (1) is equivalent to the following: there exists a G-invariant exact hessian metric $\gamma \in \Omega^1(M;T^*)$ such that $0 = [\gamma]$ in $H^1(M;T^*)_G$. When $\gamma \in \Omega^1(M;T^*)$ is a G-invariant exact hessian metric, $[\gamma] = 0$ in $H^1(M;T^*)$ by definition, however $[\gamma]$ in $H^1(M;T^*)_G$ is not necessarily zero. We can show that if $H^0(M;T^*) = 0$ and $[\gamma] = 0$ in $H^1(M;T^*)$, then $[\gamma] = 0$ in $H^1(M;T^*)_G$.

Applying Koszul's theorem, we obtain the following immediately.

Lemma 1. *Let* M *be a connected affine manifold. Then*

(1) *Suppose that a Lie group* G *acts on* M *affinely and transitively and* $H^1(M;T^*)_G = (0)$. *If there exists a* G*-invariant hessian metric on* M *then* M *is hyperbolic.*

(2) *Suppose that* M *is compact and* $H^1(M;T^*) = (0)$. *If there exists a hessian metric on* M *then* M *is hyperbolic.*

3. In this section we shall give an example of a compact affine manifold which does not admit a hessian metric. Let H be the quotient space of $\mathbb{R}^n - (0)$ by the discrete group $\Gamma = \{2^k \cdot id \in GL(n;\mathbb{R}) : k \in \mathbb{Z}\}$, called the *Hopf manifold.* In a natural way H is provided with an affine structure. Now let $\tilde{G} = \{t \cdot id \in GL(n;\mathbb{R}) : t \in \mathbb{R}^+\}$ and $G = \tilde{G}/\Gamma$. Then G is a compact Lie group acting on the affine manifold H affinely. Thus we have the isomorphism $H^p(H;\otimes^s_t T(H)) \cong H^p(H:\otimes^s_t T(H))_G$. The following vanishing theorem has been obtained originally by Koszul [3]. His theorem is more general and the proof is based on the harmonic integral.

Theorem 2. *Let* H *be the* n*-dimensional Hopf manifold. Then we have*

(1) $H^p(H,\otimes^s_t T(H)) = (0)$ *if* $p \geqq 0$ *and* $s \neq t$,

(2) H *does not admit a hessian metric if* $n \geqq 2$.

Proof. Let Φ be a d-closed form in $\Omega^p(H;\otimes^s_t T(H))_G$ and $\tilde{E} = \Sigma_i x^i \, \partial/\partial x^i$ the infinitesimal affine vector field on $\mathbb{R}^n - (o)$. Since the vector field $\tilde{E}$ is invariant by Γ, $\tilde{E}$ defines an infinitesimal affine vector field E on H. The p-form Φ can be written as follows; locally

$$\Phi = \sum_{I,J} c_{I,J} \otimes \partial/\partial x^I \otimes dx^J$$

where $c_{I,J}$ is an ordinary p-form, $\partial/\partial x^I \otimes dx^J$ is the (s,t)-tensor field with the multi-index (I,J). When we denote by L_E the Lie derivative defined by the vector field E on M, we have

$$L_E(\partial/\partial x^I) = -s \, \partial/\partial x^I \quad , \qquad L_E(dx^J) = t \, dx^J \, .$$

Since Φ is G-invariant, $L_E\Phi = 0$. On the other hand, we have the formula $L_E = di_E + i_E d$ where i_E denotes the substitution operation by the vector field E on $\Omega^p(H)$. Therefore

$$\mathbf{d}i_E\Phi = (\mathbf{d}i_E + i_E\mathbf{d})\Phi = \Sigma\, ((di_E + i_E d)c_{I,J}) \otimes \partial/\partial x^I \otimes dx^J$$

$$= \Sigma\, (L_E c_{I,J}) \otimes \partial/\partial x^I \otimes dx^J$$

$$= L_E\Phi - \Sigma\, c_{I,J} \otimes L_E(\partial/\partial x^I) \otimes dx^J$$

$$- \Sigma\, c_{I,J} \otimes \partial/\partial x^I \otimes L_E(dx^J) = (t-s)\Phi$$

and hence if $t \neq s$, Φ is $\mathbf{d}$-exact. This shows the first part of the theorem. When $n \geqq 2$, H is not hyperbolic as an affine manifold, thus Lemma 1 implies the second part of the theorem, since we have $H^1(H;T^*) = (0)$.

4. Let $\mathfrak{g}$ be a finite dimensional Lie algebra over the real field $\mathbb{R}$. $\mathfrak{g}$ is called a *left symmetric algebra* with ρ if $\mathfrak{g}$ admits a Lie algebra representation of $\mathfrak{g}$ over $\mathfrak{g}$ such that $[X,Y] = \rho(X)Y - \rho(Y)X$ for $X, Y \in \mathfrak{g}$ [5]. A multiplication in $\mathfrak{g}$ will be then defined as follows: $XY = \rho(X)Y$ for $X,Y \in \mathfrak{g}$. A Lie algebra $\mathfrak{g}$ is called an *A-Lie algebra* when $\mathfrak{g}$ has also an associative algebra structure such that $[X,Y] = X\cdot Y - Y\cdot X$ for $X,Y \in \mathfrak{g}$ [7]. Then an A-Lie algebra $\mathfrak{g}$ can be regarded as a left symmetric algebra in a natural way: $\rho(X)Y = X\cdot Y$ for $X,Y \in \mathfrak{g}$. Let $\mathfrak{g}$ be a left symmetric algebra with ρ and G a connected Lie group with Lie algebra $\mathfrak{g}$. Then there exists a unique left invariant affine structure D on G such that $D_X Y = \rho(X)Y$ for $X,Y \in \mathfrak{g}$. Here $\mathfrak{g}$ is the Lie algebra of all *left invariant* vector fields on G. Moreover if $\mathfrak{g}$ is an A-Lie algebra then the affine structure D on G defined above is bi-invariant. A left symmetric algebra $\mathfrak{g}$ is said to be *convex* (resp. *hyperbolic*) if a connected Lie group G with Lie algebra $\mathfrak{g}$ is convex (resp. hyperbolic) as an affine manifold. The representation ρ induces a representation of $\mathfrak{g}$ over $\otimes^s_t\mathfrak{g} = (\otimes^s\mathfrak{g}) \otimes (\otimes^t\mathfrak{g}^*)$, which will be also denoted by ρ. Thus we have the following complex

$$\{C^*(\mathfrak{g};\otimes^s_t\mathfrak{g}),\ d_\rho\} : C^p(\mathfrak{g};\otimes^s_t\mathfrak{g}) = \Lambda^p\mathfrak{g}^*\otimes(\otimes^s_t\mathfrak{g}) \quad ,$$

$$d_\rho c(x_0,\ldots,x_p) = \sum_{i=0}^{p} (-1)^i \rho(x_i)(c(x_0,\ldots,\hat{x}_1,\ldots,x_p))$$

$$+ \sum_{i<j} (-1)^{i+j} c([x_i,x_j],x_0,\ldots,\hat{x}_i,\ldots,\hat{x}_j,\ldots,x_p)$$

for $c \in C^p(\mathfrak{g};\otimes_t^s\mathfrak{g})$ and $x_0,\ldots,x_p \in \mathfrak{g}$.

Then we have natural isomorphisms of complexes and their cohomology groups;

$$\{C^*(\mathfrak{g};\otimes_t^s\mathfrak{g}), d\rho\} \cong \{\Omega^*(G;\otimes_t^s T(G))_G, d\!\!I\} ,$$

$$H^p(\mathfrak{g};\otimes_t^s\mathfrak{g}) \cong H^p(G;\otimes_t^s(G))_G .$$

When G is connected and compact, these cohomology groups are isomorphic to $H^p(G;\otimes_t^s T(G))$.

Theorem 3. *Let* $\mathfrak{g}$ *be a left symmetric algebra. If there exist* $E \in \mathfrak{g}$ *such that* $EX = XE = X$ *for each* $X \in \mathfrak{g}$, *then* $H^p(\mathfrak{g};\otimes_t^s\mathfrak{g}) = (0)$ *for* $p \geqq 0$ *and* $s \neq t$.

Proof. Suppose that $c \in C^p(\mathfrak{g};\otimes_t^s\mathfrak{g})$ and $d_\rho c = 0$. Then $\theta_\rho(E)c = d_\rho i(E)c + i(E)d_\rho c = d_\rho i(E)c$. On the other hand c can be written as follows:

$$c = \sum_i \gamma_i \otimes U_i \otimes V_i$$

where

$$\gamma_i \in \Lambda^p\mathfrak{g}^*, \qquad U_i \in \otimes^s\mathfrak{g} \qquad \text{and} \qquad V_i \in \otimes^t\mathfrak{g}^* .$$

Then $\theta_\rho(E)c = \Sigma_i\ (\theta(E)\gamma_i) \otimes U_i \otimes V_i + \gamma_i \otimes \rho(E)U_i \otimes V_i + \gamma_i \otimes U_i \otimes \rho(E)V_i$. Since E is the identity of the algebra $\mathfrak{g}$, $\rho(E)U = sU$ for $U \in \otimes^s g$ and $\rho(E)V = -tV$ for $V \in \otimes^t\ *$, and more $\theta(E)\gamma = 0$ for $\gamma \in \Lambda^p g^*$. Therefore $\theta_\rho(E)c = (s-t)c$. Hence if $p \geqq 0$ and $s \neq t$, then $c = (s-t)^{-1} d_\rho i(E)c$, which shows c is d_ρ-exact. This completes the proof.

Let $\mathfrak{g}$ be a left symmetric algebra and G a Lie group with Lie algebra $\mathfrak{g}$. Then we have the following canonical identification:

$$\begin{array}{ccccc}
\otimes^2\mathfrak{g}^* & = & \Gamma(G;\otimes^2T^*)_G & \subset & \Gamma(G;\otimes^2T^*) \\
\updownarrow\cong & & \updownarrow\cong & & \updownarrow\cong \\
C^1(\mathfrak{g};\mathfrak{g}^*) & = & \Omega^1(G;T^*)_G & \subset & \Omega^1(G;T^*)
\end{array}$$

where we consider only the left action of G on G.
In particular, there is a one to one correspondence between an inner product on $\mathfrak{g}$ and a left invariant Riemannian metric on G. They will be identified and denoted by the same letter. Thus an inner product h on $\mathfrak{g}$ is called a *hessian* metric on $\mathfrak{g}$ if h is a left invariant hessian metric on G. If h is an inner product on $\mathfrak{g}$ and $\gamma \in C^1(\mathfrak{g};\mathfrak{g}^*)$ corresponds to h then h is a hessian metric on $\mathfrak{g}$ if and only if $d_\rho\gamma = 0$. This is equivalent to saying that $h(X,YZ) - h(Y,XZ) = h([X,Y],Z)$ for X,Y and $Z \in \mathfrak{g}$. Therefore a hessian metric on $\mathfrak{g}$ defines a cohomology class $[\gamma]$ in $H^1(\mathfrak{g};\mathfrak{g}^*)$. A hessian metric h on $\mathfrak{g}$ is said to be d_ρ-*exact* if $\mathfrak{g}$ is d_ρ-exact; that is, $[\gamma] = 0$ in $H^1(\mathfrak{g};\mathfrak{g}^*)$. If a hessian metric on $\mathfrak{g}$ is d_ρ-exact, then the left invariant hessian metric on G is exact. However, the converse does not hold in general.

Proposition 4. *Let* $\mathfrak{g}$ *be a left symmetric algebra. If there exists nonzero* $X \in \mathfrak{g}$ *such that* $X^2 = 0$, *then* $\mathfrak{g}$ *admits no exact hessian metric.*

Proof. Let $\gamma \in C^1(\mathfrak{g};\mathfrak{g}^*)$ correspond to an exact hessian metric on $\mathfrak{g}$. Then there exists $\eta \in C^0(\mathfrak{g};\mathfrak{g}^*)$ such that $\gamma = d_\rho\eta$. Thus $0 < h(X,X) = \gamma(X)X = (d_\rho\eta(X))(X) = (\rho(X)\eta)(X) = -\eta(\rho(X)X) = -\eta(X^2) = 0$. This is a contradiction.

Theorem 5. *Let* $\mathfrak{g}$ *be a left symmetric algebra. Then*

(1) $\mathfrak{g}$ *is hyperbolic if and only if there exists a* d_ρ*-exact hessian metric on* $\mathfrak{g}$.

(2) *If* $H^1(\mathfrak{g};\mathfrak{g}^*) = (0)$ *and there exists a hessian metric on* $\mathfrak{g}$, *then* $\mathfrak{g}$ *is hyperbolic.*

(3) *If* $H^1(\mathfrak{g};\mathfrak{g}^*) = (0)$ *and there exists nonzero* $X \in \mathfrak{g}$ *such that* $X^2 = 0$, *then* $\mathfrak{g}$ *has no hessian metric.*

Proof. By Koszul's theorem and Remark 1, $\mathfrak{g}$ is hyperbolic if and only if there exists a left invariant hessian metric $\gamma \in \Omega^1(G;T^*)_G$ on G such that $[\gamma] = 0$ in $H^1(G;T^*)_G$. When $\gamma \in \Omega^1(G;T^*)_G$ is regarded as an element of $C^1(\mathfrak{g};\mathfrak{g}^*)$, $[\gamma] = 0$ in $H^1(G;T^*)_G$ if and only if

$[\gamma]=0$ in $H^1(\mathfrak{g};\mathfrak{g}^*)$. This shows (1). Let h be a hessian metric on $\mathfrak{g}$ and $\gamma \in C^1(\mathfrak{g};\mathfrak{g}^*)$ correspond to h. Since $H^1(\mathfrak{g};\mathfrak{g}^*)=0$, necessarily γ is d_ρ-exact. Thus (1) implies that M is hyperbolic, which shows (2). (3) follows from Proposition 4 and (2).

5. In [6] Shima determines the structure of a left symmetric algebra with a hessian metric. In this section we consider a similar problem for an A-Lie algebra. We denote by K the real numbers field $\mathbb{R}$, the complex numbers field $\mathbb{C}$ or the quaternion $\mathbb{H}$. In a natural way $\mathfrak{gl}(n;K)$, the set of all K-matrices of degree n, is regarded as an A-Lie algebra over $\mathbb{R}$. The group $GL(n;K)$ of all invertible elements of $\mathfrak{gl}(n;K)$ is an A-Lie group with Lie algebra $\mathfrak{gl}(n;K)$. We define $\mathfrak{a}(p,q)$ and $A(p,q)$ as follows:

$$\mathfrak{a}(p,q) = \left\{ \begin{pmatrix} a_1 & & & 0 & 0 \\ & \ddots & & & \\ & & a_p & & \\ & & & & b_1 \\ 0 & & & 0 & \vdots \\ & & & & b_q \\ 0 & & & 0 & 0 \end{pmatrix} : a_i, b_j \in \mathbb{R} \right\},$$

$$A(p,q) = \left\{ \begin{pmatrix} a_1 & & & & & 0 & 0 \\ & \ddots & & & & & \\ & & a_p & & & & \\ & & & 1 & & & b_1 \\ 0 & & & & \ddots & & \vdots \\ & & & & & 1 & b_q \\ 0 & & & & & 0 & 0 \end{pmatrix} : \begin{matrix} a_i > 0 \\ b_j \in \mathbb{R} \end{matrix} \right\}.$$

Then $\mathfrak{a}(p,q)$ is an associative subalgebra of $\mathfrak{gl}(p+q+1;K)$ and $A(p,q)$ is an A-Lie group with Lie algebra $\mathfrak{a}(p,q)$ and provided with the affine structure on $A(p;q)$ as an A-Lie group, $A(p,q)$ is affinely diffeomorphic to $(\mathbb{R}^+)^p \times \mathbb{R}^q$ where $\mathbb{R}^+$ is the open subset of all positive real numbers in $\mathbb{R}$ with the induced affine structure. In particular, the affine manifold $A(p,q)$ is convex and $A(p,q)$ is hyperbolic if and only if $q=0$.

Let $\mathfrak{g}$ be an A-Lie algebra with a hessian metric h. Regarded as an associative algebra, $\mathfrak{g}$ has a decomposition $\mathfrak{g} = \mathfrak{r} + \mathfrak{s}$ (semi-direct sum) where $\mathfrak{r}$ is the radical (the maximal nilpotent ideal) of the associative algebra $\mathfrak{g}$ and $\mathfrak{s}$ is a semisimple associative sub-algebra of $\mathfrak{g}$ [7]. Since h is a hessian metric on $\mathfrak{g}$, the restrictions of h to $\mathfrak{r}$ and $\mathfrak{s}$ are also hessian metrics on $\mathfrak{r}$ and $\mathfrak{s}$. We denote by $\mathfrak{g}^k$ the ideal of $\mathfrak{g}$ generated by $\{X_1 \ldots X_k : X_i \in \mathfrak{g}\}$.

Lemma. *Let* $\mathfrak{g}$ *be an A-Lie algebra with a hessian metric* h. *Then*

(1) *If* $\mathfrak{g}$ *is nilpotent as an associative algebra then* $\mathfrak{g}^2 = (0)$,

(2) *If* $\mathfrak{g}$ *is semisimple as an associative algebra then* $\mathfrak{g}$ *is isomorphic to* $\mathfrak{a}(p,0)$ *where* $p = \dim \mathfrak{g}$.

Proof. (1) Since $\mathfrak{g}$ is nilpotent, there exists an integer $k \geqq 1$ such that $\mathfrak{g}^k \neq (0)$ and $\mathfrak{g}^{k+1} = (0)$. Suppose that $k \geqq 2$ and $0 \neq X_1 \ldots X_k \in \mathfrak{g}^k$. Then let $X = X_1 \ldots X_k$, $Y = X_1$ and $Z = X_2 \ldots X_k$. Then $0 < h(X,X) = h(X,YZ) = h(Y,XZ) + h([X,Y],Z) = 0$ since $XZ = XY = YX = 0$. Thus $k = 1$ and hence $\mathfrak{g}^2 = (0)$.

(2) By Wedderburn's theorem, a semisimple associative algebra over $\mathbb{R}$ is a direct sum of $\mathfrak{gl}(n;K)$ and hence it contains the identity. Thus when $\mathfrak{g}$ is semisimple we have $H^1(\mathfrak{g};\mathfrak{g}^*) = (0)$ by Theorem 3. By Theorem 5, $\mathfrak{g}$ is hyperbolic. On the other hand, the affine structure on $GL(n;K)$ as an A-Lie group is the one on $GL(n;K)$ induced from K^{n^2} as an open subset. Therefore the affine manifold $GL(n;K)$ is hyperbolic if and only if $n = 1$ and $K = \mathbb{R}$. Thus $\mathfrak{g}$ is isomorphic to a direct sum of $\mathfrak{gl}(1;\mathbb{R})$, which is $\mathfrak{a}(p,0)$.

Theorem 6. *Let* $\mathfrak{g}$ *be an A-Lie algebra. Then* $\mathfrak{g}$ *admits a hessian metric if and only if* $\mathfrak{g}$ *is isomorphic to* $\mathfrak{a}(p,q)$ *as an A-Lie algebra for some* $p, q \geqq 0$. *Particularly* $\mathfrak{g}$ *is a commutative associative algebra and convex. Moreover* $\mathfrak{g}$ *is hyperbolic if and only if* $\mathfrak{g}$ *is isomorphic to* $\mathfrak{a}(p,0)$ *as an A-Lie algebra for some* $p \geqq 0$.

Proof. Let $\mathfrak{g} = \mathfrak{r} + \mathfrak{s}$ be a decomposition of $\mathfrak{g}$ as an associative algebra as above. Suppose that $\mathfrak{g}$ admits a hessian metric h. Then $\mathfrak{r}$ and $\mathfrak{s}$ are A-Lie algebras with hessian metrics and hence it follows from the above lemma that $\mathfrak{r}^2 = (0)$ and $\mathfrak{s}$ is isomorphic to $\mathfrak{a}(p,0)$ for some $p \geqq 0$. We shall prove that $\mathfrak{r} \cdot \mathfrak{s} = \mathfrak{s} \cdot \mathfrak{r} = (0)$. Let $U \in \mathfrak{s}$ and $V \in \mathfrak{r}$ such that $U^2 = U \neq 0$. Putting $X = UVU$, $Y = U$ and $Z = VU$, we have $h(UVU,UVU) = h(X,YZ) = h(Y,XZ) + h([X,Y],Z) = 0$ since X and Z are in $\mathfrak{r}$, $\mathfrak{r}^2 = (0)$ and $[X,Y] = XY - YX = 0$. Therefore $UVU = 0$. Now let $X = VU$,

$Y = V$ and $Z = U$. Then $h(VU,VU) = h(X,YZ) = h(Y,XZ) + h([X,Y],Z) = h(V,VU)$ since X and Y are in $\mathfrak{r}$ and $\mathfrak{r}^2 = (0)$. On the other hand, similarly $h(UV,VU) = h(V,UVU) + h([UV,V],U) = 0$ since $\mathfrak{r}^2 = (0)$. And $h(UV,VU) = h(VU,UV) = h(U,VUV) + h([VU,U],V) = h(VU,V)$. Therefore combining these three equalities, we have $h(VU,VU) = 0$ and hence $VU = 0$. We shall show that $UV = 0$. $h(UV,UV) = h(U,UVV) + h([UV,U],V) = -h(UV,V)$ since $U^2 = U$ and $UVU = 0$. On the other hand $h(V,UV) = h(U,VV) + h([V,U],V) = -h(UV,V)$ since $\mathfrak{r}^2 = (0)$ and $VU = 0$. Thus $h(UV,V) = 0$ and hence $h(UV,UV) = 0$. This shows that $UV = 0$. Since $\mathfrak{s}$ is a direct sum of $\mathfrak{gl}(1;\mathbb{R})$, $\mathfrak{s}$ is generated by $\{U \in \mathfrak{s} : U^2 = U\}$. Therefore we have that $\mathfrak{r}\cdot\mathfrak{s} = \mathfrak{s}\cdot\mathfrak{r} = (0)$ and $\mathfrak{g} = \mathfrak{r} \oplus \mathfrak{s}$ (direct sum). Since $\mathfrak{r}^2 = (0)$, $\mathfrak{r}$ is isomorphic to $\mathfrak{a}(0,q)$ for some $q \geqq 0$. Therefore $\mathfrak{g}$ is isomorphic to $\mathfrak{a}(p,q)$. The converse is obvious. The rest of Theroem can be proved immediately.

<u>Corollary</u>. *Let* G *be a compact* A-*Lie group. If* G, *provided with the affine structure, admits a hessian metric, then the universal covering group of* G *is isomorphic to* $A(p,q)$ *as an* A-*Lie group for some* p, $q \geqq 0$.

<u>Proof</u>. Suppose that $\gamma \in \Omega(G;T^*)$ is a hessian metric on G. Then we define a Riemannian metric $\zeta \in \Omega^1(G;T^*)_G$ as follows;

$$\zeta = \int_{a\in G} T_a^{\#}(\gamma)\, da$$

where da denotes the invariant volume element of G. Then $\mathbf{d}\zeta = 0$ since $\mathbf{d}\, T_a^{\#} = T_a^{\#}\mathbf{d}$ for each $a \in G$. Therefore ζ is a left invariant hessian metric on G. The A-Lie algebra of G admits a hessian metric and hence $A(p,q)$ is the universal covering group of G, by the above theorem.

6. Let M be a homogeneous affine manifold whose underlying manifold is the two dimensional real torus. Then M is affinely diffeomorphic to a compact two dimensional A-Lie group whose A-Lie algebra is one of the following A-Lie subalgebras of $\mathfrak{gl}(3;\mathbb{R})$ [4].

$$(1,1)\ \begin{pmatrix} a & b & 0 \\ 0 & a & 0 \\ 0 & 0 & 0 \end{pmatrix}, \qquad (1,2)\ \begin{pmatrix} a & 0 & 0 \\ 0 & d & 0 \\ 0 & 0 & 0 \end{pmatrix}, \qquad (1,3)\ \begin{pmatrix} a & b & 0 \\ -b & a & 0 \\ 0 & 0 & 0 \end{pmatrix}$$

$$(\mathrm{II})\ \begin{pmatrix} a & 0 & 0 \\ 0 & 0 & v \\ 0 & 0 & 0 \end{pmatrix}, \qquad (\mathrm{III},1)\ \begin{pmatrix} 0 & b & u \\ 0 & 0 & b \\ 0 & 0 & 0 \end{pmatrix}, \qquad (\mathrm{III},2)\ \begin{pmatrix} 0 & 0 & u \\ 0 & 0 & v \\ 0 & 0 & 0 \end{pmatrix} .$$

Theorem 7. *Let* M *be a homogeneous affine manifold whose underlying manifold is the two-dimensional torus. Then* M *admits a hessian metric if and only if* M *is affinely diffeomorphic to a compact* A-*Lie group whose* A-*Lie algebra is one of type* (I,2), (II) *or* (III,2) *above.*

Proof. We may assume that M is affinely diffeomorphic to an A-Lie group G whose A-Lie algebra is one of the above. Then it follows from the Corollary to Theorem 6 that G admits a hessian metric if and only if the A-Lie algebra of G is one of (I,2), (II) or (III,2).

Remark 2. Recently Shima proved that a compact hessian affine manifold is convex. This implies the nonexistence of a hessian metric on the Hopf manifold if the dimension $\geqq 2$. It should be remarked that the converse of Shima's theorem does not hold. The A-Lie group of type (I,1) is convex and the A-Lie group of type (III,1) is affinely complete; however neither of these admits a hessian metric.

6. References

[1] W. Greub, S. Halperin, and R. Vanstone, *Connection, Curvature and Cohomology*, Vol. II, Academic Press, New York and London, 1973.

[2] J.L. Koszul, "Variétés localement plates et convexité," *Osaka J. Math.* 2 (1965), 285-290.

[3] J.L. Koszul, "Déformations de connexions localement plates," *Ann. Inst. Fourier, Grenoble* 18, 1 (1968), 103-114.

[4] T. Nagano and K. Yagi, "The affine structures on the real two-torus (1)," *Osaka J. Math.* 11 (1974), 181-210.

[5] H. Shima, "On certain locally flat homogeneous manifolds of solvable Lie groups," *Osaka J. Math.* 13 (1976), 213-229.

[6] H. Shima, "Homogeneous hessian manifolds," in *Manifolds and Lie Groups, Papers in Honor of Yozô Matsushima*, Progress in Mathematics, Vol. 14, Birkhäuser, Boston, Basel, Stuttgart, 1981, 385-392.

[7] K. Yagi, "On compact homogeneous affine manifolds," *Osaka J. Math.* 7 (1970), 457-475.

Osaka University
Toyonaka, Osaka 560, Japan

(Received January 12, 1981)

Progress in Mathematics

Edited by J. Coates and S. Helgason

Progress in Physics

Edited by A. Jaffe and D. Ruelle

- A collection of research-oriented monographs, reports, notes arising from lectures or seminars
- Quickly published concurrent with research
- Easily accessible through international distribution facilities
- Reasonably priced
- Reporting research developments combining original results with an expository treatment of the particular subject area
- A contribution to the international scientific community: for colleagues and for graduate students who are seeking current information and directions in their graduate and post-graduate work.

Manuscripts

Manuscripts should be no less than 100 and preferably no more than 500 pages in length.

They are reproduced by a photographic process and therefore must be typed with extreme care. Symbols not on the typewriter should be inserted by hand in indelible black ink. Corrections to the typescript should be made by pasting in the new text or painting out errors with white correction fluid.

The typescript is reduced slightly (75%) in size during reproduction; best results will not be obtained unless the text on any one page is kept within the overall limit of 6x9½ in (16x24 cm). On request, the publisher will supply special paper with the typing area outlined.

Manuscripts should be sent to the editors or directly to:
Birkhäuser Boston, Inc., P.O. Box 2007, Cambridge,
Massachusetts 02139

PROGRESS IN MATHEMATICS

Already published

PM 1 Quadratic Forms in Infinite-Dimensional Vector Spaces
Herbert Gross
ISBN 3-7643-1111-8, 431 pages, $22.00, paperback

PM 2 Singularités des systèmes différentiels de Gauss-Manin
Frédéric Pham
ISBN 3-7643-3002-3, 339 pages, $18.00, paperback

PM 3 Vector Bundles on Complex Projective Spaces
C. Okonek, M. Schneider, H. Spindler
ISBN 3-7643-3000-7, 389 pages, $20.00, paperback

PM4 Complex Approximation, Proceedings, Quebec, Canada, July 3-8, 1978
Edited by Bernard Aupetit
ISBN 3-7643-3004-X, 128 pages, $10.00, paperback

PM 5 The Radon Transform
Sigurdur Helgason
ISBN 3-7643-3006-6, 202 pages, $14.00, paperback

PM 6 The Weil Representation, Maslov Index and Theta Series
Gérard Lion, Michèle Vergne
ISBN 3-7643-3007-4, 345 pages, $18.00, paperback

PM 7 Vector Bundles and Differential Equations
Proceedings, Nice, France, June 12-17, 1979
Edited by André Hirschowitz
ISBN 3-7643-3022-8, 255 pages, $14.00, paperback

PM 8 Dynamical Systems, C.I.M.E. Lectures, Bressanone, Italy, June 1978
John Guckenheimer, Jürgen Moser, Sheldon E. Newhouse
ISBN 3-7643-3024-4, 300 pages, $14.00, paperback

PM 9 Linear Algebraic Groups
T. A. Springer
ISBN 3-7643-3029-5, 304 pages, $18.00, hardcover

PM10 Ergodic Theory and Dynamical Systems I
A. Katok
ISBN 3-7643-3036-8, 352 pages, $20.00, hardcover

PM11 18th Scandinavian Congress of Mathematicians, Aarhus, Denmark, 1980
Edited by Erik Balslev
ISBN 3-7643-3034-6, 528 pages, $26.00, hardcover

PM12 Séminaire de Théorie des Nombres, Paris 1979-80
Edited by Marie-José Bertin
ISBN 3-7643-3035-X, 408 pages, $22.00, hardcover

PM13 Topics in Harmonic Analysis on Homogeneous Spaces
Sigurdur Helgason
ISBN 3-7643-3051-1, 142 pages, $12.00, hardcover

PM14 Manifolds and Lie Groups, Papers in Honor of Yozô Matsushima
Edited by J. Hano, A. Marimoto, S. Murakami, K. Okamoto, and H. Ozeki
ISBN 3-7643-3053-8, 480 pages, $35.00, hardcover

PM15 Representations of Real Reductive Lie Groups
David A. Vogan, Jr.
ISBN 3-7643-3037-6, 771 pages, $35.00, hardcover

PM16 Rational Homotopy Theory and Differential Forms
Phillip A. Griffiths, John W. Morgan
ISBN 3-7643-3041-4, 264 pages, $16.00, hardcover

PROGRESS IN PHYSICS
Already published

PPh1 Iterated Maps on the Interval as Dynamical Systems
Pierre Collet and Jean-Pierre Eckmann
ISBN 3-7643-3026-0, 256 pages, $16.00 hardcover

PPh2 Vortices and Monopoles, Structure of Static Gauge Theories
Arthur Jaffe and Clifford Taubes
ISBN 3-7643-3025-2, 275 pages, $16.00 hardcover

PPh3 Mathematics and Physics
Yu. I. Manin
ISBN 3-7643-3027-9, 111 pages, $10.00 hardcover